Lecture Notes
in Control and Information Sciences 193

Editor: M. Thoma

Alan S. I. Zinober (Ed.)

Variable Structure and Lyapunov Control

Springer-Verlag Berlin Heidelberg GmbH

Editor

Alan S. I. Zinober, PhD
Department of Applied and Computational Mathematics, University of Sheffield,
Sheffield S10 2TN, UK

ISBN 978-3-540-19869-7 ISBN 978-3-540-39325-2 (eBook)
DOI 10.1007/978-3-540-39325-2

British Library Cataloguing in Publication Data
A catalogue record for this book is available from the British Library

Originally published by Springer-Verlag Berlin Heidelberg New York in 1994

Typesetting: Camera ready by editor

69/3830-543210 Printed on acid-free paper

To

Brenda, Madeleine, Rebecca,

Cathy and Vicky

Preface

Mathematical models of actual systems contain uncertainty terms which model the designer's lack of knowledge about parameter values and disturbances. Such poorly known quantities may be assumed to be constant or time-varying. Uncertainties arising from imperfect knowledge of system inputs and inaccuracies in the mathematical modelling itself, contribute to performance degradation of the feedback control system. In self-tuning and other stochastic adaptive control systems, the parameter values and disturbances are constantly monitored using on-line identification algorithms, and appropriate adaptive globally stable controllers are implemented. These schemes, however, are costly and result in additional complexity. Simplicity and reliability are not features which should be sacrificed in a control system.

In direct contrast to these adaptive controllers, the deterministic control of uncertain time-varying systems proposes the use of straightforward *fixed nonlinear* feedback control functions, which operate effectively over a specified magnitude range of system parameter variations and disturbances, without any on-line identification of the system parameters. An immediate advantage of such an approach is that no statistical information of the system variations is required to yield the desired dynamic behaviour, and robustness is achieved, not in an average sense, but for all possible values of the underlying uncertainty. The deterministic approach thus contrasts sharply with many other adaptive control schemes, which require global parameter convergence properties. Furthermore, if the parameter variations satisfy certain matching conditions, complete insensitivity to system variations can be achieved. The main areas of application of deterministic control of uncertain systems include electric motor drives, robotics, flight control, space systems, power electronics, chemical processes, automotive control systems and magnetic levitation.

The two main approaches to deterministic control of uncertain systems are the Sliding Mode Control technique using a special behaviour of variable structure systems called the sliding regime; the second approach is generically known as the Lyapunov control design technique. The outstanding feature of these controllers is their excellent robustness and invariance properties.

The essential property of Variable Structure Control (VSC) is that the discontinuous feedback control switches on one or more manifolds in the state space. Thus the structure of the feedback system is altered or switched as the state crosses each discontinuity surface. Sliding motion occurs when the system state repeatedly crosses and immediately re-crosses a switching surface, because all motion in the neighbourhood of the manifold is directed inwards towards the manifold. Following an initial trajectory onto the switching (sliding) surfaces, the system state is constrained to lie upon these surfaces and is said to be in the *sliding mode*. In the sliding mode the system is totally invariant to a class of matched disturbances and parameter variations with known upper and lower bounds; the decoupled system dynamics then being wholly described by the reduced order dynamics of the selected sliding surfaces. In the sliding mode the control element has high (theoretically infinite) gain, while the control actually passed onto the plant takes finite values.

The discontinuous controller can be replaced in many practical applications with continuous nonlinear control which yields a dynamic response arbitrarily close to the discontinuous controller, but without undesirable chatter motion. Following an initial trajectory onto the switching (sliding) surfaces, the system state is constrained to lie in a neighbourhood of these surfaces.

VSC with a sliding mode was first studied intensively in the 1960's by Russian authors, notably Emel'yanov and Utkin, although early work was also done by Flügge-Lotz in the 1950's. In recent years the subject has attracted the attention of numerous researchers. This is reflected in learned journals, books, technical sessions at control conferences and workshops. The basic interest in the technique stems because of its applicability to linear and nonlinear dynamical systems as well as to systems with delays and distributed parameters. VSC is particularly well suited to the deterministic control of uncertain control systems. Some of the major interests have been the use of VSC and allied techniques in model-following and model reference adaptive control, tracking control and observer systems.

In the 1970's research work consolidated the linear scalar case and some attempts had been made at solving the more complex multivariable control problem. The introduction of the geometric approach to linear systems theory was rapidly translated into a general technique which allowed the solution in full generality of the sliding mode control of linear multivariable systems. However, some basic problems still remained to be solved. Most notably, the state observation problem for perturbed linear systems, needed a solution from the viewpoint of deterministic uncertainty. The 1980's witnessed the emergence of the initial steps of a general theory for nonlinear systems, most notably, the differential geometric approach for the study of nonlinear systems structure.

The theoretical results for smooth systems were rapidly translated into a more intuitive theory, while there has been more rigorous formulation of sliding mode control for nonlinear systems. The theory has now been extended to distributed parameter systems described by linear partial differential equations and delay differential systems. The sliding mode control of discrete time systems for linear and nonlinear systems remained largely unexplored until recently. Important contributions in the area of adaptation and identification of dynamical systems using sliding mode control, were made towards the end of the last decade. A user-friendly CAD design package is now available in the MATLAB environment; thus allowing the control designer who is not expert in VSC to straightforwardly design and simulate sliding mode controllers.

Recent research is beginning to consolidate nonlinear systems theory from both a geometric and an algebraic viewpoint. The algebraic approach to cast linear and nonlinear systems in a unified framework has been researched only recently, and the implications of non-traditional state space representations for dynamical systems has yielded interesting emerging consequences in sliding mode control theory.

Lyapunov control follows the approach of early research workers such as Leitmann, Corless, Gutman, Palmor and Ryan. Using a Lyapunov function and specified magnitude bounds on the uncertainties, a nonlinear control law is developed to ensure uniform ultimate boundedness of the closed-loop feedback trajectory to achieve sufficient accuracy. The resulting controller is a discontinuous control function, with generally continuous control in a boundary layer in the neighbourhood of the switching surface. The boundary layer control prevents the excitation of high-frequency unmodelled parasitic dynamics. Controllers have been devised for numerous types of system for many different classes of uncertainty. The control is designed using a Lyapunov design approach and allows for a range of expected system variation.

The chapters in this book cover the whole spectrum of Variable Structure and Lyapunov Control research and design. After an introductory chapter on the theoretical and practical design of multivariable VSC systems, there are chapters covering numerous aspects of VSC including novel mathematical approaches exploring a differential algebraic approach for the sliding mode control of nonlinear single-input single-output systems and module theory for the study of sliding modes for multivariable linear systems; robust control for systems with matched and unmatched uncertainty; a frequency domain design approach; discrete-time control; observer-control systems; the control of uncertain infinite-dimensional systems; and model-following control systems. Using the Lyapunov approach universal adaptive nonlinear feedback controllers are developed, and quadratic Lyapunov techniques are reviewed. Applications presented in detail include automobile fuel-injection control, magnetic levitation and industrial robotics.

The editor wishes to thank all the authors for their diligent cooperation in preparing their manuscripts in LaTeX, which has allowed efficient and speedy text processing using the international electronic mail network. In particular I wish to thank Hebertt Sira-Ramírez and Sarah Spurgeon for their scholarly advice and careful review of many of the chapters. Numerous other reviewers have also assisted in the preparation of this monograph. Mike Piff has patiently allowed unlimited access to his vast personal knowledge of LaTeX, while Madeleine Floy, Judith Smith and Cerys Morgan have provided efficient administrative and secretarial assistance. Finally I wish to thank my family for their encouragement and patience during the preparation of the book.

University of Sheffield Alan Zinober

August 1993

Table of Contents

List of Contributors

Francesco Amato
Dipartimento di Informatica e Sistemistica
Università degli Studi di Napoli Federico II
via Claudio 21
Napoli 80125
Italy

Professor Giorgio Bartolini
Dipartimento di Informatica Sistemistica e Telematica
Universita di Genova
Via Opera Pia 11A
Genova 16145
Italy

Professor Dong-Il Dan Cho
Department of Control and Instrumentation Engineering
Seoul National University
Seoul 171-742
Korea

Professor Martin Corless
School of Aeronautics and Astronautics
Purdue University
West Lafayette
Indiana 47907
USA

Professor Boris Curk
Faculty of Technical Sciences
University of Maribor
Smetanova 17
62000 Maribor
Slovenia

Raymond Davies
Department of Engineering
University of Leicester
University Road
Leicester LE1 7RH
UK

Professor Ahmet Denker
Department of Electrical and Electronic Engineering
Bogazici University
80815 Bebek
Istanbul
Turkey

Professor Sergey V Drakunov
The Ohio State University
2015 Neil Avenue
Columbus
Ohio 43210
USA

Christopher Edwards
Department of Engineering
University of Leicester
University Road
Leicester LE1 7RH
UK

Professor Antonella Ferrara
Dipartimento di Informatica Sistemistica e Telematica
Universita di Genova
Via Opera Pia 11A
Genova 16145
Italy

Professor Franco Garofalo
Dipartimento di Informatica e Sistemistica
Università degli Studi di Napoli Federico II
via Claudio 21
Napoli 80125
Italy

Professor Luigi Glielmo
Dipartimento di Informatica e Sistemistica
Università degli Studi di Napoli Federico II
via Claudio 21
Napoli 80125
Italy

Dr David P Goodall
School of Mathematical and Information Sciences
Coventry University
Priory Street
Coventry CV1 5FB
UK

Professor Jože Harnik
Faculty of Technical Sciences
University of Maribor
Smetanova 17
62000 Maribor
Slovenia

Professor Hideki Hashimoto
Institute of Industrial Science
University of Tokyo
7-22-1 Roppongi
Manato-ku
Tokyo 106
Japan

Professor Karel Jezernik
Faculty of Technical Sciences
University of Maribor
Smetanova 17
62000 Maribor
Slovenia

Dr Efthimios Kappos
Dept of Applied and Computational Mathematics
University of Sheffield
Sheffield S10 2TN
UK

Professor Okyay Kaynak
Department of Electrical Engineering
Bogazici University
P.K.2
Bebek
Istanbul 80815
Turkey

Yusuke Konno
Institute of Industrial Science
University of Tokyo
7-22-1 Roppongi
Manato-ku
Tokyo 106
Japan

Professor Ümit Özgüner
The Ohio State University
2015 Neil Avenue
Columbus
Ohio 43210
USA

Dr Eugene P Ryan
School of Mathematical Sciences
University of Bath
Claverton Down
Bath BA2 7AY
UK

Professor Hebertt Sira-Ramírez
Departamento Sistemas de Control
Escuela de Ingeniería de Sistemas
Facultad de Ingeniería U.L.A.
Avenida Tulio Febres Cordero
Universidad de Los Andes
Mérida 5101
Venezuela

Dr Sarah Spurgeon
Department of Engineering
University of Leicester
University Road
Leicester LE1 7RH
UK

Dr Alexander A Stotsky
Institute for Problems of Mechanical Engineering
Academy of Sciences of Russia
Lensoveta Street 57-32
196143 St Petersburg
Russia

Professor V I Utkin
Discontinuous Control Systems Laboratory
Institute of Control Sciences
Russian Academy of Sciences
Profsoyuznaya 65
GSP-312 Moscow
Russia

Dr Xinghuo Yu
Department of Mathematics and Computing
University of Central Queensland
Rockhampton
Queensland 4702
Australia

Dr Alan S I Zinober
Department of Applied and Computational Mathematics
University of Sheffield
Sheffield S10 2TN
UK

1. An Introduction to Sliding Mode Variable Structure Control

Alan S.I. Zinober

1.1 Introduction

The main features of the sliding mode and the associated feedback control law of Variable Structure Control (VSC) systems will be summarized in this chapter. Some of the important features have already been summarized in the Preface.

Variable Structure Control (VSC) is a well-known solution to the problem of the deterministic control of uncertain systems, since it yields invariance to a class of parameter variations (Draženović 1969, Utkin 1977, 1978 and 1992, Utkin and Yang 1978, DeCarlo et al 1988, Zinober 1990). The characterizing feature of VSC is sliding motion, which occurs when the system state repeatedly crosses certain subspaces, or sliding hyperplanes, in the state space. A VSC controller may comprise nonlinear and linear parts, and has been well studied in the literature.

Numerous practical applications of VSC sliding control have been reported in the literature. These include aircraft flight control (Spurgeon et al 1990), helicopters flight control, spacecraft flight control, ship steering, turbogenerators, electrical drives, overhead cranes, industrial furnace control (see Chapter 3), electrical power systems (see Chapter 8), robot manipulators (see Chapters 17 and 18), automobile fuel injection (see Chapter 16) and magnetic levitation (see Chapter 16).

The methods outlined below yield sliding hyperplanes by various approaches including complete and partial eigenstructure assignment, and reduction of the sensitivity to unmatched parameter variations. The design of the necessary sliding hyperplanes and control law may be readily achieved using the user-friendly VSC Toolbox programmed in the MATLAB environment.

One method of hyperplane design is to specify null space eigenvalues within the left-hand half-plane for the reduced order equivalent system, which are associated with the sliding hyperplanes, and to design the control to yield these eigenvalues (Dorling and Zinober 1986). Additionally one may wish to specify fully (or partially) the eigenvectors corresponding to the closed-loop eigenvalues. There exists the additional possibility of reducing the sensitivity of the specified eigenvalues to unmatched parameter uncertainty (Dorling and Zinober 1988).

An alternative design approach is to specify some region in the left-hand half-plane within which these eigenvalues must lie. Regions studied include the disc, vertical strip and damping sector in the left-hand half-plane. The methods for ensuring that the eigenvalues will lie in the required region, involve the solution of certain matrix Riccati equations (Woodham and Zinober 1990, 1991a, 1991b, 1993).

After presenting the underlying theory of the sliding mode, we shall describe some of the techniques relating to the design of the sliding hyperplanes. For completeness we also present a suitable control law to ensure the attainment of the sliding mode. A straightforward scalar illustrative example is presented. We describe briefly the quadratic performance approach (Utkin and Yang 1978), and then consider eigenstructure assignment and the mixed eigenstructure and sensitivity reduction problem. The sliding hyperplanes for a Remotely Piloted Vehicle are designed to illustrate the theory.

1.2 Regulator System

As our basic control system we shall consider the uncertain regulator

$$\dot{x}(t) = [A + \Delta A(t)]x(t) + [B + \Delta B(t)]u(t) + f(x, u, t) \tag{1.1}$$

where x is an n-vector of states and u is an m-vector of controls. It is assumed that $n > m$, that B is of full rank m and that the pair (A, B) is controllable. The matrices ΔA and ΔB represent the variations and uncertainties in the plant parameters and the control interface respectively. f represents uncertain time-varying additive terms. It is assumed further that the parameter uncertainties and disturbances are matched, occurring only on the control channels, i.e. $\mathcal{R}(B) = \mathcal{R}([B, \Delta B])$ (where $\mathcal{R}(\cdot)$ denotes the range space); and that rank $[B + \Delta B(t)] = m$ for all $t \geq 0$. This implies that for suitable choice of limiting values of the control, one can achieve *total invariance* to parameter variations and disturbances (Draženović 1969).

The overall aim of VSC is to drive the system state from an initial condition $x(0)$ to the state space origin as $t \to \infty$. The jth component u_j $(j = 1, \ldots, m)$ of the state feedback control vector $u(x)$ has a discontinuity on the jth switching surface which is a hyperplane M_j passing through the state origin. Defining the hyperplanes by

$$M_j = \{x : c_j x = 0\}, \quad (j = 1, 2, \ldots, m) \tag{1.2}$$

(where c_j is a row n-vector), the sliding mode occurs when the state lies in M_j for all j, i.e. in the sliding subspace

$$\mathcal{M} = \bigcap_{j=1}^{m} M_j \tag{1.3}$$

In practice the control discontinuity may be replaced by a *soft* nonlinearity to reduce chattering (Burton and Zinober 1986).

1.3 Model-Following Control System

Model-following control systems are very widely used in practice and VSC can be designed in a manner very similar to the regulator system (Zinober et al 1982). Consider the system

$$\begin{aligned}\dot{x}(t) &= [A + \Delta A(t)]x(t) + [B + \Delta B(t)]u(t) + f(x, u) \\ \dot{x}_m(t) &= A_m x_m(t) + B_m r(t) \qquad (1.4)\end{aligned}$$

where the first equation (as in (1.1)) describes the actual plant, and the second equation is the model plant with x_m an n-vector of model states and r a vector of reference inputs. It is desired that the actual plant states follow the model states. The error

$$e(t) = x_m(t) - x(t) \qquad (1.5)$$

should be forced to zero as time $t \to \infty$ by suitable choice of the control u.

Subject to the matrices A, B, ΔA, ΔB, A_m and B_m satisfying certain structural and matching properties (Landau 1979), we can achieve the desired objective with suitable control. The error model satisfies

$$\dot{e}(t) = A_m e(t) + [(A_m - A)x(t) - f + B_m r(t)] - Bu(t) \qquad (1.6)$$

and, subject to certain matching conditions (see Spurgeon et al 1990), the model equations with suitable linear control components, reduce to

$$\dot{e}(t) = A_m e(t) - Bu(t) \qquad (1.7)$$

The VSC of this error system may be readily designed, using the techniques previously described, by associating e with x in earlier sections. The sliding hyperplanes are now in the error state space.

Further details and examples of practical time-varying and nonlinear avionics systems are given in Spurgeon et al (1990).

1.4 The Sliding Mode

When considering the synthesis of the sliding hyperplanes, it is sufficient to study the *ideal* regulator system, without uncertainties and disturbances, given by

$$\dot{x}(t) = Ax(t) + Bu(t) \qquad (1.8)$$

Matched uncertainties are handled by suitable choice of the control function. From (1.2) the sliding mode satisfies

$$s = Cx(t) = 0 \ , \qquad t \geq t_s \qquad (1.9)$$

where t_s is the time when the sliding subspace is reached, and C is an $m \times n$ matrix. Differentiating equation (1.9) with respect to time, and substituting for $\dot{x}(t)$ from (1.8) gives

$$C\dot{x}(t) = CAx(t) + CBu(t) = 0 \ , \qquad t \geq t_s \tag{1.10}$$

Equation (1.10) may be rearranged to give

$$CBu(t) = -CAx(t) \tag{1.11}$$

The hyperplane matrix C is selected so that $|CB| \neq 0$, and therefore the product CB is invertible. Hence (1.11) may be rearranged to give the following expression for the *equivalent control*

$$u_{eq}(t) = -(CB)^{-1}CAx(t) = -Kx(t) \tag{1.12}$$

where $u_{eq}(t)$ is the linear open-loop control which would force the trajectory to remain in the null space of C, during sliding. Substituting for $u_{eq}(t)$ from equation (1.12) into (1.8) gives

$$\begin{aligned} \dot{x}(t) &= \{I - B(CB)^{-1}C\}Ax(t) \ , \qquad t \geq t_s & (1.13) \\ &= (A - BK)x(t) & (1.14) \end{aligned}$$

which is the system equation for the closed-loop system dynamics during sliding.

This motion is independent of the actual nonlinear control and depends only on the choice of C, which determines the matrix K. The purpose of the control u is to drive the state into the sliding subspace $\mathcal{M}$, and thereafter to maintain it within the subspace $\mathcal{M}$.

The convergence of the state vector to the origin is ensured by suitable choice of the feedback matrix K. The determination of the matrix K or alternatively, the determination of the matrix C defining the subspace $\mathcal{M}$, may be achieved without prior knowledge of the form of the control vector u. (The reverse is not true). The null space of C, $\mathcal{N}(C)$, and the range space of B, $\mathcal{R}(B)$, are, under the hypotheses given earlier, complementary subspaces, so $\mathcal{N}(C) \cap \mathcal{R}(B) = \{\emptyset\}$. Since motion lies entirely within $\mathcal{N}(C)$ during the ideal sliding mode, the dynamic behaviour of the system during sliding is unaffected by the controls, as they act only within $\mathcal{R}(B)$. The development of the theory and design principles is simplified by using a particular canonical form for the system, which is closely related to the Kalman canonical form for a multivariable linear system.

By assumption the matrix B has full rank m; so there exists an orthogonal $n \times n$ transformation matrix T such that

$$TB = \begin{pmatrix} 0 \\ B_2 \end{pmatrix} \tag{1.15}$$

where B_2 is $m \times m$ and nonsingular. The orthogonality restriction is imposed on T for reasons of numerical stability, and to remove the problem of inverting T when transforming back to the original system. The transformed state is $y = Tx$, and the state equation becomes

$$\dot{y}(t) = TAT^T y(t) + TBu(t) \tag{1.16}$$

The sliding condition is $CT^Ty(t) = 0, \ \forall \, t \geq t_s$. If the transformed state y is now partitioned as

$$y^T = (y_1^T \quad y_2^T); \quad y_1 \in \mathcal{R}^{n-m}, \quad y_2 \in \mathcal{R}^m \tag{1.17}$$

and the matrices $\mathcal{T}A\mathcal{T}^T$, $\mathcal{T}B$ and $C\mathcal{T}^T$ are partitioned accordingly, then equation (1.16) may be written as the following pair of equations

$$\begin{aligned} \dot{y}_1(t) &= A_{11}y_1(t) + A_{12}y_2(t) \\ \dot{y}_2(t) &= A_{21}y_1(t) + A_{22}y_2(t) + B_2u(t) \end{aligned} \tag{1.18}$$

The sliding condition becomes

$$C_1y_1(t) + C_2y_2(t) = 0 \,, \quad t \geq t_s \tag{1.19}$$

where

$$\mathcal{T}A\mathcal{T}^T = \begin{pmatrix} A_{11} & A_{12} \\ A_{21} & A_{22} \end{pmatrix}, \quad C\mathcal{T}^T = (C_1 \quad C_2) \tag{1.20}$$

and C_2 is nonsingular because CB is nonsingular.

This canonical form is central to hyperplane design methods and it plays a significant role in the solution of the reachability problem, i.e. the determination of the control form ensuring the attainment of the sliding mode in $\mathcal{M}$ (Utkin and Yang 1978, Dorling and Zinober 1986).

Equation (1.19) defining the sliding mode is equivalent to

$$y_2(t) = -Fy_1(t) \tag{1.21}$$

where the $m \times (n-m)$ matrix F is defined by

$$F = C_2^{-1}C_1 \tag{1.22}$$

so that in the sliding mode y_2 is related linearly to y_1. The sliding mode satisfies equation (1.21) and

$$\dot{y}_1 = A_{11}y_1(t) + A_{12}y_2(t) \tag{1.23}$$

This represents an $(n-m)$th order system in which y_2 plays the role of a state feedback control. So we get

$$\dot{y}_1(t) = (A_{11} - A_{12}F)y_1(t) \tag{1.24}$$

which is known as the reduced order equivalent system. The design of a stable sliding mode such that $y \to 0$ as $t \to \infty$, requires the determination of the gain matrix F such that $A_{11} - A_{12}F$ has $n-m$ left-hand half-plane eigenvalues.

1.5 Nonlinear Feedback Control

Once the sliding hyperplanes have been selected, attention must be turned to solving the reachability problem. This involves the selection of a state feedback control function $u : \mathcal{R}^n \to \mathcal{R}^m$ which will drive the state x into $\mathcal{N}(C)$

and thereafter maintain it within this subspace. There is a virtually unlimited number of possible forms for this control function, the only essential features of the form chosen being discountinuity on one or more subspaces containing $\mathcal{N}(C)$. In general the variable structure control law consists of two additive parts ; a linear control law u_ℓ and a nonlinear part u_n, which are added to form u. The linear control is merely a state feedback controller

$$u_\ell(x) = Lx \tag{1.25}$$

while the nonlinear feedback controller u_n incorporates the discountinuous elements of the control law. Consider here the unit vector control

$$u_n(x) = \rho\frac{s}{\|s\|} = \rho\frac{Cx}{\|Cx\|} \ , \quad \rho > 0 \tag{1.26}$$

in the form (Ryan and Corless 1984)

$$u(x) = Lx + \rho\frac{Nx}{\| Mx \|} \tag{1.27}$$

where the null spaces of N, M and C are coincident : $\mathcal{N}(N) = \mathcal{N}(M) = \mathcal{N}(C)$. Starting from the transformed state y, we form a second transformation $\mathcal{T}_2 : \mathcal{R}^n \to \mathcal{R}^n$ such that

$$z = \mathcal{T}_2 y \tag{1.28}$$

where

$$\mathcal{T}_2 = \begin{pmatrix} I_{n-m} & 0 \\ F & I_m \end{pmatrix} \tag{1.29}$$

The matrix T_2 is clearly nonsingular, with inverse

$$\mathcal{T}_2^{-1} = \begin{pmatrix} I_{n-m} & 0 \\ -F & I_m \end{pmatrix} \tag{1.30}$$

Partitioning $z^T = (z_1^T \quad z_2^T)$ with $z_1 \in \mathcal{R}^{n-m}$ and $z_2 \in \mathcal{R}^m$

$$z_1 = y_1; \quad z_2 = Fy_1 + y_2 \tag{1.31}$$

from which it is clear that the conditions $s \equiv 0$ and $z_2 \equiv 0$ are equivalent (in the sense that the points of the state space at which $s = 0$ are precisely the points at which $z_2 = 0$). The transformed system equations become

$$\dot{z}_1 = \Sigma z_1 + A_{12} z_2 \tag{1.32}$$

$$\dot{z}_2 = \Theta z_1 + \Phi z_2 + B_2 u \tag{1.33}$$

where

$$\begin{aligned} \Sigma &= A_{11} - A_{12}F \\ \Theta &= F\Sigma - A_{22}F + A_{21} \\ \Phi &= FA_{12} + A_{22} \end{aligned} \tag{1.34}$$

In order to attain the sliding mode it is necessary to force z_2 and $\dot{z}_2$ to become identically zero. Define the linear part of the control to be

$$u_\ell(z) = -B_2^{-1}\{\Theta z_1 + (\Phi - \Phi_*)z_2\} \tag{1.35}$$

where Φ_* is any $m \times m$ matrix with left-hand half-plane eigenvalues. Transforming back into the original state space (x-space) gives

$$L = -B_2^{-1}(\Theta \quad \Phi - \Phi_*)T_2T \tag{1.36}$$

The linear control law u_ℓ drives the state component z_2 to zero asymptotically; to attain $\mathcal{N}(C)$ in finite time, the nonlinear control component u_n is required. This nonlinear control must be discontinuous whenever $z_2 = 0$, and continuous elsewhere. Letting P_2 denote the positive definite unique solution of the Lyapunov equation

$$P_2\Phi_* + \Phi_*^T P_2 = -I_m \tag{1.37}$$

then $P_2z_2 = 0$ if and only if $z_2 = 0$, and we may take

$$u_n(z) = -\rho\frac{B_2^{-1}P_2z_2}{\| P_2z_2 \|}, \quad z_2 \neq 0 \tag{1.38}$$

where $\rho > 0$ is a scalar parameter to be selected by the designer. When $z_2 = 0$, u_n may be arbitrarily defined as any function satistying $\| u_n \| \leq \rho$. Expressing the control in x-space, we have

$$N = -B_2^{-1}(0 \quad P_2)T_2T \tag{1.39}$$

$$M = (0 \quad P_2)T_2T \tag{1.40}$$

For the more general system (1.1) in which disturbances and uncertainties are present, a similar control structure may be employed. However, in this case the scalar ρ of (1.38) is replaced by a time-varying state-dependent function incorporating two design parameters γ_1, γ_2, upon which the time to reach $\mathcal{N}(C)$ also depends (Ryan and Corless 1984).

Discontinuous control produces chatter motion in the neighbourhood of the sliding surface. In may practical applications this cannot be tolerated. There are numerous techniques to "smooth" the control function. Perhaps the most straightforward smoothed continuous nonlinear control, which eliminates the chatter motion, is (see, for example, Burton and Zinober (1986) and Zinober (1990))

$$u(x) = Lx + \rho\frac{Nx}{\|Mx\| + \delta}, \qquad \delta > 0^+ \tag{1.41}$$

1.6 Second-Order Example

To illustrate some of the main ideas consider the simple scalar double-integrator plant

$$\ddot{x}(t) = bu(t) + f(t)$$

with the positive parameter b uncertain but taking known bounded maximum and minimum values, and $f(t)$ a disturbance. Here the sliding subspace will be a one-dimensional space, a straight line through the state origin

$$s = gx + \dot{x} = 0$$

During sliding for $t > t_s$ we require

$$\lim_{s \to 0^-} \dot{s} > 0 \qquad \text{and} \qquad \lim_{s \to 0+} \dot{s} < 0$$

and then

$$s = 0 \qquad \text{and} \qquad \dot{s} = 0$$

i.e. the state remains on the sliding surface. Then

$$\dot{s} = gx + \dot{x} = 0$$

which yields the dynamics of a reduced *first order* system, i.e. $n - m = 1$. So

$$\dot{x} = -gx$$

with eigenvalue $-g$ and

$$x(t) = x(t_s)e^{-gt}$$

So one obtains *exactly* the closed-loop eigenvalue $-g$ by specifying the sliding line (1.6). The dynamics in the sliding mode are independent of the parameter b.

The discontinuous control

$$u = \rho \frac{s}{|s|}$$

can maintain sliding motion on $s = 0$ within a bounded region of the state origin, for a range of values of b with the precise value of b not required to be known (Utkin 1977, 1978, 1992). The equivalent control which theoretically can maintain the state on the sliding line, is the linear control

$$u_{eq} = -\frac{gx}{b}$$

To achieve the sliding mode with this linear control would require *exact* knowledge of b; unlike for the case of nonlinear control.

Smooth nonlinear control has the form

$$u = \rho \frac{s}{|s| + \delta}$$

Simulation results are presented in Figs. 1.1 and 1.2 for discontinuous ($\delta = 0$) and smooth control ($\delta = 0.01$) with $\rho = 1$. The state trajectories are very similar. During the sliding mode the smooth control is equal to the equivalent control u_{eq}, which is included in the control graph of Fig. 1.1. Note the elimination of chatter when using the smooth control. The invariance of the system to a matched disturbance function $f(t)$ is demonstrated in Fig. 1.3 for the case of smooth control ($\delta = 0.01$).

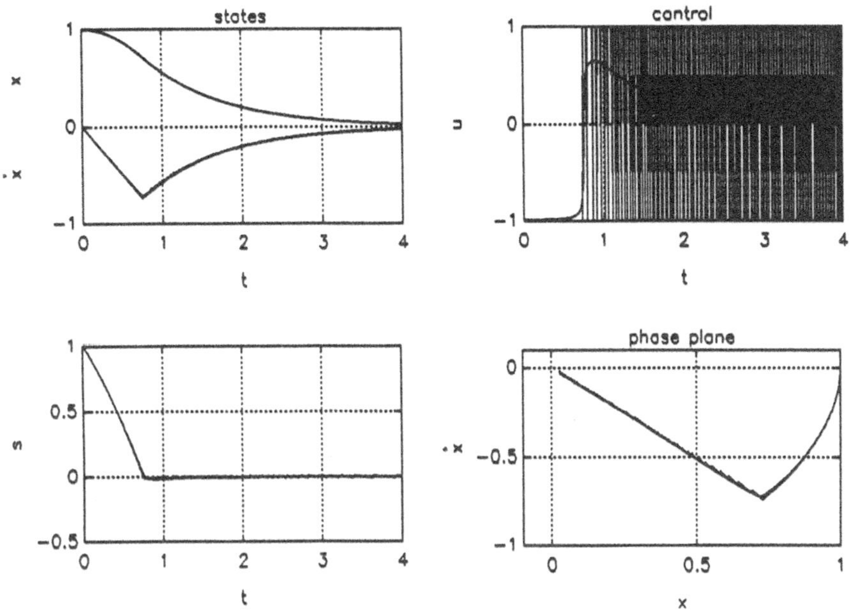

Fig. 1.1. Double integrator plant with discontinuous control

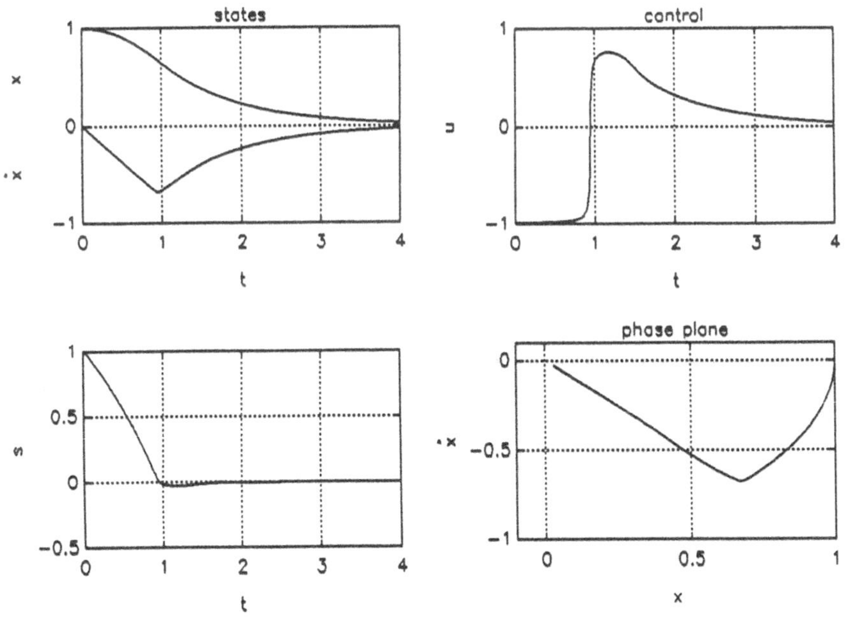

Fig. 1.2. Double integrator plant with smooth control

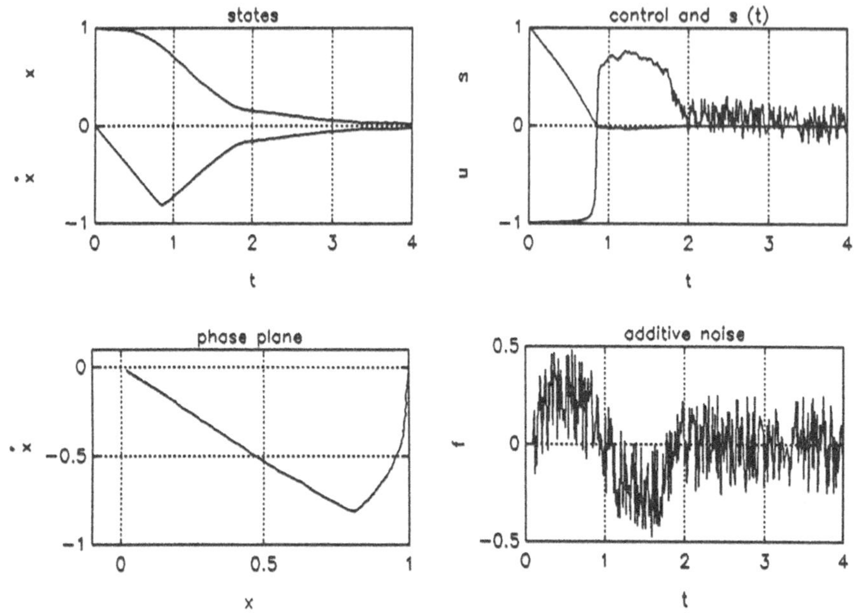

Fig. 1.3. Double integrator plant with smooth control and disturbance

1.7 Quadratic Performance

One way to design the sliding hyperplanes is by minimizing the quadratic performance

$$J = \frac{1}{2}\int_{t_s}^{\infty} x^T Q x \,\mathrm{d}t \tag{1.42}$$

where the matrix $Q > 0$ is positive definite symmetric and t_s is the time of attaining the sliding mode. Partitioning the product compatibly with y, and defining

$$\hat{Q} = Q_{11} - Q_{12}Q_{22}^{-1}Q_{21} \tag{1.43}$$

$$\hat{A} = A_{11} - A_{12}Q_{22}^{-1}Q_{21} \tag{1.44}$$

$$v(t) = y_2(t) + Q_{22}^{-1}Q_{21}y_1(t) \tag{1.45}$$

this problem may be restated as: minimize

$$J(v) = \frac{1}{2}\int_{t_s}^{\infty} \left\{ y_1^T(t)\hat{Q}y_1(t) + v^T(t)Q_{22}v(t) \right\} \mathrm{d}t \tag{1.46}$$

subject to

$$\dot{y}_1(t) = \hat{A}y_1(t) + A_{12}v(t) \tag{1.47}$$

which has the form of the standard linear quadratic optimal regulator problem (Utkin and Yang 1978).

The controllability of (A, B) is sufficient to ensure the controllability of $(\hat{A}, A_{12})$. Moreover, the positivity condition on Q ensures that $Q_{22} > 0$ (so that Q_{22}^{-1} exists) and that $\hat{Q} > 0$. Thus a positive-definite unique solution P is guaranteed for the algebraic matrix Riccati equation

$$P\hat{A} + \hat{A}^T P - PA_{12}Q_{22}^{-1}A_{12}^T P + \hat{Q} = 0 \tag{1.48}$$

associated with the problem (1.46), and the optimal control v is given by

$$v(t) = -Q_{22}^{-1}A_{12}^T P y_1(t) \tag{1.49}$$

Using (1.45) this may be transformed to give

$$y_2(t) = -Q_{22}^{-1}\left(Q_{21} + A_{12}^T P\right) y_1(t) = -F y_1(t) \tag{1.50}$$

and F is readily determined once the matrix Riccati equation (1.48) has been solved.

1.8 Eigenstructure Assignment

For the multiple input case Utkin and Yang (1978) have shown that the pair (A_{11}, A_{12}) is controllable and that eigenvalue assignment in (1.24) is therefore feasible. It is well known, however, that the assignment of eigenvalues of an nth order m-input system requires only n of the nm d.o.f. available in choosing the feedback gain matrix (Shah et al 1975). The remaining $n(m-1)$ d.o.f. may be utilized in some other way; in particular, by partially assigning the eigenvectors. For convenience it is assumed here that the nonzero sliding mode eigenvalues are distinct from each other and from the eigenvalues of A_{11}. Suppose that the sliding mode has commenced on $\mathcal{N}(C)$. Then

$$\dot{x}(t) = (A - BK)x(t) \tag{1.51}$$

where K is defined by (1.12). During the sliding mode x must remain in $\mathcal{N}(C)$, so that

$$C(A - BK) = 0 \Longleftrightarrow \mathcal{R}(A - BK) \subseteq \mathcal{N}(C) \tag{1.52}$$

Let $\{\lambda_i : i = 1, ..., n\}$ be the eigenvalues of $A - BK$ with corresponding eigenvectors v_i. Then (1.52) implies that

$$C(A - BK)v_i = \lambda_i C v_i = 0 \tag{1.53}$$

so that either λ_i is zero or $v_i \in \mathcal{N}(C)$. Now $A_{eq} = A - BK$ has precisely m zero-valued eigenvalues, so let $\Lambda = \{\lambda_i : i = 1, ..., n-m\}$ be the nonzero distinct eigenvalues. Specifying the corresponding eigenvectors $\{v_i : i = 1, \ldots, n-m\}$ fixes the null space of C, since dim $\mathcal{N}(C) = n - m$. However, C is not uniquely determined, because

$$CV = 0 \ , \qquad V = (v_1 \dots v_{n-m}) \tag{1.54}$$

has m^2 degrees of freedom. Defining

$$W = \begin{pmatrix} W_1 \\ W_2 \end{pmatrix} = \mathcal{T}V \tag{1.55}$$

with the partitioning of W compatible with that of y, (1.54) becomes

$$\begin{aligned} 0 = C\mathcal{T}^T\mathcal{T}V &= \begin{pmatrix} C_1 & C_2 \end{pmatrix} \begin{pmatrix} W_1 \\ W_2 \end{pmatrix} \\ &= C_2 \begin{pmatrix} F & I_m \end{pmatrix} \begin{pmatrix} W_1 \\ W_2 \end{pmatrix} \end{aligned} \tag{1.56}$$

giving the equation

$$FW_1 = -W_2 \tag{1.57}$$

The eigenvectors v_i of $A - BK$ are not generally freely assignable. Shah et al (1975) have shown that at most m elements of an eigenvector may be assigned arbitrarily, after which the remaining $n-m$ elements are fully determined by the assigned elements. Consider the assignable subspace corresponding to a given eigenvalue (Klein and Moore 1977). It has been shown by Sinswat and Fallside (1977) that this assignable subspace for an eigenvalue λ_i may be characterized as the null space of the $n \times n$ matrix $H(\lambda_i)$ defined by

$$\begin{aligned} H(\lambda_i) &= (I_n - BB^+)(A - \lambda_i I_n) \\ B^+ &= (B^T B)^{-1} B^T \end{aligned} \tag{1.58}$$

which follows from the requirements that $(A - \lambda_i I_n)v_i$ must lie in $\mathcal{R}(B)$. The transformation matrix $\mathcal{T}$ is nonsingular, so

$$H(\lambda_i)v = 0 \Longleftrightarrow \mathcal{T}H(\lambda_i)\mathcal{T}^T\mathcal{T}v = 0 \tag{1.59}$$

and

$$\begin{aligned} \mathcal{T}H(\lambda_i)\mathcal{T}^T &= (I_n - \mathcal{T}BB^+\mathcal{T}^T)(\mathcal{T}A\mathcal{T}^T - \lambda_i I_n) \\ &= \begin{pmatrix} A_{11} - \lambda_i I_{n-m} & A_{12} \\ 0 & 0 \end{pmatrix} \end{aligned} \tag{1.60}$$

Therefore an arbitrary vector v lies within $\mathcal{N}(H(\lambda_i))$ if and only if $w = \mathcal{T}v \in \mathcal{N}(H^*(\lambda_i))$ where

$$H^*(\lambda_i) = (A_{11} - \lambda_i I_{n-m} \qquad A_{12}) \tag{1.61}$$

Note that $H^*(\lambda_i)$ has dimensions $(n-m) \times n$, and therefore requires less storage space than the original $n \times n$ matrix $H(\lambda_i)$. Moreover, $H^*(\lambda_i)$ provides clarification of the number of degrees of freedom available in assigning the eigenvector corresponding to λ_i, for if $w = \mathcal{T}v$ is partitioned compatibly with y, then $w \in \mathcal{N}(H^*(\lambda_i))$ implies

$$(A_{11} - \lambda_i I_{n-m})w_1 = -A_{12}w_2 \tag{1.62}$$

from which it is clear that fixing the m elements of w_2 uniquely determines w_1, and hence v. Note also that the requirement that W_1 must have linearly independent columns, is a further restriction on the assignable eigenvectors arising from the requirement that the reduced-order system should have distinct eigenvalues and hence linearly independent eigenvectors.

The concept of assignable subspaces is applied to sliding mode design as follows. The designer selects the desired elements of the closed-loop eigenvector v_i corresponding to a nonzero sliding mode eigenvalue λ_i.

If r $(1 \leq r \leq m)$ elements of v_i are specified, the remaining $n-r$ elements are determined directly by solving $H^*(\lambda_i)Tv_i = 0$, taking the minimum norm solution if $r < m$.

If more than m but less than n elements are specified, $m < r < n$, a quadratic programming problem needs to be solved, i.e. determine an assignable eigenvector using a least squares fit to the specified elements whilst minimizing the contribution of the remaining elements.

If all of the n elements of v_i are specified, the assignability of the vector is tested by transforming it and applying (1.61). If this result is nonzero, v_i must be modified to give the *closest* assignable eigenvector, which is found by projection into the current assignable subspace $\mathcal{N}(H^*(\lambda_i)T)$.

1.9 Sensitivity Reduction

When the matching criterion does not hold, VSC will not yield total invariance to all the parameter uncertainties (Utkin 1977). It may be useful to attempt to minimize the sensitivity of the location of the closed-loop eigenvalues to unmatched parameter variations. We can use any remaining degrees of freedom to select unspecified elements of the eigenvectors so as to minimize the sensitivity. An algorithm for sliding hyperplane design has been described (Dorling and Zinober 1988), incorporating the algorithm of Kautsky and Nichols (1983). The algorithm yields a near minimum value for the spectral condition number, $\kappa(V)$, using an iterative algorithm, which minimizes a related conditioning measure κ_c.

In the MATLAB VSC Toolbox an additional algorithm has been included which combines the previous eigenstructure assignment techniques with the sensitivity reduction approach. After computing s $(s < n-m)$ eigenvectors according to the specified criteria, the remaining $n-m-s$ eigenvectors are determined using the iterative sensitivity reduction approach. Available degrees of freedom are used to select unspecified eigenvectors so as to minimize the measure κ_c.

1.10 Eigenvalue Assignment in a Region

1.10.1 Eigenvalue Assignment in a Sector

Of the eigenvalue regional assignment problems, perhaps the most useful practically is the placing of all the closed-loop eigenvalues within a specified damping sector or cone in the left-hand half-plane (Woodham and Zinober 1993). Define a region bounded by a line making an angle θ with the imaginary axis, and crossing the real axis at α, where α is any negative real number; and the reflection of this line in the real axis. The angle θ is measured in an anti-clockwise direction from the imaginary axis, and lies between 0° and 90°. We want to determine the equivalent state feedback matrix F (1.22), such that all the eigenvalues of the closed-loop system lie within the required region. The region is bounded by the lines

$$y \sin\theta + (x - \alpha)\cos\theta = 0 \tag{1.63}$$

$$y \sin\theta - (x - \alpha)\cos\theta = 0 \tag{1.64}$$

The region we are considering is to the left of these lines, and excludes the origin, so we require

$$y \sin\theta + (x - \alpha)\cos\theta < 0 \tag{1.65}$$

and

$$y \sin\theta - (x - \alpha)\cos\theta > 0 \tag{1.66}$$

Consider the matrix equation

$$e^{j\theta} A^* P + e^{-j\theta} PA - 2\alpha P \cos\theta = -Q \tag{1.67}$$

where Q is an arbitrary positive definite matrix and * denotes the complex conjugate transpose. Let λ and ν be an eigenvalue and the corresponding right eigenvector of A, so $A\nu = \lambda\nu$ and $\nu^* A^* = \bar{\lambda}\nu^*$. Premultiply (1.67) by ν^* and postmultiply it by ν to give

$$e^{j\theta} \nu^* A^* P\nu + e^{-j\theta} \nu^* PA\nu - 2\alpha\nu^* P\nu \cos\theta = -\nu^* Q\nu \tag{1.68}$$

Substituting for $A\nu$ and $\nu^* A^*$, and rearranging gives

$$\nu^* P\nu \left(e^{j\theta}\bar{\lambda} + e^{-j\theta}\lambda - 2\alpha\cos\theta \right) = -\nu^* Q\nu \tag{1.69}$$

Let $\lambda = x + jy$, so $\bar{\lambda} = x - jy$. Substituting into (1.69) gives

$$2\left((x - \alpha)\cos\theta + y\sin\theta \right) \nu^* P\nu = -\nu^* Q\nu \tag{1.70}$$

Since Q is positive definite and we require P to be positive definite, it follows that

$$(x - \alpha)\cos\theta + y\sin\theta < 0 \tag{1.71}$$

In other words, if there exists a positive definite solution P to (1.67), all the eigenvalues of the matrix A lie to the left of the line (1.71).

For the sliding mode we require the $(n-m)$ left-hand half-plane closed-loop eigenvalues of the reduced order equivalent system $(A_{11} - A_{12}F)$ to lie within the specified region. It has been shown (Woodham and Zinober 1993) that the feedback matrix F can be determined by solving the complex matrix Riccati equation

$$\begin{aligned} e^{j\theta}(A_{11} - \alpha I)^* P + e^{-j\theta} P(A_{11} - \alpha I) \\ - PA_{12}R^{-1}A_{12}^T P = -Q \end{aligned} \tag{1.72}$$

with R a positive definite $m \times m$ matrix. The real matrix F, is given by

$$F = R^{-1}A_{12}^T \hat{P} \tag{1.73}$$

where

$$\hat{P}_{ij} = \sqrt{P_{ij}\bar{P}_{ij}} \tag{1.74}$$

The choice of the weighting matrix R has an effect on the positioning of the eigenvalues within the region (Woodham and Zinober 1993).

1.10.2 Eigenvalue Assignment in a Disc

The technique of placing all the closed-loop eigenvalues of a system within a specified circular disc with centre $-\alpha + j0$ and radius r has been adapted for use with a VSC system. In this case the $n - m$ closed-loop eigenvalues of the reduced order equivalent sliding system are required to be placed within the specified disc.

Furuta and Kim (1987) have studied the standard linear regulator problem (1.1) with linear feedback $u = Gx$. Consider the matrix equation

$$-\alpha A^* P - \alpha PA + A^* PA + (\alpha^2 - r^2)P = -Q \tag{1.75}$$

where Q is an arbitrary positive definite matrix, * denotes the conjugate transpose of a matrix, and α and r are scalars. Let λ and v respectively be an eigenvalue and right eigenvector of A, then

$$Av = \lambda v \quad \text{and} \quad v^* A^* = \bar{\lambda} v^* \tag{1.76}$$

Premultiplying (1.75) by v^*, postmultiplying by v and using (1.76) gives

$$\{-\alpha(\bar{\lambda} + \lambda) + |\lambda|^2 + \alpha^2 - r^2\} v^* P v = -v^* Q v \tag{1.77}$$

Let $\lambda = x + jy$ with $\bar{\lambda} = x - jy$. Then (1.77) becomes

$$\{(x - \alpha)^2 + y^2 - r^2\} v^* P v = -v^* Q v \tag{1.78}$$

Since Q is positive definite, and we require P to be positive definite, it follows that

$$(x - \alpha)^2 + y^2 - r^2 < 0 \tag{1.79}$$

So all the eigenvalues of A will lie within the disc with centre $-\alpha + j0$ and radius r, if there exists a positive definite solution P of (1.75) (see Furuta and Kim 1987 for *only if* proof). In this case the eigenvalues of $A+BG$ are required to lie within a disc of radius r and centre $\alpha + j0$, so (1.75) becomes

$$\begin{aligned} -\alpha(A+BG)^{*}P - \alpha P(A+BG) \\ + (A+BG)^{*}P(A+BG) \\ + (\alpha^2 - r^2)P = -Q \end{aligned} \tag{1.80}$$

with

$$G = -(r^2 R + B^T P B)^{-1} B^T P (A - \alpha I) \tag{1.81}$$

where R is an arbitrary positive definite symmetric matrix.

For the sliding mode design using the above framework the $(n-m)$ left-hand half-plane eigenvalues of the $(n-m)$th reduced order system with $\Psi = A_{11} - A_{12}F$ are to be placed in a specified disc. Ψ is of the form $\mathcal{A} + \mathcal{B}F$ with $\mathcal{A} = A_{11}$, $\mathcal{B} = -A_{12}$ and F the *feedback* matrix.

Once r and α have been assigned, (1.80) is solved for P. The choice of the two arbitrary matrices Q and R affects the placement of the eigenvalues within the specified disc (Woodham 1991, Woodham and Zinober 1991, 1991b). If R is chosen to be diag$\{r_1, r_2, .., r_m\}$ and the linear control is $u = \mathcal{K}Gx$ where

$$\mathcal{K} = \mathrm{diag}\{k_1, k_2, \ldots, k_m\} \tag{1.82}$$

then the gain margin is given by

$$\frac{1}{1+a_i} < k_i < \frac{1}{1-a_i}, \quad i = 1, 2, \ldots, m \tag{1.83}$$

where

$$a_i = \{r^2 r_i / (r^2 r_i + \lambda_{\max})\}^{\frac{1}{2}} \tag{1.84}$$

and $\lambda_{\max}$ is the maximum eigenvalue of $B^T PB$. As $r \to 0$ the a_i approach zero, and the gain margin, which indicates the degree of stability, decreases. Then all the poles of the closed-loop linear system are assigned to the same point and the robustness of the solution may be weak. Thus the choice of the R matrix affects the robustness of the solution.

1.10.3 Eigenvalue Assignment in a Vertical Strip

The problem of placing all the closed-loop eigenvalues of a system within a vertical strip (Juang et al 1989) has been extended for use with the sliding mode. Consider again the general system (1.1) and two positive real numbers h_1 and h_2 with $h_2 > h_1$ which specify the open vertical strip $[-h_2, -h_1]$ on the negative real axis.

Define the matrix

$$\tilde{A} = A + h_1 I \tag{1.85}$$

Suppose that

$$u = -rBR^{-1}B^TP \tag{1.86}$$

where R is an $m \times m$ positive definite symmetric matrix, P is the solution of the Riccati equation

$$PBR^{-1}B^TP - \tilde{A}^TP - P\tilde{A} = 0 \tag{1.87}$$

and the constant gain r is chosen to be

$$r = 0.5 + (h_2 - h_1)/(2\mathrm{Tr}(\tilde{A}^+)) \tag{1.88}$$

where $\mathrm{Tr}(\tilde{A}^+)$ is the sum of the positive eigenvalues of $\tilde{A}$. Then the resulting closed-loop system is

$$\dot{x}(t) = (A - rBK)x(t) \tag{1.89}$$

If $h_2 > \max\{|Re\,\lambda_i|\}$ for all i, where λ_i are the eigenvalues of A, then the eigenvalues of $(A - rBK)$ will all lie within the vertical strip $[-h_2, -h_1]$.

In our case the $n - m$ eigenvalues of $A_{11} - A_{12}F$ are required to be placed within the above vertical strip. It is not possible to move the original eigenvalues (those of A_{11}) towards the right-hand half-plane, so the value of h_2 is limited by the eigenvalues of A_{11}. Having selected h_1, the matrix $\tilde{A}$ is computed. The Riccati equation (1.87) is then solved to give the matrix P. Then $\mathrm{Tr}(\tilde{A}^+)$ and $\max\{|Re\,\lambda_i|\}, i = 1, \ldots, n - m$, are computed, and h_2 is chosen within the limits stated above. Finally the F matrix is computed, and the eigenvalues of $(A_{11} - A_{12}F)$ will be located within the specified vertical strip (Woodham 1991, Woodham and Zinober 1991a, 1991b).

1.11 Example: Remotely Piloted Vehicle

To illustrate some of the design techniques for the sliding hyperplanes we consider the following example of a remotely piloted vehicle (RPV) (Safonov et al 1981, Safonov and Chiang 1988)

$$\dot{x} = Ax + Bu \tag{1.90}$$

where

$$A = \begin{bmatrix} -.0257 & -36.6170 & -18.8970 & -32.0900 & 3.2509 & -.7626 \\ .0001 & -1.8997 & .9831 & -.0007 & -.1708 & -.0050 \\ .0123 & 11.7200 & -2.6316 & .0009 & -31.6040 & 22.3960 \\ 0 & 0 & 1.0000 & 0 & 0 & 0 \\ 0 & 0 & 0 & 0 & -30.0000 & 0 \\ 0 & 0 & 0 & 0 & 0 & -30.0000 \end{bmatrix} \tag{1.91}$$

$$B = \begin{bmatrix} 0 & 0 \\ 0 & 0 \\ 0 & 0 \\ 0 & 0 \\ 30 & 0 \\ 0 & 30 \end{bmatrix} \tag{1.92}$$

$n = 6$ and $m = 2$, and

$$T = \begin{bmatrix} -1 & 0 & 0 & 0 & 0 & 0 \\ 0 & 0 & 0 & 1 & 0 & 0 \\ 0 & 0 & 1 & 0 & 0 & 0 \\ 0 & -1 & 0 & 0 & 0 & 0 \\ 0 & 0 & 0 & 0 & -1 & 0 \\ 0 & 0 & 0 & 0 & 0 & -1 \end{bmatrix} \tag{1.93}$$

$$A_{11} = \begin{bmatrix} -.0257 & 32.0900 & 18.8970 & -36.6170 \\ 0 & 0 & 1.0000 & 0 \\ -.0123 & .0009 & -2.6316 & -11.7200 \\ .0001 & .0007 & -.9831 & -1.8997 \end{bmatrix} \tag{1.94}$$

$$A_{12} = \begin{bmatrix} 3.2509 & -.7626 \\ 0 & 0 \\ 31.6040 & -22.3960 \\ -.1708 & -.0050 \end{bmatrix} \tag{1.95}$$

(i) We consider first the quadratic performance approach. For $Q = I_n$ the VSC Toolbox yields

$$F = \begin{bmatrix} .8640 & 1.7736 & 1.1102 & -2.1252 \\ -.4995 & -1.1616 & -.7489 & 1.2822 \end{bmatrix} \tag{1.96}$$

with resulting eigenvalues

$$\Lambda = \{-31.8682,\ -22.4504,\ -4.9609,\ -.6838\} \tag{1.97}$$

For $Q = \mathrm{diag}(1, 5, 10, 15, 20, 25)$

$$F = \begin{bmatrix} 0.1943 & 1.0389 & .7154 & -1.2218 \\ -.0967 & -.5734 & -.3959 & .6235 \end{bmatrix} \tag{1.98}$$

with $\Lambda = \{-26.3189,\ -4.9716 \pm 2.1072i,\ -.6823\}$.

(ii) Suppose that we wish to place the eigenvalues of the sliding hyperplanes at the $n - m$ $(= 4)$ specified locations in the left-hand half-plane of the complex plane

$$\Lambda = \{-1,\ -2,\ -3,\ -4\} \tag{1.99}$$

Then

$$F = \begin{bmatrix} .7481 & 13.1149 & 8.6078 & -13.8084 \\ 1.0580 & 18.3794 & 12.0859 & -18.9744 \end{bmatrix} \tag{1.100}$$

and

$$C = \begin{bmatrix} -.7481 & 13.8084 & 8.6078 & 13.1149 & -1.0000 & 0 \\ -1.05807 & 18.9744 & 12.0859 & 18.3794 & 0 & -1.0000 \end{bmatrix} \tag{1.101}$$

(iii) For our example, with $m = 2$, we can specify, for a given eigenvalue, at most two desired elements of an eigenvector v_d, and obtain an exact solution. We shall use the symbol $\star$ to indicate an unspecified eigenvector element. We consider the eigenvalue $\lambda = -1$.
For $v_{d_1} = (1, 0, \star, \star, \star, \star)^T$ we obtain

$$v_1 = (1, 0, -.0388, .0388, -.2140, -.3053)^T \tag{1.102}$$

For $v_{d_2} = (1, 0, -0.5, \star, \star, \star)^T$ we obtain

$$v_2 = (1.0040, -.1914, -.3979, .3979, -1.2345, -1.6714)^T \tag{1.103}$$

For $v_{d_3} = (1, 0, -.5, .5, \star, \star)^T$ we obtain

$$v_3 = (1.0045, -.2153, -.4426, .4426, -1.3615, -1.8414)^T \tag{1.104}$$

With the specified eigenvector v_{d_1} we obtain as sliding mode eigenvalues $\Lambda = \{-1,\ -2,\ -3,\ -4\}$ using the sensitivity reduction technique for assigning the remaining eigenvectors, the reduced order (y_1-space) eigenvector matrix

$$W_1 = \begin{bmatrix} -.9985 & .1440 & .1007 & -.9353 \\ .0387 & .4301 & -.3113 & .0009 \\ -.0387 & -.8601 & .9338 & -.0035 \\ 0 & .2334 & -.1446 & -.3539 \end{bmatrix} \tag{1.105}$$

and

$$F = \begin{bmatrix} .3812 & 13.1623 & 8.8499 & -12.8361 \\ .5349 & 18.3054 & 12.3844 & -17.5811 \end{bmatrix} \tag{1.106}$$

(iv) We next assign the eigenvalues of the sliding mode to lie in a sector. Suppose $\alpha = -1$ and $\theta = 0$, with $Q = 10I_4$ and $R = I_2$, we get

$$F = \begin{bmatrix} 2.5601 & 13.4619 & 2.7937 & 3.5231 \\ -1.1207 & -13.5987 & -1.9521 & -4.6892 \end{bmatrix} \tag{1.107}$$

with the eigenvalues

$$\Lambda = \{-121.9553,\ -16.6076,\ -5.4937,\ -1.1107\} \tag{1.108}$$

With $Q = 6I_4$ and $R = I_2$ we obtain

$$\Lambda = \{-94.0677,\ -17.0100,\ -5.3790,\ -1.0104\} \tag{1.109}$$

while $Q = qI_4$, $q \leq 5$ does not place the eigenvalues in the appropriate region.
(v) We can assign the eigenvalues to lie within a disc. For the disc with centre -2 and radius 1 we obtain

$$F = \begin{bmatrix} .4005 & 8.2867 & 7.4507 & -7.6527 \\ .5661 & 11.5734 & 10.4775 & -10.2811 \end{bmatrix} \tag{1.110}$$

and eigenvalues

$$\{-2.3479, \ -1.2518, \ -2.0021 \pm .0044i\} \tag{1.111}$$

(vi) To achieve sufficiently fast motion to the sliding surface we should select the range space dynamics to be suitably fast. Here we select the 2 ($= m$) range space eigenvalues to be -2.5, -3.5, i.e.

$$\Phi^* = \begin{bmatrix} -2.5 & 0 \\ 0 & -3.5 \end{bmatrix} \tag{1.112}$$

For the case (1.105) in (iii) we obtain

$$L = \begin{bmatrix} -.0278 & 4.1795 & 1.0607 & 1.5045 & -8.5208 & 6.6143 \\ -.0568 & 6.4289 & 1.8817 & 2.7077 & -13.2046 & 10.1394 \end{bmatrix} \tag{1.113}$$

$$M = \begin{bmatrix} -.0762 & 2.5672 & 1.7700 & 2.6325 & -.2000 & 0 \\ -.0764 & 2.5116 & 1.7692 & 2.6151 & 0 & -.1429 \end{bmatrix} \tag{1.114}$$

$$N = \begin{bmatrix} -.0025 & .0856 & .0590 & .0877 & -.0067 & 0 \\ -.0025 & .0837 & .0590 & .0872 & 0 & -.0048 \end{bmatrix} \tag{1.115}$$

in the feedback control law (1.27).

1.12 Conclusions

After introducing the concept of the sliding mode some sliding mode design approaches have been described. They are applicable to regulator and model-following systems, and also to tracking problems with suitable modifications. These and other algorithms have been incorporated into a CAD VSC Toolbox in the MATLAB environment. This user-friendly Toolbox (available from the author) allows the control designer to synthesize and simulate the sliding hyperplanes and the feedback control law of a VSC system in a straightforward manner using a wide variety of techniques. The MATLAB platform provides powerful high-level matrix arithmetic and graphical routines to be easily accessed by the user either in a program or keyboard input mode.

1.13 Acknowledgement

The author acknowledges support under the Science and Engineering Research Council grant GR/E46943.

References

Burton, J.A., Zinober, A.S.I. 1986, Continuous approximation of variable structure control. *Int. J. Systems Science* **17**, 876–885

DeCarlo, R.A., Żak, S.H., Matthews, G.P. 1988, Variable structure control of nonlinear multivariable systems: a tutorial. *Proc IEEE* **26**, 1139–1144

Dorling, C.M., Zinober, A.S.I. 1986, Two Approaches to Hyperplane Design in Multivariable Variable Structure Control Systems. *International Journal of Control* **44**, 65–82

Dorling, C.M., Zinober, A.S.I. 1988, Robust hyperplane design in multivariable variable structure control systems. *International Journal of Control* **48**, 2043–2054

Draženović, B. 1969, The invariance conditions in variable structure systems. *Automatica* **5**, 287–295

Furuta, K., Kim, S.B. 1987, Pole Assignment in a Specified Disc. *IEEE Transactions on Automatic Control* **AC-32**, 423–427

Juang, Y-T., Hong, Z-C., Wang, Y-T. 1989, Robustness of Pole Assignment in a Specified Region. *IEEE Transactions on Automatic Control* **AC-34**, 758–760

Kautsky J., Nichols, N.K., Van Dooren, P. 1985, Robust Pole Assignment in Linear State Feedback. *International Journal of Control* 41, 1129–1155

Klein, G., Moore, B.C. 1977, Eigenvalue generalised eigenvector assignment with state feedback. *IEEE Transactions on Automatic Control* **AC–22**, 140–141

Landau, I.D. 1979, *Adaptive control: The model reference approach*, M. Dekker, New York

Ryan, E.P., Corless, M. 1984, Ultimate boundedness and asymptotic stability of a class of uncertain dynamical systems via continuous and discontinuous feedback control. *IMA J Math Control Information* **1**, 223–242

Safonov, M.G., Chiang, R.Y. 1988, CACSD using the state-space L_∞ theory – A design example. *IEEE Transactions on Automatic Control* **AC-33**, 477–479

Safonov, M.G., Laub, A.J., Hartmann, G.L. 1981, Feedback properties of multivariable systems: The role and use of the return difference matrix. *IEEE Transactions on Automatic Control* **AC-26**, 47–65

Shah, S.L., Fisher, D.G., Seborg, D.E. 1975, Eigenvalue/eigenvector assignment for multivariable systems and further results output feedback control. *Electron Letters* **11**, 388–389

Sinswat, V., Fallside, F. 1977, Eigenvalue/eigenvector assignment by state-feedback. *International Journal of Control* **26**, 389–403

Spurgeon, S.K., Yew, M.K., Zinober, A.S.I., Patton, R.J. 1990, Model-following control of time-varying and nonlinear avionics systems, in *Deterministic control of uncertain systems*, ed. Zinober, A.S.I., Peter Peregrinus Press, 96–114

Utkin, V.I. 1977, Variable structure systems with sliding mode. *IEEE Transactions on Automatic Control* **AC-22**, 212–222

Utkin, V.I. 1978, *Sliding Modes and Their Application in Variable Structure Systems*, MIR, Moscow

Utkin, V.I. 1992, *Sliding Modes in Control Optimization*, Springer-Verlag, Berlin

Utkin, V.I., Yang, K.D. 1978, Methods for Constructing Discontinuity Planes in Multidimensional Variable Structure Systems. *Autom. Remote Control* **39**, 1466–1470

Woodham, C.A. 1991, *Eigenvalue Placement for Variable Structure Control Systems*, Ph D Thesis, University of Sheffield

Woodham, C.A., Zinober, A.S.I. 1990, New Design Techniques for the Sliding Mode. *Proc IEEE International Workshop on VSS and their Applications*, Sarajevo, 220–231

Woodham, C.A., Zinober, A.S.I. 1991a, Eigenvalue Assignment for the Sliding Hyperplanes, *Proc IEE Control Conference*, Edinburgh, 982–988

Woodham, C.A., Zinober, A.S.I. 1991b, Robust Eigenvalue Assignment Techniques for the Sliding Mode, *IFAC Symposium on Control System Design*, Zurich, 529–533

Woodham, C.A., Zinober, A.S.I. 1993, Eigenvalue placement in a specified sector for variable structure control systems, *International Journal of Control* **57**, 1021–1037

Zinober, A.S.I. (editor), 1990, *Deterministic control of uncertain systems*, Peter Peregrinus Press, London

Zinober, A.S.I., El-Ghezawi, O. M. E., Billings, S. A. 1982, Multivariable variable-structure adaptive model-following control systems. *Proc IEE* **129D**, 6–12

2. An Algebraic Approach to Sliding Mode Control

Hebertt Sira-Ramírez

2.1 Introduction

Recent developments in nonlinear systems theory propose the use of differential algebra for the conceptual formulation, clear understanding and definitive solution of long standing problems in the discipline of automatic control. Fundamental contributions in this area are due to Fliess (1986, 1987, 1988a, 1988b, 1989a, 1989b) while some other work has been independently presented by Pommaret (1983, 1986). Similar developments have resulted in a complete restatement of linear systems theory using the theory of Modules (see Fliess (1990c)).

In this chapter implications of the differential algebraic approach for the sliding mode control of *nonlinear* single-input single-output systems are reviewed. We also explore the implications of using module theory in the treatment of sliding modes for the case of (multivariable) *linear* systems.

Formalization of sliding mode control theory, within the framework of differential algebra and module theory, represents a theoretical need. All the basic elements of the theory are recovered from this viewpoint, and some fundamental limitations of the traditional approach are therefore removed.

For instance, input-dependent sliding surfaces are seen to arise naturally from this new approach. These manifolds are shown to lead to continuous, rather than bang-bang, inputs and chatter-free sliding regimes. Independence of the dimension of the desired ideal sliding dynamics with respect to that of the underlying plant, is also an immediate consequence of the proposed approach. A relationship linking controllability of a nonlinear system and the possibility of creating *higher order* sliding regimes is also established using differential algebra. The implications of the module theoretic approach to sliding regimes in linear systems seem to be multiple. Clear connections with decouplability, nonminimum phase problems, and the irrelevance of matching conditions from an input-output viewpoint, are but a few of the theoretical advantages with far reaching practical implications.

The first contribution using differential algebraic results in sliding mode control was given by Fliess and Messager (1990). These results were later extended and applied in several case studies by Sira-Ramírez et al (1992), Sira-Ramírez and Lischinsky-Arenas (1991) and Sira-Ramírez (1992a, 1992b, 1992c, 1993). Recent papers dealing with the multivariable linear systems case are those of Fliess and Messager (1991) and Fliess and Sira-Ramírez (1993). Extensions to pulse-width-modulation, and pulse frequency modulation control strategies may also be found in Sira-Ramírez (1992d, 1992e). Some of these results, obtained for sliding mode control, can be related to ideas presented by Emelyanov (1987, 1990) in his *binary systems* formulation of control problems. In Emelyanov's work, however, the basic developments are not drawn from differential algebra. The algebraic approach to sliding regimes in perturbed linear systems was studied by Fliess and Sira-Ramírez (1993a, 1993b). The theory is presented here in a tutorial fashion with a number of illustrative examples.

Section 2.2 is devoted to general background definitions used in the differential algebraic approach to nonlinear systems theory. Section 2.3 presents some of the fundamental implications of this new trend to sliding mode control analysis and synthesis. As a self-contained counterpart of the results for nonlinear systems, Sect. 2.4 is devoted to present the module theoretic approach to sliding mode control in linear systems. Sect. 2.5 contains some conclusions and suggestions for further work.

2.2 Basic Background to Differential Algebra

In this section we present in a tutorial fashion some of the basic background to differential algebra which is needed for the study of nonlinear dynamical systems. The results are gathered from Fliess's numerous contributions with little or no modification. Further details are found in Fliess (1988a, 1989a).

2.2.1 Basic Definitions

Definition 2.1 *An ordinary differential field K is a commutative field in which a single operation, denoted by "d/dt" or ".", called derivation, is defined, which satisfies the usual rules:* $\mathrm{d}(ab + c)/\mathrm{dt} = (\mathrm{d}a/\mathrm{dt})b + a(\mathrm{d}b/\mathrm{dt}) + \mathrm{d}c/\mathrm{dt}$ *for any a, b and c in K. If all elements c in K satisfy* $\mathrm{d}c/\mathrm{dt} = 0$, *then K is said to be a field of constants.*

Example 2.2 The field $\mathbb{R}$ of real numbers, with the operation of time differentiation $\mathrm{d}/\mathrm{d}t$, trivially constitutes a differential field, which is a field of constants. The field of rational functions in t with coefficients in $\mathbb{R}$, denoted by $\mathbb{R}(t)$, is a differential field with respect to time derivation. $\mathbb{R}(x)$ is also a differential field for any differentiable indeterminate x.

Definition 2.3 *Given a differential field L which contains K, we say L is a differential field extension of K, and denote it by L/K, if the derivation in K is a restriction of that defined in L.*

Example 2.4 $\mathbb{R}(t)/\mathbb{R}$ is a differential field extension over the set of real numbers. The differential field $\mathbb{R}(t)/Q(t)$ is also a differential field extension over the field $Q(t)$ of all rational functions in t with coefficients in the set of rational numbers Q. Similarly, the field $C(t)$ of rational functions in t with complex coefficients, is a differential field extension of, both $\mathbb{R}(t)$ and of $Q(t)$. Evidently, $C(t)/Q$ and $C(t)/C$ are also differential field extensions.

In the following developments u is considered to be a scalar differential indeterminate and k stands for an ordinary differential field with derivation denoted by $\mathrm{d}/\mathrm{d}t$.

Definition 2.5 *By $k\langle u\rangle$, we denote the differential field generated by u over the ground field k. i.e., the smallest differential field containing both k and u. This field is clearly the intersection of all differential fields which contain the union of k and u.*

Example 2.6 Consider the field of all possible rational expressions in u and its time derivatives, with coefficients in $\mathbb{R}$. This differential field is $\mathbb{R}\langle u\rangle$. A typical element in $\mathbb{R}\langle u\rangle$ may be

$$\frac{u^{(3)}}{u^2} + \frac{3u^2\dot{u} + \pi(\ddot{u})^{-1}u^4 - 1.02(\dot{u})^3}{\sqrt{7}u^{(5)} + u} - 5^{0.2}u \tag{2.1}$$

Example 2.7 Let $x_1, \ldots, x_n$ be differential indeterminates. Consider the differential field $k\langle u\rangle$. One may then extend $k\langle u\rangle$ to a differential field K containing all possible rational expressions in the variables $x_1, \ldots, x_n$, and their time derivatives, with coefficients in $k\langle u\rangle$. For instance, a typical element in $K/\mathbb{R}\langle u\rangle$ may be

$$\frac{u^2\left({x_1}^3\right)^2 - \frac{1}{\log(\pi)}\left(\frac{2\ddot{u}+u^3u^{(7)}}{1+\pi u}\right)\left(\dot{x}_5\right)^2 x_6 + x_2}{\sqrt[5]{5}x_3x_4\left(\ddot{x}_1\right)^3 + u^{(3)} - e^2\dot{u}x_2} \tag{2.2}$$

A differential field K, like the one just described, is addressed as a *finitely generated field extension* over $\mathbb{R}\langle u\rangle$. In general, K does not coincide with $\mathbb{R}\langle u, x\rangle$ and it is somewhat larger since we find in K some other variables, like e.g. outputs, which may not be in $\mathbb{R}\langle u, x\rangle/\mathbb{R}\langle u\rangle$.

Definition 2.8 *Any element of a differential field extension, say L/K, has only two possible characterizations. Either it satisfies an algebraic differential equation with coefficients in K, or it does not. In the first case, the element is said to be differentially algebraic over K, otherwise it is said to be differentially transcendental over K. If the property of being differentially algebraic is shared by all elements in L, then L is said to be a differentially algebraic extension of K. If, on the contrary, there is at least one element in L which is differentially transcendent over K, then L is said to be a differentially transcendent extension of K.*

Example 2.9 Consider $k\langle u\rangle$, with k being a constant field. If x is an element which satisfies, $\dot{x} - ax - u = 0$, then x is differentially algebraic over $k\langle u\rangle$. However, since no further qualifications have been given, u is differentially transcendent over k.

Definition 2.10 *A differential transcendence basis of L/K is the largest set of elements in L which do not satisfy any algebraic differential equation with coefficients in K, i.e. they are not differentially K-algebraically dependent. A non-differential transcendence basis of L/K is constituted by the largest set of elements in L which do not satisfy any algebraic differential equation with coefficients in K. The number of elements constituting a differential transcendence basis is called the differential transcendence degree, and denoted by diff tr d°. The (non-differential) transcendence degree (tr d°) refers to the cardinality of a non-differential transcendence basis.*

Example 2.11 In the previous example the differential field extension $k\langle x, u\rangle/k\langle u\rangle$ is algebraic over $k\langle u\rangle$, but, on the other hand, $k\langle u\rangle/k$ is differentially transcendent over k, with u being the differential transcendence basis. Note that x is transcendent over $k\langle u\rangle$ as it does not satisfy any algebraic equation, but does satisfy a differential one. Hence, x is a non-differential transcendence basis of $k\langle x, u\rangle/k\langle u\rangle$. Evidently, *diff tr d* $^\circ k\langle x, u\rangle/k\langle u\rangle = 0$, and *tr d* $^\circ k\langle x, u\rangle/k\langle u\rangle = 1$

Theorem 2.12 *A finitely generated differential extension L/K is differentially algebraic if, and only if its (non-differential) transcendence degree is finite.*

Proof. See Kolchin (1973).

Definition 2.13 *A dynamics is defined as a finitely generated differentially algebraic extension $K/k\langle u\rangle$ of the differential field $k\langle u\rangle$.*

The input u is regarded as an *independent* indeterminate. This means that u is a *differentially transcendent* element of K/k, i.e. u does not satisfy any algebraic differential equation with coefficients in k. It is easy to see, that if u is a differential transcendent element of $k\langle u\rangle$, then it is also a differential transcendence element of $K/k\langle u\rangle$.

The following result is quite basic:

Proposition 2.14 *Suppose $x = (x_1, x_2, ..., x_n)$ is a non-differential transcendence basis of $K/k\langle u\rangle$, then, the derivatives $\mathrm{d}x_i/\mathrm{d}t; (i = 1, \ldots, n)$ are $k\langle u\rangle$ -algebraically dependent on the components of x.*

Proof. This is immediate.

One of the consequences of all these results, discussed by Fliess (1990a) is that a more general and natural representation of nonlinear systems requires *implicit* algebraic differential equations. Indeed, from the preceeding proposition, it follows that there exist exactly n polynomial differential equations with coefficients in k, of the form

$$P_i\left(\dot{x}_i, x, u, \dot{u}, \ldots, u^{(\alpha)}\right) = 0 \ ; \qquad i = 1, \ldots, n \tag{2.3}$$

implicitly describing the controlled dynamics with the inclusion of input time derivatives up to order α.

It has been shown by Fliess and Hassler (1990) that such implicit representations are not entirely unusual in physical examples. The more traditional form of the state equations, known as *normal form*, is recovered in a local fashion, under the assumption that such polynomials locally satisfy the following rank condition

$$\operatorname{rank}\begin{bmatrix} \frac{\partial P_1}{\partial \dot{x}_1} & 0 & \cdots & 0 \\ 0 & \frac{\partial P_2}{\partial \dot{x}_2} & \ddots & \vdots \\ 0 & 0 & \cdots & \frac{\partial P_n}{\partial \dot{x}_n} \end{bmatrix} = n \tag{2.4}$$

The time derivatives of the x_i's may then be solved locally

$$\dot{x}_i = p_i(x, u, \dot{u}, \ldots, u^{(\alpha)}) = 0 \quad i = 1, \ldots, n \tag{2.5}$$

It should be pointed out that even if (2.3) is in polynomial form, it may happen that (2.5) is not. The representation (2.5) is known as the *Generalized State Representation* of a nonlinear dynamics.

2.2.2 Fliess's Generalized Controller Canonical Forms

The following theorem constitutes a direct application of the theorem of the *differential primitive element* which may be found in Kolchin (1973). This theorem plays a fundamental role in the study of systems dynamics from the differential algebraic approach (Fliess 1990a).

Theorem 2.15 *Let $K/k\langle u\rangle$ be a dynamics. Then, there exists an element $\xi \in K$ such that $K = k\langle u, \xi\rangle$ i.e., such that K is the smallest field generated by the indeterminates u and ξ.*

Proof. See Fliess (1990a).

The (non-differential) transcendence degree n of $K/k\langle u\rangle$ is the smallest integer n such that $\xi^{(n)}$ is $k\langle u\rangle$-algebraically dependent on ξ, $\mathrm{d}\xi/\mathrm{d}t, \ldots, \mathrm{d}^{(n-1)}\xi/\mathrm{d}t^{(n-1)}$. We let $q_1 = \xi$, $q_2 = \mathrm{d}\xi/\mathrm{d}t, \ldots$, $q_n = \mathrm{d}^{(n-1)}\xi/\mathrm{d}t^{(n-1)}$. It follows that $q = (q_1, \ldots, q_n)$ also qualifies as a (non-differential) transcendence basis of $K/k\langle u\rangle$.Hence, one obtains a nonlinear generalization of the controller canonical form, known as the *Global Generalized Controller Canonical Form* (GGCCF)

$$\begin{aligned} \dot{q}_1 &= q_2 \\ \dot{q}_2 &= q_3 \end{aligned}$$

$$\vdots \qquad (2.6)$$

$$C(\dot{q}_n, q, u, \dot{u}, \ldots, u^{(\alpha)}) = 0$$

where C is a polynomial with coefficients in k. If one can solve locally for the time derivative of q_n in the last equation of 2.6, one obtains locally an explicit system of first order differential equations, known as the *Local Generalized Controller Canonical Form* (LGCCF)

$$\begin{aligned} \dot{q}_1 &= q_2 \\ \dot{q}_2 &= q_3 \\ &\vdots \\ \dot{q}_n &= c(q, u, \dot{u}, \ldots, u^{(\alpha)}) \end{aligned} \qquad (2.7)$$

Remark. We assume throughout that $\alpha \geq 1$, i.e. the input u explicitly appears before the n-th derivative of the differential primitive element. The case $\alpha = 0$ corresponds to that of exactly linearizable systems under state coordinate transformations and static state feedback. One may still obtain the same smoothing effect of dynamical sliding mode controllers which we shall derive in this article, by considering arbitrary *prolongations* of the input space (i.e. addition of integrations before the input signal). This is accomplished by successively considering the *extended system* (Nijmeijer and Van der Schaft 1990), and proceeding to use the same differential primitive element yielding the LGCCF of the original system.

Example 2.16 Consider the second order system
$\dot{x}_1 = x_2 + u$, $\dot{x}_2 = u$. Then one may consider $\xi = x_1$ as a differential primitive element. In this case the GCCF of the system is simply $\dot{\xi}_1 = \xi_2$, $\dot{\xi}_2 = u + \dot{u}$

2.2.3 Input-Output Systems

Definition 2.17 *(Fliess 1988) Let k be a differential ground field and let u be a differential transcendent element over k. A single input-single output system consists of*

(i) a given input u

(ii) an output y, belonging to a universal differential field extension U, such that y is differentially algebraic over the differential field $k\langle u\rangle$, which denotes the smallest differential field containing, both k and u.

Remark. An input-output system may be viewed as a finitely generated differential field extension $k\langle y, u\rangle / k\langle u\rangle$. The differential field $k\langle y, u\rangle$ is, hence, differentially algebraic over $k\langle u\rangle$, i.e. y satisfies an algebraic differential equation with coefficients in $k\langle u\rangle$.

Definition 2.18 *Let $k\{y,u\}$ denote the differential ring generated by y and u and let U be a universal differential field. A differential homomorphism $\psi : k\{y,u\} \mapsto U$ is defined as a homomorphism which commutes with the derivation defined on $k\{y,u\}$, i.e.*

$$\forall \ \theta \in k\{u,y\}, \qquad \psi\left(\dot{\theta}\right) = \frac{d}{dt}\psi(\theta) \tag{2.8}$$

Definition 2.19 *A differential k-specialization of the differential ring $k\{y,u\}$ is a differential homomorphism $\psi : k\{y,u\} \mapsto U$, taking $k\{y,u\}$ into the universal differential field U, which leaves the elements of the ground field k invariant, i.e.*

$$\forall \ a \in k, \quad \psi(a) = a \tag{2.9}$$

The differential transcendence degree of the extension over k, of the differential quotient field $Q(\psi(k\{y,u\}))$, is nonnegative and it is never higher than the differential transcendence degree of $k\langle u\rangle/k$ (i.e. $\textit{diff tr d}^{\,\circ}Q(\psi(k\{y,u\})/k \leq \textit{diff tr d}^{\,\circ}k\langle u\rangle/k = 1$). One frequently takes ψ as the identity mapping.

Remark. Differential specializations have been found to have a crucial relevance in the definition of the zero dynamics (Fliess 1990b). Indeed, consider the input-output system $k\langle y,u\rangle/k\langle u\rangle$. Let J be the largest differential subfield of $k < u,y >$ which contains $k\langle y\rangle$ and such that $J/k\langle y\rangle$ is differentially algebraic. Notice that J is not, in general, equal to $k\langle y,u\rangle$, unless the system is left invertible. Consider now the differential homomorphism $\psi : k\{y,u\} \mapsto U$, such that $\psi(y) = 0$. Hence, $\psi(y^{(\beta)}) = 0$, for all $\beta \geq 1$. It follows that $\psi(k\{y,u\}) = k\{u\}$ and the quotient field $Q(\psi(k\{y,u\}))/k$ coincides with the differential field extension $k\langle u\rangle/k$. Extend now the corresponding differential specialization ψ to the differential field J, in a trivial manner, and obtain a *smaller* differential field J^*. The specialized extension J^*/k , which is evidently differentially algebraic, is called the *zero dynamics.*

In the language of differential algebra, feedback is also accounted for, in all generality, by means of differential specializations (Fliess 1989a). This most appealing way of treating the fundamental concept of control theory is stated as follows:

Definition 2.20 *A closed-loop control is a differential k-specialization ψ : $k\{y,u\} \mapsto U$ such that $\textit{diff tr d}^{\,\circ}Q(\psi(k\{y,u\}))/k = 0$. We refer to such feedback loops as pure feedback loops. In such a case, the specialized elements $\psi(u)$, $\psi(y)$ satisfy an ordinary algebraic differential equation. Whenever $\textit{diff tr d}^{\,\circ}k\langle\psi(y)\rangle/k$ is zero, the closed-loop is said to be degenerate .*

We are mainly interested in those cases for which $\textit{diff tr d}^{\,\circ}Q(\psi(k\{y,u\}))/k = 0$. However, let v be a scalar differential transcendent element of $k\langle v\rangle/k$, such that $\psi(u)$, $\psi(y)$ are differentially algebraic over $k\langle v\rangle$. Then, if *diff tr d* $^\circ Q(\psi(k\{y,v\}))/k = 1$, the underlying differential specialization ψ leads to a regular feedback loop with an (independent) external input v (Fliess 1987).

Definition 2.21 *An input-output system $k\langle y,u\rangle/k\langle u\rangle$ is invertible if u is differentially algebraic over $k\langle y\rangle$, i.e. if diff tr d $^\circ k\langle y,u\rangle/k\langle y\rangle = 0$. It is easy to see that every nontrivial single-intput single-output system is always invertible.*

2.3 A Differential Algebraic Approach to Sliding Mode Control of Nonlinear Systems

In this section we present some applications of the results of the differential algebraic approach, proposed by Fliess for the study of control systems, to characterize in full generality, sliding mode control of nonlinear systems.

2.3.1 Differential Algebra and Sliding Mode Control of Nonlinear Dynamical Systems

Consider a (nonlinear) dynamics $K/k\langle u\rangle$. Furthermore, let, $\zeta = (\zeta_1, ..., \zeta_n)$ be a non-differential transcendence baisis for K, i.e. the transcendence degree of $K/k\langle u\rangle$ is then assumed to be n.

Definition 2.22 *A sliding surface candidate is any non k-algebraic element σ of $K/k\langle u\rangle$ such that its time derivative $d\sigma/dt$ is $k\langle u\rangle$-algebraically dependent on ζ , i.e. there exists a polynomial S over k such that*

$$S(\dot{\sigma}, \zeta, u, \dot{u}, \ldots, u^{(\alpha)}) = 0 \tag{2.10}$$

Remark. In the traditional definition of the sliding mode for systems in *Kalman form* with state ζ, the time derivative of the sliding surface was required to be only algebraically dependent on ζ and u. Hence, all the resulting sliding mode controllers were necessarily static. One can generalize this definition using differential algebra. The differential algebraic approach naturally points to the possibilities of *dynamical sliding mode controllers* specially in the case of nonlinear systems, where elimination of input derivatives from the system model may not be possible at all (see Fliess and Hasler (1990) for a physical example).

Proposition 2.23 *The element σ in $K/k\langle u\rangle$ is a sliding surface candidate if it is k-algebraically dependent on all the elements of a transcendence basis ζ.*

Proof. The time derivative of σ is k-algebraically dependent on the derivatives of every element in the transcendence basis ζ. Therefore, $d\zeta/dt$ is $k\langle u\rangle$-algebraically dependent on ζ

The condition in the above proposition is clearly not necessary as σ may well be k-algebraically dependent only on some elements of the transcendence basis ζ, and still have $d\sigma/dt$ being $k\langle u\rangle$-algebraically dependent on ζ. Imposing on σ a ***discontinuous sliding dynamics*** of the form

$$\dot{\sigma} = -W\text{sign } \sigma \tag{2.11}$$

one obtains from (2.10) an ***implicit dynamical sliding mode controller*** given by

$$S(-W\text{sign}(\sigma), \zeta, u, \dot{u}, \ldots, u^{(\alpha)}) = 0 \tag{2.12}$$

which is an implicit time-varying discontinuous ordinary differential equation for the control input u. The two *structures* associated with the underlying variable structure control system are represented by the following pair of implicit dynamical controllers

$$\begin{array}{lll} S(-W, \zeta, u, \dot{u}, \ldots, u^{(\alpha)}) & = 0 & \text{for } \sigma > 0 \\ S(W, \zeta, u, \dot{u}, \ldots, u^{(\alpha)}) & = 0 & \text{for } \sigma < 0 \end{array} \tag{2.13}$$

each one valid, respectively, on one of the *regions* $\sigma > 0$ and $\sigma < 0$. Precisely when $\sigma = 0$ neither of the control structures is valid. One then ideally characterizes the motions by formally assuming $\sigma = 0$ and $d\sigma/dt = 0$ in (2.10). We formally define the ***equivalent control dynamics*** as the dynamical state feedback control law obtained by letting $d\sigma/dt$ become zero in (2.12), and consider the resulting implicit differential equation for the equivlent control, here denoted by u_{eq}

$$S(0, \zeta, u_{eq}, \dot{u}_{eq}, \ldots, u_{eq}^{(\alpha)}) = 0 \tag{2.14}$$

According to the initial conditions of the state ζ and the control input and its derivatives, one obtains in general, $\sigma =$ *constant.* Hence, the sliding motion ideally taking place on $\sigma = 0$ may be viewed as a particular case of the motions of the system obtained by means of the equivalent control.

Note that whenever $\partial S/\partial\dot{\sigma} \neq 0$, one locally obtains from the implicit equation (2.10)

$$\dot{\sigma} = s(\zeta, u, \dot{u}, \ldots, u^{(\alpha)}) \tag{2.15}$$

The corresponding dynamical sliding mode feedback controller, satisfying (2.11), is given by

$$s(\zeta, u, \dot{u}, \ldots, u^{(\alpha)}) = -W\text{sign } \sigma \tag{2.16}$$

Furthermore, if $\partial\sigma/\partial u^{(\alpha)} \neq 0$, one obtains locally a time-varying state space representation for the dynamical sliding mode controller (2.16) in ***normal*** form

$$\begin{aligned}\dot{u}_1 &= u_2\\ \dot{u}_2 &= u_3\\ &\vdots\\ \dot{u}_\alpha &= \theta(u_1,\ldots,u_\alpha,\zeta,W\text{sign }\sigma)\end{aligned} \tag{2.17}$$

All discontinuities arising from the bang-bang control policy (2.11) are seen to be confined to the highest derivative of the control input through the nonlinear function θ. The output u of the dynamical controller is clearly the outcome of α integrations performed on such a discontinuous function θ and for this reason u is, generically speaking, sufficiently continuous.

2.3.2 Dynamical Sliding Regimes Based on Fliess's GCCF

The general results on canonical forms for nonlinear systems, presented in Sect. 2.2, have an immediate consequence in the definition of sliding surfaces for stabilization and tracking problems. We explore the stabilization problem below.

Consider a system of the form (2.7) and the following sliding surface coordinate function, expressed in terms of the generalized phase coordinates q

$$\sigma = c_1 q_1 + c_2 q_2 + \cdots + c_{n-1} q_{n-1} + q_n \tag{2.18}$$

where the scalar coefficients c_i $(i = 1, \cdots, n-1)$ are chosen in such a manner that the polynomial

$$p(s) = c_1 + c_2 s + \cdots + c_{n-1} s^{n-2} + s^{n-1} \tag{2.19}$$

in the complex variable s, is Hurwitz. Imposing on the sliding surface coordinate function σ the discontinuous dynamics (2.11), then the trajectories of σ are seen to exhibit, within finite time T given by $T = W^{-1}|s(0)|$, a sliding regime on $\sigma = 0$. Substituting in (2.11) the expression (2.18) for σ, and using (2.7), one obtains after some straightforward algebraic manipulations, the implicit dynamical sliding mode controller

$$\begin{aligned}c(q,u,\dot{u},\ldots,u^{(\alpha)}) &= c_{n-1}\dot{\sigma} + c_1 c_{n-1} q_1 + (c_2 c_{n-1} - c_1) q_2 + \cdots\\ &\quad + (c_{n-2}c_{n-1} - c_{n-3}) q_{n-2} + (c_{n-1}^2 - c_{n-2}) q_{n-1}\\ &\quad - W\text{sign }\sigma\\ &= -c_1 q_2 - c_2 q_3 - \cdots - c_{n-2} q_{n-1} - c_{n-1} q_n\\ &\quad - W\text{sign }\sigma\end{aligned} \tag{2.20}$$

Evidently, under ideal sliding conditions $\sigma = 0$, the variable q_n no longer qualifies as a state variable for the system since it is expressible as a linear combination of the remanining states and, hence, q_n is no longer a non-differentially transcendent element of the field extension K. The ideal (autonomous) closed-loop dynamics may then be expressed in terms of a *reduced* non-differential transcendence basis of K/k which only includes the remaining $n-1$ phase coordinates associated with the original differential primitive element. This leads to the ideal sliding dynamics

$$\begin{aligned} \dot{q}_1 &= q_2 \\ \dot{q}_2 &= q_3 \\ &\vdots \\ \dot{q}_{n-1} &= -c_1 q_1 - c_2 q_2 - \cdots - c_{n-2} q_{n-2} - c_{n-1} q_{n-1} \end{aligned} \tag{2.21}$$

The characteristic polynomial of (2.21) is evidently given by (2.19) and hence the (reduced) autonomous closed-loop dynamics is asymptotically stable to zero. Note that, by virtue of (2.18), the condition $\sigma = 0$ holds, and due to the asymptotic stability of (2.21), the variable q_n also tends to zero in an asymptotically stable fashion. The equivalent control, denoted by u_{eq}, is a ***virtual*** feedback control action achieving ideally smooth evolution of the system on the constraining sliding surface $\sigma = 0$, provided initial conditions are precisely set on such a switching surface. The equivalent control is formally obtained from the condition $d\sigma/dt = 0$, i.e.

$$\begin{aligned} c(q, u, \dot{u}_{eq}, \ldots, u_{eq}^{(\alpha)}) &= c_1 c_{n-1} q_1 + (c_2 c_{n-1} - c_1) q_2 + \cdots \\ &\quad + (c_{n-2} c_{n-1} - c_{n-3}) q_{n-2} + (c_{n-1}^2 - c_{n-2}) q_{n-1} \end{aligned} \tag{2.22}$$

Since q asymptotically converges to zero, the solutions of the above time-varying implicit differential equation, describing the evolution of the equivalent control, asymptotically approach the solutions of the following autonomous implicit differential equation

$$c(0, u, \dot{u}, \ldots, u^{(\alpha)}) = 0 \tag{2.23}$$

Equation (2.23) constitutes the *zero dynamics* (Fliess 1990b) associated with the problem of zeroing the differential primitive element, considered now as an (auxiliary) output of the system. Note that (2.23) may also be regarded as the zero dynamics associated with the zeroing of the sliding surface coordinate function σ. If (2.23) locally asymptotically approaches a constant equilibrium point $u = U$, then the system is said to be *locally minimum phase* around such an equilibrium point, otherwise the system is said to be ***non-minimum phase***. The equivalent control is, thus, locally asymptotically stable to U, whenever the underlying input-output system is minimum phase.

One may be tempted to postulate, for the sake of physical realizability of the sliding mode controller, that a sliding mode control strategy is properly defined whenever the zero dynamics associated with the system is constituted by an asymptotically stable motion towards equilibrium. In other words, the input-output system should be minimum phase. It must be pointed out, however, that non-minimum phase systems might make perfect physical sense and that, in some instances, instability of a certain state variable or input does not necessarily imply disastrous effects on the controlled system (for an example of this frequently overlooked fact, see Sira-Ramírez (1991, 1993)).

2.3.3 Some Formalizations of Sliding Mode Control for Input-Output Nonlinear Systems

Definition 2.24 *Consider a differential k-specialization ϕ, mapping $k\{y\} \mapsto U$, such that diff tr d $^{\circ}Q(\phi(k\{y\}))/k = 0$. The elements $\sigma \in Q(\phi(k\{y\}))/k$ are referred to as ideal sliding dynamics , or sliding surfaces. Note that $Q(\phi(k\{y\}))/k = k\langle\phi(y)\rangle/k$. We will be using the identity map for the mapping ϕ from now on. A sliding surface σ is, therefore, directly taken from the specialized extension $k\langle y\rangle/k$, as $\sigma = 0$.*

Definition 2.25 *Let σ be an element of $k\langle y\rangle/k$ such that $\sigma = 0$ represents a desirable ideal sliding dynamics. An equivalent control, corresponding to σ, is said to exist for the system $k\langle y,u\rangle/k\langle u\rangle$, if there exists a differential k-specialization $\psi : k\{y,u\} \mapsto U$, which represents a pure feedback loop, such that $d\sigma/dt$ is identically zero. A sliding regime is said to exist on $\sigma = 0$ if $\sigma \in k\langle\psi(y)\rangle/k$ and diff tr $d^{o}k\langle\psi(y)\rangle/k = 0$. The differential k-specialization $\psi : k\{y,u\} \mapsto U$, may be computed, in principle, from the condition $d\sigma/dt = 0$.*

Sliding mode control thus leads to a very special class of degenerate feedback in which the resulting closed-loop system ideally satisfies a preselected autonomous algebraic differential equation. Note that, in this setting and at least for single-input single-output systems, the order of the highest derivative of the output y in the differential equation representing the ideal sliding dynamics, is not necessarily restricted to be smaller than the highest order of the derivative of y in the differential equation defining the input-output system. The following helps to formalize this issue.

Definition 2.26 *An element r in the differential field $k\langle y\rangle/k$ is said to be a prolongation of an element $\rho \in k\langle y\rangle/k$, if r is obtained by a finite number of time differentiations performed on ρ, i.e. if there exist a nonnegative integer, L such that $r = \rho^{(L)}$. The integer L, of required differentiations, is called the length of the prolongation. Similarly, given an input-output system $k\langle y,u\rangle/k\langle u\rangle$ a prolonged system is obtained by straightforward differentiation of the input-output relation (Nijmeijer and Van der Schaft 1990). All prolongations of an input-output system rest in the differential field extension: $k\langle y,u\rangle/k\langle u\rangle$.*

Proposition 2.27 *Let $k\langle y,u\rangle/k\langle u\rangle$ be an invertible system, then any prolongation of the system, of finite length, is also invertible.*

Proof. It is easy to see that *diff tr d* $^{\circ}k\langle y\rangle/k$ is invariant with respect to prolongations.

Theorem 2.28 *Modulo singularities in the actual computation of the required control input, and the need for suitable prolongations, the equivalent control always exists for a given element $\sigma \in k\langle y\rangle$.*

Proof. The result is obviously true from the fact that the single input-single output system $k\langle y, u\rangle / k\langle u\rangle$ is trivially invertible, modulo the possible local singularities.

Example 2.29 Consider the first order input-output system $\dot{y} = u$ and the asymptotically stable second order ideal sliding dynamics $\sigma = \ddot{y} + 2\xi\omega_n\dot{y} + \omega_n^2 y = 0, \xi > 0,\ \omega_n > 0$. The dynamical feedback (equivalent) controller $\ddot{u} = -2\xi\omega_n\dot{u} - \omega_n^2 u$, obtained from $\dot{\sigma} = \ddot{u} + 2\xi\omega_n\dot{u} + \omega_n^2 u = 0$, defines the equivalent control for arbitrary initial conditions in u.

Remark. We have defined sliding motions in a quite general and relaxed sense. Essentially, we have required only that the ideal (autonomous) sliding dynamics be synthesizable, in principle, by pure feedback. The process of actually achieving a sliding regime on such a desirable autonomous dynamics may then be carried out through discontinous or continuous (e.g. high gain) feedback control of a static or dynamic nature. Owing to the generally local nature of the invertibility of a given system, as well as the possible presence of singularities, it may actually happen that finding well-defined discontinuous or continuous feedback policies, which eventually result in closed-loop compliance with the ideal sliding dynamics, may not be possible at all due to singularities.

Consider now a regular feedback loop with an external input v, obtained from the differential k-specializations ϕ^+, and ϕ^- mapping $k\{y\} \mapsto U$, such that

$$\textit{diff tr } d^\circ\ Q(\phi^+(k\{y\}))/k = \textit{diff tr } d^\circ\ Q(\phi^-(k\{y\}))/k = 1 \tag{2.24}$$

In particular, let the external input v be obtained as $v = -W\mathrm{sign}(\sigma)$. The controlled elements $\sigma \in Q(\phi^+(k\{y\}))/k$ and $\sigma \in Q(\phi^-(k\{y\}))/k$ are referred to as *controlled motions towards sliding*, and the differential specializations ϕ^+ and ϕ^- constitute the sliding mode control strategy .

Example 2.30 Consider again the single integrator system with a higher order sliding surface. A sliding regime is achieved on $\sigma = 0$ in finite time by imposing on σ the discontinuous dynamics $d\sigma/dt = -W\mathrm{sign}\ \sigma$, i.e. $\dot{\sigma} = y^{(3)} + 2\xi\omega_n\ddot{y} + \omega_n^2\dot{y} = -W\mathrm{sign}\ \left(\ddot{y} + 2\xi\omega_n\dot{y} + \omega_n^2 y\right)$. Using suitably prolonged system equations, one obtains the dynamical sliding mode controller $\ddot{u} = -2\xi\omega_n\dot{u} - \omega_n^2 u - W\mathrm{sign}\left(\dot{u} + 2\xi\omega_n u + \omega_n^2 y\right)$.

2.3.4 An Alternative Definition of the Equivalent Control Dynamics

One may generate a differential algebraic extension of $k\langle u\rangle$ by adjoining the sliding surface element σ to u, and considering $k\langle u, \sigma\rangle$ as an input-output system. The differential field extension $k\langle u, \sigma\rangle / k\langle u\rangle$ is indeed an input-output system, or, more precisely, an input-sliding surface system. The element σ is then a non-differential transcendence element of the field extension $k\langle u, \sigma\rangle / k\langle u\rangle$. It therefore satisfies an algebraic differential equation with coefficients in $k\langle u\rangle$. This means that there exists a polynomial with coefficients in k such that

$$P(\sigma, \dot{\sigma}, \ldots, \sigma^{(\rho)}, u, \dot{u}, \ldots, u^{(\gamma)}) = 0 \tag{2.25}$$

where we have implicitly assumed that ρ is the smallest integer such that $d^{\rho}\sigma / dt^{\rho}$ is algebraically dependent upon $\sigma, \dot{\sigma}, \ldots, \sigma^{(\rho)}, u, \dot{u}, \ldots, u^{(\gamma)}$. This general characterization of sliding surface coordinate functions has not been clearly established in the sliding mode control literature. Obtaining a differential equation for the sliding surface coordinate σ, which is independent of the system state, has direct implications for the area of *higher order* sliding motions (see Chang (1991)), for a second order sliding motion example) and some recent developments in *binary control systems.* We will explore only the first issue in Section 2.3.5. A state-independent implicit definition of the *equivalent control dynamics* can then be immediately obtained from (2.25) by setting σ and its time derivatives to zero

$$P(0, 0, \ldots, 0, u, \dot{u}, \ldots, u^{(\gamma)}) = 0 \tag{2.26}$$

2.3.5 Higher Order Sliding Regimes

Recently some effort has been devoted to the smoothing of system responses to sliding mode control policies through so called *higher order* sliding regimes. *Binary control systems*, as applied to variable structure control, are also geared towards obtaining asymptotic convergence towards the sliding surface, in a manner that avoids control input chattering through integration. These two develpments are also closely related to the differential algebraic approach. In the following paragraphs we explain in complete generality how the same ideas may be formally derived from differential algebra.

Consider (2.25) with σ as an output and rewrite in the following *Global Generalized Observability Canonical Form* (GGOCF) (Fliess 1990a)

$$\begin{aligned} \dot{\sigma}_1 &= \sigma_2 \\ \dot{\sigma}_2 &= \sigma_3 \\ &\vdots \\ P(\sigma_1, \quad \ldots &, \sigma_\rho, \dot{\sigma}_\rho, u, \dot{u}, \ldots, u^{(\gamma)}) = 0 \end{aligned} \tag{2.27}$$

As before, an explicit LGOCF can be obtained for the element σ whenever $\partial P / \partial \dot{\sigma}_\rho \neq 0$

$$\begin{aligned} \dot{\sigma}_1 &= \sigma_2 \\ \dot{\sigma}_2 &= \sigma_3 \\ &\vdots \\ \dot{\sigma}_\rho &= p\left(\sigma_1, \ldots, \sigma_\rho, u, \dot{u}, \ldots, u^{(\gamma)}\right) \end{aligned} \tag{2.28}$$

Definition 2.31 *An element σ of the dynamics $K/k\langle u\rangle$ admits a ρ-th order sliding regime if the GOCF (2.29) associated with σ is ρ-th order.*

One defines a ρ-th order *sliding surface candidate* as any arbitrary (algebraic) function of σ and its time derivatives up to $(r-1)$-st order. For obvious reasons the most convenient type of function is represented by a suitable linear combination of σ and its time derivatives, which achieves stabilization

$$s = m_1\sigma_1 + m_2\sigma_2 + \cdots + m_{n-1}\sigma_{\rho-1} + \sigma_\rho \tag{2.29}$$

First-order sliding motion is then imposed on this linear combination of generalized phase variables by means of the discontinuous sliding mode dynamics

$$\dot{s} = -M\text{sign } s \tag{2.30}$$

This policy results in the implicit dynamical higher order sliding mode controller

$$\begin{aligned} p(\sigma_1, \ldots, \sigma_\rho, u, \dot{u}, \ldots, u^{(\gamma)}) &= -m_1\sigma_2 - m_2\sigma_3 - \cdots - m_{\rho-2}\sigma_{\rho-1} - m_{\rho-1}\sigma_\rho \\ &\quad -M\text{sign }(s) \end{aligned} \tag{2.31}$$

As previously discussed, s goes to zero in finite time and, provided the coefficients in (2.29) are properly chosen, an ideal asymptotically stable motion can be then obtained for s, which is governed by the autonomous linear dynamics

$$\begin{aligned} \dot{\sigma}_1 &= \sigma_2 \\ \dot{\sigma}_2 &= \sigma_3 \\ &\vdots \\ \dot{\sigma}_{\rho-1} &= -m_1\sigma_1 - \cdots - m_{\rho-1}\sigma_{\rho-1} \end{aligned} \tag{2.32}$$

2.3.6 Sliding Regimes in Controllable Nonlinear Systems

The differentially algebraic closure of the ground field k in the dynamics K is defined as the differential field κ, where $K \supseteq \kappa \supseteq k$, consisting of the elements of K which are differentially algebraic over k. The field k is differentially algebraically closed if and only if $k = \kappa$.

The following definition is taken from Fliess (1991) (see also Pommaret (1991)).

Definition 2.32 *The dynamics $K/k\langle u\rangle$ is said to be algebraically controllable if and only if the ground field k is differentially algebraically closed in K.*

Algebraic controllability implies that all elements of K are necessarily influenced by the input u, since they satisfy a differential equation which is not independent of u and possibly some of its time derivatives.

Proposition 2.33 *A higher order sliding regime can be created for any element s of the dynamics $K/k\langle u\rangle$ if and only if $K/k\langle u\rangle$ is controllable.*

Proof. Sufficiency is obvious from the fact that s satisfies a differential equation with coefficients in $k\langle u\rangle$. For the necessity of the condition, suppose, contrary to what is asserted, that $K/k\langle u\rangle$ is not controllable, but that a higher order sliding regime can be created on any element of the differential field extension $K/k\langle u\rangle$. Since k is not differentially algebraically closed, there are elements in K, which belong to a differential field κ containing k, which satisfy differential equations with coefficients in k. Clearly these elements are not related to the control input u through differential equations. It follows that a higher order sliding regime cannot be imposed on such elements. A contradiction is established.

In this more relaxed notion of sliding regime, one may say that sliding mode behaviour can be imposed on any element of the dynamics of the system, if and only if the system is controllable. The characterization of sliding mode existence through controllability, is a direct consequence of the differential algebraic approach.

2.4 A Module Theoretic Approach to Sliding Modes in Linear Systems

The particularization of the differential algebraic approach to the case of linear systems applies the notion of *Modules of Khäler differentials.* This theory establishes far reaching properties of the linearized version of the system to those of the underlying nonlinear system (see Fliess (1991) for details). It turns out that, in its own right, the theory of linear systems can be handled in a self contained manner, from the theory of modules over rings of finite linear differential operators. This approach discards the need to relate the linear system to some linearizability properties of an underlying nonlinear system generating it, which operates in the vicinity of an equilibrium point. Due to the wide spread knowledge about linear systems, this latter approach is preferred in the presentation that follows.

In this section we address the algebraic approach to sliding mode control of linear systems. We first provide some background definitions of the relevant topics in algebra. The reader is referred to the book by Adkins and Weintraub (1992) for a fundamental background. We shall be closely following the work of Fliess (1990c) for the portion containing background material on the applications of module thoery to linear systems. The algebraic approach to sliding mode control is taken from Fliess and Sira-Ramírez (1993a, 1993b).

Definition 2.34 *A ring $(R, +, \cdot)$ is a set R with two binary operations*

$$+ \quad : R \rightarrow R(addition)$$

$$\cdot \quad : R \to R(multiplication)$$

such that $(R, +)$ *is an abelian group with a* zero. *Multiplication and addition satisfy the usual properties of associativity and distributivity.*

Here we shall be dealing only with commutative rings with *identity.*

Example 2.35

The set $2Z$ of *even* integers is a ring without an identity. The set of all square $n \times n$ matrices defined over the field of real numbers. The set of all polynomials in an indeterminate x

Definition 2.36 *Let R be an arbitrary ring with identity. A left R-module is an abelian group M together with a scalar multiplication map*

$$\cdot \ : R \times M \to M$$

which satisfies the following axioms $\forall a, b \in R, \ m, n \in M$

$$\begin{aligned}
&a(m+n) = am + an \\
&(a+b)m = am + bm \\
&(ab)m = a(bm) \\
&1m = m.
\end{aligned}$$

Example 2.37

Let F be a *field*, then an F-module V is called a *vector space* over F. Let R be an arbitrary ring. The set of matrices $M_{m,n}(R)$ is a left R-module via left scalar multiplication of matrices. Any subgroup $N \subset M$ which is closed under scalar multiplication by elements in R is itself a module, called a *submodule* of M.

If $S \subset M$, then $[S]$ denotes the intersection of all submodules of M containing S. We may say that $[S]$ is the "smallest" submodule, with respect to inclusion, containing the set S. The submodule $[S]$ is also called the *submodule of M generated by S.*

Definition 2.38 *M is* fintely generated *if* $M = [S]$ *for some finite subset S of M. The elements of S are called the "generators" of M. The rank of a module M is the cardinality of the minimal set of generators of M in S.*

We denote by $k\left[\frac{d}{dt}\right]$ the *ring of finite linear differential operators.* These are operators of the following form

$$\sum_{finite} a_\alpha \frac{\mathrm{d}^\alpha}{\mathrm{d}t^\alpha}, \quad a_\alpha \in k$$

The ring $k\left[\frac{\mathrm{d}}{\mathrm{dt}}\right]$ is commutative if, and only if, k is a field of constants. We will be primarily concerned with rings of linear differential operators with real coefficients. This necessarily restricts the class of problems treated to linear, time-invariant, systems. The results, however, can be extended to time-varying systems by using rings defined over *principal ideal domains* (see Fliess (1990c)).

Definition 2.39 *Let M be a left $k\left[\frac{\mathrm{d}}{\mathrm{dt}}\right]$-module. An element $m \in M$ is said to be* torsion *if and only if there exists $\pi \in k\left[\frac{\mathrm{d}}{\mathrm{dt}}\right]$, $\pi \neq 0$, such that $\pi m = 0$ i.e. m satisfies a linear differential equation with coefficients in k.*

Definition 2.40 *A module T such that all its elements are torsion is said to be a torsion module.*

Definition 2.41 *A finite set of elements in a $k[\frac{\mathrm{d}}{\mathrm{dt}}]$-module M constitutes a* basis *if every element in the module may be uniquely expressed as a $k[\frac{\mathrm{d}}{\mathrm{dt}}]$-linear combination of such elements. A module M is said to be* free *if it has a basis.*

Proposition 2.42 *Let M be a finitely generated left $k\left[\frac{\mathrm{d}}{\mathrm{dt}}\right]$-module. M is torsion if and only if the dimension of M as a k-vector space is finite .*

Definition 2.43 *The set of all torsion elements of a module M is a submodule T called the* torsion submodule *of M.*

Definition 2.44 *A module M is said to be free if and only if its torsion submodule is trivial.*

Theorem 2.45 *Any finitely generated left $k\left[\frac{\mathrm{d}}{\mathrm{dt}}\right]$-module M can be decomposed into a direct sum*

$$M = T \oplus \Phi$$

where T is the torsion submodule and Φ is a free submodule.

2.4.1 Quotient Modules

Let M be an R-module and let $N \subset M$ be a submodule of M, then N is a subgroup of the abelian group M and we can form the *quotient group* M/N as the set of all *cosets*

$$M/N = \{m + N \; ; \; \text{for } \; m \in M\} \tag{2.33}$$

They evidently accept the operation of addition as a well defined (commutative) operation

$$(m + N) + (p + N) = (m + p) + N$$

The elements $m+N$ of M/N can now be endowed with an R-module structure by defining scalar products in a manner inherited from M, namely,

$$a(m+N) = am + N \; ; \; \forall a \in R \text{ and } m \in M$$

The elements $m' = m(\text{mod} N)$ are called the *residues* of M in M/N. The map $M \to M/N$ taking $m \mapsto m' = m + N$ is called the *canonical projection*.

2.4.2 Linear Systems and Modules

Linear systems enjoy a particularly appealing characterization from the algebraic viewpoint. This has been long recognized since the work of Kalman (1970). More recently Fliess (1990c) has provided a rather different approach to such characterization, which still uses modules but in a different context. In this section we follow the work of Fliess (1990c) with little or no modifications.

Definition 2.46 *A linear system is a finitely generated left* $k\left[\frac{\mathrm{d}}{\mathrm{dt}}\right]$*-module* Λ.

Example 2.47 (Fliess, 1990c) Consider a system Σ as a finite set of quantities $w = (w_1, \ldots, w_q)$ which are related by a set of homogeneous linear differential equations over k.

Let

$$E_\alpha\left(w_i^{(v_j)}\right) = \sum_{finite} a_{\alpha,i,j} w_i^{(v_j)} = 0, \; (a_{\alpha,i,j} \in k)$$

Consider the left $k\left[\frac{\mathrm{d}}{\mathrm{dt}}\right]$-module $\mathcal{F}$ spanned by $\overline{w} = (\overline{w}_1, \ldots, \overline{w}_q)$ and let $\Xi \subset \mathcal{F}$ be the submodule spanned by

$$e_\alpha = E_\alpha\left(\overline{w}_i^{(v_j)}\right) = \sum_{finite} a_{\alpha,i,j} \overline{w}_i^{(v_j)}, \; (a_{\alpha,i,j} \in k)$$

The quotient module $\Lambda = \mathcal{F}/\Xi$ is the module corresponding to the system. It is easy to see that the canonical image (residue) of w in $\mathcal{F}/\Xi$ satisfies the system equations.

2.4.3 Unperturbed Linear Dynamics

Definition 2.48 *A linear dynamics* $\mathcal{D}$ *is a linear system* $\mathcal{D}$ *where we distinguish a finite set of quantities, called the inputs* $u = (u_1, \ldots, u_m)$*, such that the module* $\mathcal{D}/[u]$ *is torsion.*

The set of inputs u are said to be *independent* if and only if $[u]$ is a *free* module. An output vector $y = (y_1, \ldots, y_p)$ is a finite set of elements in $\mathcal{D}$.

Example 2.49 (Fliess 1990c) Consider the single input single output system

$$a(\frac{d}{dt})y = b(\frac{d}{dt})u \qquad a, b \in k[\frac{d}{dt}] \ , \ a \neq 0$$

Take as the free left $k[\frac{d}{dt}]$-module $\mathcal{F} = [\overline{u}, \overline{y}]$ spanned by $\overline{u}$, $\overline{y}$. Let $\Xi \subset \mathcal{F}$ be the submodule spanned by $a(\frac{d}{dt})\overline{y} - b(\frac{d}{dt})\overline{u}$. The quotient module $\mathcal{D} = \mathcal{F}/\Xi$ is the system module. Let u, y be the *residues* of $\overline{u}, \overline{y}$ in $\mathcal{D}$. Then u, y satisfy the system equations. If we let $\underline{y}$ be the residue of y in $\mathcal{D}/[u]$, then $\underline{y}$ satisfies $a(\frac{d}{dt})\underline{y} = 0$, which is torsion.

2.4.4 Controllability

Definition 2.50 *A linear system is said to be controllable if and only if its associated module Λ is* free.

Example 2.51 The system given by $\dot{w}_1 = w_2$ is controllable since its associated module is not torsion.

Definition 2.52 *A linear dynamics $\mathcal{D}$, with input u, is said to be* controllable *if and only if the associated linear system is controllable.*

Example 2.53 The linear dynamics $\dot{x}_1 = u$ is controllable, since its associated linear system is described by a free module. The module decomposition $\mathcal{D} = \Phi \oplus T$ shows that a system is controllable if and only if T is trivial.

Example 2.54 The linear system $\dot{w}_1 = w_2$; $\dot{w}_2 = -w_2$ is uncontrollable since its associated module can be decomposed as $[w_1] \oplus [w_2]$ with $[w_2]$ being evidently torsion.

2.4.5 Observability

Definition 2.55 *A linear dynamics $\mathcal{D}$ with input u and output y, is said to be observable if and only if $\mathcal{D} = [u, y]$. The quotient module $\mathcal{D}/[u, y]$ is trivial.*

Example 2.56 The linear dynamics $\dot{x}_1 = x_2$; $\dot{x}_2 = u$; $y = x_1$ is observable since $x_1 = y$; $x_2 = \dot{y}$.

If the system is unobservable then $[u, y] \subset \mathcal{D}$ and the quotient module $\mathcal{D}/[u, y]$ is torsion.

Example 2.57 The linear dynamics $\dot{x}_1 = x_1$; $\dot{x}_2 = u$; $y = x_2$ is unobservable since $x_1 \notin [u, y]$ and the residues $\overline{x}_1\overline{x}_2$ in the quotient module $\mathcal{D}/[u, y]$ satisfy $\dot{\overline{x}}_1 - \overline{x}_1 = 0$ and $\dot{\overline{x}}_2 = 0$ which is torsion but nontrivial.

2.4.6 Linear Perturbed Dynamics

Here we will introduce the basic elements that allow us to treat sliding mode control of perturbed linear systems from an algebraic viewpoint. The basic developments and details may also be found in Fliess and Sira-Ramírez (1993b)

Definition 2.58 *A linear perturbed dynamics $\overline{\mathcal{D}}$ is a module where we distinguish a control input vector $\overline{u} = (\overline{u}_1, \ldots, \overline{u}_m)$ and perturbation inputs $\overline{\xi} = (\overline{\xi}_1, \ldots, \overline{\xi}_m)$ such that*

$$\overline{\mathcal{D}}/[\overline{u}, \overline{\xi}] = \textit{torsion}.$$

Consider the canonical epimorphism

$$\phi : \overline{\mathcal{D}} \rightarrow \overline{\mathcal{D}}/[\overline{\xi}] = \mathcal{D}$$

Since $[\overline{u}] \cap [\overline{\xi}] = 0$, then $\phi \mid_{[\overline{u}]}$ and $\phi \mid_{[\overline{\xi}]}$ are isomorphisms, i.e.

$$[\overline{u}] \simeq [u] \quad ; \quad [\overline{\xi}] \simeq [\xi]$$

This means that we should not distinguish between "perturbed" and "unperturbed" versions of the control input (i.e. between $\overline{u}$ and u), nor between similar versions of the perturbation input ($\overline{\xi}$ and ξ). Since $\mathcal{D}/[u]$ is torsion, we call $\mathcal{D}$ the unperturbed linear dynamics with u being the unperturbed control.

Control and perturbation inputs are not assumed to interact, thus the condition

$$[\overline{\xi}] \cap [\overline{u}] = \{0\}$$

appears to be quite natural. It will be assumed furthermore assumed that $[\overline{u}]$ is free. This means that we are essentially considering linear systems with unrestricted control inputs. Note, however, that perturbations are not necessarily independent in the sense that they might indeed satisfy some (linear unknown) set of differential equations. For this reason we assume here that $[\overline{\xi}]$ is not necessarily free, i.e. it may be torsion. It is reasonable to assume that the unperturbed version of the system, $\mathcal{D}$ is controllable, i.e. $\mathcal{D}$ is free. Regulation of uncontrollable systems is only possible in quite limited and unrealistic cases.

2.4.7 A Module-Theoretic Characterization of Sliding Regimes

The work presented here is taken from Fliess and Sira-Ramírez (1993a), where an algebraic characterization of sliding regimes is presented in terms of module theory.

Definition 2.59 *Let $\overline{\mathcal{D}}$ be a linear perturbed dynamics, such that $\mathcal{D}$ is controllable. We define a submodule $\overline{S}$ of $\overline{\mathcal{D}}$ as a sliding submodule if the following conditions holds*

(i) The sliding module does not contain elements which are driven exclusively by the perturbations. This condition is synthesized by $[\overline{S}] \cap [\overline{\xi}] = 0$

(ii) The canonical image S *of* $\overline{S}$ *in* $\mathcal{D} = \overline{\mathcal{D}}/[\overline{\xi}]$ *is a rank* m *free submodule, i.e. the quotient module*

$$\mathcal{D}/S \text{ is torsion.}$$

This condition means that all the control effort is spent in making the system behave as elements that are found in S*.*

It is convenient to assume that the unperturbed version of the system is observable; $\mathcal{D} = [u, y]$*. This guarantees that elements in the sliding module* S *may be obtained, if necessary, from asymptotic estimation procedures.*

$\mathcal{D}/S$ is the unperturbed (residual) *sliding dynamics* while $\overline{\mathcal{D}}/\overline{S}$ is the *perturbed sliding dynamics*. The canonical image of $\overline{u}$ in $\overline{\mathcal{D}}/\overline{S}$ is the *perturbed equivalent control*, denoted by $\overline{u}_{eq}$. The canonical image of u on $\mathcal{D}/S$ is addressed simply as the *equivalent control*, u_{eq}. Note that $\overline{u}_{eq}$ generally depends on the perturbation inputs $\overline{\xi}$, while u_{eq}, is perturbation independent.

Example 2.60 Consider the linear perturbed dynamics $\dot{\overline{y}} = \overline{u} + \overline{\xi}$. In this case $\overline{\mathcal{D}} = [\tilde{u}, \tilde{y}, \tilde{\xi}]/[\tilde{e}]$, with $\tilde{e} = \dot{\tilde{y}} - \tilde{u} - \tilde{\xi}$. The module $\overline{\mathcal{D}}/[\overline{u}, \overline{\xi}] = torsion$ and $\mathcal{D}$ is rank 1, with u acting as a basis. $\mathcal{D}$ is also controllable. The condition $\dot{\overline{y}} = -\overline{y}$ may be regarded as a *desirable* asymptotically stable dynamics. Consider $\overline{S} = [\overline{s}] = [\overline{y} + \overline{u}]$. It is easy to see that $\overline{S} \subset \overline{\mathcal{D}}$ with rank $S = 1$, while $\overline{S} \cap [\overline{\xi}] = 0$. Finally, the residue $\underline{y}$ of y in $\mathcal{D}/[y + u]$ satisfies : $\dot{\underline{y}} = -\underline{y}$, which is torsion. Note that the unperturbed equivalent control satisfies $\dot{u}_{eq} + u_{eq} = 0$, while the perturbed equivalent control satisfies $\dot{\overline{u}}_{eq} + \overline{u}_{eq} = -\overline{\xi}$.

2.4.8 The Switching Strategy

Let $z = (z_1, \ldots, z_m)$ be a *basis* of S and $\overline{z} = (\overline{z}_1, \ldots, \overline{z}_m)$ be a *basis* of $\overline{S}$. The basis z is the image of $\overline{z}$ under $\phi|_{\overline{S}}$. The input-output system relating u to z is *right* and *left invertible*, and hence *decouplable*. Therefore the multivariable case reduces to the single-input single-output case. The basis z (resp. $\overline{z}$) is unique up to a constant factor.

Example 2.61 Consider the previous example, $\dot{\overline{y}} = \overline{u} + \overline{\xi}$, with sliding module S generated by $s = u + y$. The element $z = u + y$ is a basis for S, while $\overline{z} = \overline{u} + \overline{y}$ is a basis for $\overline{S}$. The relation between z and u is trivially invertible. A switching strategy is obtained by condsidering $\dot{z} = -W \text{sign} z$, with $W > 0$ a sufficiently large constant. This choice results in the discontinuous controller, $\dot{u} + u = -W$ sign $(u + y)$. The response of the perturbed basis to the synthesized controller is governed by $\dot{\overline{z}} = \overline{\xi} - W$ sign $\overline{z}$.

2.4.9 Relations with Minimum Phase Systems and Dynamical Feedback

Definition 2.62 *Let $[u, S]$ stand for the module generated by u and S. The sliding module S is said to be minimum phase if and only if one of the following conditions are satisfied*

(i) $[u] = S$

(ii) If $[u] \not\subset S$ *then the endomorphism* τ*, defined as*
$\tau : [u, S]/S \rightarrow [u, S]/S,$ *has eigenvalues with negative real parts.*

The first condition means that the elements of the vector u can be expressed as a (decoupled) $k[\frac{d}{dt}]$-linear combination of the basis elements in S. The second condition means that some Hurwitz differential polynomial associated with u can be expressed as a decoupled $k[\frac{d}{dt}]$ linear combination of the basis elements in S.

Example 2.63 In the previous example the basis z for S was taken to be $z = u + y$ and evidently $[u] \not\subset S$, since u is not expressible as a $k[\frac{d}{dt}]$ linear combination of z. Definitely $[u] \subset [u, S] = [u, z]$ since $\dot{z} = \dot{u} + u$. The residue $\underline{u}$ of u in $[u, z]/[z]$ satisfies the linear system equation $\dot{\underline{u}} + \underline{u} = 0$ and therefore the sliding module is minimum phase.

2.4.10 Non-Minimum Phase Case

Let S be non-minimum phase. One may replace z by some other output $\sigma \in \mathcal{D}$, which is for instance a basis of $[u, z]$ and such that the transfer function relating u and σ is minimum phase.

It is easy to see, due to linearity, that the convergence of σ ensures that of z. Thus the minimum phase case is recovered. If the resulting numerator of the transfer function, relating σ and u, is not constant, then switchings will be taken by the highest order derivative of the control signal. This gives naturally the possibility of smoothed sliding mode controllers (see Sira-Ramirez (1992a, 1992c, 1993)).

2.4.11 Some Illustrations

Example 2.64 Consider the perturbed linear dynamics, $\dot{\overline{y}} = \overline{u} + \xi$, and the (desired) unperturbed *second* order dynamics given by $\ddot{\overline{y}} + 2\zeta\omega_n\dot{\overline{y}} + \omega_n^2\overline{y} = 0$. Consider the sliding module $S \subset \mathcal{D}$, generated by $z = \dot{u} + 2\zeta\omega_n u + \omega_n^2 y$. The element z is a basis for S and $\overline{z} = \dot{\overline{u}} + 2\zeta\omega_n\overline{u} + \omega_n^2\overline{y}$ is a basis for $\overline{S}$. The residue $\underline{y}$ of y in $\mathcal{D}/S$ satisfies the relation $\ddot{\underline{y}} + 2\zeta\omega_n\dot{\underline{y}} + \omega_n^2\underline{y} = 0$, which is certainly torsion and asymptotically stable to zero.

Evidently $[u] \not\subset [z]$. In order to obtain the necessary inclusion, consider the module $[u, z]$. Here one finds that the relationship between u and the basis element z for S, is given by $\dot{z} = \ddot{u} + 2\zeta\omega_n\dot{u} + \omega_n^2 u$ Taking the quotient $[u, z]/[z]$, one is left with the torsion system $\ddot{\underline{u}} + 2\zeta\omega_n\dot{\underline{u}} + \omega_n^2\underline{u} = 0$.

The linear map associated to $\frac{\mathrm{d}}{\mathrm{dt}}$ is represented by the matrix

$$\mathcal{T} = \begin{bmatrix} 0 & 1 \\ -\omega_n^2 & -2\zeta\omega_n \end{bmatrix}$$

which has eigenvalues with negative real parts. The sliding module S is therefore minimum phase.

Let W be a positive constant parameter. A dynamical sliding mode controller, which is robust with respect to ξ, is given by

$$\ddot{\overline{u}} + 2\zeta\omega_n\dot{\overline{u}} + \omega_n^2\overline{u} = -W\mathrm{sign}(\dot{\overline{u}} + 2\zeta\omega_n\overline{u} + \omega_n^2\overline{y}).$$

Use of the proposed dynamical switching strategy on the system leads to the following regulated dynamics for $\overline{z}$,

$$\dot{\overline{z}} = \ddot{\xi} + 2\zeta\omega_n\dot{\xi} + \omega_n^2\xi - W\mathrm{sign}\ \overline{z}.$$

For sufficiently high values of the gain parameter W, the element $\overline{z}$ goes to zero in finite time, and the desired (torsion) dynamics is achieved.

Example 2.65 Consider the nonminimum phase system $\ddot{\overline{y}} + 2\zeta\omega_n\dot{\overline{y}} + \omega_n^2\overline{y} = \dot{\overline{u}} - \beta\overline{u} + \xi$, (with $\beta > 0$), and the desired dynamics $\dot{\overline{y}} + \alpha\overline{y} = 0 \;\; ; \;\; \alpha > 0$. Evidently, $z = \dot{y} + \alpha y$ is a basis for the sliding submodule S, and $z = 0$ is deemed to be desirable.

However, as before, $[u] \not\subset S$. The relationship between z and u is readily obtained as $\ddot{u} + (\alpha - \beta)\dot{u} - \alpha\beta u = \ddot{z} + 2\zeta\omega_n\dot{z} + \omega_n^2 z$. The canonical image $\underline{u}$ of u in $[u, z]/[z]$ leads to the following unstable (torsion) dynamics $\ddot{\underline{u}} + (\alpha - \beta)\dot{\underline{u}} - \alpha\beta\underline{u} = (\frac{\mathrm{d}}{\mathrm{dt}} + \alpha)(\frac{\mathrm{d}}{\mathrm{dt}} - \beta)\underline{u} = 0$. The sliding module is therefore nonminimum phase.

Take a new basis σ of S such that $\dot{\sigma} = \beta\sigma + \alpha\dot{y} + \alpha y$. Note that $z = \dot{\sigma} - \beta\sigma$ and $\overline{z} = \dot{\overline{\sigma}} - \beta\overline{\sigma}$. One now has $\ddot{\sigma} + 2\zeta\omega_n\dot{\sigma} + \omega_n^2\sigma = \dot{u} + \alpha u$. The residue of u in $[u, \sigma]/[\sigma]$ satisfies $\dot{\underline{u}} + \alpha\underline{u} = 0$, and the sliding module is now minimum phase.

A robust dynamical sliding mode controller may now be synthesized which guarantees asymptotic convergence of $\overline{\sigma}$ to zero, and hence of $\overline{z}$ to zero. The desired unforced dynamics is, therefore, asymptotically attainable by means of dynamical sliding modes.

2.5 Conclusions

The differential algebraic approach to system dynamics provides both theoretical and practical grounds for the development of the sliding mode control of nonlinear dynamical systems. More general classes of sliding surfaces, which include inputs and possibly their time derivatives, have been shown naturally to allow for chatter-free sliding mode controllers of dynamical nature. Although equivalent smoothing effects can be similarly obtained by simply resorting to appropriate system extensions or prolongations of the input space, the theoretical simplicity and conceptual advantages stemming from the differential algebraic approach, bestow new possibilities for the broader area of discontinuous feedback control. For instance, the same smoothing effects and theoretical richness can be used for the appropriate formulation and study of many potential application areas based on pulse-width-modulated control strategies (Sira-Ramírez 1992d). The less explored pulse-frequency-modulated control techniques have also been shown to benefit from this new approach (Sira-Ramírez 1992e, Sira-Ramírez and Llanes-Santiago 1992). Possible extensions of the theory to nonlinear multivariable systems, and to infinite dimensional systems such as delay differential systems and systems described by partial differential equations, deserve attention.

Module Theory recovers and generalizes all known results of sliding mode control of linear multivariable systems. A more relaxed concept of sliding motions evolve in this context, as any desirable output dynamics is synthesizable by minimum phase sliding mode control. This statement is independent of the order of the desired dynamics. Generalizations demonstrate, for instance, that *matching conditions* are linked to particular state space realizations, but they have no further meaning from a general viewpoint. This fact has also been corroborated in recent developments in sliding observers (see Sira-Ramírez and Spurgeon (1993)). Multivariable sliding mode control problems have been shown to be always reducible to single-input single output problems in a natural manner.

Nonminimum phase problems have been shown to be handled by a suitable change of the output variable, whenever possible. The practical implications of this result seem to be multiple (see also Benvenuti et al (1992)). Extension of the results here presented to the case of time varying linear systems requires non-conmutative algebra.

An exciting area in which the algebraic approach may be used to full advantage is the area of sliding mode observers for linear systems. An interesting area rest on the extension of sliding mode theory from an algebraic viewpoint, to nonlinear multivariable sytems. The results so far seem to indicate that the class of systems to which the theory can be extended without unforseen complications is constrained to the class of *flat systems* (see Fliess et al (1991)).

References

Adkins, W. A., Weintraub, S.H. 1992, *Algebra : An approach via module theory*, Springer-Verlag, New York

Chang, L.W. 1991, A versatile sliding control with a second-order sliding condition. *Proc American Control Conference,* , Boston 54–55

Benvenuti, L., Di Benedetto, M. D., Grizzle, J. W. 1992, *Approximate output tracking for nonlinear non-minimum phase systems with applications to flight control*, Report CGR–92–20, Michigan Control Group Reports. University of Michigan, Ann Arbor, Michigan

Emelyanov, S.V. 1987, *Binary control systems*, MIR, Moscow

Emelyanov, S.V. 1990, The principle of duality, new types of feedback, variable structure and binary control, *Proc IEEE Int. Workshop on Variable Structure Systems and their Applications*, Sarajevo, 1–10

Fliess, M. 1986, A note on the invertibility of nonlinear input-output differential systems. *Systems and Control Letters* **8**, 147–151

Fliess, M. 1987, Nonlinear control theory and differential algebra: Some illustrative examples. *Proc IFAC, 10th Triennial World Congress*, Munich, 103–107

Fliess, M. 1988a, Nonlinear control theory and differential algebra, in *Modelling and adaptive control*, Byrnes, Ch. I. Kurzhanski, A., Lect. Notes in Contr. and Inform. Sci., **105**, Springer-Verlag, New York, 134–145

Fliess, M. 1988b, Généralisation non linéaire de la forme canonique de commande et linarisation par bouclage. *C.R. Acad. Sci. Paris* **I-308** , 377–379

Fliess, M. 1989a, Automatique et corps différentieles. *Forum Mathematicum* **1**, 227–238

Fliess, M. 1989b, Generalized linear systems with lumped or distributed parameters and differential vector spaces. *International Journal of Control* **49**, 1989–1999

Fliess, M. 1990a, Generalized controller canonical forms for linear and nonlinear dynamics. *IEEE Transactions on Automatic Control* **35**, 994–1001

Fliess, M. 1990b, What the Kalman state variable representaion is good for. *Proc IEEE Conference on Decision and Control,* **3**, Honolulu, 1282–1287

Fliess M. 1990c, Some basic structural properties of generalized linear systems *Systems and Control Letters* **15** 391–396.

Fliess, M. 1991, Controllability revisited, in *Mathematical System Theory : The Influence of R.E. Kalman*, ed. Antoulas, A.C., Springer-Verlag, New York, 463–474

Fliess, M., Hassler, M. 1990, Questioning the classical state-space description via circuit examples, in *Mathematical Theory of Networks and Systems*, eds. Kaashoek, M.A., Ram, A.C.M., van Schuppen, J.H., Progress in Systems and Control Theory, Birkhauser, Boston

Fliess, M., Lévine, J., Rouchon, P. 1991, A simplified approach of crane control via generalized state-space model. *Proc IEEE Conference on Decision and Control,* **1**, Brighton, England, 736–741

Fliess, M., Messager, F. 1990, Vers une stabilisation non lineaire discontinue, in *Analysis and Optimization of Systems*, eds. Bensoussan, A., Lions, J.L., Lect. Notes Contr. Inform. Sci., **144**, Springer-Verlag, New York, 778–787

Fliess, M., Messager, F. 1991, Sur la commande en régime glissant. *C. R. Acad. Sci. Paris* **I-313**, 951–956

Fliess, M., Sira-Ramírez, H. 1993a, Regimes glissants, structures variables linéaires et modules. *C.R. Acad. Sci. Paris*, submitted for publication

Fliess, M., Sira-Ramírez, H. 1993b. A Module Theoretic Approach to Sliding Mode Control in Linear Systems, *Proc IEEE Conference on Decision and Control,* , submitted for publication

Kalman, R., Falb, P., Arbib, M. 1970, *Topics in Mathematical Systems Theory*, McGraw-Hill, New York

Kolchin, E.R. 1973, *Differential algbebra and algebraic groups*, Academic Press, New York

Nijmeijer, H., Van der Schaft, A. 1990, *Nonlinear dynamical control systems*, Springer-Verlag, New York

Pommaret, J.F. 1983, *Differential galois theory*, Gordon and Breach, New York

Pommaret, J.F. 1986, Géométrie différentielle algébrique et théorie du contrôle. *C.R. Acad. Sci. Paris* **I-302**, 547–550

Sira-Ramírez, H. 1991, Dynamical feedback strategies in aerospace systems control: A differential algebraic approach. *Proc First European Control Conference*, Grenoble, 2238–2243

Sira-Ramírez, H. 1993, Dynamical variable structure control strategies in asymptotic output tracking problems. *IEEE Transactions on Automatic Control* , to appear

Sira-Ramírez, H. 1992a, Asymptotic output stabilization for nonlinear systems via dynamical variable structure control. *Dynamics and Control* **2**, 45–58

Sira-Ramírez, H. 1992b, The differential algebraic approach in nonlinear dynamical feedback controlled landing maneuvers. *IEEE Transactions on Automatic Control* **AC-37**, 1173–1180

Sira-Ramírez, H. 1992c, Dynamical sliding mode control strategies in the regulation of nonlinear chemical processes. *International Journal of Control* **56**, 1–21

Sira-Ramírez, H. 1992d, Dynamical pulse width modulation control of nonlinear systems. *Systems and Control Letters* **18**, 223–231.

Sira-Ramírez, H. 1992e, Dynamical discontinuous feedback control in nonlinear systems. *Proc IFAC Nonlinear Control Systems Conference*, Burdeaux, 471–476

Sira-Ramírez, H. 1993, A Differential Algebraic Approach to Sliding Mode Control of Nonlinear Systems. *International Journal of Control* **57**, 1039–1061

Sira-Ramírez, H., Ahmad, S., Zribi, M. 1992, Dynamical feedback control of robotic manipulators with joint flexibility. *IEEE Transactions on Systems Man and Cybernetics* **22**, 736–747

Sira-Ramírez, H., Lischinsky-Arenas, P. 1991, The differential algebraic approach in nonlinear dynamical compensator design for dc-to-dc power converters. *International Journal of Control* **54**, 111–134

Sira-Ramírez, H., Llanes-Santiago, O. 1992, An extended system approach to dynamical pulse-frequency-modulation control of nonlinear systems. *Proc IEEE Conference on Decision and Control,* **1**, Tucson, 2376–2380

Sira-Ramírez, H., Spurgeon, S.K. 1993, On the robust design of sliding observers for linear systems, submitted for publication

3. Robust Tracking with a Sliding Mode

Raymond Davies, Christopher Edwards and Sarah K. Spurgeon

3.1 Introduction

The system analyst represents the salient features of a given physical process using a mathematical model. Any such model, whether derived from first principles using the laws of physics or developed using system identification techniques, will contain uncertainties due to modelling assumptions, lack of precise knowledge of system data and external effects all of which may vary in both time and space.

One of the possible tools available for control system design and analysis of such uncertain dynamical systems involves the evocation of a deterministic approach. Within this category of design tools, the two main approaches are *Variable Structure Control* (VSC), particularly with a sliding mode, and Lyapunov control. Historically, VSC is characterized by a control structure which is switched as the system state crosses specified discontinuity surfaces in the state-space and the sliding mode describes the particular case when, following a preliminary motion onto the switching surfaces, the system state is constrained to lie upon the surfaces. The approach exhibits the well known property of total invariance to all matched uncertainty when sliding. Further, in the presence of only matched uncertainty, the system's dynamic behaviour when in the sliding mode, will be wholly described by the chosen switching surfaces.

The major practical disadvantage of this approach is the fundamental requirement of a discontinuous control structure. This has resulted in the development of continuous approximations to the discontinuous elements, see for example Burton and Zinober (1986), and also the use of boundary layer techniques (Slotine 1984). It should be noted that such approximations result in a continuous motion within a bounded region of the sliding surfaces and not a true sliding mode. For the case of a dynamic system containing only matched uncertainty, such approximations to the required discontinuous control action will consequently induce some sensitivity to the uncertainty contribution during sliding which will, in turn, affect the ideal dynamic behaviour prescribed by the switching surfaces. It is seen that for the case of problems containing matched uncertainty, where the sliding philosophy is particularly appropriate, implementation considerations result in motion about rather than constrained to lie within the sliding surfaces.

Many physical systems contain both matched and unmatched uncertainty. A second disadvantage of the traditional sliding mode approach to design is that unmatched contributions are not formally considered. For example, it can be shown, that for the case of motion constrained to the sliding surface, the dynamic behaviour when sliding will vary as a function of the unmatched uncertainty. Ryan and Corless (1984) use a Lyapunov approach to develop a continuous nonlinear controller which incorporates consideration of unmatched uncertainty contributions. The freedom to deal with unmatched uncertainty is obtained by considering the goal of motion about rather than constrained to prescribed sliding surfaces as the start point for the design procedure.

It has already been seen that although from the theoretical point of view a traditional sliding mode design uses a discontinuous control strategy to ensure motion lies on the prescribed discontinuity surfaces in the sliding mode, this requirement has to be relaxed for practical implementation; the consequence is motion about the switching surfaces. The Ryan and Corless (1984) approach recognises this fact and exploits the freedom thus provided to incorporate additional robustness considerations at the design stage. Bounded motion about the nominal sliding mode dynamic in the presence of bounded matched and unmatched uncertainty is the result. Although intuitively appealing and theoretically elegant, the original results are very conservative. The uncertainty class considered requires a relatively small upper bound to be placed upon the matched and unmatched uncertainty contributions. This has been found to restrict the practical viability of the results. Spurgeon and Davies (1993) have investigated the possibility of restricting the uncertainty class for which the work was originally considered. It has been shown that a subclass of that considered by Ryan and Corless (1984) is sufficiently general to cover a broader class of engineering applications and reduce the conservativeness of the results.

This work develops this practical control design methodology to incorporate a demand following requirement. Section 3.2 formulates the problem and defines the associated uncertainty class. The design of the sliding manifold and an assessment of its properties is presented in Sect. 3.3. Section 3.4 defines the associated nonlinear control structure which is shown to produce bounded motion about the ideal sliding mode dynamic which has been specified by the choice of sliding manifold. Section 3.5 considers the application of the proposed nonlinear tracking strategy to the design of a temperature control scheme for an industrial furnace.

3.2 Problem Formulation

Consider an uncertain dynamical system of the form

$$\dot{x}(t) = Ax(t) + Bu(t) + F(t, x, u) \tag{3.1}$$

with output

$$y(t) = \gamma^T x(t) + h(t, x) \tag{3.2}$$

where $x \in \mathbb{R}^n$, $u \in \mathbb{R}^m$, $y \in \mathbb{R}^p$, $p \leq m$, $m \leq n$. The known matrix pair (A, B) defining the nominal linear system is assumed controllable with B of full rank. It is assumed in the theoretical development that the system states are available to the controller and so an observability requirement is not necessary. The output $y(t)$ merely represents those linear combinations of system states which are required to track the prescribed reference signals. As might be expected, an overall controllability requirement for tracking is required and this will be developed in Sect. 3.3. The unknown functions $F(\cdot,\cdot,\cdot) : \mathbb{R} \times \mathbb{R}^n \times \mathbb{R}^m \to \mathbb{R}^n$ and $h(\cdot,\cdot) : \mathbb{R} \times \mathbb{R}^n \to \mathbb{R}^p$ model uncertainties in the system and output respectively. For ease of exposition, it is assumed that $F \in \mathcal{F}$, a known class of functions whereby the matched and unmatched uncertainty components can be decomposed in the form

$$F = f(t,x) + g(t,x,u) \tag{3.3}$$

$$\begin{array}{lcl} f(\cdot,\cdot) : \mathbb{R} \times \mathbb{R}^n & \to & (\operatorname{im} B)^{\perp} \\ g(\cdot,\cdot,\cdot) : \mathbb{R} \times \mathbb{R}^n \times \mathbb{R}^m & \to & \operatorname{im} B \end{array}$$

where $\operatorname{im}(\cdot)$ denotes the range of $(\cdot)$ and f, g and h are Carathéodory functions[1]. It will be shown later in this section that the uncertainty function $h(t,x)$ appears as unmatched uncertainty in an augmented system containing the states (3.1). This function $h(t,x)$ is also assumed to belong to a known class of functions which will be denoted $\mathcal{H}$. The matched and unmatched components of each $F(t,x,u) \in \mathcal{F}$ and the function $h(t,x) \in \mathcal{H}$ are to be expressed in the form

$$\begin{aligned} f(t,x) &= F_1(t,x)x + F_2(t,x) \\ g(t,x,u) &= G_1(t,x,u)u + G_2(t,x) \\ h(t,x) &= H_1(t,x)x + H_2(t,x) \end{aligned} \tag{3.4}$$

where

$$\begin{array}{ll} \|F_1(t,x)\| < K_{F_1} \; , & \|F_2(t,x)\| < K_{F_2} \\ \|G_1(t,x,u)\| < K_{G_1} \; , & \|G_2(t,x)\| < K_{G_2} \\ \|H_1(t,x)\| < K_{H_1} \; , & \|H_2(t,x)\| < K_{H_2} \end{array} \tag{3.5}$$

Here $\|\cdot\|$ denotes the 2-norm, K_{F_1}, K_{F_2}, K_{G_1}, K_{G_2}, K_{H_1} and K_{H_2} are known constants. Equations (3.4) and (3.5) imply

$$\begin{aligned} \|f(t,x)\| &\le K_{F_1}\|x\| + K_{F_2} \\ \|g(t,x,u)\| &\le K_{G_1}\|u\| + K_{G_2} \\ \|h(t,x)\| &\le K_{H_1}\|x\| + K_{H_2}. \end{aligned} \tag{3.6}$$

In the original work, Ryan and Corless (1984) considered the problem of state regulation with the following uncertainty class

[1] A function $f(t,x)$ is a Carathéodory function if for all $t \in \mathbb{R}$ it is continuous in $x \in \mathbb{R}$, for all x it is Lebesgue measurable in t, and is dominated by a locally integrable function in t.

$$\begin{aligned} \|f(t,x)\| &\le \bar{K}_{F_1}\|x\| + \bar{K}_{F_2} \\ \|g(t,x,u)\| &\le \bar{K}_{G_1}\|u\| + \alpha(t,x) \end{aligned} \tag{3.7}$$

where $\bar{K}_{F_1}$, $\bar{K}_{F_2}$ and $\bar{K}_{G_1}$ are known constants and $\alpha(t,x)$ is a known strongly Carathéodory function[2]. However, the structure (3.4) was not imposed. This work will show that the imposition of (3.4) enables the development of an associated nonlinear controller which is applicable to systems of the form (3.1), (3.2) where the norm bounds can have significantly larger values than would be allowable without the added structural constraints. Further, it will be seen that the additional structural constraints are sufficiently general to cover most practical applications. Consideration of the system (3.1), (3.2) with uncertainty class (3.4), (3.5), (3.6) thus provides a framework for practical robust control design and analysis. The design objective requires that the chosen output $y(t)$ asymptotically tracks a reference signal $w(t)$ which is assumed to be a vector whose entries are piecewise constant. In order to circumvent problems at the points of discontinuity, define the tracking demand $W(t)$ by

$$\dot{W} = \Gamma W + w(t) \tag{3.8}$$

where $\Gamma \in \mathbb{R}^{p\times p}$ is a stable matrix. This necessitates consideration of the derivative of the output uncertainty $h(\cdot,\cdot)$:

$$\dot{h}(t,x) = \frac{\partial h}{\partial x}\dot{x} + \frac{\partial h}{\partial t} \tag{3.9}$$

where for all $h \in \mathcal{H}$ it is assumed that

$$\left\|\frac{\partial h}{\partial x}\right\| \le K_{H_3}\ , \qquad \left\|\frac{\partial h}{\partial t}\right\| \le K_{H_4}$$

so that

$$\left\|\dot{h}(t,x)\right\| \le K_{H_3}\|\dot{x}\| + K_{H_4} \tag{3.10}$$

The tracking approach to be adopted is to employ integral action, Bühler (1991). This is particulary suitable for the temperature control problem to be considered in Sect. 3.5 where integral action is traditionally employed within the industry. To introduce this integral action into the system (3.1), (3.2) define states $x_R \in \mathbb{R}^p$ and a nonsingular matrix $T_i \in \mathbb{R}^{p\times p}$ such that

$$\dot{x}_R = T_i^{-1}\left(W(t) - y(t)\right) \tag{3.11}$$

where T_i can be considered a design parameter representing the traditional integrator time constant. The design approach to be adopted is the traditional two stage process associated with the design of sliding mode control schemes. The first stage is concerned with the design of a sliding manifold $\mathcal{S}$, dependent upon $W(t)$, upon which desired behaviour can be guaranteed for the nominal system (3.1) in the presence of the matched uncertainty, $g(t,x,u)$, alone.

[2]A Carathéodory function $f(t,x)$ is said to be strongly Carathéodory if $\|f(t,x)\| \le m$ where m is constant.

The second stage concerns the development of a continuous nonlinear control strategy which will ensure the attractivity of a bounded neighbourhood of $\mathcal{S}$ thus ensuring bounded motion about the desired class (3.7), (3.10) and hence asymptotic tracking of $W(t)$. For this development, an augmented state-space is considered where

$$\tilde{x} = \begin{bmatrix} x_R^T & x^T \end{bmatrix}^T \tag{3.12}$$

The state equations of the augmented system are readily expressed from (3.1), (3.3) and (3.11) as

$$\dot{\tilde{x}}(t) = \begin{bmatrix} 0 & -T_i^{-1}\gamma^T \\ 0 & A \end{bmatrix} \tilde{x}(t) + \begin{bmatrix} 0 \\ B \end{bmatrix} u(t) + \begin{bmatrix} T_i^{-1} \\ 0 \end{bmatrix} W(t) + \begin{bmatrix} -T_i^{-1}h \\ F \end{bmatrix} \tag{3.13}$$

As was stated earlier, the output uncertainty function $h(t,x)$ appears as wholly unmatched uncertainty. In addition no component of the demand signal $W(t)$ is matched.

The development of an appropriate sliding manifold $\mathcal{S}$ will now be explored.

3.3 Design of the Sliding Manifold

The switching function is defined to be of the form

$$S(\tilde{x}, W) = \tilde{C}\tilde{x}(t) - C_W W(t) \tag{3.14}$$

where $C_W \in \mathbb{R}^{m \times p}$ is an arbitrary, possibly zero, matrix which influences only the steady-state value of the integral states x_R and $\tilde{C} \in \mathbb{R}^{m \times (n+p)}$ is defined by

$$\tilde{C} = \begin{bmatrix} C_R & C \end{bmatrix} \tag{3.15}$$

with $C_R \in \mathbb{R}^{m \times p}$ and $C \in \mathbb{R}^{m \times n}$. The criteria for selection of $\tilde{C}$ and its role in prescribing the system dynamic performance will now be investigated.

An orthogonal transformation $\tilde{T} \in \mathbb{R}^{(n+p) \times (n+p)}$ is first introduced, as in Dorling and Zinober (1990), to partition the matched and unmatched uncertainty. Let $\bar{T} \in \mathbb{R}^{n \times n}$ be an orthogonal matrix satisfying

$$\bar{T}B = \begin{bmatrix} 0 \\ B_2 \end{bmatrix}, \quad \bar{T} = \begin{bmatrix} T \\ T' \end{bmatrix} \tag{3.16}$$

where $B_2 \in \mathbb{R}^{m \times m}$, $T \in \mathbb{R}^{(n-m) \times n}$ and $T' \in \mathbb{R}^{m \times n}$. Then let

$$\tilde{T} = \begin{bmatrix} I_p & 0 \\ 0 & \bar{T} \end{bmatrix} \tag{3.17}$$

and define a first coordinate transformation by

$$\hat{x} = \tilde{T}\tilde{x} \tag{3.18}$$

The system representation becomes

$$\dot{\hat{x}} = \begin{bmatrix} 0 & -T_i^{-1}\gamma_1^T & -T_i^{-1}\gamma_2^T \\ 0 & A_{11} & A_{12} \\ 0 & A_{21} & A_{22} \end{bmatrix} \hat{x} + \begin{bmatrix} 0 \\ 0 \\ B_2 \end{bmatrix} u + \begin{bmatrix} T_i^{-1} \\ 0 \\ 0 \end{bmatrix} W(t) + \begin{bmatrix} -T_i^{-1}h \\ Tf \\ T'g \end{bmatrix} \tag{3.19}$$

where

$$\bar{T}A\bar{T}^T = \begin{bmatrix} A_{11} & A_{12} \\ A_{21} & A_{22} \end{bmatrix} \tag{3.20}$$

with $A_{11} \in \mathbb{R}^{(n-m)\times(n-m)}$, $A_{12} \in \mathbb{R}^{(n-m)\times m}$, $A_{21} \in \mathbb{R}^{m\times(n-m)}$, $A_{22} \in \mathbb{R}^{m\times m}$ and

$$\gamma^T\bar{T}^T = \begin{bmatrix} \gamma_1^T & \gamma_2^T \end{bmatrix} \tag{3.21}$$

with $\gamma_1 \in \mathbb{R}^{p\times(n-m)}$, $\gamma_2 \in \mathbb{R}^{p\times m}$. In order to group the matched and unmatched uncertainty contributions a convenient block partition structure is imposed on (3.19) whereby

$$\begin{bmatrix} \begin{bmatrix} 0 & -T_i^{-1}\gamma_1^T \\ 0 & A_{11} \end{bmatrix} & \begin{bmatrix} -T_i^{-1}\gamma_2^T \\ A_{12} \end{bmatrix} \\ \begin{bmatrix} 0 & A_{21} \end{bmatrix} & A_{22} \end{bmatrix} = \begin{bmatrix} \bar{A}_{11} & \bar{A}_{12} \\ \bar{A}_{21} & \bar{A}_{22} \end{bmatrix} \tag{3.22}$$

Here $\bar{A}_{11} \in \mathbb{R}^{(p+n-m)\times(p+n-m)}$, $\bar{A}_{12} \in \mathbb{R}^{(p+n-m)\times m}$, $\bar{A}_{21} \in \mathbb{R}^{m\times(p+n-m)}$ and $\bar{A}_{22} \in \mathbb{R}^{m\times m}$. Transforming and compatibly partitioning the switching function (3.14), yields

$$S(\hat{x}, W) = \begin{bmatrix} \begin{bmatrix} C_R & C_1 \end{bmatrix} & C_2 \end{bmatrix} \hat{x}(t) - C_W W(t) \tag{3.23}$$

where

$$C\bar{T}^T = \begin{bmatrix} C_1 & C_2 \end{bmatrix} \tag{3.24}$$

with $C_1 \in \mathbb{R}^{m\times(n-m)}$ and $C_2 \in \mathbb{R}^{m\times m}$. Partition the transformed state vector so that

$$\hat{x}(t) = \begin{bmatrix} \hat{x}_1(t) \\ \hat{x}_2(t) \end{bmatrix} \tag{3.25}$$

where $\hat{x}_1 \in \mathbb{R}^{p+n-m}$ and $\hat{x}_2 \in \mathbb{R}^m$. Applying the sliding condition to the switching function (3.23) then yields

$$0 = \begin{bmatrix} C_R & C_1 \end{bmatrix} \hat{x}_1(t) + C_2\hat{x}_2(t) - C_W W(t) \tag{3.26}$$

Assuming the parameters of the switching function are chosen so that C_2 is nonsingular, equation (3.26) defines the sliding manifold to be

$$\mathcal{S} = \{ (\hat{x}_1, \hat{x}_2) : \hat{x}_2 = -C_2^{-1} \begin{bmatrix} C_R & C_1 \end{bmatrix} \hat{x}_1(t) + C_2^{-1}C_W W(t) \} \tag{3.27}$$

it follows that m of the states in (3.19) may be expressed in terms of the remaining $n - m + p$ states and the tracking demand during the traditional sliding mode phase. Specify C_R and C_1 by

$$C_2 M = \begin{bmatrix} C_R & C_1 \end{bmatrix} \tag{3.28}$$

The role of M is seen to be that of a full-state feedback prescribing the dynamics of the nominal $(\bar{A}_{11}, \bar{A}_{12})$ subsystem which in turn specifies the desired dynamic behaviour of the full system during the sliding mode. It should be noted that controllability of the original (A, B) pair guarantees controllability of the (A_{11}, A_{12}) pair but is not sufficient to guarantee controllability of the $(\bar{A}_{11}, \bar{A}_{12})$ pair. Before proceeding with the selection of the switching surface parameters it is thus necessary to check the controllability of the $(\bar{A}_{11}, \bar{A}_{12})$ pair. From the controllability of the (A, B) pair it follows that the pair $(\bar{A}_{11}, \bar{A}_{12})$ is controllable provided

$$\operatorname{rank}\begin{bmatrix} \gamma_1^T & \gamma_2^T \\ A_{11} & A_{12} \end{bmatrix} = n - m + p \tag{3.29}$$

Note that the condition $p \leq m$ imposed initially is a necessary but not sufficient condition for controllability. Any robust linear design procedure such as that proposed by Kautsky et al (1985) may then be used to determine an appropriate feedback gain matrix M which can then be used to determine C_R and C_1 using (3.28).

The switching function design framework has been outlined. The properties of system motion constrained to the sliding manifold $\mathcal{S}$ will now be explored; these developments assume the existence of an appropriate controller. In order to facilitate the analysis a final transformation matrix $\hat{T} \in \mathrm{IR}^{(n+p)\times(n+p)}$ is defined by

$$\hat{T} = \begin{bmatrix} I_{n-m+p} & 0 \\ M & I_m \end{bmatrix} \tag{3.30}$$

with associated state transformation

$$\begin{bmatrix} \xi \\ \phi \end{bmatrix} = \hat{T}\hat{x} \tag{3.31}$$

where $\xi \in \mathrm{IR}^{p+n-m}$ and $\phi \in \mathrm{IR}^m$. The transformed augmented system becomes

$$\begin{aligned} \begin{bmatrix} \dot{\xi}(t) \\ \dot{\phi}(t) \end{bmatrix} &= \begin{bmatrix} \Sigma & \bar{A}_{12} \\ \Theta & \Omega \end{bmatrix} \begin{bmatrix} \xi(t) \\ \phi(t) \end{bmatrix} + \begin{bmatrix} 0 \\ B_2 \end{bmatrix} u(t) \\ &+ \begin{bmatrix} \begin{bmatrix} I_p \\ 0 \end{bmatrix} \\ M_1 \end{bmatrix} T_i^{-1} W(t) + \begin{bmatrix} \begin{bmatrix} -T_i^{-1} h \\ Tf \end{bmatrix} \\ -M_1 T_i^{-1} h + M_2 T f + T' g \end{bmatrix} \end{aligned} \tag{3.32}$$

where

$$\begin{aligned} \Sigma &= \bar{A}_{11} - \bar{A}_{12} M \\ \Theta &= M\Sigma + \bar{A}_{21} - \bar{A}_{22} M \\ \Omega &= M\bar{A}_{12} + \bar{A}_{22} \\ M &= \begin{bmatrix} M_1 & M_2 \end{bmatrix} \end{aligned} \tag{3.33}$$

with $M_1 \in \mathrm{IR}^{m\times p}$ and $M_2 \in \mathrm{IR}^{m\times(n-m)}$. In terms of the (ξ, ϕ) coordinate system the sliding manifold (3.27) may be defined by

$$\mathcal{S} = \left\{(\xi, \phi) : \phi = C_2^{-1} C_W W\right\}. \tag{3.34}$$

For system motion constrained to the sliding manifold (3.34), (3.32) reduces to

$$\dot{\xi}(t) = \Sigma\,\xi(t) + \left(\bar{A}_{12} C_2^{-1} C_W + \begin{bmatrix} T_i^{-1} \\ 0 \end{bmatrix}\right) W(t) + \begin{bmatrix} -T_i^{-1} h \\ T f \end{bmatrix} \tag{3.35}$$

It is already seen that the choice of M determines the nominal dynamics and that the sliding system is completely insensitive to the matched uncertainty, $g(t, x, u)$. The ideal sliding mode performance is given by

$$\dot{\xi}_m(t) = \Sigma\,\xi_m(t) + \left(\bar{A}_{12} C_2^{-1} C_W + \begin{bmatrix} T_i^{-1} \\ 0 \end{bmatrix}\right) W(t) \tag{3.36}$$

The effect of unmatched contributions $h(t, x)$ and $f(t, x)$ upon the actual sliding dynamics (3.35) must now be addressed. A Lyapunov argument similar to that used by Ryan and Corless (1984) will be utilised. Let

$$\tilde{\Sigma} = \Sigma + \Delta\Sigma \tag{3.37}$$
$$\tilde{A}_{12} = \bar{A}_{12} + \Delta\bar{A}_{12} \tag{3.38}$$

where

$$\begin{bmatrix} \Delta\Sigma & \Delta\bar{A}_{12} \end{bmatrix} = \begin{bmatrix} -T_i^{-1} H_1 \\ T F_1 \end{bmatrix} \tilde{T}^T \begin{bmatrix} \begin{bmatrix} 0 & I_{n-m} \\ & -M \end{bmatrix} & \begin{matrix} 0 \\ I_m \end{matrix} \end{bmatrix} \tag{3.39}$$

Define P_1 as the unique, symmetric positive definite solution to the Lyapunov equation

$$P_1 \Sigma + \Sigma^T P_1 + I_{n-m+p} = 0. \tag{3.40}$$

Assumption 3.1 *There exists a real scalar parameter* ν *satisfying* $0 < \nu \leq \bar{\nu}$ *where*

$$\bar{\nu} = -\sup_{F_1, H_1} \lambda_{\max}\left(P_1 \tilde{\Sigma} + \tilde{\Sigma}^T P_1\right) \tag{3.41}$$

and $\lambda_{\max}(\cdot)$ *denotes the maximum eigenvalue. The properties of motion constrained to* $\mathcal{S}$ *are expressed in terms of the following result.*

Theorem 3.2 *If Assumption 3.1 holds, then*
(i) The system is globally uniformly ultimately bounded, as defined by Ryan and Corless (1984), with respect to the ellipsoid

$$E_1(r_1) = \left\{\xi : \frac{1}{2}\xi^T P_1 \xi \leq r_1\right\} \tag{3.42}$$

where

$$r_1 = \varepsilon_1 + \frac{2\left\|P_1\right\|^2}{\nu^2}\left(K_1 + K_2 \|W\|_{\max}\right)^2 \tag{3.43}$$

and

$$K_1 = \sup_{F_2,H_2} \left\| P_1^{\frac{1}{2}} \begin{bmatrix} -T_i^{-1}H_2 \\ TF_2 \end{bmatrix} \right\| \tag{3.44}$$

$$K_2 = \sup_{F_1,H_1} \left\| P_1^{\frac{1}{2}} \left(\tilde{A}_{12}C_2^{-1}C_W + \begin{bmatrix} T_i^{-1} \\ 0 \end{bmatrix} \right) \right\| \tag{3.45}$$

with $\|W\|_{\max}$ *denoting the maximum norm of all possible demand signals* W *and* $\varepsilon_1 > 0$ *is an arbitrary small constant.*
(ii) If the deviation of the system state (3.35) from the ideal sliding mode dynamics (3.36) is given by

$$\Delta\xi(t) = \xi(t) - \xi_m(t) \tag{3.46}$$

with $\Delta\xi(t_0) = 0$, *then the deviation from the ideal sliding motion is bounded with respect to the ellipsoid* $E_1(r_2)$ *where*

$$r_2 = \begin{cases} 2\|P_1\|^2 \left(K_1 + K_3\|W\|_{\max} + K_4 \left\| P_1^{\frac{1}{2}}\xi(t_0) \right\| \right)^2 & \text{if } \xi(t_0) \notin E_1(r_1) \\ 2\|P_1\|^2 \left(K_1 + K_3\|W\|_{\max} + K_4\sqrt{2r_1} \right)^2 & \text{if } \xi(t_0) \in E_1(r_1) \end{cases} \tag{3.47}$$

with

$$K_3 = \sup_{F_1,H_1} \left\| P_1^{\frac{1}{2}}\, \Delta\tilde{A}_{12}\, C_2^{-1}C_W \right\| \tag{3.48}$$

$$K_4 = \sup_{F_1,H_1} \left\| P_1^{\frac{1}{2}}\, \Delta\Sigma\, P_1^{-\frac{1}{2}} \right\| \tag{3.49}$$

i.e.

$$\Delta\xi(t) \in E_1(r_2) \;\; \forall t \geq t_0$$

Proof.
(i) Consider the following positive definite Lyapunov function candidate

$$V_1(\xi) = \frac{1}{2}\xi^T P_1 \xi \tag{3.50}$$

where P_1 is defined in (3.40). Along any solution

$$\dot{V}_1(\xi) = \xi^T P_1 \left(\tilde{\Sigma}\xi + \tilde{A}_{12}\phi + \begin{bmatrix} I_p \\ 0 \end{bmatrix} T_i^{-1}W + \begin{bmatrix} -T_i^{-1}H_2 \\ TF_2 \end{bmatrix} \right) \tag{3.51}$$

Here the uncertainty structure (3.4) has been exploited to develop the linear pertubation matrices defined in (3.37)-(3.39). Applying the quadratic stability criterion (3.41), see Barmish (1983) and Khargonekar et al (1990), yields

$$\dot{V}_1(\xi) \leq -\frac{\nu}{2}\|\xi\|^2 + \xi^T P_1 \left(\tilde{A}_{12}\phi + \begin{bmatrix} I_p \\ 0 \end{bmatrix} T_i^{-1}W + \begin{bmatrix} -T_i^{-1}H_2 \\ TF_2 \end{bmatrix} \right) \tag{3.52}$$

For motion constrained to $\mathcal{S}$, knowledge of W determines ϕ from (3.34). It follows that

$$\begin{aligned}\dot{V}_1(\xi) &\leq -\nu V_1(\xi)\,\|P_1\|^{-1} + \sqrt{2V_1(\xi)}\left\{\left\|P_1^{\frac{1}{2}}\begin{bmatrix} -T_i^{-1}H_2 \\ TF_2 \end{bmatrix}\right\|\right.\\ &\quad \left. + \left\|P_1^{\frac{1}{2}}\left(\tilde{A}_{12}C_2^{-1}C_W + \begin{bmatrix} T_i^{-1} \\ 0 \end{bmatrix}\right)\right\|\,\|W\|\right\} \qquad (3.53)\end{aligned}$$

With r_1, K_1 and K_2 defined in (3.43)-(3.45) it is true that $\dot{V}_1(\xi) < 0$ if $V_1(\xi) > r_1 - \varepsilon$ and the states ultimately enter the ellipsoid (3.42). Global uniform ultimate boundedness with respect to $E_1(r_1)$ is therefore proved.

(ii) To investigate the deviation from the ideal sliding mode dynamics consider this time $V_1(\Delta\xi)$. Along any solution

$$\begin{aligned}\dot{V}_1(\Delta\xi) &= \Delta\xi^T P_1\left(\Sigma\,\Delta\xi + \Delta\Sigma\,\xi + \Delta\bar{A}_{12}\,C_2^{-1}C_W W + \begin{bmatrix} -T_i^{-1}H_2 \\ TF_2 \end{bmatrix}\right)\\ &\leq -\frac{1}{2}\|\Delta\xi\|^2 + \|\Delta\xi^T P_1\,\Delta\Sigma\,\xi\| + \|\Delta\xi^T P_1\,\Delta\bar{A}_{12}\,C_2^{-1}C_W W\|\\ &\quad + \left\|\Delta\xi^T P_1\begin{bmatrix} -T_i^{-1}H_2 \\ TF_2 \end{bmatrix}\right\|\\ &\leq -V_1(\Delta\xi)\,\|P_1\|^{-1} + \sqrt{2V_1(\xi)}\left\{\left\|P_1^{\frac{1}{2}}\,\Delta\Sigma\,P_1^{-\frac{1}{2}}\right\|\,\left\|P_1^{\frac{1}{2}}\xi\right\|\right.\\ &\quad \left. + \left\|P_1^{\frac{1}{2}}\,\Delta\bar{A}_{12}\,C_2^{-1}C_W W\right\| + \left\|P_1^{\frac{1}{2}}\begin{bmatrix} -T_i^{-1}H_2 \\ TF_2 \end{bmatrix}\right\|\right\} \qquad (3.54)\end{aligned}$$

Clearly

$$\left\|P_1^{\frac{1}{2}}\xi(t)\right\| \leq \begin{cases} \left\|P_1^{\frac{1}{2}}\xi(t_0)\right\| & \text{if } \xi(t_0) \notin E_1(r_1) \\ \sqrt{2r_1} & \text{if } \xi(t_0) \in E_1(r_1) \end{cases} \qquad (3.55)$$

With r_2, K_1, K_3 and K_4 defined in (3.44), (3.47), (3.48) and (3.49), it follows that the deviation from ideal model motion is bounded with respect to the ellipsoid $E_1(r_2)$. □

Theorem 3.2 relates solely to system motion constrained to $\mathcal{S}$. It demonstrates that the system exhibits the well known invariance to matched uncertainty. Further, it is shown that under particular conditions the deviation from the ideal sliding mode dynamics is bounded in the presence of the given unmatched uncertainty class. The control effort required to ensure $\mathcal{S}$ or a neighbourhood of $\mathcal{S}$ is reached must now be developed.

3.4 Nonlinear Controller Development and Associated Tracking Properties

The discontinuous control action ensuring motion constrained to $\mathcal{S}$ which is traditionally employed for the design of sliding mode controllers is undesirable from the practical point of view. Here, a continuous nonlinear controller will be employed to ensure the system (3.32) reaches a bounded region containing $\mathcal{S}$. Ideally, motion is close to $\mathcal{S}$ to ensure insensitivity to matched uncertainty and as close as possible to the ideal dynamics prescribed by the choice of sliding surfaces. These ideal dynamics are, of course, affected by both matched and unmatched uncertainty in this formulation.

A control strategy of the form

$$u(\xi, \phi, W, \dot{W}) = u_{\mathrm{L}}(\xi, \phi, W, \dot{W}) + u_{\mathrm{NL}}(\xi, \phi, W) \tag{3.56}$$

is hypothesized where u_{L} is a linear feedback effort and u_{NL} is a nonlinear continuous control component. The linear feedback term is defined as follows

$$\begin{aligned} u_{\mathrm{L}}(\xi, \phi, W, \dot{W}) &= -B_2^{-1} \left[\begin{array}{cc} \Theta & \Omega - \Omega^* \end{array} \right] \left[\begin{array}{c} \xi(t) \\ \phi(t) \end{array} \right] \\ &\quad - B_2^{-1} \left(\Omega^* C_2^{-1} C_W + M_1 T_i^{-1} \right) W + B_2^{-1} C_2^{-1} C_W \dot{W} \end{aligned} \tag{3.57}$$

where $\Omega^* \in \mathbb{R}^{m \times m}$ is a design matrix whose stable eigenvalues contribute to the rate of decay of the range space states, ϕ, into the neighbourhood of $\mathcal{S}$. The nonlinear control contribution has the form

$$u_{\mathrm{NL}}(\xi, \phi, W) = -\varrho \frac{B_2^{-1} P_2 \left(\phi(t) - C_2^{-1} C_W W(t)\right)}{\left\| P_2 \left(\phi(t) - C_2^{-1} C_W W(t)\right) \right\| + \delta} \tag{3.58}$$

Here P_2 is the unique symmetric positive definite solution of the Lyapunov equation

$$P_2 \Omega^* + (\Omega^*)^T P_2 + I_m = 0 \tag{3.59}$$

and $\delta > 0$ is a constant whose role is to smooth the otherwise discontinuous nonlinear control action.

Assumption 3.3 *There exists a real scalar parameter σ satisfying $0 < \sigma \leq \bar{\sigma}$ where*

$$\bar{\sigma} = \inf_{G_1} \lambda_{\min} \left[I_m + T' \frac{G_1}{2} B_2^{-1} + (B_2^{-1})^T \frac{G_1^T}{2} (T')^T \right] \tag{3.60}$$

and $\lambda_{\min}(\cdot)$ denotes the minimum eigenvalue.

With Assumption 3.3 the function $\varrho(t, \xi, \phi)$ is then given by

$$\varrho = \frac{\gamma_1}{\sigma} \left(\eta(t, \xi, \phi) + \gamma_2 \right) \tag{3.61}$$

with $\gamma_1 \geq 1$, $\gamma_2 > 0$. There is some choice available for the parameter $\eta(t,\xi,\phi)$ subject to the satisfaction of the following condition

$$\begin{aligned}\eta(t,\xi,\phi) \;\geq\; & \Big\| -M_1 T_i^{-1} h(t,x) + M_2 T f(t,x) \\ & + T'\left(G_1(t,x,u)u_L(\xi,\phi,W,\dot{W}) + G_2(t,x)\right)\Big\| \end{aligned} \tag{3.62}$$

Using the imposed structural constraints from (3.4) and repeatedly applying the properties of a vector norm yields the following expression for $\eta(t,\xi,\phi)$

$$\begin{aligned}\eta(t,\xi,\phi) \;=\; & \left(K_{F_1}\|M_2\| + K_{H_1}\left\|M_1 T_i^{-1}\right\|\right)\|x\| + K_{F_2}\|M_2\| \\ & + K_{G_1}\left\|u_L(\xi,\phi,W,\dot{W})\right\| + K_{G_2} + K_{H_2}\left\|M_1 T_i^{-1}\right\| \end{aligned} \tag{3.63}$$

This particular choice has been found to be appropriate for all application studies considered to date. With the control strategy presented in (3.56)-(3.62), the following holds:

Theorem 3.4
(i) The uncertain system (3.32) is globally uniformly ultimately bounded with respect to the subspace $\mathcal{N}$ where $\mathcal{S} \subset \mathcal{N}$ and

$$\mathcal{N} = \{(\xi,\phi) : V_2(\phi,W) \leq r_3\} \tag{3.64}$$

with

$$V_2(\phi,W) = \frac{1}{2}\left(\phi - C_2^{-1}C_W W\right)^T P_2\left(\phi - C_2^{-1}C_W W\right) \tag{3.65}$$

and

$$r_3 = \left[\frac{\delta}{\gamma_1 - 1}\right]^2 \frac{1}{2\lambda_{\min}(P_2)} \tag{3.66}$$

(ii) If $(\xi(t_0),\phi(t_0)) \notin \mathcal{N}$ then the time T_1 required to reach $\mathcal{N}$ satisfies

$$T_1 \leq \frac{1}{\gamma_2}\sqrt{\frac{2}{\lambda_{\min}(P_2)}}\left(\sqrt{V_2(\phi(t_0))} - \sqrt{r_3}\right) \tag{3.67}$$

but if $(\xi(t_0),\phi(t_0)) \in \mathcal{N}$ then $(\xi,\phi) \in \mathcal{N}$ $\forall t \geq t_0$.
(iii) If the motion is constrained to $\mathcal{N}$, the states $\xi(t)$ are ultimately bounded with respect to the ellipsoid $E_1(r_4)$ where

$$r_4 = \varepsilon_2 + \frac{2\|P_1\|^2}{\nu^2}\left(K_1 + K_2\|W\|_{\max} + K_5\sqrt{2r_3}\right)^2 \tag{3.68}$$

$$K_5 = \sup_{F_1,H_1}\left\|P_1^{\frac{1}{2}}\tilde{A}_{12}P_2^{-\frac{1}{2}}\right\| \tag{3.69}$$

with $\varepsilon_2 > 0$ an arbitrarily small positive constant. In addition, the deviation from the ideal sliding mode dynamic behaviour is bounded with respect to the ellipsoid $E_1(r_5)$ where

$$
r_5 = \begin{cases} 2\|P_1\|^2 \left(K_1 + K_3\|W\|_{\max} + K_4 \left\|P_1^{\frac{1}{2}}\xi(t_0)\right\| + K_5\sqrt{2r_3}\right)^2 \\ \quad \textit{if } \xi(t_0) \notin E_1(r_4) \\ 2\|P_1\|^2 \left(K_1 + K_3\|W\|_{\max} + K_4\sqrt{2r_4} + K_5\sqrt{2r_3}\right)^2 \\ \quad \textit{if } \xi(t_0) \in E_1(r_4). \end{cases} \tag{3.70}
$$

Proof. Although the choice of the nonlinear control component is different from that employed by Ryan and Corless (1984), the above result can be proved using a very similar theoretical approach. Much of the detail is therefore omitted below.

(i) With the proposed control strategy, the closed-loop dynamics may be expressed by

$$
\begin{aligned}
\begin{bmatrix} \dot{\xi}(t) \\ \dot{\phi}(t) \end{bmatrix} &= \begin{bmatrix} \tilde{\Sigma} & \tilde{A}_{12} \\ 0 & \Omega^* \end{bmatrix} \begin{bmatrix} \xi(t) \\ \phi(t) \end{bmatrix} + \begin{bmatrix} \begin{bmatrix} I_p \\ 0 \end{bmatrix} T_i^{-1} \\ C_2^{-1}C_W\Gamma - \Omega^* C_2^{-1} C_W \end{bmatrix} W(t) \\
&+ \begin{bmatrix} 0 \\ C_2^{-1}C_W \end{bmatrix} w - \begin{bmatrix} \begin{bmatrix} -T_i^{-1}H_2 \\ TF_2 \end{bmatrix} \\ -M_1T_i^{-1}h + M_2Tf + T'g \end{bmatrix} \\
&+ \begin{bmatrix} 0 \\ I_m \end{bmatrix} \varrho \frac{P_2\left(\phi - C_2^{-1}C_W W\right)}{\left\|P_2\left(\phi - C_2^{-1}C_W W\right)\right\| + \delta}
\end{aligned} \tag{3.71}
$$

Along any solution of the Lyapunov function (3.65)

$$
\begin{aligned}
\dot{V}_2(\phi, W) &= \left(\phi - C_2^{-1}C_W W\right)^T P_2 \Big[\Omega^* \left(\phi - C_2^{-1}C_W W\right) \\
&- \varrho \frac{P_2\left(\phi - C_2^{-1}C_W W\right)}{\left\|P_2\left(\phi - C_2^{-1}C_W W\right)\right\| + \delta} + \left(-M_1T_i^{-1}h + M_2Tf + T'g\right)\Big]
\end{aligned} \tag{3.72}
$$

Substituting from (3.4), let

$$
-M_1T_i^{-1}h + M_2Tf + T'g = T'G_1(t,x,u)u_{\mathrm{NL}}(\xi,\phi,W) + \gamma(t,\xi,\phi,W,\dot{W}) \tag{3.73}
$$

where

$$
\begin{aligned}
\gamma(t,\xi,\phi,W,\dot{W}) &= -M_1T_i^{-1}h + M_2Tf \\
&+ T'\left(G_1(t,x,u)u_{\mathrm{NL}}(\xi,\phi,W,\dot{W}) + G_2(t,x)\right)
\end{aligned} \tag{3.74}
$$

Using the expression (3.73) in (3.72)

$$
\begin{aligned}
\dot{V}_2(\phi, W) &= \left(\phi - C_2^{-1}C_W W\right)^T P_2 \Big[\Omega^* \left(\phi - C_2^{-1}C_W W\right) \\
&- \left(I_m + T'G_1B_2^{-1}\right) u_{\mathrm{NL}}(\xi,\phi,W) + \gamma(t,\xi,\phi,W,\dot{W})\Big]
\end{aligned} \tag{3.75}
$$

With ϱ, σ and η as defined in (3.61), (3.62) it follows that

$$\dot{V}_2(\phi, W) \leq -\frac{1}{2}\left\|\phi - C_2^{-1}C_W W\right\|^2 - \gamma_2\left\|P_2\left(\phi - C_2^{-1}C_W W\right)\right\| - \left\|P_2\left(\phi - C_2^{-1}C_W W\right)\right\|\sigma\left(u_{\mathrm{NL}}(\xi,\phi,W) - \frac{\varrho}{\gamma_1}\right) \quad (3.76)$$

Considering the structure of u_{NL}, (3.58), it is seen that $\dot{V}_2(\phi, W) < 0$ if

$$\left\|P_2\left(\phi - C_2^{-1}C_W W\right)\right\| \geq \frac{\delta}{\gamma_1 - 1}. \quad (3.77)$$

Verification of (i) follows directly.
(ii) Note that if $\dot{x} \leq -ax - b\sqrt{x}$, then the time taken for x to move from x_0 to x_1 is given by

$$T \leq \frac{2}{a}\ln\left(\frac{a\sqrt{x_0}+b}{a\sqrt{x_1}+b}\right) \leq \frac{2}{b}\left(\sqrt{x_0} - \sqrt{x_1}\right) \quad (3.78)$$

Result (ii) follows from this observation.
(iii) Consider now motion constrained to $\mathcal{N}$ when the following constraint on the state ϕ holds

$$\frac{1}{2}\left(\phi - C_2^{-1}C_W W\right)^T P_2\left(\phi - C_2^{-1}C_W W\right) \leq r_3 \quad (3.79)$$

Taking again the Lyapunov function candidate (3.50) it follows that

$$\dot{V}_1(\xi) \leq -\nu V_1(\xi)\left\|P_1\right\|^{-1} + \sqrt{2V_1(\xi)}\left\{\left\|P_1^{\frac{1}{2}}\tilde{A}_{12}P_2^{-\frac{1}{2}}\right\|\sqrt{2r_3} + \left\|P_1^{\frac{1}{2}}T_i^{-1}W\right\| + \left\|P_1^{\frac{1}{2}}\begin{bmatrix} -T_i^{-1}H_2 \\ TF_2 \end{bmatrix}\right\|\right\} \quad (3.80)$$

With r_4, K_1, K_3 and K_5 from (3.68), (3.44), (3.48) and (3.69) respectively, it follows that $\dot{V}_1(\xi) < 0$ if $V_1(\xi) > r_4 - \varepsilon_2$ and ultimate boundedness with respect to $E_1(r_4)$ is proved. To investigate the deviation from the ideal dynamic behaviour prescribed by the choice of switching surfaces, consider again the Lyapunov function candidate $V_1(\Delta\xi)$. Along any solution

$$\dot{V}_1(\Delta\xi) = \Delta\xi^T P_1\left\{\Sigma\,\Delta\xi + \bar{A}_{12}\left(\phi - C_2^{-1}C_W W\right) + \begin{bmatrix} -T_i^{-1}h \\ Tf \end{bmatrix}\right\} \quad (3.81)$$

$$\leq -\frac{1}{2}\left\|\Delta\xi\right\|^2 + \Delta\xi^T P_1\left\{\tilde{A}_{12}\left(\phi - C_2^{-1}C_W W\right) + \Delta\Sigma\,\xi - \Delta\bar{A}_{12}C_2^{-1}C_W W + \begin{bmatrix} -T_i^{-1}H_2 \\ TF_2 \end{bmatrix}\right\} \quad (3.82)$$

Again

$$\left\|P_1^{\frac{1}{2}}\xi(t)\right\| \leq \begin{cases} \left\|P_1^{\frac{1}{2}}\xi(t_0)\right\| & \text{if } \xi(t_0) \notin E_1(r_4) \\ \sqrt{2r_4} & \text{if } \xi(t_0) \in E_1(r_4) \end{cases} \quad (3.83)$$

With r_5, K_1, K_3, K_4 and K_5 defined in (3.70), (3.44), (3.48), (3.49), (3.69), it follows that the deviation from ideal model motion is bounded with respect to the ellipsoid $E_1(r_5)$. □

The controller (3.56)–(3.63) may be conveniently expressed in terms of the system state (3.12) and measurement (3.2) by using the inverse transformations (3.18) and (3.31)

$$u_{\mathrm{L}}(\tilde{x}, W, \dot{W}) = L\tilde{x}(t) + L_W W(t) + L_{\dot{W}} \dot{W}(t) \tag{3.84}$$

with

$$\begin{aligned} L &= -B_2^{-1} \left[\ \Theta \quad \Omega - \Omega^* \ \right] \hat{T}\tilde{T} && (3.85) \\ L_W &= -B_2^{-1} \left(\Omega^* C_2^{-1} C_W + M_1 T_i^{-1} \right) && (3.86) \\ L_{\dot{W}} &= B_2^{-1} C_2^{-1} C_W && (3.87) \end{aligned}$$

and

$$u_{\mathrm{NL}}(\tilde{x}, W) = \varrho \frac{N(\tilde{x}, W)}{\|M(\tilde{x}, W)\| + \delta} \tag{3.88}$$

where

$$\begin{aligned} N(\tilde{x}, W) &= -B_2^{-1} P_2 C_2^{-1} S(\tilde{x}, W) && (3.89) \\ M(\tilde{x}, W) &= P_2 C_2^{-1} S(\tilde{x}, W) && (3.90) \end{aligned}$$

it has been hypothesized that the controller detailed above, with appropriately selected parameters, provides a robust tracking performance. For completeness, this tracking performance will now be explored. In the absence of uncertainty, it follows from (3.71) that the following relationships hold in the steady state:

$$\begin{aligned} 0 &= \Gamma W + w \\ 0 &= \Sigma \xi + \bar{A}_{12} \phi + \begin{bmatrix} T_i^{-1} \\ 0 \end{bmatrix} W \\ 0 &= \Omega^* \left(\phi - C_2^{-1} C_W W \right) + B_2 u_{\mathrm{NL}}(\xi, \phi, W) \end{aligned} \tag{3.91}$$

In order to investigate the tracking error in the presence of uncertainty, define

$$\begin{aligned} \bar{W} &= W + \Gamma^{-1} w \\ \bar{\xi} &= -\Sigma^{-1} \left[\bar{A}_{12} C_2^{-1} C_W + \begin{bmatrix} T_i^{-1} \\ 0 \end{bmatrix} \right] \Gamma^{-1} w \\ \bar{\phi} &= \phi + C_2^{-1} C_W \Gamma^{-1} w \end{aligned} \tag{3.92}$$

where it is assumed for the purpose of this analysis that Σ is nonsingular. The closed-loop dynamics of the states defined in (3.92) are determined by

$$\begin{bmatrix} \dot{\bar{\xi}} \\ \dot{\bar{\phi}} \end{bmatrix} = \begin{bmatrix} \tilde{\Sigma} & \tilde{A}_{12} \\ 0 & \Omega^* \end{bmatrix} \begin{bmatrix} \bar{\xi} \\ \bar{\phi} \end{bmatrix} + \begin{bmatrix} \begin{bmatrix} T_i^{-1} \\ 0 \end{bmatrix} + \bar{A}_{12} C_2^{-1} C_W \\ C_2^{-1} C_W \Gamma - \Omega^* C_2^{-1} C_W \end{bmatrix} \bar{W} + \ldots$$

$$\ldots + \begin{bmatrix} \Delta\Sigma\,\Sigma^{-1}\left(\bar{A}_{12}C_2^{-1}C_W + \begin{bmatrix} T_i^{-1} \\ 0 \end{bmatrix} - \Delta\bar{A}_{12}C_2^{-1}C_W\right) \\ 0 \end{bmatrix} \Gamma^{-1}w$$

$$+ \begin{bmatrix} \begin{bmatrix} -T_i^{-1}H_2 \\ TF_2 \end{bmatrix} \\ -M_1T_i^{-1}h + M_2Tf + T'g \end{bmatrix} + \begin{bmatrix} 0 \\ B_2 \end{bmatrix} u_{\mathrm{NL}}(\bar{\xi},\bar{\phi},\bar{W}) \quad (3.93)$$

In addition, a bound on the tracking error between the chosen output $y(t)$ and the tracking demand $W(t)$ as defined in (3.11) will be derived.

Theorem 3.5
(i) The $\bar{\phi}$, $\bar{W}$ states remain within the ellipsoid

$$E_2(r_6) = \left\{(\bar{\xi},\bar{\phi}) \,:\, V_2(\bar{\phi},\bar{W}) \le r_6\right\} \quad \forall\, t \ge t_0$$

and ultimately enter the ellipsoid

$$E_2(r_3) = \left\{(\bar{\xi},\bar{\phi}) \,:\, V_2(\bar{\phi},\bar{W}) \le r_3\right\}$$

where

$$r_6 = \max\left\{V_2(\bar{\phi}(t_0),\bar{W}(t_0))\,,\ r_3\right\}$$

and V_2 and r_3 are as previously defined in (3.65), (3.66).
(ii) The states $\bar{\xi}(t)$ remain within the ellipsoid $E_1(r_7)$ $\forall t \ge t_0$ and ultimately enter the ellipsoid $E_1(r_8)$ where

$$\begin{aligned} r_7 &= \max\left\{V_1(\bar{\xi}(t_0))\,,\ r_8 - \varepsilon_3\right\} \\ r_8 &= \varepsilon_3 + \frac{2\,\|P_1\|^2}{\nu^2}\left[K_1 + K_3\sqrt{2r_3} + K_6\right]^2 \end{aligned}$$

with

$$K_6 = \sup_{F_1,H_1,w}\left\| P_1^{\frac{1}{2}}\left\{\Delta\Sigma\,\Sigma^{-1}\left(\bar{A}_{12}\,C_2^{-1}C_W + \begin{bmatrix} T_i^{-1} \\ 0 \end{bmatrix} - \Delta\bar{A}_{12}\,C_2^{-1}C_W\right)\Gamma^{-1}w\right\}\right\|$$

and $\varepsilon_3 > 0$ an arbitrarily small constant.
(iii) Let P_R be the unique, positive definite solution to the Lyapunov equation

$$P_R\Gamma + (\Gamma')^T P_R + I_p = 0$$

where Γ is as defined in (3.8). The $\dot{x}_R$ states remain within the ellipsoid

$$E_3(r_9) = \left\{\dot{x}_R \,:\, \frac{1}{2}\dot{x}_R^T P_R \dot{x}_R \le r_9\right\}$$

and ultimately enter the ellipsoid $E_3(r_{10})$ where

$$\begin{aligned} r_9 &= \max\left\{V_R(\dot{x}_R(t_0))\,,\ r_{11}\right\} \\ r_{10} &= \varepsilon_4 + 2\left(\|P_R\|\,K_8\right)^2 \\ r_{11} &= 2\left(\|P_R\|\,K_7\right)^2 \end{aligned}$$

and

$$K_7 = \sup_{(\bar\xi,\bar\phi,\bar W)\in\mathcal{M}_1} \Bigg\| P_R^{\frac{1}{2}} \Bigg\{ -T_i^{-1} \left[\ \Gamma T_i \quad (\gamma^T + H_3)\bar T^T\ \right] \tilde T^{-1} \begin{bmatrix} \tilde\Sigma & \tilde A_{12} \\ 0 & \Omega^* \end{bmatrix} \begin{bmatrix} \bar\xi \\ \bar\phi \end{bmatrix}$$
$$- T_i^{-1}\left(\gamma^T + \frac{\partial h}{\partial x}\right)(T')^T\left(-M_1 T_i^{-1} - \Omega^* C_2^{-1} C_W + C_2^{-1} C_W \Gamma\right)\bar W$$
$$- T_i^{-1}\left[\ \Gamma T_i \quad (\gamma^T + \tfrac{\partial h}{\partial x})\bar T^T\ \right]\begin{bmatrix} I_{n-m+p} \\ -M \end{bmatrix}\{\Delta\Sigma\,\Sigma^{-1}\,(\bar A_{12} C_2^{-1} C_W$$
$$+ \begin{bmatrix} T_i^{-1} \\ 0 \end{bmatrix}\Bigg) - \Delta\bar A_{12} C_2^{-1} C_W\Bigg\}\Gamma^{-1} w + T_i^{-1}\Bigg\{\Gamma H_2 - \frac{\partial h}{\partial t}$$
$$- \left(\gamma^T + \frac{\partial h}{\partial x}\right)\bar T^T\begin{bmatrix} T F_2 \\ -M_1 T_i^{-1} H_1 x + M_2 T F_1 x + T' g \end{bmatrix}\Bigg\}$$
$$- T_i^{-1}\left(\gamma^T + \frac{\partial h}{\partial x}\right)(T')^T B_2 u_{\mathrm{NL}}(\bar\xi,\bar\phi,\bar W)\Bigg\}\Bigg\|$$

and

$$K_8 = \sup_{(\bar\xi,\bar\phi,\bar W)\in\mathcal{M}_2} \|\cdot\|$$

in which $\|\cdot\|$ *denotes the norm of the function defined in* K_7 *and* $\mathcal{M}_1$, $\mathcal{M}_2$ *denote the sets*

$$\begin{aligned} \mathcal{M}_1 &= \{(\bar\xi,\bar\phi,\bar W) : (\bar\phi,\bar W)\in E_2(r_6),\ \bar\xi\in E_1(r_7)\} \\ \mathcal{M}_2 &= \{(\bar\xi,\bar\phi,\bar W) : (\bar\phi,\bar W)\in E_2(r_3),\ \bar\xi\in E_1(r_8)\} \end{aligned}$$

Proof.
(i) Follows directly from Theorem 3.4, part (i).
(ii) Follows from applying the procedure (3.50)–(3.52) to the $\bar\xi, \dot{\bar\xi}$ pair and noting that constraint (3.79) applies to the $\bar\phi$, $\bar W$ states from (i).
(iii) Consider the Lyapunov function candidate

$$V_R = \frac{1}{2}\dot x_R^T P_R \dot x_R.$$

Differentiate (3.11) noting that for the system, (3.13), subject to the transformations (3.18) and (3.31), the following identity holds

$$(\gamma^T + H_1)\left[\ 0 \quad \bar T^T\ \right]\tilde T^{-1} = -\left[\ T_i \quad 0\ \right]\tilde T^{-1}\begin{bmatrix} \tilde\Sigma & \tilde A_{12} \\ 0 & \Omega^* \end{bmatrix}.$$

Expressing $\ddot x_R$ in terms of the $\bar\xi$, $\bar\phi$, $\bar W$ states yields the required result. □

For the nominal system in the absence of uncertainty where an appropriate choice of nonlinear control component is a zero control effort, it is interesting to note that $K_7 = 0$. The $\dot{x}_R$ states thus ultimately enter the ellipsoid $E_3(\varepsilon_4)$ where $\varepsilon_4 > 0$ is an arbitrarily small constant and thus asymptotic tracking is achieved.

A case study will now be presented in order to illustrate the practical viability of the theoretical results developed in this paper. The design of a temperature control scheme for an industrial furnace is considered. Particular attention will be paid to the engineering design criteria which can be used to select the free parameters present in the proposed tracking methodology.

3.5 Design Example: Temperature Control of an Industrial Furnace

The heating plant considered in this section is of the design shown schematically in Fig. 3.1 and may be considered as a gas filled enclosure, bounded by insulating surfaces and containing a heat sink. Heat input is supplied by a single burner located in one end wall and the combustion products are evacuated through a flue in the roof.

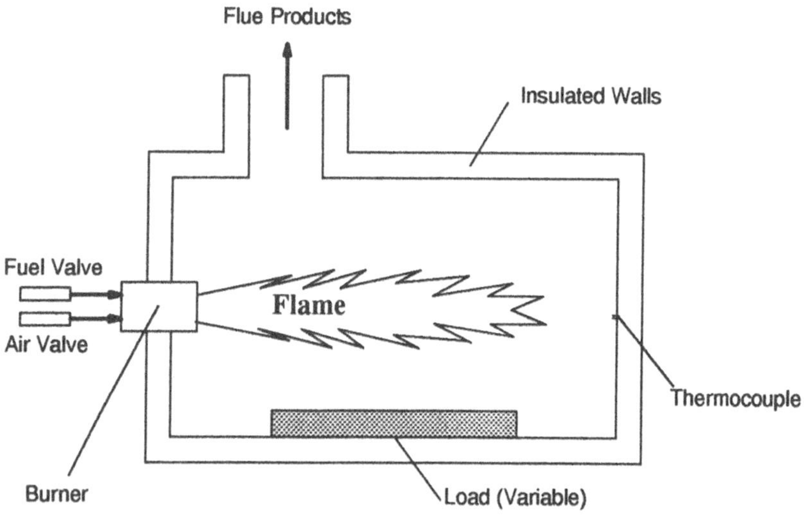

Fig. 3.1. Schematic of the box furnace considered

The control problem considered in this section is the manipulation of the fuel flow rate so that the temperature at some point in the furnace adheres to some temperature/time profile. It is assumed that a controller for the air flow valve already exists, which, for any given fuel flow adjusts the air flow rate to ensure good combustion efficiency and an appropriate concentration of oxygen in the flue products.

Heat transmission within high temperature heating plant is principally by radiation in a participating medium. This is governed essentially by fourth power laws and so is inherently nonlinear, added to which are the nonlinearities associated with the flow valves. This, together with the disturbances caused by changes in the load and alterations in the desired operating points make the control strategy outlined earlier with its inherent robustness properties attractive for such an uncertain system.

Using a system identification package several different transfer functions have been obtained from different plant trials. The transfer functions are all in the following form

$$G(s) = \frac{(b_1 s + b_0)e^{-ds}}{a_2 s^2 + a_1 s + a_0} \tag{3.94}$$

where the time delay is of the order 10 seconds. Using a Padé approximation for the time delay, state space realizations of order 3 have been generated for the different transfer functions. One of these realizations has been selected as the nominal system and the nonlinear controller together with a linear observer designed around this system. To examine the controller's robustness a different realization was used to drive the observer, whose states were used to calculate the control action for the different system.

The design matrices T_i^{-1} and C_2 are both scalars for the furnace realizations considered and are both set to unity in the example that follows. The dynamics of the sliding mode, i.e. the spectrum of Σ is set to be $\{-0.03, -0.035, -0.025\}$ by appropriate choice of M. These poles give dynamics that are marginally faster than the slowest pole of the open loop system. The scalar C_W influences the steady state value of the integral states and can be chosen arbitrarily. However it can be chosen so in the absence of disturbances at steady state then $x_R = 0$ (assuming that ideal sliding motion has been obtained). The scalar Ω^* has been assigned the value -0.1. From these values the state feedback matrix L and the feed-forward gain matrices L_W and $L_{\dot{W}}$ can be calculated from equations (3.86) and (3.87) and the nonlinear components $M(\tilde{x}, W)$, $N(\tilde{x}, W)$ from (3.89) and (3.90).

For the single input-single output case considered the design parameter Γ associated with the tracking demand dynamics (3.8) is also a scalar. Fig. 3.2 below shows the responses of a nominal furnace model for values of Γ in the range $(-0.010, -0.025)$ which can readily be seen to affect the rise-time/ over-shoot. For this application, the response with a fast rise time with minimal over-shoot would be regarded as the most acceptable. In the simulations that follow the value of Γ associated with this (-0.01) will be used.

On account of the system identification approach adopted, the bounds on the nonlinear/uncertain functions $f(t, x)$ and $g(t, x, u)$ cannot be computed directly. Consequently K_{F_1}, K_{F_2}, K_{G_1} and K_{G_2} have been assigned reasonable values employing information obtained from experiments using known disturbance matrices F_1, F_2, G_1 and G_2. For simplicity in the example that follows it has been assumed that $h(t, x) \equiv 0$. The nonlinear design parameters γ_1 and γ_2 and the smoothing parameter δ have been set to 1.1, 0.05 and 0.01 respectively. Linear simulation results are shown in Figs. 3.3 and 3.4.

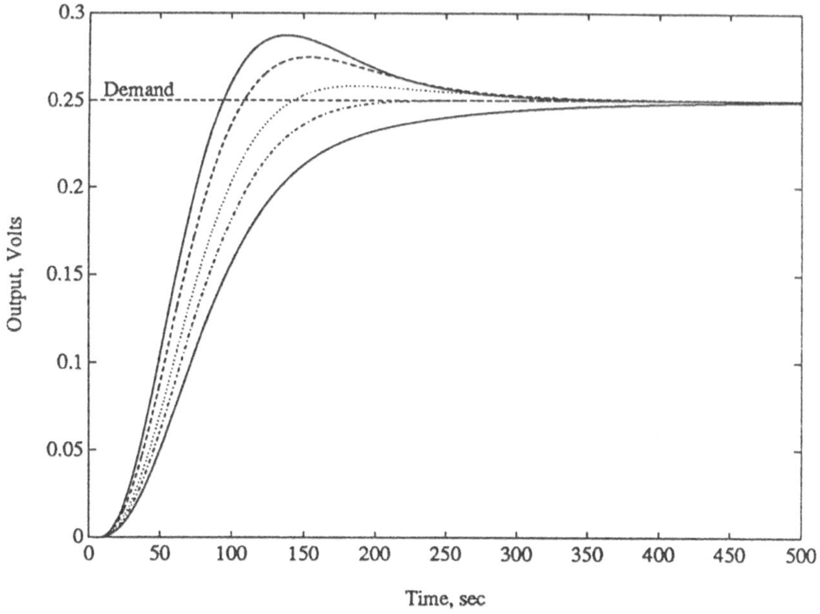

Fig. 3.2. Responses for different demand dynamics

The established technique for the mathematical modelling of industrial furnaces is the *Zone Method* further details of which can be found in the book by Rhine and Tucker (1991). This approach basically involves the breaking up of all enclosure surfaces and volumes into a patch-work of sub-surfaces and sub-volumes which are termed *zones*. Each zone must be small enough to be considered isothermal. Radiation exchange factors are then calculated for every possible pair of zones, and the integro-differential equations governing radiative heat transfer are reduced to algebraic equations, which can be solved numerically. A well validated nonlinear dynamic model of this type is available to this work. However, first it should be noted that this tracking controller requires complete state information; in practice, measurement of such internal states is not possible and therefore an appropriate state estimation policy is required. Following the success of the linear simulation results which were obtained using the proposed sliding mode control policy, a robust sliding observer as developed by Edwards and Spurgeon (1993) is employed. Fig. ?? shows the performance of the nonlinear furnace simulation when controlled by the proposed strategy in conjunction with the sliding observer. The heating plant under consideration needs to operate over a wide temperature range maintaining close tracking of a specified trajectory. Such a typical demand signal has been used for this simulation test. Visually, perfect tracking is obtained. Fig. ?? shows the associated control signal which is seen to be very smooth.

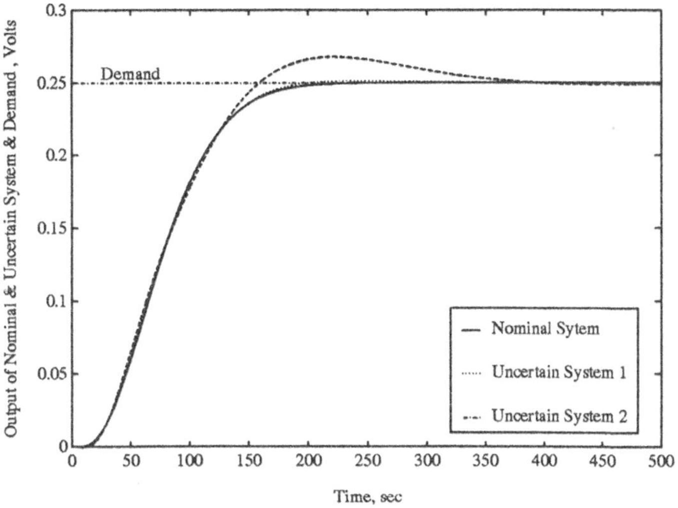

Fig. 3.3. System outputs from different linear models

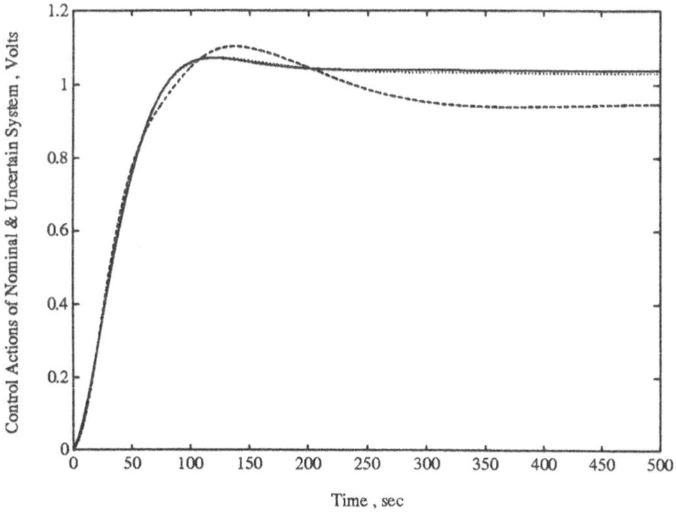

Fig. 3.4. Corresponding control actions from different linear models

3.6 Conclusions

It is well known that a problem formulation containing only matched uncertainty can be forced to attain a sliding mode and exhibit the precise nominal dynamic which is defined by the choice of switching surface. This paper has formulated a nonlinear control strategy which will prescribe bounded motion about an ideal sliding mode dynamic for an uncertainty set including both matched and unmatched uncertainty which can be readily applied to engineering problems. A tracking requirement has been successfully incorporated into the methodology. The results have been illustrated by considering the design of a temperature controller for an industrial furnace.

3.7 Acknowledgements

Financial support from the UK Science and Engineering Research Council (Grant Reference GR/H23368) and the provision of a Research Scholarship by British Gas PLC are gratefully acknowledged.

References

Barmish, B.R. 1983, Stabilization of uncertain systems via linear control. *IEEE Transactions on Automatic Control* **28**, 848–850

Bühler, H. 1991, Sliding mode control with switching command devices. in *Deterministic Control of Uncertain Systems*, ed. Zinober, A.S.I., Peter Peregrinus, London, 27–51

Burton, J.A., Zinober, A.S.I. 1986, Continuous approximation of variable structure control. *International Journal of Systems Science* **17**, 875–885

Dorling, C.M., Zinober, A.S.I. 1990, Hyperplane design and CAD of variable structure control systems. in *Deterministic Control of Uncertain Systems*, ed. Zinober, A.S.I., Peter Peregrinus London, 52–79

Edwards, C., Spurgeon, S.K. 1993, On the development of discontinuous observers. *International Journal of Control* , to appear

Kautsky, J., Nichols, N.K., Van Dooren, P. 1985, Robust pole assignment in linear state feedback. *International Journal of Control* **41**, 1129–1155

Khargonekar, P.P., Petersen, R., Zhou, K. 1990, Robust stabilization of uncertain linear systems: Quadratic stabilizability and H^{∞} control theory. *IEEE Transactions on Automatic Control* **35**, 356–361

Rhine, J.M., Tucker, R.J. 1991, Modelling of gas fired furnaces and boiler and other industrial heating processes, McGraw-Hill, New York, Chapters 13 and 14

Ryan, E.P., Corless, M. 1984, Ultimate boundedness and asymptotic stability of a class of uncertain dynamical systems via continuous and discontinuous feedback control. *IMA Journal of Mathematics and Control Information* **1**, 223–242

Slotine, J.J. 1984, Sliding controller design for nonlinear systems. *International Journal of Control* **53**, 163–179

Spurgeon, S.K., Davies, R. 1993, A nonlinear control strategy for robust sliding mode performance in the presence of unmatched uncertainty. *International Journal of Control* **57**, 1107-1123

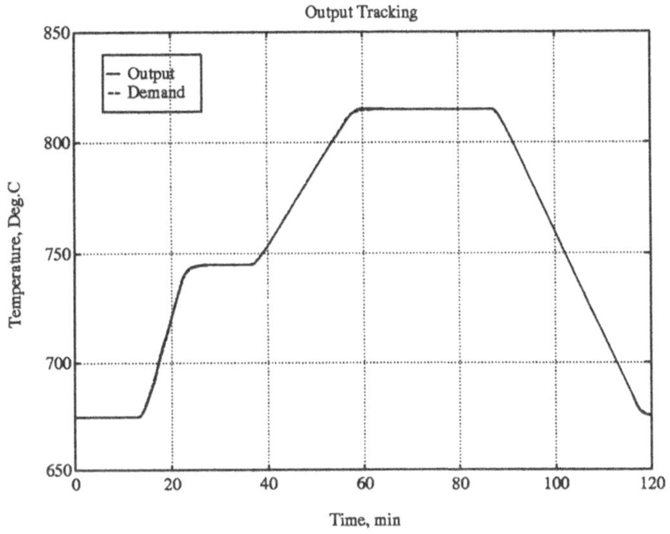

Fig. 3.5. Tracking performance achieved with the nonlinear furnace model

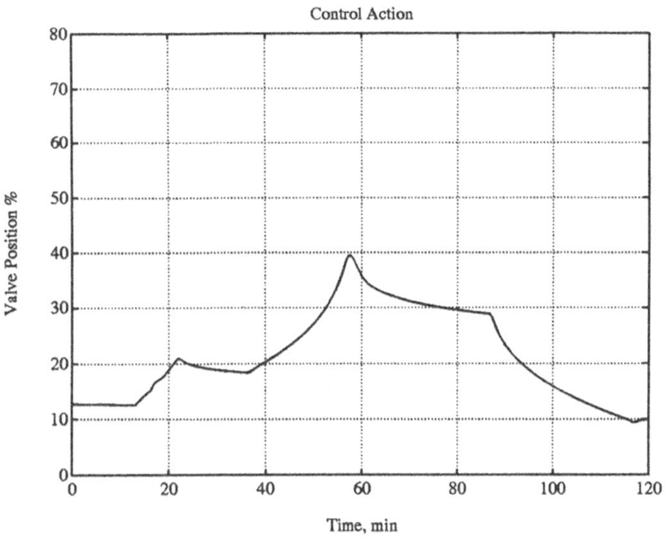

Fig. 3.6. Control action applied to the nonlinear furnace model

4. Sliding Surface Design in the Frequency Domain

Hideki Hashimoto and Yusuke Konno

4.1 Introduction

A new method for the sliding surface design of variable structure control (VSC) systems using the frequency criteria of H^∞ control theory, is presented. H^∞ theory is a well known control technique used to suppress high frequency modes of the controlled plant, i.e. loop shaping. The robust performance of sliding mode control has been confirmed by practical experiments (Young 1993). It is well known that nonlinearities and plant parameter uncertainties can be suppressed by proper design of a sliding controller.

Here we treat the control of plants with high frequency resonance modes. Usually we cannot obtain exact models of physical systems Sliding mode control can be applied to the plants with uncertainties, but, if we design fast sliding mode dynamics in such plants to improve the transient response, high frequency control inputs excite resonance modes and may cause undesired vibration. The new design method is proposed to satisfy two conflicting requirements: fast response and vibration suppression.

Our proposed design introduces additional states to construct the so-called generalized plant so that the control input does not excite the high frequency resonance modes. The usual feedback sliding mode design requires all the plant states. This means that there is no freedom to suppress the high frequency component of control input. Otherwise we have to use a dynamical filter to attenuate the high frequency gain of the closed-loop system. The design technique provides for the shaping of the closed-loop transfer function in the sliding mode by using H^∞ theory.

Young and Özgüner (1990) described an approach to suppress the high frequency component of the input by using frequency shaped LQ design. Using this method, an appropriate frequency dependent weight function $R(\omega)$ is selected and high frequency control inputs are penalized. The frequency shaped LQ method is closely related to H^∞ control theory in terms of using frequency weights. However, the H^∞ method specifies the frequency response of the closed loop directly.

In this chapter the concept of the frequency shaped sliding mode using H^∞ control theory, is introduced for uncertain systems. This approach achieves frequency shaping sliding mode much more easily than frequency shape LQ design. In Sect. 4.2 we discuss the design of the sliding surface using the LQ optimal method and describe the frequency shaped LQ approach. Section 4.3 briefly discusses H^2/H^∞ optimal control and the new design method is introduced.

Then in Sect. 4.4 the design method is applied to an elastic joint manipulator and the efficiency of the method is demonstrated.

4.2 Sliding Mode using the LQ Approach

A linear control law based on a quadratic cost function is well known as Linear Quadratic (LQ) optimal control. Utkin and Young (1978) have applied this method to the synthesis of the sliding mode. This method was extended to frequency shaped sliding mode by introducing frequency depending weights in Young and Özgüner (1990).

4.2.1 Linear Quadratic Optimal Sliding Mode

We consider the following linear time invariant (LTI) system

$$\dot{x} = Ax + Bu \tag{4.1}$$

where $A \in R^{n \times n}$ and $B \in R^{n \times m}$. The cost functional to be minimized is

$$J = \int_{t_s}^{\infty} x^T Q x \, \mathrm{dt} \tag{4.2}$$

where t_s is the time at which sliding mode begins and Q is a symmetric positive definite matrix. Using the state variable transformation T

$$T^{-1}B = \begin{bmatrix} 0 \\ B_2 \end{bmatrix} \tag{4.3}$$

(4.1) and (4.2) can be rewritten as

$$\frac{\mathrm{d}}{\mathrm{dt}} \begin{bmatrix} x_1 \\ x_2 \end{bmatrix} = \begin{bmatrix} A_{11} & A_{12} \\ A_{21} & A_{22} \end{bmatrix} \begin{bmatrix} x_1 \\ x_2 \end{bmatrix} + \begin{bmatrix} 0 \\ B_2 \end{bmatrix} u \tag{4.4}$$

$$J = \int_{t_s}^{\infty} (x_1^T Q_{11} x_1 + 2 x_1^T Q_{12} x_2 + x_2^T Q_{22} x_2) \, \mathrm{dt} \tag{4.5}$$

where $x_1 \in R^{n-m}$ and $x_2 \in R^m$.

The sliding surface $\sigma = 0$ of the sliding mode can be determined so as to minimize the cost functional (4.5). This problem can be regarded as a linear state feedback control design for the following subsystem

$$\dot{x}_1 = A_{11} x_1 + A_{12} x_2 \tag{4.6}$$

with the cost functional (4.5). In (4.6) x_2 is considered to be the input of the subsystem, and the state feedback controller $x_2 = K x_1$ for this subsystem gives the sliding surface of the total system, namely $\sigma = x_2 - K x_1 = 0$. For simplicity we assume $Q_{12} = Q_{21}^T = 0$. The optimal sliding surface is given by

$$\sigma = x_2 + Q_{22}^{-1} A_{12}^T P x_1 = 0 \ , \qquad K = -Q_{22}^{-1} A_{12}^T P \tag{4.7}$$

where $P > 0$ is a unique solution of following Riccati equation

$$PA + A^T P - PA_{12} Q_{22}^{-1} A_{12}^T P + Q_{11} = 0. \tag{4.8}$$

and $K = -Q_{22}^{-1} A_{12}^T P$. For the existence of the solutions of the Riccati equation (4.8), the pair (A_{11}, A_{12}) must be controllable, the pair $(Q_{11}^{(1/2)}, A_{11})$ must be observable, $Q_{11} \geq 0$ has to be semi-positive and $Q_{22} > 0$ (Utkin and Young 1978).

4.2.2 Frequency Shaped LQ Approach

The frequency shaped LQ approach is based on frequency dependent weights. The cost function (4.5) can be written in the frequency domain using Perseval's theorem as

$$J = \frac{1}{2\pi} \int_{-\infty}^{\infty} (x_1^T(j\omega) Q_{11} x_1(j\omega) + x_2^T(j\omega) Q_{22} x_2(j\omega)) \, d\omega \tag{4.9}$$

In the frequency domain a frequency dependent weight matrix $Q_{22}(j\omega)$ is introduced so that control inputs for certain frequencies can be amplified or suppressed. We can choose Q_{22} to yield the reduction of high frequency control inputs to the subsystem (4.6). This approach is realized using state space representation. The frequency dependent weight $Q_{22}(j\omega)$ must be a rational function of ω^2 to yield a solution to the problem (Gupta 1980). The transfer function matrix $W_2(s)$ is defined as

$$Q_{22}(j\omega) = W_2(j\omega)^* W_2(j\omega) \tag{4.10}$$

where $W_2(s)^*$ stands for the conjugate transpose of $W_2(s)$. The frequency shaped input $\tilde{u}$ is given by

$$\tilde{u} = W_2(s) x_2 \ . \tag{4.11}$$

$W_2(s)$ has the following state space representation

$$\begin{aligned} \dot{x}_{w2} &= A_{w2} x_{w2} + B_{w2} x_2 \\ \tilde{u} &= C_{w2} x_{w2} + D_{w2} x_2 \end{aligned} \tag{4.12}$$

Then the cost functional (4.9) can be rewritten as

$$\begin{aligned} J &= \frac{1}{2\pi} \int_{-\infty}^{\infty} (x_1^*(j\omega) Q_{11} x_1(j\omega) + (W_2(j\omega) x_2(j\omega))^* W_2(j\omega) x_2(j\omega)) \, d\omega \\ &= \int_{t_s}^{\infty} (x_1^T(t) Q_{11} x_1(t) + \tilde{u}(t)^T \tilde{u}(t)) \, dt \end{aligned} \tag{4.13}$$

We introduce the following extended plant

$$
\begin{aligned}
\dot{x}_e &= A_e x_e + B_e \ , \quad x_e = \begin{bmatrix} x_{w2} \\ x_1 \end{bmatrix} \\
A_e &= \mathrm{diag}\,(A_{w2}, A_{11}) \ , \quad B_e = \begin{bmatrix} B_{w2} \\ A_{12} \end{bmatrix} \\
Q_e &= \mathrm{diag}\,(C_{w2}^T C_{w2}, Q_{11}) \ , \quad N_e = \begin{bmatrix} C_{w2}^T D_{w2} \\ 0 \end{bmatrix} \\
R_e &= D_{w2}^T D_{w2}
\end{aligned}
\tag{4.14}
$$

and then the cost functional (4.13) is

$$
J = \int_{t_s}^{\infty} \left(x_e^T Q_e x_e + 2 x_e^T N_e x_2 + x_2^T R_e x_2 \right) \, dt \tag{4.15}
$$

Minimization of this cost function with cross term between state and control input is achieved by solving the Riccati equation

$$
P_e A_e + A_e^T P_e - (P_e B_e + N_e) R_e^{-1} (B_e^T P_e + N_e^T) + Q_e = 0 \tag{4.16}
$$

The optimal sliding surface is, using the solution of 4.16,

$$
\sigma = x_2 + R_e^{-1} (B_e^T P_e + N_e^T) x_e \tag{4.17}
$$

4.3 H^2/H^∞ approach

Linear control theory has developed rapidly especially in the field of robust control. H^∞ optimal control theory is an excellent result of this development. H^∞ control has a close relation to Linear Quadratic Gaussian (LQG) control including the frequency shaped case, which is covered by H^2 control methods. This section introduces H^2/H^∞ control methods and then develops the theory of the optimal sliding mode based on H^∞ control.

4.3.1 H^2/H^∞ Optimal Control

The control goal is formulated through a norm minimization of the generalized plant, where H^2 and H^∞ norms are used to formulate the cost function. If $G(s)$ is a stable transfer matrix in the frequency domain, then the H^2 and H^∞ norms are

$$
\|G(s)\|_2 = \left(\frac{1}{2\pi} \int_{-\infty}^{\infty} \mathrm{trace}\,[G(j\omega)^* G(j\omega)] \, d\omega \right)^{1/2} \tag{4.18}
$$

$$
\|G(s)\|_\infty = \sup_{\omega} \sigma_{max}[G(j\omega)] \tag{4.19}
$$

To measure performance using these norms, the generalized plant G and controller K as shown in Fig. 4.1 are used. The norms between ω and z are used to measure performance. The generalized plant consists of the controlled plant and the frequency dependent weights which penalize control action for high frequencies and the state error for low frequencies. The signal w contains all external inputs including disturbances, noise, and references. The outputs z and y are the measured variables and u is the control input. We can easily derive the controller which minimizes $||G_{zw}||_2$ by solving two Riccati equations; however, the exact solution globally minimizing $||G_{zw}||_\infty$ is not obtained, rather a sub-optimal solution can be obtained using both state space and transfer function formulations. The sub-optimal problem is formulated as finding a controller with an upper bound for the H^∞ norm as

$$||G_{zw}(s)||_\infty \leq 1 \tag{4.20}$$

The closed-loop is internally stable, i.e. the four transfer matrices of the closed loop system from $[w \quad u]$ to $[z \quad y]$ are asymptotically stable.

There are several ways to solve the H^∞ sub-optimal problem. The method based on the transfer function matrix yields the controller through Youla parametrization (Francis 1987). This method is suitable for the output feedback case. In the state space approach an elegant solution is given by Doyle et al (1989) and Zhou and Khargonekar (1988) with the same formulation as LQG design.

4.3.2 Generalized Plant Structure

If we consider the subsystem (4.6), the cost function for the frequency shaped LQ design contains both the state vector and control input cost terms. The generalized plant of this case is shown in Fig. 4.2, where the output vector z contains the weighted state variable $Q_{11}^{1/2}x_1$ and frequency weighted control input $W_2(s)x_2$. The exogenous signal w which is assumed to be white noise, excites all the state variables including W_2.

In the H^∞ case, we define the error signal e between the reference inputs r and the subsystem outputs $y_1 = C_1x_1$ for tracking error measurement, instead

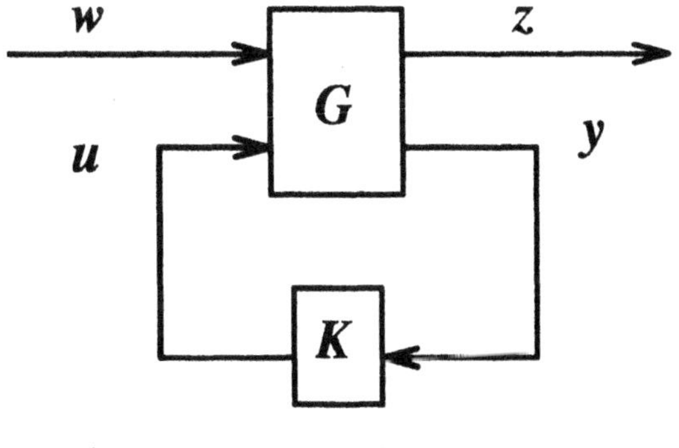

Fig. 4.1. Generalized plant and controller

of Q_{11} in the H^2 case.

$$e = r - y_1 = r - C_1 x_1 \tag{4.21}$$

The input to $W_1(s)$ in Fig. 4.3 is the tracking error e which is expected to be small at low frequencies. Therefore $W_1(s)$ should have a low-pass characteristic which penalizes low frequency error. It is straightforward to assign the characteristics of the tracking error behaviour in the frequency domain.

4.3.3 Controller Solution

In the state space formulation the H^∞ sub-optimal controller is given by the solution of two Riccati equations, as in the H^2 case (Doyle et al 1989, Zhou and Khargonekar 1988). If we can use all the state variables and all the external inputs of the generalized plant (the so-called Full Information problem), the class of controller K includes the case of a constant matrix. The scheme which Doyle et al (1989) proposed is famous for its simplicity, but has some restrictive conditions for the construction of the generalized plant. In the following we study the H^∞ controller proposed by Zhou and Khargonekar (1988). It is slightly more complicated than Doyle's method.

The generalized plant is

$$\begin{aligned} \dot{x} &= Ax + B_1 w + B_2 z \\ z &= C_1 x + D_{12} u + D_{11} w. \end{aligned} \tag{4.22}$$

For the generalized plant (4.22), the state feedback H^∞ controller can be obtained as follows. If rank $(D_{12}) = k > 0$, select any U and Σ which satisfy the decomposition form

$$D_{12} = U\Sigma \ , \quad D_{12} : p_1 \times m_2 \ , \quad U : p_1 \times k \ , \quad \Sigma : k \times m_2 \tag{4.23}$$

where p_1 and m_2 are dependent on z and u. Then the matrices Φ_F, R and Ξ_F are defined as

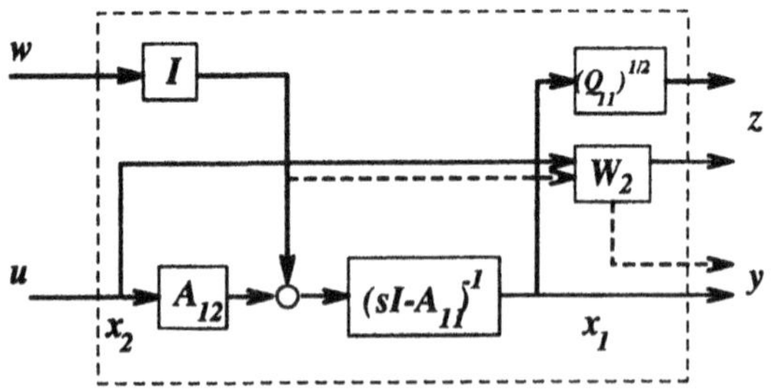

Fig. 4.2. Generalized plant (LQ design)

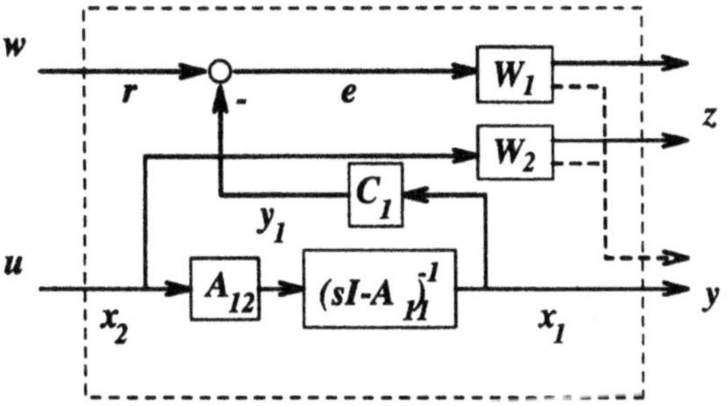

Fig. 4.3. Generalized plant (H^∞ design)

$$\begin{aligned}\Phi_F &= I-(\Sigma^+\Sigma)^T\\ R &= I+D_{11}(\gamma^2 I-D_{11}^T D_{11})^{-1}D_{11}^T\\ \Xi_F &= \Sigma^T(\Sigma\Sigma^T)^{-1}(U^T RU)^{-1}(\Sigma\Sigma^T)^{-1}\Sigma\end{aligned}$$

where Σ^+ denotes the generalized inverse of Σ. If $D_{12}=0$ then we can choose U and Σ as

$$\Phi_F = I\ , \qquad \Xi_F = 0 \tag{4.24}$$

Other matrices are defined as

$$\begin{aligned}A_F &= A+B_1(\gamma^2 I-D_{11}^T D_{11})^{-1}D_{11}^T C_1\\ B_F &= B_2+B_1(\gamma^2 I-D_{11}^T D_{11})^{-1}D_{11}^T D_{12}\\ C_F &= \{I+D_{11}(\gamma^2 I-D_{11}^T D_{11})^{-1}D_{11}^T\}^{(1/2)}C_1\\ D_F &= B_1(\gamma^2 I-D_{11}^T D_{11})^{-(1/2)}\\ F_F &= \{I+D_{11}(\gamma^2 I-D_{11}^T D_{11})^{-1}D_{11}^T\}^{(1/2)}D_{12}\end{aligned}$$

Theorem 4.1 *If and only if the H^∞ sub-optimal controller satisfying (4.20) exists, the following two conditions are met*

(i) $\gamma^2 I-D_{11}^T D_{11}$ *is positive definite*

(ii) The Riccati equation

$$\begin{aligned}&(A_F-B_F\Xi_F F_F^T C_F)^T P+P(A_F-B_F\Xi_F F_F^T C_F)\\ &\quad +PD_F D_F^T P-PB_F\Xi_F B_F^T P-\frac{1}{\epsilon}PB_F\Phi_F^T\Phi_F B_F^T P\\ &\quad +C_F^T(I-F_F\Xi_F F_F^T)C_F+\epsilon I=0\end{aligned} \tag{4.25}$$

has a positive definite solution for sufficiently small ϵ.

Then the state feedback gain K is

$$K=-\left\{\frac{1}{2\epsilon}\Phi_F^T\Phi_F+\Xi_F\right\}B_F^T P-\Xi_F F_F^T C_F \tag{4.26}$$

Proof. See Zhou and Khargonekar (1988).

In the design of the sliding mode, H^∞ control theory is applied to the generalized plant in Fig. 4.3. The states are x_1 and x_w; the vector x_w consists of the states $W_1(s)$ and $W_2(s)$; and x_2 is the input of the generalized plant. The sliding surface is

$$\sigma = x_2 - K \begin{bmatrix} x_1 \\ x_w \end{bmatrix} \tag{4.27}$$

where the feedback gain K is obtained from Theorem 4.1.

4.4 Simulation

Here we apply the proposed H^∞ frequency shaped sliding mode to a flexible manipulator joint.

4.4.1 Plant Model

The elastic joint manipulator shown in Fig. 4.4 has two inertias and a payload as nonlinear disturbance. This plant has fourth order dynamics represented in

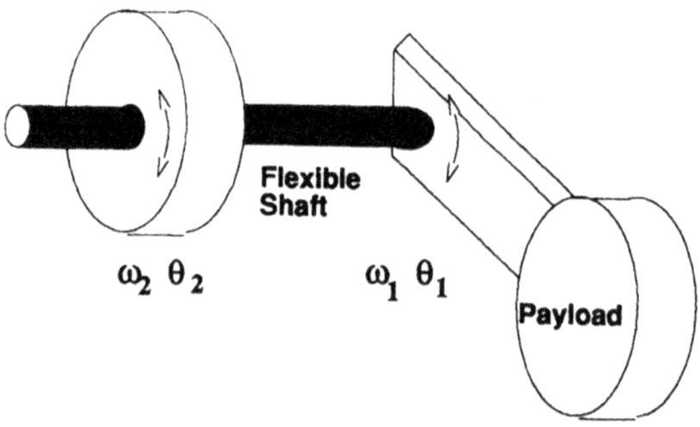

Fig. 4.4. Elastic joint manipulator

state space by

$$\dot{x} = Ax + Bu + f \tag{4.28}$$

where

$$\begin{aligned} x &= \begin{bmatrix} \theta_1 & \dot{\theta}_1 & \theta_2 & \dot{\theta}_2 \end{bmatrix}^T \\ A &= \begin{bmatrix} 0 & 1 & 0 & 0 \\ -k/I & -d/I & k/I & d/I \\ 0 & 0 & 0 & 1 \\ k/J & d/J & -k/J & -d/J \end{bmatrix} \\ B &= \begin{bmatrix} 0 & 0 & 0 & 1/J \end{bmatrix}^T \end{aligned}$$

$$f = \begin{bmatrix} 0 & -M*g*L\sin(\theta_1)/I & 0 & 0 \end{bmatrix}^T$$

θ_1	payload position (rad)
θ_2	motor position (rad)
I	payload inertia (Kgm^2)
J	motor inertia (Kgm^2)
k	shaft stiffness (Nm/rad)
d	shaft damping (Nm/rad/s)
M	payload weight (Kg)
g	gravitational constant (m/s^2)
L	arm length (m)

The control variable is θ_1 and we assume θ_2 and $\dot{\theta}_2$ to be observable. In this design example the plant parameters are chosen to have the values

$I = 0.2\,\text{Kgm}^2$	$J = 0.8\,\text{Kgm}^2$
$k = 1000$ Nm/rad	$d = 1$ Nm/rad/s
$M = 1$ Kg	$g = 9.8\,\text{m/s}^2$
$L = 0.3$ m	

This plant has a resonance mode at $\omega_r = 80$ rad/s.

4.4.2 Controller Design

In the design of the switching surface, we ignore disturbance and link flexibility. This assumption implies the reduction of the plant dynamics ($\theta_1 = \theta_2$), so the plant has unmodelled dynamics. The subsystem (4.6) used for design is

$$\dot{x}_1 = x_2 \tag{4.29}$$

where $x_1 = \theta_2$ and $x_2 = \dot{\theta}_2$. The weighting function $W_1(s)$ determines the command response because the input to $W_1(s)$ is the error signal, i.e. command − output. We choose $W_1(s)$ to be

$$W_1(s) = \frac{s+100}{10s+1} \tag{4.30}$$

which yields a tracking error specification below 0.1 Hz of less than 40 dB. The weighting function $W_2(s)$ is chosen to reduce high frequency input to the subsystem. In this case, the input of the subsystem is the angular velocity ($\dot{\theta}_2$). Reduction of any high frequency input components is desirable for vibration suppression.

We must remember the important relation between $W_1(s)$ and $W_2(s)$, i.e. fast command response and low frequency control input cannot be realized at the same time. Cutoff frequencies of $W_1(s)$ and $W_2(s)$ must not be close together for existence of the H^∞ sub-optimal controller. Considering the above restriction, we select $W_2(s)$ as

$$W_2(s) = \left(\frac{10s + g}{s + 10g}\right)^2 \tag{4.31}$$

In (4.31) g determines the cutoff frequency of $W_2(s)$. We find that the minimum value of g for the existence of an iterative solution of the Riccati equation (4.25), is 9.8. The order of the sliding surface is 4, i.e. x_1 and x_w are 2-vectors.

4.4.3 Simulation Results

Figure 4.5 shows the desired step response. This plot was computed using H^∞ linear feedback of the subsystem (4.29) and we consider this as the ideal sliding mode with no model uncertainty. Figure 4.6 shows the plant with the

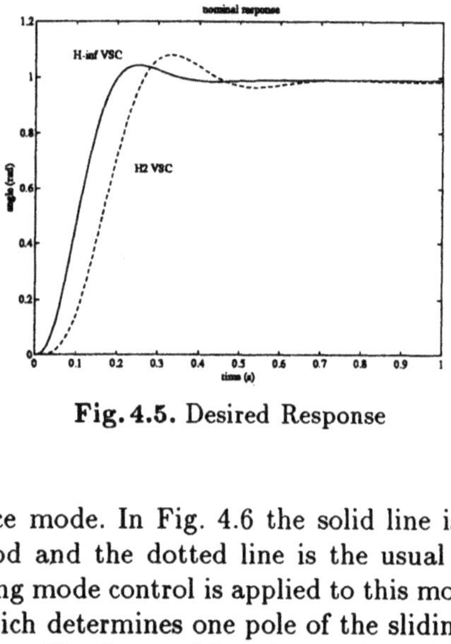

Fig. 4.5. Desired Response

unmodelled resonance mode. In Fig. 4.6 the solid line is the response of the proposed H^∞ method and the dotted line is the usual sliding mode design. When the usual sliding mode control is applied to this model, we have only one design parameter which determines one pole of the sliding mode dynamics for a second order system. We choose this pole equal to the slowest pole desired in the frequency shaped sliding mode, i.e. $\sigma = \dot{\theta}_2 + 12.3\theta_2)$. The results show that the frequency shaped sliding mode successfully suppresses vibration and maintains a good transient response. Gravity is considered as a nonlinearity in Fig. 4.7. Invariance of the sliding mode is also maintained for the frequency shaped case.

4.5 Conclusions

We have proposed a new method of sliding surface design using the frequency domain. Through the simulation of an elastic joint manipulator, the efficiency of the approach has been demonstrated. For the H^∞ norm it is easy to assign the sliding mode dynamics from a frequency specification of the reference response. If we have a priori information of any resonance modes, we obtain

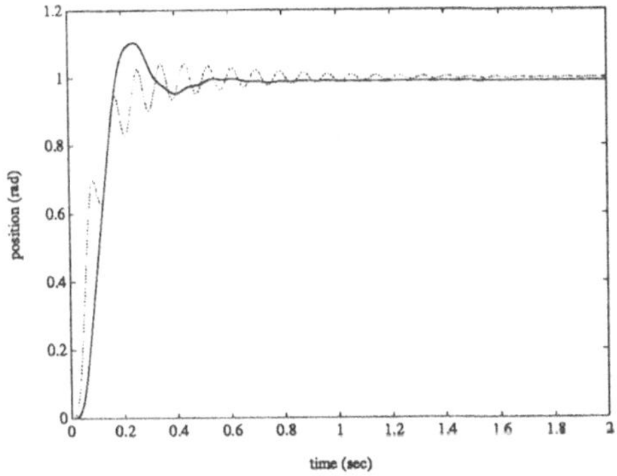

Fig. 4.6. Simulation result (no gravity)

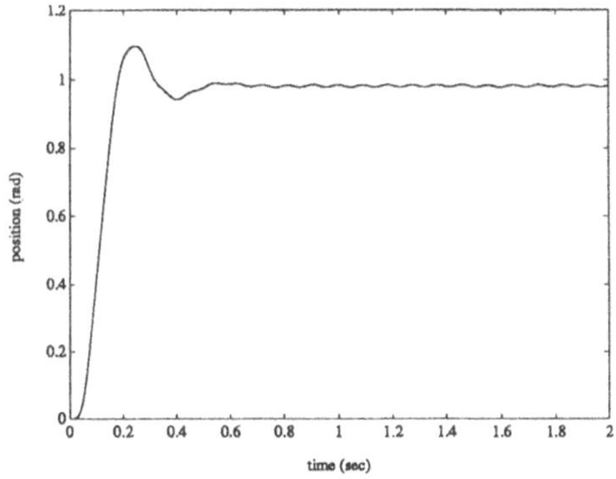

Fig. 4.7. Simulation result (with gravity)

better performance using an observer which estimates the resonance modes. However, the problem of optimal observer design has yet to be fully solved.

References

Doyle, J.C., Glover, K., Khargonekar, P., Francis, B.A. 1989, State-Space Solutions to Standard H_2 and H^∞ Control Problems. *IEEE Transactions on Automatic Control* **AC-34**, 831–847

Francis, B.A. 1987, *A Course in H^∞ Control Theory*, Lecture Notes in Control and Information Sciences, **88**, Springer-Verlag, Berlin

Gupta, N.K. 1980, Frequency-shaped cost functionals: Extension of linear-quadratic-gaussian design. *Journal of Guidance and Control* **3**, 529–535

Utkin, V.I., Young, K.D. 1978, Methods for Constructing Discontinuous Planes in Multidimensional Variable Structure Systems. *Automation and Remote Control* **31**, 1466–1470

Young, K.K. 1993, *Variable Structure Control for Robotics and Aerospace Systems*, Elsevier Science

Young, K.D., Özgüner, Ü. 1990, Frequency Shaped Sliding Mode Synthesis. *International Workshop on VSS and their Applications*, Sarajevo

Zhou, K., Khargonekar, P.P. 1988, An Algebraic Riccati Equation Approach to H^∞ Optimization. *Systems and Control Letters* **11-2**, 85–91

5. Sliding Mode Control in Discrete-Time and Difference Systems

Vadim I. Utkin

5.1 Introduction

Sliding mode control has been widely used because of its robustness properties, and the ability to decouple high dimensional systems into a set of independent subproblems of lower dimension (Utkin 1992). Thus far the theory of sliding mode control has been developed mainly for finite dimensional continuous-time systems described by ordinary differential equations. Recently several papers have been published on sliding modes in distributed parameter systems described by partial differential equations and ordinary differential equations in Banach spaces (Utkin and Orlov 1990). The sliding mode is generated by means of discountinuities in the controls on a manifold in the state space. The discontinuity manifold $\mathcal{S}$, consisting of state trajectories, is attained from initial conditions in a finite time interval. From a mathematical point of view it should be emphasized that on $\mathcal{S}$ the Cauchy problem does not have a unique solution for $t < 0$. In other words, a shift operator establishing correspondence between the states at two different time instants is not invertible at points in the sliding manifold. Indeed, any point where the sliding mode exists, may be reached along a sliding trajectory in $\mathcal{S}$ or by a trajectory from outside $\mathcal{S}$.

For discrete-time systems the concept of sliding needs to be clarified, since discontinuous control does not enable generation of motion in an arbitrary manifold, and results in chattering or oscillations at the boundary layer at the sampling frequency (Kotta 1989). There are many different approaches to the design of discrete-time sliding mode control, associated with motion in the manifold's boundary layer, with width of the sampling-interval order (Milosavljević 1985, Spurgeon 1991, Sarpturk 1987, Furuta 1990).

Generally in continuous-time systems with continuous control, the manifold consisting of state trajectories can be reached only asymptotically. In contrast to continuous-time systems, in discrete-time systems with continuous control, motion may exist with state trajectories in some manifold with a finite time interval preceding this motion (Drakunov and Utkin 1992). So the motion may be called the "sliding mode". Moreover, in contrast to continuous-time systems the shift operator in discrete-time systems is not invertible. In discrete-time systems the continuous operator in the system equation, which matches a system state from one sampling instant into the next state, is a shift operator. If the sliding mode occurs, state trajectories are in a manifold of lower dimension than that of the original system. This means that the inverse of the shift operator does not exist since it transforms a domain of full dimension into another domain in a manifold of lower dimension (Gantmacher 1959).

The similarity of the sliding mode in continuous and discrete-time systems has been established in terms of the shift operator. The concept of the "sliding mode" for dynamic systems of general type, can be represented by a shift operator (Drakunov and Utkin 1992).

Sect. 5.2 is dedicated to the basic concepts of sliding modes in dynamic systems. In later sections the further development of sliding mode control design methods is presented for discrete-time linear finite and infinite-dimensional systems, the design of finite observers, and the control for systems with delays and differential-difference systems. An example illustrates sliding mode control of the longitudinal oscillations of a one-dimensional flexible bar.

5.2 Semi-Group Systems and Sliding Mode Definition

Consider a finite-dimensional continuous-time system

$$\dot{x} = f(t,x) + B(t,x)u \tag{5.1}$$

with $x \in \mathbb{R}^n$, $u \in \mathbb{R}^m$. In sliding mode control the control components u_i have discontinuities on the surfaces $\sigma_i = \{x : s_i(t,x) = 0\}$ in the state space, i.e.

$$u_i(t,x) = \begin{cases} u_i^+(t,x) & \text{if} \quad s_i(t,x) > 0 \\ u_i^-(t,x) & \text{if} \quad s_i(t,x) < 0 \end{cases} \tag{5.2}$$

for $i = 1, 2, \ldots, m$ where $u_i^+(t,x)$ and $u_i^-(t,x)$ are continuous functions. Starting from some time instant t_0 every state trajectory belongs to the intersection of the surfaces σ_i for $i = 1, 2, \ldots, m$. The motion along this manifold (termed the sliding mode) is described by $(n-m)$-th order system equations. To derive the sliding mode equation the original control should be substituted by the so-called "equivalent control" u_{eq}, consisting of the solution to the equation $\dot{s} = 0$ with respect to the control. For the simplest example with $n = m = 1$

$$\dot{x} = f(x) + u \ , \qquad u = -M \operatorname{sign} x \tag{5.3}$$

with $M > f_0$, $f_0 = \sup |f(x)|$, the sliding mode arises in the "manifold" $x = 0$, at least for $t \geq x(0)/(M - f_0)$ (see Fig. 5.2). Digital computer implementation of the control (5.3) with a sampling interval δ leads to oscillations at finite frequency (see Fig. 5.2). This example illustrates the chattering problem which arises in systems with discontinuous control implemented digitally. Since within a sampling interval a control value is constant, the switching frequency cannot exceed the sampling frequency. Now suppose that for any constant u the solution to (5.6) may be found, i.e. $x(t) = F(x(0), u)$. For $t = 0$, the control $u(x(0), \delta)$ may be chosen so that $x(\delta) = 0$, which means that $x((k+1)\delta) = 0$ with the control $u(x(k\delta), \delta)$. So, in the discrete-time system

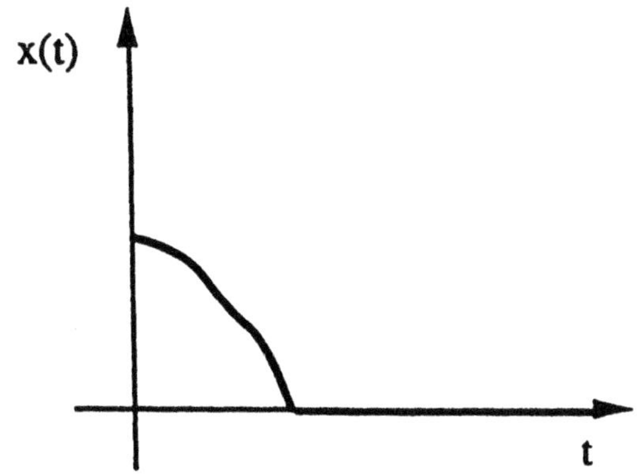

Fig. 5.1. Ideal sliding in continuous-time system

$$x_{k+1} = F(x_k, u_k) \ , \qquad u_k = u(x_k) = u(x(k\delta), \delta) \tag{5.4}$$

the value $x_k = 0$ for $k \geq 1$. Since $F(x(0)), \delta, u)$ tends to $x(0)$ as $\delta \to 0$, the function $u(x(0), \delta)$ may exceed the control bounds $u_{\max}$ or $u_{\min}$. As a result the control represented in Fig. 5.2, steers x_k to zero only after a finite number of steps (see Fig. 5.2). Thus the manifold $x = 0$ is reached after a finite time interval, and thereafter the state remains in the manifold. Similarly to continuous-time systems, this motion may be referred to as the "discrete-time sliding mode". Note that the sliding mode is generated in the discrete-time system with a *continuous* control.

The above first order example clarifies the definition of the term "discrete-time sliding mode" introduced by Drakunov and Utkin (1992) for an arbitrary finite-dimensional discrete-time system.

Definition 5.1 *In the discrete-time dynamic system*

$$x(k+1) = F(x(k)) \qquad x \in \mathbb{R}^n \tag{5.5}$$

a discrete-time sliding mode takes place on the subset $\mathcal{M}$ of the manifold $\sigma = \{x : s(x) = 0\}$, $s \in \mathbb{R}^m$ $(m < n)$, if there exists an open neighbourhood U of this subset such that for $x \in U$ it follows that $s(F(x)) \in M$ (see Fig. 5.2).

In contrast to continuous-time systems the sliding mode may arise in discrete-time systems with a continuous function on the right hand side of system equations.

The similarity of the two types of sliding modes is that the families of state space transformations representing the closed-loop systems are semi-groups rather then groups, since the inverse transformation values for the states in the sliding manifold are not unique.

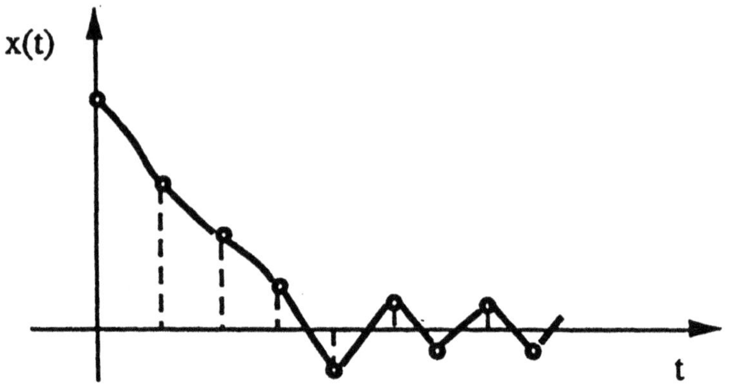

Fig. 5.2. Discrete-time system with discontinuous control

The families of transformations

$$F(t, t_0, \cdot) : \mathcal{X} \rightarrow \mathcal{X} \tag{5.6}$$

with t_0, $t \in T$, $t_0 \leq t$ ($T \in \mathbb{R}$ or $\mathbb{N}$ to embrace continuous and discrete-time cases) is the most general description of dynamic systems in the metric space $\mathcal{X}$. F is a continuous function of x satisfying the semi-group condition $F(t, t_1, F(t_1, t_0, x_0)) = F(t, t_0, x_0)$ for every $t_0 \leq t_1 \leq t$, $x_0 \in \mathcal{X}$ and $F(t, t_1, x) = x$ for every $t \in T$, $x \in \mathcal{X}$. If F corresponds to a system of ordinary differential equations with the existence and uniqueness of the Cauchy problem solution, then for all $t_0 \leq t$, $x \in \mathcal{X}$ the transformation F is invertible and its inverse $F^{-1}(t, t_0, x)$ is equal to $F(t_0, t, x)$. This means that the family $\{F(t_0, t, x)\}_{t_0, t \in T}$ is a group.

We will now develop the concept of the sliding mode further. Consider dynamic systems determined by the shift operator (5.6). This class embraces continuous-time ordinary differential and difference-differential equations, or more generally hereditary equations, which contain hysteresis loops, delays, etc. The above property, established for continuous and discrete-time systems, namely the violation of the group condition in the sliding manifold, is the core idea for the formulation of the concept of the sliding mode in dynamical systems (5.6).

Definition 5.2 *A point x in the state space $\mathcal{X}$ of a dynamic system in the family of semi-group transformations $\{F(t, t_0, \cdot)\}_{t_0 \leq t}$ is said to be a sliding mode point at the time instant $t \in T$, if for every $t_0 \in T$, $t_0 < t$ the transformation $F(t, t_0, \cdot)$ is not invertible at this point (and the equation $F(t, t_0, \cdot) = x$ has more than one solution ξ). A set $\Sigma \subset T \times \mathcal{X}$ in the state space is a sliding mode set, if for every $(t, x) \in \Sigma$, x is a sliding mode point at the time instant t.*

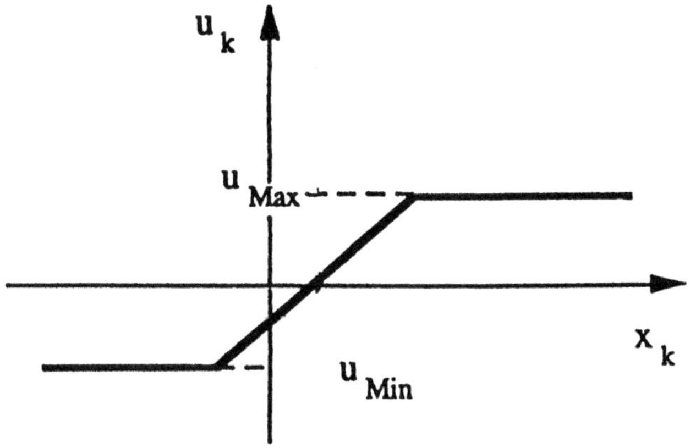

Fig. 5.3. Control in discrete-time system with sliding mode

Design procedures based on the above concept of the sliding modes in dynamic systems, will be developed below. First we confine ourselves to "invalid" problems which are nontraditional in sliding mode theory, but which may be interepreted in terms of Definition 5.2, and will serve as illustrative examples.

Example 5.3 Consider the system

$$\phi(x)\dot{x} = f(t,x) \;, \qquad x \in \mathbb{R}^n \tag{5.7}$$

(where f is Lipshitz and ϕ is a smooth scalar valued function) which cannot be presented in the Cauchy form on the manifold $\phi(x) = 0$. In the vicinity of the set

$$U = \{x : \phi(x) = 0\} \bigcap \{x : \nabla\phi(x)\; f(t,x) < -\phi_0\}$$

with $\phi_0 > 0$ (see Fig. 5.3) the time derivative of the function $V = \phi^2(x)$

$$\dot{V} = 2\phi\; \nabla\phi\; \dot{x} = 2\nabla\phi\; f < -2\phi_0 \tag{5.8}$$

is negative which means that $\phi(x)$ vanishes in a finite time interval. After reaching the surface $\phi(x) = 0$ further state trajectories remain on the surface. Any point of U may be reached from both of the domains $\phi(x) > 0$ and $\phi(x) < 0$ and along trajectories in the surface $\phi(x) = 0$. According to Definition 5.2 the set U is the sliding mode set of the system (5.7).

Example 5.4 Let the control for the system (5.1) have the form

$$u = -(GB)^{-1}Gf - s/||s||^{\frac{1}{2}} \;, \qquad G = \partial s/\partial x \tag{5.9}$$

The control is a continuous function of the state but the feedback system does not satisfy the uniqueness solution condition since a Lipschitz constant does not exist in the vicinity of $s = 0$.

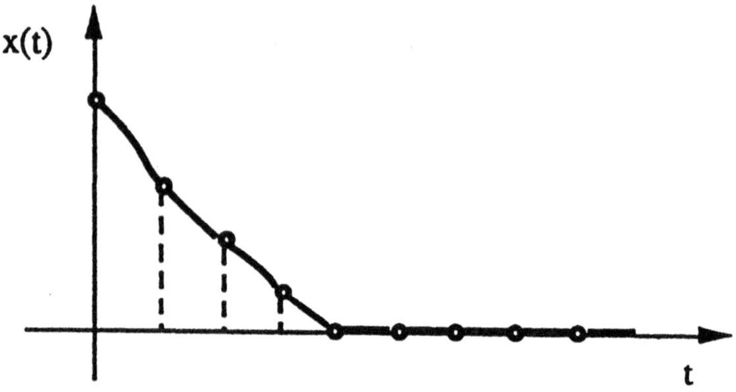

Fig. 5.4. Sliding mode in discrete-time system

The projection onto the subspace s is governed by the equation

$$\dot{s} = -s/||s||^{\frac{1}{2}} \tag{5.10}$$

and the time-derivative of the Lyapunov function $V = \frac{1}{2}s^T s$ is

$$\dot{V} = -(2V)^{\frac{3}{4}} \tag{5.11}$$

The solution to (5.11) $V(t) = (V_0^{\frac{1}{4}} - t/2)^4/2$ decays, and for the initial conditions $V_0 = \frac{1}{2}s(x_0)^T s(x_0)$ vanishes for $t \geq 2(2V_0)^{\frac{1}{4}}$. Hence, after a finite time inverval the state reaches the manifold $s(x) = 0$ and further trajectories are confined to this manifold. Again, according to Definition 5.2 the system motion may be refered to as the sliding mode.

Example 5.5 Another example is the singularly perturbed system

$$\dot{x} = f(t, x, z) \tag{5.12}$$
$$\mu\dot{z} = g(t, x, z) \tag{5.13}$$

where $\mu > 0$ is a small parameter, $x \in \mathbb{R}^n$ represents "slow" and $z \in \mathbb{R}^m$ "fast" variables. Under the conditions of Tihonov's Theorem (Kokotović et al 1976) (relating to the asymptotic stability of the "fast" subsystem (5.13)) the motion of (5.12) and (5.13) with initial states x_0, z_0 tending to zero when $t > 0$, is represented by the equations

$$\begin{aligned} \dot{x}^*(t) &= f(t, x^*, \phi(t, x^*)) \\ z^*(t) &= \phi(t, x^*(t)) \end{aligned} \tag{5.14}$$

with $x^*(0) = x_0$, $z^*(0) = \phi(0, x_0)$, where $\phi(t, x)$ is the solution of the algebraic equation $g(t, x, \phi) = 0$. Equation (5.14) describes the behaviour of the dynamic

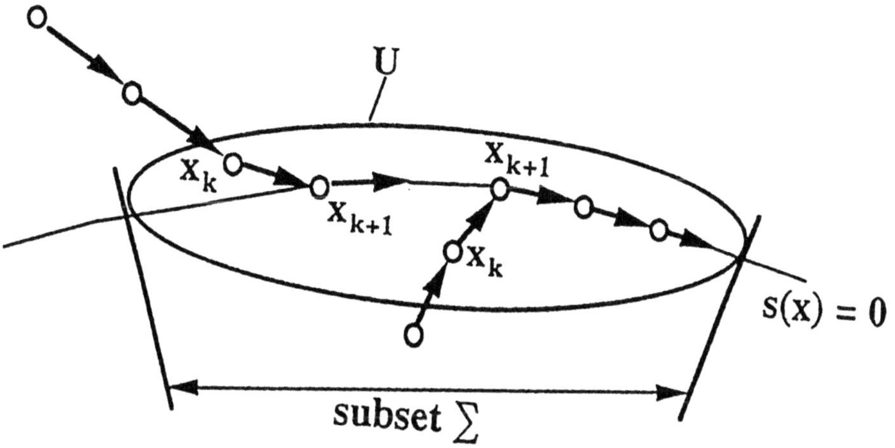

Fig. 5.5. Definition of discrete-time sliding mode control

system when the state jumps from the point (x_0, z_0) to $(x^*(0), z^*(0))$ instantly and then proceeds along a trajectory in the manifold $g(t, x, z) = 0$. Similarly to the previous examples this manifold is the sliding mode set, though there is no sliding motion in the system (5.12) and (5.13) for any $\mu \neq 0$.

5.3 Discrete-time Sliding Mode Control in Linear Systems

This section deals with discrete-time sliding mode control of linear time-invariant continuous-time plants. Let us assume that a sliding manifold is linear for a discrete-time system $x_{k+1} = F(x_k)$, i.e. $s_k = Cx_k$, $C \in \mathbb{R}^{m \times n}$. Following from Definition 5.1 the sliding mode existence condition has the form

$$s_{k+1} = C(F(x_k)) = 0 \tag{5.15}$$

for any $x_k \in \mathcal{M}$. To design a descrete-time controller based on the condition (5.15) the continuous-time equations should be transformed into discrete-time form.

Consider the linear time-invariant system

$$\dot{x} = Ax + Bu + Df(t) \tag{5.16}$$

where $x \in \mathbb{R}^m$, $u \in \mathbb{R}^m$, $f(t)$ is the reference input, and A, B and D are constants, with a digital controller. This can be written as

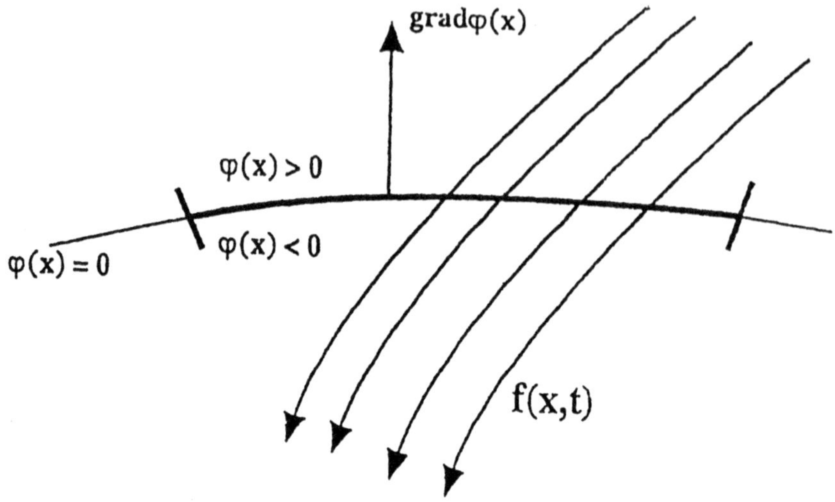

Fig. 5.6. Vector field in system with sliding mode

$$x_{k+1} = A^* x_k + B^* u_k + D^* f_k \tag{5.17}$$

where

$$A^* = e^{A\delta}\,, \quad B^* = \int_0^\delta e^{A(\delta-\tau)} B \,\mathrm{d}\tau\,, \quad D^* = \int_0^\delta e^{A(\delta-\tau)} D \,\mathrm{d}\tau$$

and the reference input $f(t)$ is assumed to be constant within a sampling interval, i.e. $f(t) = f_k$. In accordance with (5.15), the discrete-time sliding mode exists if the matrix CB^* has an inverse and the control u_k is designed as the solution of

$$s_{k+1} = CA^* x_k + CD^* f_k + CB^* u_k = 0 \tag{5.18}$$

i.e.

$$u_k = -(CB^*)^{-1}(CA^* x_k + CD^* f_k) \tag{5.19}$$

By analogy with continuous-time systems, the control law (5.19) yielding motion in the manifold will be referred to as the "equivalent control". To reveal the structure of the equivalent control, let us represent it as the sum of two linear functions

$$u_{k_{eq}} = (CB^*)^{-1} s_k - (CB^*)^{-1}[(CA^* - C)x_k + CD^* f_k] \tag{5.20}$$

and

$$s_{k+1} = s_k + (CA^* - C)x_k + CD^* f_k + CB^* u_k \tag{5.21}$$

As in the first order example considered above, $u_{k_{eq}}$ tends to infinity with $\delta \to 0$ for $s_k \neq 0$, since $(CB^*)^{-1} \to \infty$ while $(CB^*)^{-1}(CA^* - C)$ and $(CB^*)^{-1}CD^*$ take finite values. So the bounds for the control should be taken into account.

Suppose that the control can vary within the domain $||u_k|| \leq u_0$ and

$$||(CB^*)^{-1}||.||(CA^* - C)x_k + CD^* f_k|| < u_0 \tag{5.22}$$

The control

$$u = \begin{cases} u_{k_{eq}} & \text{if} \quad ||u_{k_{eq}}|| \leq u_0 \\ -u_0 \dfrac{u_{k_{eq}}}{||u_{k_{eq}}||} & \text{if} \quad ||u_{k_{eq}}|| > u_0 \end{cases} \tag{5.23}$$

is in the admissible domain. From (5.21)-(5.23) it follows that for

$$s_{k+1} = (s_k + (CA^* - C)x_k + CD^* f_k)\left(1 - \frac{u_0}{||u_{k_{eq}}||}\right) \ , \ 1 - \frac{u_0}{||u_{k_{eq}}||} > 0 \tag{5.24}$$

Then

$$\begin{aligned} ||s_{k+1}|| &= ||(s_k + (CA^* - C)x_k + CD^* f_k)|| \left(1 - \frac{u_0}{||u_{k_{eq}}||}\right) \\ &\leq ||s_k|| + ||(CA^* - C)x_k + CD^* f_k|| \\ &\quad - \frac{u_0 ||s_k + (CA^* - C)x_k + CD^* f_k||}{||(CB^*)^{-1}||.||s_k + (CA^* - C)x_k + CD^* f_k||} \\ &\leq ||s_k|| + ||(CA^* - C)x_k + CD^* f_k|| - \frac{-u_0}{||(CB^*)^{-1}||} \\ &< ||s_k|| \end{aligned}$$

Hence $||s_k||$ decreases monotonously and, after a finite number of steps, $u_{k_{eq}}$ belongs to the admissible domain according to (5.20) and (5.22) $||u_{k_{eq}}|| \leq u_0$ for $s = 0$), which is a δ-order boundary layer of the manifold $s = 0$. In compliance with the definition of $u_{k_{eq}}$, the subsequent part of the state trajectory will be in the manifold and discrete-time sliding mode will take place. Note that the boundary layer is the neighbourhood $\mathcal{M}$ in terms of Definition 5.1.

As mentioned previously, a discontinuous control in a discrete-time system results in chattering in the boundary layer of the discontinuity manifold. The control (5.23) provides the motion in the manifold $s = 0$. Similarly to the case of continuous-time systems, the equation $s = Cx = 0$ enables the reduction of the system order, and the desired dynamics of the sliding mode, governed by the $(n-m)$-th order system, can be designed by appropriate choice of matrix C in the sliding mode equation. Complete information on the plant parameters is needed for inplementation of the control (5.23). Suppose now that the system operates under uncertainty conditions: the matrices A and D and the input f are unknown and may vary in some ranges such that the condition (5.22) holds. Similarly to (5.23) the control

$$u_k = \begin{cases} -(CB^*)^{-1}s_k & \text{if} \quad ||(CB^*)^{-1}s_k|| \leq u_0 \\ -u_0 \dfrac{(CB^*)^{-1}s_k}{||(CB^*)^{-1}s_k||} & \text{if} \quad u_{k_{eq}} > u_0 \end{cases} \tag{5.25}$$

is in the admissible domain but it does not depend on the plant parameters and input. From (5.21), (5.22) and (5.25) it follows that for $||(CB^*)^{-1}s_k|| > u_0$

$$s_{k+1} = s_k(1 - u_0/||(CB^*)^{-1}s_k||) + (CA^* - C)x_k + CD^* f_k \tag{5.26}$$

for $1 - u_0/||(CB^*)^{-1}s_k|| > 0$, and

$$\begin{aligned} ||s_{k+1}|| &\le ||s_k||(1 - u_0/||(CB^*)^{-1}s_k||) + ||(CA^* - C)x_k + CD^* f_k|| \\ &= ||s_k|| - u_0||s_k||/||(CB^*)^{-1}s_k|| + ||(CA^* - C)x_k + CD^* f_k|| \\ &\le ||s_k|| - u_0/||(CB^*)^{-1}|| + ||(CA^* - C)x_k + CD^* f_k|| \\ &< ||s_k|| \end{aligned}$$

Hence, as for the complete information case, the value of $||s_k||$ decreases monotonically and, after a finite number of steps, $(CB^*)^{-1}s_k$ will belong to the admissible control domain. Following (5.21) $s_{k+1} = (CA^* - C)x_k + CD^* f_k$. Since the matrices $CA^* - C$ and CD^* are of δ-order, the subsequent system motion will be in a δ-order boundary layer of the manifold $s = 0$ while the control values are in the admissible domain. This result should be expected for systems operating under uncertainty conditions, since we dealing with an open-loop system within a sampling interval. In contrast to discrete-time systems with discontinuous control, the systems with control (5.25) are free of chatter.

5.4 Discrete-Time Sliding Modes in Infinite-Dimensional Systems

Recent research has been oriented towards the generalization of the sliding mode control concept for the case of infinite-dimensional systems. The main reason hindering such generalization is unboundedness of the operators in the system equations.

Mathematical models, embracing the majority of infinite-dimensional processes of modern technology, are differential equations in Banach space

$$\dot{x} = Ax + Bu \tag{5.27}$$

where A is a linear operator, generally unbounded, x and u are elements of Banach spaces B_x and B_u, and B is a linear bounded operator. The control has discontinuities in some manifold $s = Cx = 0$, $s \in B_s$, with bounded operator C, transforming B_x into B_s.

For real physical processes the solution to the corresponding differential equation is a bounded function of t. Formalization of this condition usually means that the operator A generates the analytical semi-group $U(t)$. The latter is a bounded operator and

$$x(t) = U(t)x(0) \tag{5.28}$$

is a solution to (5.27) with $u = 0$. The above condition holds if the operator $-A$ (or $-A + \lambda_0 I$ for some λ_0) is strongly positive (Utkin and Orlov 1990). For this class of operators, Lyapunov functional method for generating sliding modes on the manifold $s = 0$ have been developed, and validity of the equivalent control method has been substantiated, in spite of the unboundness of the operators.

Generalization of the discrete-time sliding mode approach to infinite dimensional systems has proved to be simplier, since the operators determining the dependence between $x(t+\delta)$ and $x(t)$, i.e.

$$x(t+\delta) = U(\delta)x(t) + \int_0^\delta U(\delta - t)Bu(t)\,\mathrm{d}t \tag{5.29}$$

are bounded. Correspondingly, for digital control $u(t) = u_k = \text{const}$, for $\delta k < t < \delta(k+1)$, all the operators in the discrete-time equation are bounded

$$x_{k+1} = A^*x + B^*u\ , \qquad A^* = U(\delta)\ , \qquad B^* = \int_0^\delta U(\delta - t)B\,\mathrm{d}t \tag{5.30}$$

For the operator CB^* to be invertible and $(CB^*)^{-1}$ bounded, the design procedure below finite dimensional systems. In the sense of Definition 5.1, the control (5.22),(5.23) coincides with that developed in Sect. 5.3 for control (5.22),(5.23) provides the sliding mode in the manifold $s = 0$ after a finite number of steps.

Suppose that the Banach space B can be represented as the sum

$$B_x = X_1 \oplus X_2 \tag{5.31}$$

where the subspace

$$X_1 = \ker C = \{x \in B_x, Cx_1 = 0\} \tag{5.32}$$

and X_2 is a space belonging to B_x. Then the sliding mode equation (or (5.30) with $u_k = u_{k_{eq}}$)

$$x_{k+1} = A^*x_k - B^*(CB^*)^{-1}CA^*x_k \tag{5.33}$$

can be written as a system of two equations with bounded operators A_{ij} $(i, j = 1, 2)$ depending upon A^*, B^* and C, i.e.

$$\begin{aligned} x_1(k+1) &= A_{11}x_1(k) + A_{12}x_2(k) \\ x_2(k+1) &= A_{21}x_1(k) + A_{22}x_2(k) \end{aligned} \tag{5.34}$$

If the sliding mode occurs, (5.31), (5.32) imply that $x_2 = 0$ and

$$x_1(k+1) = A_{11}x_1(k) \tag{5.35}$$

The sliding mode dynamics (5.35) can be influenced by a proper choice of operator C in the equation of manifold.

Example 5.6 Let us consider the discrete-time control of an infinite dimensional plant governed by the heat equation

$$\begin{aligned}\frac{\partial Q(y,t)}{\partial t} &= \frac{\partial^2 Q(y,t)}{\partial y^2} + u(y,t) \qquad t>0, \quad 0<y<1 \\ Q(0,t) &= Q(1,t)=0, \quad t>0, \quad Q(y,0)=Q(y)\end{aligned} \tag{5.36}$$

which describes the temperature field $Q(y,t)$ of a one-dimensional bar with distributed heat source $u(y,t)$ as the control. If the solution $Q(y,t)$ is interpreted as a function of time with values in the space of double differentiable functions, then (5.36) can be viewed as the equation analogous to the differential equation (5.27) in the Banach space $L_2(0,1)$ with operator $A = \partial^2/\partial y^2$ and $B = 1$. The solution has the form

$$\begin{aligned}Q(y,t) &= U(t)Q_0(y) + \int_0^t U(t-\tau)u(y,\tau)\,\mathrm{d}t \\ &= 2\sum_{n=1}^{\infty} \mathrm{e}^{-n^2\pi^2 t}\left[\int_0^1 Q_0(\xi)\sin n\pi\xi\,\mathrm{d}\xi\right]\sin n\pi y \\ &\quad + 2\sum_{n=1}^{\infty}\int_0^t \mathrm{e}^{-n^2\pi^2(t-\tau)} \times \\ &\qquad \left[\int_0^1 u(\xi,\tau)\sin n\pi\xi\,\mathrm{d}\xi\right]\mathrm{d}\tau\sin n\pi y\end{aligned} \tag{5.37}$$

Bearing in mind that

$$Q_0(y) = \sum_{n=1}^{\infty} q_n(0)\sin n\pi y\ , \qquad q_n(0) = \int_0^t Q_0(\xi)\sin n\pi\xi\,\mathrm{d}\xi \tag{5.38}$$

$$u(y,t) = \sum_{n=1}^{\infty} u_n(t)\sin n\pi y\ , \qquad u_n(t) = 2\int_0^t u(\xi,t)\sin n\pi\xi\,\mathrm{d}\xi \tag{5.39}$$

the validity of (5.37) as a solution to (5.36) may be checked by direct substitution. The semi-group $U(t)$ in (5.37) is analytical with the upper estimate $\|U(t)\| \le e^{-\pi^2 t}$, $t>0$.

Suppose that the control is a discrete-time function of y, namely $u(k) = u(y,k)$ for $\delta k < t < \delta(k+1)$. Then

$$Q(y,k+1) = \sum_{n=1}^{\infty}\left[e^{-n^2\pi^2\delta}q_n(k) + \left(\int_0^\delta e^{-n^2\pi^2(\delta-\tau)}\,\mathrm{d}\tau\right)u_n(k)\right]\sin n\pi y \tag{5.40}$$

where $q_n(k)$ and $u_n(k)$ may be found similarly to (5.38) and (5.39).

The equivalent control yielding discrete-time sliding mode on the manifold $Q(y)=0$ can be found from (5.40) and (5.39) as

$$u(y,k) = \sum_{n=1}^{\infty} u_{n_{eq}}(k)\sin(n\pi y) \tag{5.41}$$

where

$$u_{n_{eq}}(k) = -\frac{n^2\pi^2}{e^{n^2\pi^2\delta}-1} q_n(k) \tag{5.42}$$

The series (5.41) with finite values $q_n(k)$ converges but may have high enough values with a small sampling interval δ and exceed the constraints posed on the control

$$||u||_{L_2} = \left[\sum_{n=1}^{\infty} u_n^2\right]^{\frac{1}{2}} \leq u_0 \tag{5.43}$$

Then, similarly to (5.23), the expression for the control action (5.41), (5.42) is modified so that the control varies only within the admissible domain (5.43), i.e.

$$u_n(k) = \begin{cases} u_{n_{eq}} & \text{if } ||u_{eq}|| \leq u_0 \\ -\dfrac{n^2\pi^2}{e^{n^2\pi^2\delta}-1} q_n(k) \dfrac{u_0}{||u_{eq}||} & \text{if } ||u_{eq}|| > u_0 \end{cases} \tag{5.44}$$

and

$$\begin{aligned} q_n(k+1) &= q_n(k)e^{-n^2\pi^2\delta} + \int_0^{\delta} e^{-n^2\pi^2(\delta-\tau)}\, d\tau\, u_n(k) \\ &= e^{-n^2\pi^2\delta}\left[1 - \frac{u_0}{||u_{eq}||}\right] q_n(k) \end{aligned}$$

for $||u_{eq}|| > u_0$. This indicates the convergence of the $q_n(k)$ and $||u_{eq}||$ to zero, so that, after N (finite) steps, the inequality $||u_{eq}(y, N)|| \leq u_0$ holds, and the sliding mode exists on the manifold $Q(y) = 0$ for $k > N$. So, discrete-time distributed control is designed in the form

$$u(y,k) = \sum_{n=1}^{\infty} u_n(k) \sin n\pi y \tag{5.45}$$

where the components of the control $u_n(k)$ vary in accordance with (5.44). As in finite dimensional systems, sliding motion arises with control which is a continuous function of the system state.

It should be noted that we can also consider the situation in the presence of a term $a(y)f(t)$, which plays the role of a reference signal. This means that, in (5.40), we should replace $u_n(k)$ with $u_n(k) + a_n f(k)$. The coefficients in the second term are constant.

5.5 Sliding Modes in Systems with Delays

The subject of the present and subsequent sections is design methods for systems governed by difference and differential-difference equations. These types of equations may serve as mathematical models for dynamic systems with delays and distributed systems with finite dimensional inputs and outputs. It will be shown that, in terms of sliding mode control, a deadbeat observer (Kwakernaak and Sivan 1972) may be designed for continuous-time linear systems.

Consider the system of differential-difference equations in the block form

$$\dot{x}(t) = A_{11}x(t) + A_{12}z(t) \tag{5.46}$$

$$z(t) = A_{21}x(t-\tau) + A_{22}z(t-\tau) + B_0u(t-\tau) \tag{5.47}$$

where $x \in \mathbb{R}^n$, $z \in \mathbb{R}^k$ and $u \in \mathbb{R}^m$. The pair (A_{11}, A_{12}) is assumed to be controllable, and the difference system (5.47) invertible with output $\tilde{A}_{12}z(t)$ ($\tilde{A}_{12}$ consists of basic rows of A_{12}). The block-control form of the controllable system has been described by Drakunov et al (1990a, 1990b) for systems of ordinary differential equations. The design yields quasi-control $z(t)$ to be assigned equal to $z^*(x)$, and gives the desired motion in the first block (5.46). The control is used to fulfill the condition $z \to z^*$. The stabilization problem for the subsystem (5.46) can be solved within the class of dynamic systems with sliding modes.

Within the framework of sliding mode control algorithms the above stabilization approach leads to a two stage procedure. The first step, design of the discontinuous quasi-control $z = z^*(x)$, enforces the asymptotically stable sliding mode along the manifold $\sigma = \{x : S(x) = 0\}$ in the subsystem (5.46). In the second step the control $u = u(x, z)$ is designed so that $\sigma_0 = \{x : z - z^*(x) = 0\}$ is a sliding manifold for differential-difference system (5.46), (5.47). The sliding mode exists on the intersection $\sigma \bigcap \sigma_0$.

Example 5.7 Consider a time-invariant linear system with a delay in the input variable

$$\dot{x} = Ax + Bu(t-\tau) \tag{5.48}$$

where $x \in \mathbb{R}^n$, $u \in \mathbb{R}^m$, $t > 0$ and the initial conditions are $x(0) = x_0$, $u(\xi) = u_0(\xi)$, $-\tau < \xi < 0$. This system may be presented in a the differential-difference block form with quasi-control $z(t) = u(t-\tau)$, $A_{11} = A$, $A_{12} = B$, $A_{21} = A_{22} = 0$ and $B_0 = I$. Let us suppose that there is a smooth function $S(x) = (s_1(x), \ldots, s_k(x))$ with values in R^k, and a discontinuous quasi-control $z^* \in \mathbb{R}^k$ with components

$$z_i^*(x(t)) = \begin{cases} z_i^+(x(t)) & \text{if} \quad s_i(x(t)) > 0 \\ z_i^-(x(t)) & \text{if} \quad s_i(x(t)) < 0 \end{cases} \qquad i = 1, 2, \ldots, k \tag{5.49}$$

such that every state trajectory after a finite time belongs to the intersection of the surfaces $\sigma_i = \{x : s_i(x) = 0\}$. and thereafter the sliding mode exists. The quasi-control $z(t)$ should be equal to $z^*(x(t))$. If we assign

$$u(t) = z^*(x(t+\tau)) \tag{5.50}$$

the sliding takes place on the manifold $s_0 = \{x : z - z^*(x) = 0\}$. The values of $x(t+\tau)$ can be extrapolated from

$$x(t+\tau) = e^{At}x(t) + \int_0^{\tau} e^{Ax}Bu(t-\xi)\,d\xi \tag{5.51}$$

The motion along the sliding manifold is described by the system

$$\dot{x}(t) = Ax(t) + Bz^*(x(t)) \tag{5.52}$$

and the sliding mode also occurs on $\bigcap_{i=1}^{k} \sigma_i$. Therefore in the system (5.48) sliding modes exist on $\sum = \bigcap_{i=0}^{k} \sigma_i$.

Note that the system (5.48) is not a finite dimensional system and the equality $u(t-\tau) = z^*(x(t))$ holds for $t \geq \tau$ which means that the sliding mode exists in the manifold σ_0 in the sense of Definition 5.2.

5.6 Finite Observers with Sliding Modes

Consider a linear time-invariant system

$$\dot{x}(t) = Ax(t) + Bu(t) \tag{5.53}$$

where $x \in \mathbb{R}^n$, $u \in \mathbb{R}^m$ and $t > 0$, with an output vector $y = Cx \in \mathbb{R}^\ell$. A conventional aproach to the problem of estimating the state $x(t)$ from the measurements y is to use an asymptotic observer of the form

$$\dot{\hat{x}} = A\hat{x} + Bu + L(y - C\hat{x}) \tag{5.54}$$

By suitable choice of the gain matrix L, with observability conditions on the pair (A, C), all the eigenvalues of the matrix $A - LC$ of the system

$$\dot{e}(t) = (A - LC)e(t) \ , \qquad e(t) = \hat{x}(t) - x(t) \tag{5.55}$$

can be assigned arbitrarily and therefore the desired rate of convergence of the estimate $\hat{x}(t)$ to $x(t)$ may be achieved.

Let us show that the observer based on the sliding mode concept in continuous-time difference system can be designed with finite-time convergence to the system state. The solution of (5.53) with $x(t-\tau)$ as the initial condition at time $t - \tau$ $(t > 0)$ is

$$x(t) = \mathrm{e}^{At} x(t-\tau) + \int_0^{\tau} \mathrm{e}^{A\xi} Bu(t-\xi)\, \mathrm{d}x \tag{5.56}$$

The system state $x(t)$ may be estimated by an observer of the form

$$\hat{x}(t) = \mathrm{e}^{A\tau} \hat{x}(t-\tau) + \int_0^{\tau} \mathrm{e}^{A\xi} Bu(t-\xi)\, \mathrm{d}x + L(y(t-\tau) - C\hat{x}(t-\tau)) \tag{5.57}$$

where $\hat{x}(t)$ is an estimate of $x(t)$. From (5.56) and (5.57) we obtain a difference equation for the estimation error

$$e(t) = (\mathrm{e}^{At} - LC)e(t-\tau) \ , \quad e(t) = \hat{x}(t) - x(t) \tag{5.58}$$

Since the pair $(\exp(At), C)$ is observable (from the observability of (A, C)), the eigenvalues of $\exp At - LC$ can be assigned arbitrarily within a unit circle by the proper choice of L so $e(t) \to 0$ as $t \to \infty$. If L is chosen so that all the eigenvalues are equal to zero, then $(\exp(At) - LC)^i = 0$ for some $i < n$ (Gantmacher 1959) and there exists $t_1 > 0$ such that $e(t) \equiv 0$ when $t \geq t_1$. We can say that according to Definition 5.2 the sliding mode occurs on the manifold $e = 0$ after time t. In the case of a scalar observation $t_1 \leq \tau n$ and e tends to zero with $\tau \to 0$ (though $L \to \infty$). Since the observer (5.57) handles delayed observations $y(t-\tau)$ it can be used for systems with delay in the observation vector. But if τ is fixed, the upper bound of the convergence time $n\tau$ cannot be made arbitrary small. This time can be made as close to τ as desired by using an observer of higher order

$$\begin{aligned}
\hat{x}(t) &= e^{A\delta}\hat{x}(t-\delta) + \int_0^\delta e^{Ax} Bu(t-\xi)\,dx + L_0(y(t-\tau) - \hat{y}_{r-1}(t-\delta)) \\
\hat{y}_1 &= C\hat{x}(t-\delta) + L_1(y(t-\tau) - \hat{y}_{r-1}(t-\delta)) \\
\dots & \quad \dots \\
\hat{y}_i &= \hat{y}_{i-1} + L_i(y(t-\tau) - \hat{y}_{r-1}(t-\delta)) \\
\dots & \quad \dots \\
\hat{y}_{r-1} &= \hat{y}_{r-2} + L_{r-1}(y(t-\tau) - \hat{y}_{r-1}(t-\delta))
\end{aligned}$$

where $\hat{y}_i(t)$ is an estimate of $y(t - i\delta)$ and $r\delta = \tau$ (r is a positive integer). The gain matrices $\tilde{L} = (\tilde{L}_1, \dots, \tilde{L}_{r-1})^T$ may be designed to provide the desired values of the eigenvalues of the system

$$\hat{\tilde{e}(t)} = (\tilde{A} - \tilde{L}\tilde{C})\tilde{e}(t-\delta) \tag{5.59}$$

where the estimate error is

$$\tilde{e}(t) = x(t) - \hat{x}(t) \;,\quad y(t-\delta) - \hat{y}_1(t) \;, \dots, y(t-(r-1)\delta) - \hat{y}_{r-1}(t))^T \tag{5.60}$$

$$\tilde{A} = \left[\begin{array}{c|ccccc}
e^{A\delta} & & & 0 & & \\
C & & & 0 & & \\
\hline
 & I_k & 0 & \dots & & \\
 & \dots & I_k & 0 & \dots & \\
0 & & & & \ddots & \ddots \\
 & & & & I_k & 0
\end{array}\right] , \qquad \tilde{C} = (0, 0, \dots, I_k) \tag{5.61}$$

If the L_i are such that all the eigenvalues of $\tilde{A} - \tilde{L}\tilde{C}$ (the pair $(\tilde{A}, \tilde{C})$ is observable) are equal to zero, the sliding mode occurs on the manifold $e = 0$, the transient time $t_1 < (n + r)\delta = n\delta + \tau$ and its upper estimate tends to τ with $\delta \to 0$.

For the system (5.53) without delay, a finite observer of lower dimension can also be designed. By using a nonsingular state transformation the system (5.56) can be represented in the block form

$$y(t) = A_{11}y(t-\tau) + A_{12}z(t-\tau) + v_1(t) \quad (5.62)$$

$$z(t) = A_{21}y(t-\tau) + A_{22}z(t-\tau) + v_2(t) \quad (5.63)$$

where $z \in \mathbb{R}^{n-k}$, and $v_1(t)$, $v_2(t)$ are known functions depending upon control variables $u(\xi)$, $\xi \in [t-\tau, t]$. Since $y(t)$ and $y(t-\tau)$ are measured, we can consider $y_1(t) = A_{12}z(t-\tau)$ as a new observation vector for the system (5.62) (the pair (A_{22}, A_{12}) is observable which follows from the observability of the pair (A, C)) and an observer of the form (5.57) can be used to estimate the state vector.

5.7 Control of Longitudinal Oscillations of a Flexible Bar

This section deals with the longitudinal oscillations in the one dimensional flexible bar with a load of mass m at the right end and a force F applied to the left end. Let $d(t,x)$ be the deviation at time t of the bar at the point which has coordinate x with respect to the left end in an unexcited state $(0 < x < \ell)$; $c(t,x)$ is an absolute coordinate of this point and $e(t)$ is the absolute coordinate of the left end of the bar (see Fig. 5.7). Then $c(t,x) = e(t) + x + d(t,x)$, and c

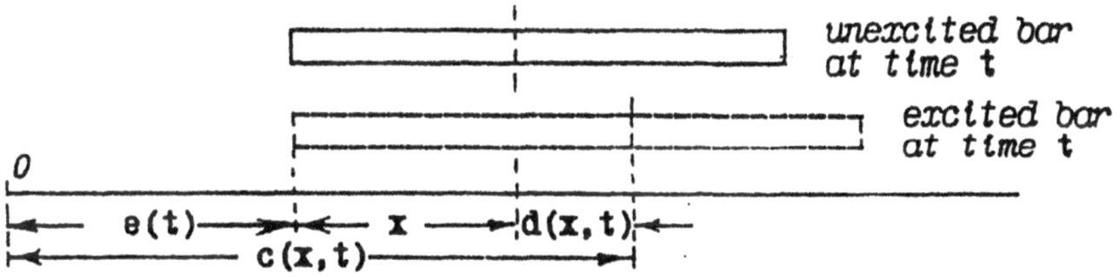

Fig. 5.7. Coordinates of moving flexible bar

and d are governed by

$$\frac{\partial^2 c(t,x)}{\partial t^2} = a^2 \frac{\partial^2 d(t,x)}{\partial x^2}$$

where a is a constant depending on the geometry and material of the bar. The boundary conditions corresponding to the control force F and the load with mass m are

$$a^2 \frac{\partial d(t,0)}{\partial x} = -F \ , \quad a^2 \frac{\partial d(t,\ell)}{\partial x} = -m \frac{\partial^2 c(t,\ell)}{\partial t^2}$$

Therefore the variable $Q(t,x) = e(t) + d(t,x)$ satisfies

$$\frac{\partial^2 Q(t,x)}{\partial t^2} = a^2 \frac{\partial^2 Q(t,x)}{\partial x^2} \tag{5.64}$$

with boundary conditions

$$a^2 \frac{\partial Q(t,0)}{\partial x} = -F \ , \quad a^2 \frac{\partial Q(t,\ell)}{\partial x} = -m \frac{\partial^2 Q(t,\ell)}{\partial t^2} \tag{5.65}$$

Consider the control F and $Q(t,\ell)$ (which differs from the position of the load by the constant ℓ), respectively as the system input $u(t)$ and output $y(t)$. Let us find the transfer function $W(p)$ via the Laplace transformation of (5.64), (5.65) with zero initial conditions

$$\begin{array}{ll} Q(0,x) = \dfrac{\partial Q(0,x)}{\partial t} = 0 \ , & p^2 \hat{Q}(p,x) = a^2 \hat{Q}''(p,x) \\ a^2 \hat{Q}^2(p,0) = -F(p) \ , & a^2 \hat{Q}'(p,\ell) = -mp^2 \hat{Q}(p,\ell) \end{array} \tag{5.66}$$

where $\hat{Q}(p,x) = \mathcal{L}Q(t,x)$ and $\hat{F}(p) = \mathcal{L}F(t)$. The solution of the boundary value problem (5.66) is

$$\hat{Q}(p,x) = \frac{(1 - mp/a)e^{-p(\ell-x)/a} + (1 + mp/a)e^{p(\ell-x)/a}}{(1 + mp/a)e^{p\ell/a} - (1 - mp/a)e^{-p\ell/a})} \hat{F}(p) \tag{5.67}$$

and $W(p)$ may be found from (5.67) with $x = \ell$

$$W(p) = \frac{2e^{-p\tau}}{ap((1 + mp/a) + (1 - mp/a)e^{-2p\tau}}$$

with $\tau = \ell/a$. The corresponding differential-difference equation may be presented in the form

$$m\ddot{y}(t) + m\dot{y}(t - 2\tau) + a\dot{y}(t) - a\dot{y}(t - 2\tau) = 2u(t - \tau)$$

Denoting $s_1(t) = y(t)$, $s_2(t) = \dot{y}(t)$, $s_3(t) = m\ddot{y}(t) + a\dot{y}(t)$ and $s_4(t) = 2as_2(t - \tau) - s_3(t - \tau)$ we obtain the motion equations

$$\dot{s}_1(t) = s_2(t) \ , \quad \dot{s}_2(t) = -\frac{a}{m} s_2(t) + \frac{1}{m} s_3(t) \tag{5.68}$$

$$s_3(t) = s_4(t - \tau) + 2u(t - \tau) \ , \quad s_4(t) = -s_3(t - \tau) + 2as_2(t - \tau) \tag{5.69}$$

The system (5.68), (5.69) may be treated as being in block-control form (Drakunov et al 1990a, 1990b) consisting of two blocks with $s_3(t)$ as a quasi-control in the first block (5.68) and the control $u(t)$ in the second block (5.69). In the block design procedure s is first designed to provide the desired eigenvalues in the first block. If $s_3(t) = k_1 s_1(t) + k_2 s_2(t)$ or $s(t) = s_3(t) - k_1 s_1(t) - k_2 s_2(t) = 0$, then a suitable choice of k_1, k_2 enables decay at the desired rate of the variable $s(t) = y(t) = Q(t,\ell)$ to be controlled. In the second step of the design procedure the control $u(t)$ is chosen to guarantee $s = 0$. The second block is governed by the difference equation (5.69) and according to the concept of sliding mode developed for discrete time systems it may also arise if some manifold consisting of trajectories is reached within a finite time interval.

Another way is to assign $s_3(t) = -M\,\mathrm{sign}\,(ks_1(t) + s_2(t))$. In this case $s(t) = s_3(t) + M\,\mathrm{sign}\,(ks_1(t) + s_2(t))$ and at first sliding occurs on the surface $s = 0$ and then on $ks_1(t) + s_2(t) = 0$. In (5.69) with control

$$u(t) = u_{eq}(t) = \frac{1}{2}(k_1 s_1(t+\tau) + k_2 s_2(t+\tau) + s_3(t-\tau) - 2as_2(t-\tau)) \quad (5.70)$$

or

$$u(t) = u_{eq}(t) = \frac{1}{2}(-M\,\mathrm{sign}\,(ks_1(t+\tau)+s_2(t+\tau))+s_3(t-\tau)-2as_2(t-\tau)) \quad (5.71)$$

the origin $s = 0$ is reached within a finite time $t < \tau$. If the control is bounded $|u(t)| \leq M$, then

$$u(t) = \begin{cases} u_{eq}(t) & \text{if} \quad |u_{eq}(t)| \leq M \\ M_1\,\mathrm{sign}\,u_{eq}(t) & \text{if} \quad |u_{eq}(t)| > M \end{cases}$$

and there exists an open domain containing the origin of the state space of the system (5.68), (5.69) so that, for all initial conditions from this domain, the sliding mode occurs along the manifold $s = 0$. The values of $s_1(t+\tau)$, $s_2(t+\tau)$ in (5.70), (5.71) can be calculated as a solution of (5.68) with known input $s_3(t)$ (right-hand side of (5.69)).

Let $\phi(\tau) = \exp(At)$ with

$$A = \begin{bmatrix} 0 & 1 \\ 0 & -\frac{a}{m} \end{bmatrix}$$

Then

$$\begin{bmatrix} s_1(t+\tau) \\ s_2(t+\tau) \end{bmatrix} = \phi(\tau)\begin{bmatrix} s_1(t) \\ s_2(t) \end{bmatrix} + \frac{1}{m}\int_t^{t+\tau} \phi(t+\tau-\xi) \times \begin{bmatrix} 0 \\ -s_3(\xi-2\tau) + 2u(\xi-\tau) - 2as_1(\xi-2\tau) \end{bmatrix} d\xi$$

The last term depends on the current value of control $u(\xi)$ and $s_3(\xi - \tau)$, $s_2(\xi - \tau)$, for $t - \tau < \xi < t$ in the τ-interval preceding t. If only $s_1(t)$ is accessible, the states s_2 and s_3 can be found using an asymptotic observer

$$\begin{aligned} \dot{\hat{s}}_1(t) &= \hat{s}_2(t) + L_1(\hat{s}_1(t) - y(t)) \\ \dot{\hat{s}}_2(t) &= -\frac{a}{m}\hat{s}_2(t) - \frac{1}{m}\hat{s}_3(t) + L_2(\hat{s}_1(t) - y(t)) \\ \dot{\hat{s}}_3(t) &= -\hat{s}_3(t-2\tau) + 2u(t-\tau) + 2a\hat{s}_1(t-2\tau) + L_3(\hat{s}_(t-2\tau) - y(t-2\tau)) \end{aligned}$$

By suitable choice of the input gains L_1, L_2, L_3 the convergence of the values of $\hat{s}_i$ to s_i with $t \to \infty$ may be achieved.

5.8 Conclusions

Wide use of digital controllers has placed onto the research agenda the generalization of sliding mode control methodology to discrete-time control systems. In the first studies, control algorithms intended for continuous- time systems were applied to discrete-time problems; resulting in chattering since the switching frequency can not exceed that of sampling. Then methods for reducing chattering were developed in many publications.

However, the fundamental question — what is the sliding mode in discrete-time systems? — was not considered. Discontinuous control in continuous-time systems may result in sliding in some manifold, while it results in chattering in discrete-time systems. The sliding mode may be originated in discrete-time systems with continuous control after a finite time interval, while any manifold with state trajectories may be reached asympotically only in continuous-time systems with continuous control (precisely speaking for systems governed by differential equations with Lipschitzian right-hand sides).

Design methods for sliding mode control for finite and infinite dimensional discrete-time and difference systems have been developed in this chapter. They enables decoupling of the overall dynamics into independent partial motions of lower dimension, and low sensitivity to system uncertainties. For all systems the motions are free of chattering, which has been the main obstacle for certain applications of discontinuous control action in systems governed by discrete and difference equations.

References

Drakunov, S.V., Izosimov, D.B., Luk'yanov A.G., Utkin V.A. and Utkin V.I. 1990a, Block control principle I. *Automation and Remote Control*, **51**, 601–609

Drakunov, S.V., Izosimov, D.B., Luk'yanov A.G., Utkin V.A. and Utkin V.I. 1990b, Block control principle II. *Automation and Remote Control*, **51**, 737–746

Drakunov, S.V., Utkin, V.I. 1990, Sliding mode in dynamic systems. *International Journal of Control* **55**, 1029–1037

Furuta, K. 1990, Sliding mode control of a discrete system. *Systems and Control Letters* , **14**, 145–152

Gantmacher, F.R. 1959, *The theory of matrices, Vol.1*, Chelsia, New York

Kokotović, P.V., O'Malley, R.B., Sannuti, P. 1976, Singular perturbations and order reduction in control theory. *Automatica* **12**, 123–132

Kotta, U. 1989, Comments on the stability of discrete-time sliding mode control systems. *IEEE Transactions on Automatic Control* **34**, 1021–1022.

Kwakernaak, H., Sivan R. 1972, *Linear optimal control systems*, Wiley Interscience, New York

Milosavljević, C. 1985, General conditions for the existence of a quasi-sliding mode on the switching hyperplane in discrete variable structure systems. *Automation and Remote Control* **46**, 679–684

Sarpturk, S.Z., Isteganopolis, Y., Kaynak O. 1987, On the stability of discrete-time sliding mode control systems. *IEEE Transactions on Automatic Control* **10** 930–932

Spurgeon, S.K. 1991, Sliding mode control design for uncertain discrete-time systems. *Proc IEEE Conference on Decision and Control,* , Brighton, England, 2136–2141

Utkin, V.I. 1992, *Sliding modes in control and optimization*, Springer-Verlag, Berlin.

Utkin, V.I., Orlov, Y.V. 1990, *Theory of infinite-dimensional control systems with sliding modes*, Nauka, Moscow (in Russian)

6. Generalized Sliding Modes for Manifold Control of Distributed Parameter Systems

Sergey Drakunov and Ümit Özgüner

6.1 Introduction

The traditional approach to control design for infinite dimensional systems is based upon the approximation of the system by a finite set of ordinary differential equations. Although standard this approach often leads to severe contradictions. An example is a simple system with delay

$$\dot{x} = u(t - \tau) \tag{6.1}$$

which if approximated by a finite dimensional system, seems to be stabilizable for any prescribed time interval, but in reality it cannot be stabilized faster than τ. This contradiction leads to unsatisfactory performance. The use of more adequate models allows us to obtain systems with better properties.

In this chapter we consider the problem of stabilizing distributed parameter systems. We base our approach on a concept which can be called *manifold control*. Its main features are as follows: the design procedure is divided into two steps. In the first step the manifold is designed in such a way that it will be integral for the closed-loop system. In the second step the control which forces the system to move along the manifold is found. The crucial point of the problem is to design the integral manifold which guarantees system stabilization. This work is based on earlier results reported by Drakunov and Utkin (1992) and Drakunov and Özgüner (1992).

Sliding mode control, which is now widely used for finite dimensional systems, can be considered as a predecessor of manifold control. It yields stable and robust closed-loop systems in the case of finite-dimensional plants, but the direct application of this design technique based on approximate models of distributed parameter systems may lead to undesirable closed-loop perfomance.

For example, in flexible structures the direct application of sliding mode control leads to to the excitation of high frequency modes neglected in the model. This phenomenon is due to the delays which are inherent in such systems. The presence of delays can be explained by wave propagation across the structure. If the control is applied to the boundary of the flexible structure, its action can influence other parts of the structure only after the wave, caused by the actuation, propagates and reaches those parts. For the case of large flexible structures these delays are not small and cannot be neglected (see Young,

Özgüner and Xu 1993). The naive application of sliding mode control ignoring these effects leads to chattering and may not be successful.

We shall consider mathematical models in the form of partial differential equations (PDE's) which allow us us to take into account the features described above and therefore design more appropriate control algorithms. The generalization of the sliding mode control concept to systems with delays and for more general dynamic systems described by semigroups of state space transformations was originally considered by Drakunov and Utkin (1991, 1992). Here we introduce the linear transformation of the state variable so as to address the problem in a simpler setting. The nondispersive wave equation is chosen as a canonical form for distributed parameter systems described by partial differential equations. Since for the many cases the nondispersive wave equation is equivalent to a system with delay, this allows the transformed system to use the control algorithms based on the manifold approach developed earlier for systems of differential-difference equations (Drakunov and Utkin 1992).

The problem of designing the control law which assigns the desired stable integral manifold to the system can be solved by using various methods, including linear techniques; the use of sliding modes makes the closed-loop system highly insensitive to external disturbances and parameter variations.

6.2 Manifold Control: Generalization of the Sliding Mode Control Concept

There are two aspects in traditional sliding mode control design: the choice of the sliding surface and synthesis of the control law for the the reduced order problem. From the point of view of dynamic system theory, the sliding surface is just a stable integral manifold of the closed-loop system, with the specific property that in the area of attraction the system state is absorbed by the manifold in finite time.

For finite dimensional closed-loop systems in $\mathbb{R}^n$ modelled as

$$\dot{x} = f(t, x) \tag{6.2}$$

such manifolds can exist only if the right hand side does not satisfy the well known Lipshitz condition

$$|f(t, x) - f(t, y)| < L|x - y| \tag{6.3}$$

which is usually required to guarantee the uniqueness of the solution both for $t \geq t_0$ and $t < t_0$.

Let $F(t; t_0, x_0)$ be a solution of the system of ordinary differential equations (6.2) with initial condition $x(t_0) = x_0$, i.e. a transition function. Then the Lipshitz condition implies that F is defined for $t \geq t_0$ and $t < t_0$. The family of state space transformations $F(t; t_0, .)$ is a group with respect to the composition operation. The inverse element to $F(t; t_0, .)$ is $F(t_0; t, .)$. For an asymptotically stable integral manifold, the trajectory initiated in its vicinity tends to, but never reaches it.

In contrast to the equations whose right hand side satisfies the Lipshitz condition, in systems with discontinuities there are integral manifolds which can be reached in finite time. Consider the system in $\mathbb{R}^n$

$$\dot{x} = f(t, x) + B(t, x)u \tag{6.4}$$

where $f(x)$, $B(x)$ are functions which satisfy the Lipshitz condition, and $u \in \mathbb{R}^m$ is discontinuous on the smooth surfaces $\{x : s_i(x) = 0\}$ $i = 1, 2, \ldots, m$ in $\mathbb{R}^n$

$$u_i(x) = \begin{cases} u_i^+(x) & \text{if } s_i(x) > 0 \\ u_i^-(x) & \text{if } s_i(x) < 0 \end{cases} \tag{6.5}$$

If the sliding mode exists on the intersection of discontinuity surfaces $\sigma = \bigcap_{i=1}^m \{x : s_i(x) = 0\}$ then

(i) σ is an integral manifold since it consists of system trajectories

(ii) the uniqueness of the inverse of the shift operator $F(t; t_0, .)$ does not hold on σ since each point on σ can be reached in at least two different ways: from outside (because of the finite time of convergence) and from points on the manifold itself.

These are characteristic features of the sliding mode and they are taken as a basis for the generalization of the sliding mode concept.

The general definition of a dynamic system in any metric space $\mathcal{X}$ (including those described by partial differential equations) utilizes a description in the form of a transition operator F similar to the one considered above. Generally $F(t; t_0, .)$ (which may represent the desired closed-loop system) is a two-parameter family of state space transformations $F(t; t_0, .) : \mathcal{X} \to \mathcal{X}$, where F satisfies the conditions

$$F(t; t_1, F(t_1; t_0, x)) = F(t; t_0, x) \tag{6.6}$$

$$F(t; t, x) = x \tag{6.7}$$

for all $t > t_1 > t_0$, $x \in \mathcal{X}$, i.e. the set of all $F(t; t_0, .)$ for $t \geq t_0$ constitutes at least a semigroup. In order for this set to be considered as a group, the uniqueness of the inverse F^{-1} operator is needed. But as we demonstrated above, this uniqueness may be violated at sliding manifolds for the finite dimensional cases. This is the reason why, for the general case, such manifolds are also called sliding manifolds.
The following definition was introduced by Drakunov and Utkin (1992).

Definition 6.1 *$x \in \mathcal{X}$ is said to be a sliding point at the time instant t if for every $t_0 < t$, the equation $F(t; t_0, \xi) = x$ has more than one solution ξ.*

This definition implies that sliding manifolds are asymptotically stable manifolds to which the system state converges in finite time from any initial condition in the area of attraction.

The underlying philosophy of the proposed approach as applied to PDE models of distributed parameter systems, is the same as in any sliding mode control design for finite dimensional systems. After representing the system in a "convenient" form, a sliding manifold is chosen and then the control is designed such that the system state reaches this manifold in finite time and then "slides" along it. The control for this case is not necessarily discontinous.

As a first example we consider control of a linear continuous-time difference system or of neutral type (Bellman and Cooke 1963)

$$x(t) = Ax(t-\tau) + Bu(t-\tau) \tag{6.8}$$

where $x \in \mathbb{R}^n$, $u \in \mathbb{R}^m$. In contrast to the ordinary differential equation case, there exists a linear control $u = Gx$ such that the closed-loop system

$$x(t) = (A+BG)x(t-\tau) \tag{6.9}$$

has stable integral manifolds σ reachable in finite time.

The state space $\mathcal{X}$ of the system (6.9) is a set of functions with values in $\mathbb{R}^n$ and defined on the interval $\theta \in (-\tau, 0]$, so it is infinite dimensional. The current state at the instant t can be interpreted as the trajectory $x(t+\theta)$ for $-\tau < \theta \leq 0$. Let there exist a nonzero matrix C such that

$$C(A+BG) = 0 \tag{6.10}$$

Consider the subset σ of $\mathcal{X}$ consisting of the functions $x(\theta)$ belonging to the null space of C

$$\sigma = \{x : Cx = 0\} \subset \mathcal{X} \tag{6.11}$$

Then σ is an integral manifold for (6.9) and every trajectory is absorbed by this manifold in finite time. The condition (6.10) implies that $\det(A+BG) = 0$ and therefore the corresponding transition function is not uniquely invertible on σ; thus it satisfies Definition 6.1 and can be called a sliding manifold. As in traditional sliding modes, the motion on the manifold is described by lower order equations and the equivalent control method can be used.

6.3 Canonical Form of the Distributed Parameter System

In the theory of time-invariant finite-dimensional systems one of the major tools of control design is the transformation of the system to different canonical forms. Analogously, we consider linear transformations of time-invariant distributed parameter system. For distributed parameter systems described by partial differential equations the nondispersive wave equation serves as a canonical form. Since for many cases the nondispersive wave equation is equivalent to the system with delay, this allows us to use control algorithms with generalized sliding for the transformed system.

6.3.1 Problem Statement

We consider the case of systems described by the partial differential equations

$$\frac{\partial^i Q(t,x)}{\partial t^i} = \mathcal{L}Q(t,x) \tag{6.12}$$

with first or second order time partial derivatives and up to Kth order partial derivatives with respect to the multidimensional spatial variable x taking values in the bounded domain $\Omega \subset \mathbb{R}^N$ with a smooth boundary $\partial\Omega$. So $i = 1$ or $i = 2$ and $\mathcal{L}$ is a differential operator of order K

$$\mathcal{L}Q(t,x) = \sum_{i=1}^{K} \sum_{k_1+\ldots+k_N=i} a_{k_1,\ldots,k_N}(x) \frac{\partial^i}{\partial^{k_1}x_1 \ldots \partial^{k_N}x_N} Q(t,x) \tag{6.13}$$

where $k_1 \geq 0, \ldots, k_N \geq 0$. The boundary conditions for the equation (6.12) are of the form

$$\Gamma Q(t,x)|_{x\in\partial\Omega} = B(x)|_{x\in\partial\Omega} u(t) \tag{6.14}$$

where Γ is the differential operator similar to (6.13) of order $K-1$, and $u \in \mathbb{R}^m$, $0 \leq m \leq K-1$.

The other case considered is when a finite dimensional control variable $u \in \mathbb{R}^m$ enters the right hand side of (6.12)

$$\frac{\partial^i Q(t,x)}{\partial t^i} = \mathcal{L}Q(t,x) + B(x)u(t) \tag{6.15}$$

For this case we consider the homogeneous boundary conditions

$$\Gamma Q(t,x)|_{x\in\partial\Omega} = 0 \tag{6.16}$$

The initial conditions are

$$Q(0,x) = Q_0(x) \ , \qquad i = 0 \tag{6.17}$$

$$Q(0,x) = Q_0(x) \ , \qquad i = 2 \tag{6.18}$$

$$\frac{\partial}{\partial t} Q(0,x) = Q_1(x) \tag{6.19}$$

The objective of the control design is to find the control which stabilizes the system. To correctly define the solution of (6.12), (6.14) or (6.15), (6.16) one needs to describe the classes of functions $a_{k_1,\ldots,k_N}(x)$, $B(x)$ and $u(t)$. Since as a result of our control design the variable $u(t)$ can be a discontionuous function, the natural class of permissible controls is L_2 on any finite interval $[0,T]$. That leads to the necessity to understand the solutions as generalized functions or distributions.

It is assumed that the corresponding boundary value problem is well posed and its solution for any $u(t) \in L_2([0,T])$ is an element of a Sobolev space $H_2^1([0,T] \times \Omega)$. According to the *Lions Trace Theorem* in $H^1(\Omega)$ (see Lions

1972) the equality (6.14) can be continuously extended in a unique manner from $C^\infty(\bar{\Omega})$ onto the Sobolev space of fractional order $H_2^{\frac{1}{2}}(\partial\Omega)$. So the natural assumption on the matrix valued function $B(x)$ is such that each of its components belong to $H_2^{\frac{1}{2}}(\partial\Omega)$. For the case of (6.15) we assume that the components of $B(x)$ belong to $L_2(\Omega)$.

The initial functions Q_0 and Q_1 are assumed to be elements of $H_2^1(\Omega)$. This problem statement embraces many physical models described by partial differential equations; e.g. dispersive and nondispersive waves, diffusions, plate vibrations.

6.3.2 Linear Transformation

We use a linear transformation of the variable Q into the variable P. Since H_2^1 is a Hilbert space, according to the Riez theorem the general form of the transformation is an inner product. So, without loss of generality we can define P as

$$P(t,\xi) = \int_\Omega \mathcal{D}(\xi,x)Q(t,x)\,\mathrm{d}x \tag{6.20}$$

where ξ is a new spatial variable. We will show that it is sufficient to consider one dimensional ξ.

Let $\mathcal{D}$ satisfy an adjoint to (6.12) or (6.15), the homogeneous boundary value problem

$$\frac{\partial^i \mathcal{D}(\xi,x)}{\partial \xi^i} = \mathcal{L}^*\mathcal{D}(\xi,x) \tag{6.21}$$

$$\Gamma^*\mathcal{D}(\xi,x)|_{x\in\partial\Omega} = 0 \tag{6.22}$$

with initial conditions

$$\mathcal{D}(0,x) = \mathcal{D}_0(x)\ , \qquad i=0 \tag{6.23}$$

$$\mathcal{D}(0,x) = \mathcal{D}_0(x)\ , \qquad i=2 \tag{6.24}$$

$$\frac{\partial}{\partial\xi}\mathcal{D}(0,x) = \mathcal{D}_1(x) \tag{6.25}$$

By using Green's formula (see Treves (1975)) it can be shown that $P(\xi,x)$ satisfies

$$\frac{\partial^i P(t,\xi)}{\partial t^i} = \frac{\partial^i P(t,\xi)}{\partial \xi^i} + \varphi(\xi)u(t) \tag{6.26}$$

where $\varphi(\xi)$ is a generalized function (distribution) defined by the trace of the adjoint variable $\mathcal{D}(\xi,x)$ or its spatial derivatives on the boundary $\partial\Omega$ (its form depends on the type of boundary operator Γ).

For $i=1$ or $i=2$ (6.26) is a hyperbolic partial differential equation which plays the role of a canonical form for the distributed parameter system. The integral transform makes it possible to split the initial control design problem into two parts; the first being the problem to find the kernel $\mathcal{D}$ of the transform

(6.21) does not depend on the control variable, so this problem can be solved off-line); and the second part is the design of the stabilizing control for the first or the second order nondispersive wave equation (6.26).

The important point here is that the transformed problem (6.26) always has only a one-dimensional spatial variable ξ, even if x in the original problem was multidimensional. Moreover, for many cases (6.26) can be written in the equivalent form of a differential-difference system to be described below.

The initial conditions for (6.26) are defined by initial conditions for the variable Q

$$P(0,\xi) = \int_{\Omega} \mathcal{D}(\xi,x)Q_0(x)\,\mathrm{d}x \tag{6.27}$$

$$\frac{\mathrm{d}}{dt}P(0,\xi) = \int_{\Omega} \mathcal{D}(\xi,x)Q_1(x)\,\mathrm{d}x \tag{6.28}$$

To define the solution of (6.26) uniquely one needs additional conditions (one for $i=1$, and two for $i=2$). The possibility to assign any $\mathcal{D}_0$ and $\mathcal{D}_1$, within the class of functions which give the nonsingular transformation $\mathcal{D}$ (satisfying assumptions of the Theorem 6.2 in Sect. 6.3.3), provides a degree of freedom for the "convenient" choice of these conditions and will be discussed in the subsequent sections for particular cases.

6.3.3 Nonsingularity of the Integral Transform

The main question concerning transformation (6.20) is its invertibility so that the stabilization of (6.26) implies the stabilization of (6.12). We will show now how the nonsingularity condition can be satisfied.

Let us assume that the operators $\mathcal{L}$ and Γ are selfadjoint

$$\mathcal{L}^* = \mathcal{L} \tag{6.29}$$

$$\Gamma^* = \Gamma \tag{6.30}$$

and that there exists a modal expansion of the solution of (6.12), (6.14) or (6.15), (6.16)

$$Q(t,x) = \sum_{k=1}^{\infty} Q_k(t)X_k(x) \tag{6.31}$$

converging in $H_1^2([0,T]\times\Omega)$, where $X_k(x)$ are the spatial modes (eigenfunctions) of the boundary value problem for differential operator $\mathcal{L}$. The functions $X_k(x)$ and the corresponding eigenvalues are defined by the equations

$$\mathcal{L}X_k(x) = \lambda_k X_k(x) \tag{6.32}$$

$$\Gamma X_k(x)|_{x\in\partial\Omega} = 0 \tag{6.33}$$

The functions $Q_k(t)$ are time modes corresponding to the eigenvalues λ_k which satisfy the equation

$$\frac{\mathrm{d}^i}{\mathrm{d}t^i} Q_k(t) + \lambda_k Q_k(t) = b_k u \tag{6.34}$$

The problem of the stabilization of $Q(t,x)$ for all $x \in \Omega$ is equivalent to the problem of stabilization of $Q_k(t)$ for all $k = 1, 2, \ldots$.

Since $\mathcal{D}$ satisfies the homogeneous boundary value problem (6.21), (6.22) and the operators $\mathcal{L}$, Γ are selfadjoint, the corresponding modal expansion for $\mathcal{D}$ has the similar form

$$\mathcal{D}(\xi, x) = \sum_{k=1}^{\infty} D_k(\xi) X_k(x) \tag{6.35}$$

where the $X_k(x)$ are the same as in (6.31). The functions $D_k(\xi)$ are the solutions of an infinite set of homogeneous equations

$$\frac{\mathrm{d}^i}{\mathrm{d}\xi^i} D_k(\xi) + \lambda_k D_k(\xi) = 0 \tag{6.36}$$

If $i = 1$ there is one inital condition for each k

$$D_k(0) = d_k \tag{6.37}$$

where

$$d_k = \frac{1}{\int_\Omega X_k^2(x)\,\mathrm{d}x} \int_\Omega \mathcal{D}_0(x) X_k(x)\,\mathrm{d}x \tag{6.38}$$

and for the case $i = 2$ two conditions

$$\mathcal{D}_k(0) = d_{0k} \tag{6.39}$$

$$\dot{\mathcal{D}}_k(0) = d_{1k} \tag{6.40}$$

where

$$d_{0k} = \frac{1}{\int_\Omega X_k^2(x)\,\mathrm{d}x} \int_\Omega \mathcal{D}_0(x) X_k(x)\,\mathrm{d}x \tag{6.41}$$

$$d_{1k} = \frac{1}{\int_\Omega X_k^2(x)\,\mathrm{d}x} \int_\Omega \mathcal{D}_1(x) X_k(x)\,\mathrm{d}x \tag{6.42}$$

These initial conditions should be chosen in such a way that the corresponding transform

$$P(t, \xi) = \int_\Omega \mathcal{D}(\xi, x) Q(t, x)\,\mathrm{d}x \tag{6.43}$$

is nonsingular, i.e. the stabilization of P will imply stabilization of Q.

Theorem 6.2 *The transformation (6.43) is nonsingular if and only if*

(i) for the case $i = 1$, for all $k = 0, 1, \dots$

$$d_k \neq 0 \tag{6.44}$$

(ii) for the case $i = 2$, for all $k = 0, 1, \dots$

$$d_{k0}^2 + d_{k1}^2 \neq 0 \tag{6.45}$$

Proof. Substituting (6.31) and (6.35) into (6.43) and using the orthogonality of $X_i(x)$ and $X_j(x)$ for $i \neq j$, we obtain

$$P(t,\xi) = \sum_{k=1}^{\infty} Q_k(t)\mathcal{D}_k(\xi) \tag{6.46}$$

If for all k the initial conditions (6.37), or (6.39) and (6.40) are nontrivial, then all $D_k(\xi)$ are nontrivial and since they are linearly independent, the stabilization of $P(t,\xi)$ implies that all $Q_k(t)$ tend to zero, when $t \to \infty$. Therefore, $Q(t,x)$ is also stabilized. On the other hand, if for some k the initial conditions (6.37) or both (6.39), (6.40), are zero, then the stabilization of $P(t,\xi)$ does not guarantee that the corresponding mode $Q_k(t)$ is stabilized since for this case the expansion (6.31) does not contain Q_k. □

We have proved that the neccesary and sufficient condition for the transform (6.43) to be nonsingular, is the requirement that all the modes must be excited in the solution of the adjoint boundary value problem which provides the kernel $\mathcal{D}$.

6.4 Manifold Control of Differential-Difference Systems

As will be seen in the examples below, the stabilization of (6.26) is often equivalent to the stabilization of differential-difference system. In this section we consider the class of such systems and provide the stabilizing control design based on manifold control with generalized sliding.

The systems under study have the block form (Drakunov et al 1990) comprized of blocks of differential equations coupled with blocks of difference equations of type (6.8). We consider two configurations

Configuration A

$$\dot{x}(t) = A_{11}x(t) + A_{12}z(t) \tag{6.47}$$
$$z(t) = A_{21}x(t-\tau) + A_{22}z(t-\tau) + B_0u(t-\tau) \tag{6.48}$$

Configuration B

$$z(t) = A_{11}z(t-\tau) + A_{12}x(t-\tau) \qquad (6.49)$$
$$\dot{x}(t) = A_{21}z(t) + A_{22}x(t) + B_0u(t) \qquad (6.50)$$

It is assumed that $x \in \mathbb{R}^{n_1}$, $z \in \mathbb{R}^{n_2}$ and $u \in \mathbb{R}^m$. A_{11}, A_{12}, A_{21}, A_{22}, B_0 are matrices of appropriate dimensions. Since A_{12} is not necessarily a full rank matrix, we can use the representation

$$A_{12} = B_1C_2 \qquad (6.51)$$

where B_1 and C_2 have full column and row rank respectively. We assume that the pairs (A_{11}, B_1) in both types of configurations are controllable and the systems (C_2, A_{22}, B_0) are invertible. Denoting $v = C_2z$ we consider v as a new control variable in (6.47) and (6.49), and as the output variable for (6.48) and (6.50). In the first block of **Configuration A**

$$\dot{x}(t) = A_{11}x(t) + B_1v(t) \qquad (6.52)$$

we use sliding mode control $v(x) = \mathrm{col}\,(v_1, \ldots, v_{n2})$ to provide stability

$$v_i(x) = \begin{cases} v_i^+(x) & \text{if } s_i(x) > 0 \\ v_i^-(x) & \text{if } s_i(x) < 0 \end{cases} \qquad (6.53)$$

The sliding manifold for this block is

$$\sigma_0 = \{x : s(x) = 0\} \qquad (6.54)$$

where $s = \mathrm{col}\,(s_1, \ldots, s_{n2})$.

Since the second block (6.48) is a difference system, under the invertibility condition it is possible to find a control u which provides the desired discontinuous function for the output $v(x) = C_2z$ (Drakunov and Utkin 1992). This equality defines another manifold in the system state space similar to (6.11)

$$\sigma_1 = \{(x, z) : v(x) - C_2z = 0\} \qquad (6.55)$$

As a result, the state slides on the intersection $\sigma = \sigma_0 \cap \sigma_1$. That means we can assign any desired rate of stability for the closed-loop system (6.47)-(6.48).

For the system in **Configuration B** the design procedure is different. On the first step we find a linear control $v = Dz(t)$ for the first block (6.49)

$$z(t) = A_{11}z(t-\tau) + B_1v(t-\tau) \qquad (6.56)$$

to establish stability for the difference system (6.49), which can be achieved by using the sliding mode control on some manifold σ_2 system as described earlier. Then the problem is to steer $s = Dz(t) - C_2x(t)$ to zero or to reach the manifold

$$\sigma_3 = \{(x, z) : s(x, z) = 0\} \qquad (6.57)$$

We solve this problem in the class of sliding mode control algorithms

$$u_i(x) = \begin{cases} u_i^+(x) & \text{if } s_i(x) > 0 \\ u_i^-(x) & \text{if } s_i(x) < 0 \end{cases} \qquad (6.58)$$

The resulting motion again will occur on the intersection $\sigma = \sigma_2 \cap \sigma_3$.

6.5 Wave Equation

As demonstrated in Sect. 6.3, the stabilization problem for a wide class of distributed parameter systems is equivalent to that of the nondispersive wave equation. In this section we show how this equation can be converted into differential-difference form and then stabilized by applying the approach developed in the previous section. For the dispersive equation the integral transform shown earlier allows us to solve the problem for the general case.

6.5.1 Suppressing Vibrations of a Flexible Rod

As a first example we study the longitudinal or torsional oscillations of a flexible rod. The control is assumed to be a force or torque applied at one end of the rod, the other end is free. Let Q be the displacement of the rod from the unexcited position. We then have the following equations for a unit rod with normalized parameters (Meirovitch 1986)

$$\frac{\partial^2 Q(t,x)}{\partial t^2} = \frac{\partial^2 Q(t,x)}{\partial x^2} \tag{6.59}$$

$$\frac{\partial Q(t,0)}{\partial x} = -u(t) \tag{6.60}$$

$$\frac{\partial Q(t,1)}{\partial x} = 0 \tag{6.61}$$

where x is the position along the rod and $u(t)$ denotes the actuation force or torque. The problem (6.59)-(6.61) has the "canonical" form (6.26) with $P = Q$, $\xi = x$ and φ a delta function

$$\varphi(\xi) = \delta(\xi) \tag{6.62}$$

Applying the Laplace transform to (6.59) and boundary conditions (6.60), (6.61) with the zero initial conditions

$$Q(0,x) = 0 \tag{6.63}$$

$$\frac{\partial}{\partial x} Q(0,x) = 0 \tag{6.64}$$

we have

$$p^2 \hat{Q}(p,x) = \hat{Q}''(p,x) \tag{6.65}$$

$$\hat{Q}'(p,0) = -\hat{u}(p) \tag{6.66}$$

$$\hat{Q}'(p,1) = 0 \tag{6.67}$$

where $\hat{Q}(p,x) = \mathcal{L}Q(t,x)$ and $\hat{u}(p) = \mathcal{L}u(t)$. The solution of this boundary value problem for the ordinary differential equation (6.65) is

$$\hat{Q}(p,x) = \frac{e^{p(x-1)} + e^{-p(x-1)}}{e^p - e^{-p}} \frac{1}{p} \hat{u}(p) \tag{6.68}$$

The solution of the stabilization problem depends greatly on what point of the rod is considered as the system output. We shall consider the free end of the rod as an output, i.e. the noncollocated actuator/sensor case

$$y(t) = Q(t,1) \tag{6.69}$$

From (6.68) we obtain

$$\hat{y}(p) = \hat{Q}(p,1) = \frac{2}{e^p - e^{-p}} \frac{1}{p} \hat{u}(p) \tag{6.70}$$

In the time domain the correspondence between $u(t)$ and $y(t)$ may be written as

$$\dot{y}(t+1) - \dot{y}(t-1) = 2u(t) \tag{6.71}$$

or

$$\dot{y}(t) - \dot{y}(t-2) = 2u(t-1) \tag{6.72}$$

We can now write this equation in the form of the differential-difference system in **Configuration A** by introducing a new variable z

$$\dot{y}(t) = z(t) \tag{6.73}$$

$$z(t) = z(t-2) + 2u(t-1) \tag{6.74}$$

The block representation of the differential-difference system simplifies the developement of the control algorithm. Considering the variable $z(t)$ in the first block (6.73) as the control, we can obtain sliding mode by assigning

$$z(t) = -\lambda \operatorname{sgn} y(t) \tag{6.75}$$

This equality is valid if

$$s(t) = z(t) + \lambda \operatorname{sgn} y(t) = 0 \tag{6.76}$$

To achieve the above, we can use the control

$$u(t) = -\frac{1}{2} z(t-1) - \frac{1}{2} \lambda \operatorname{sgn} y(t+1) \tag{6.77}$$

This control algorithm seems to be noncausal, however using an extrapolator it can in fact be realized as an operator on the current and the past values of the control variables. To demonstrate this, solve (6.73) taking $y(t)$ as the initial condition

$$y(t+1) = y(t) + \int_t^{t+1} z(\tau)\, d\tau \tag{6.78}$$

or using (6.74)

$$y(t+1) = y(t) + \int_{t-1}^{t} (z(\tau-1) + 2u(\tau))\, d\tau \tag{6.79}$$

$$= y(t) + y(t-1) - y(t-2) + 2\int_{t-1}^{t} u(\tau)\, d\tau \tag{6.80}$$

since $\dot{y}(t) = z(t)$. Substituting $y(t+1)$ from this expression into (6.77) and again using the fact that $z(t) = \dot{y}(t)$, we obtain

$$u(t) = -\frac{1}{2}\dot{y}(t-1) - \frac{1}{2}\lambda\,\mathrm{sgn}\,(y(t) + y(t-1) - y(t-2) + 2\int_{t-1}^{t} u(\tau)\,d\tau) \quad (6.81)$$

With this control the system (6.73),(6.74) and therefore (6.59) is stabilized in finite time.

Another possibility is to represent the equation (6.72) in the form of **Configuration B** as

$$y(t) = y(t-2) + 2v(t-1) \quad (6.82)$$

$$\dot{v}(t) = u(t) \quad (6.83)$$

To stabilize the difference system by means of the control variable v we need to drive the variable $s(t) = (1-\lambda)y(t-1) + 2v(t)$, where $|\lambda| < 1$, to zero. The equality $s(t) = 0$ can be considered as a sliding manifold for the second block (6.83). If the control is

$$u(t) = -\mu\,\mathrm{sgn}\,(2v(t) + (1-\lambda)y(t-1)) - (1-\lambda)\dot{y}(t-1) \quad (6.84)$$

then

$$\dot{s} = -2\mu\,\mathrm{sgn}\,s \quad (6.85)$$

Therefore we will have $s = 0$ in finite time.

6.5.2 Rod with Additional Mass

Consider the case when a unit mass is attached to the right end of the rod (Drakunov and Utkin 1992)

$$\frac{\partial^2 Q(t,x)}{\partial t^2} = \frac{\partial^2 Q(t,x)}{\partial x^2} \quad (6.86)$$

$$\frac{\partial Q(t,0)}{\partial x} = -u(t) \quad (6.87)$$

$$\frac{\partial Q(t,1)}{\partial x} = -\frac{\partial^2 Q(t,1)}{\partial t^2} \quad (6.88)$$

Again applying the Laplace transform to the equation (6.86) with boundary conditions (6.87) and (6.88), we obtain

$$p^2\hat{Q}(p,x) = \hat{Q}''(p,x) \quad (6.89)$$

$$\hat{Q}'(p,0) = -\hat{u}(p) \quad (6.90)$$

$$\hat{Q}'(p,1) = -p^2\hat{Q}(p,1) \quad (6.91)$$

The solution of this boundary value problem is

$$\hat{Q}(p,x) = \frac{(1-p)e^{p(x-1)} + (1+p)e^{-p(x-1)}}{(1+p)e^{p} - (1-p)e^{-p}} \frac{1}{p} \hat{u}(p) \tag{6.92}$$

If $Q(t,1)$ is the output variable

$$y(t) = Q(t,1) \tag{6.93}$$

then from (6.92) it follows that

$$\hat{y}(p) = \hat{Q}(p,1) = \frac{2}{(1+p)e^{p} - (1-p)e^{-p}} \frac{1}{p} \hat{u}(p) \tag{6.94}$$

The corresponding differential-difference equation is

$$\ddot{y}(t) + \ddot{y}(t-2) + \dot{y}(t) - \dot{y}(t-2) = 2u(t-1) \tag{6.95}$$

Denoting $x_1(t) = y(t)$, $x_2(t) = \dot{y}(t)$, $z_1(t) = \ddot{y}(t) + \dot{y}(t)$, $z_2(t) = 2x_2(t-1) - z_1(t-1)$ we obtain the system in the block form of **Configuration A**

$$\dot{x}_1(t) = x_2(t) \tag{6.96}$$

$$\dot{x}_2(t) = -x_2(t) + z_1(t) \tag{6.97}$$

$$z_1(t) = z_2(t-1) + 2u(t-1) \tag{6.98}$$

$$z_2(t) = -z_1(t-1) + 2x_2(t-1) \tag{6.99}$$

If

$$z_1(t) = -\mu \operatorname{sgn}(\lambda x_1(t) + x_2(t)) \tag{6.100}$$

then sliding mode occurs in the first block (6.96), (6.97) and $\dot{x}_1(t) = -\lambda x_1(t)$. Therefore x_1 tends to zero with the desired rate. The equality (6.100) will be valid if the control

$$u(t) = -\frac{1}{2} z_2(t) - \frac{1}{2} \mu \operatorname{sgn}(\lambda x_1(t+1) + x_2(t+1)) \tag{6.101}$$

is used. The values $x_1(t+1)$ and $x_2(t+1)$ can be obtained as a solution of (6.96), (6.97).

Let $\Phi(\tau) = \exp(A\tau)$, where A is a matrix of the linear system (6.96), (6.97)

$$A = \begin{bmatrix} 0 & 1 \\ 0 & -1 \end{bmatrix} \tag{6.102}$$

then

$$\begin{bmatrix} x_1(t+1) \\ x_2(t+1) \end{bmatrix} = \Phi(1) \begin{bmatrix} x_1(t) \\ x_2(t) \end{bmatrix} + \int_t^{t+1} \Phi(t+1-\tau) \begin{bmatrix} 0 \\ z_1(\tau-2) + 2u(\tau-1) - 2x_2(\tau-2) \end{bmatrix} d\tau \tag{6.103}$$

The last term depends only on the current values of control u and $x_1(\tau-1)$, $x_2(\tau-1)$ for $t-1 < \tau < t$ in the 1-interval preceeding t.

Since only $y(t) = x_1(t)$ is accessible for the measurement, an observer can be used for estimating x_2, z_1, z_2

$$\dot{\hat{x}}_1(t) = \hat{x}_2(t) + L_1(\hat{x}_1(t) - y(t)) \tag{6.104}$$
$$\dot{\hat{x}}_2(t) = -\hat{x}_2(t) + \hat{z}_1(t) + L_2(\hat{x}_1(t) - y(t)) \tag{6.105}$$
$$\hat{z}_1(t) = \hat{z}_2(t-1) + 2u(t-1) + L_3(\hat{x}_1(t) - y(t)) \tag{6.106}$$
$$\hat{z}_2(t) = -\hat{z}_1(t-1) + 2\hat{x}_2(t-1) + L_4(\hat{x}_1(t) - y(t)) \tag{6.107}$$

By a proper choice of the gains L_i we can obtain the convergence of the observer.

6.5.3 Semi-Infinite Rod with Distributed Control

Consider the semi-infinite rod with free left end and a scalar control force u distributed along it in accordance with the density function $\varphi(x)$. The equations describing the rod are (Meirovitch 1986)

$$\frac{\partial^2 Q(t,x)}{\partial t^2} = \frac{\partial^2 Q(t,x)}{\partial x^2} + \varphi(x)u(t) \tag{6.108}$$
$$\frac{\partial Q(t,0)}{\partial x} = 0 \tag{6.109}$$
$$\lim_{x\to\infty} \frac{\partial Q(t,x)}{\partial x} = 0 \tag{6.110}$$

where $0 \le x < \infty$.

It can be shown that the problem of stabilizing the output $y = Q(t, x_0)$ is also equivalent to the problem of stabilizing a set of differential-difference equations for some particular functions $\varphi(x)$; exponential, trigonometric functions or their linear combinations. When the function $\varphi(x)$ cannot be chosen within the described class, it may be possible to approximate it by these linear combinations.

Applying the Laplace transform to the equation (6.108) and boundary conditions (6.109) we obtain

$$p^2\hat{Q}(p,x) = \hat{Q}''(p,x) + \varphi(x)\hat{u}(p) \tag{6.111}$$
$$\hat{Q}'(p,0) = 0 \tag{6.112}$$
$$\lim_{x\to\infty} \hat{Q}'(p,x) = 0 \tag{6.113}$$

For

$$\varphi(x) = e^{\alpha x} \tag{6.114}$$

assuming that $\alpha < 0$, the solution of this boundary value problem is

$$\hat{Q}(p,x) = \frac{\alpha e^{-px} + p e^{\alpha x}}{p(p^2 - \alpha^2)} \hat{u}(p) \tag{6.115}$$

which for a fixed $x = \tau$ corresponds to the differential-difference equation

$$y_\tau^{(3)}(t) - \alpha^2 \dot{y}_\tau(t) = \alpha u(t-\tau) + e^{\alpha\tau}\dot{u}(t) \tag{6.116}$$

where $y_\tau(t) = Q(t,\tau)$. If $\tau = 0$

$$y(t) = Q(t,0) \tag{6.117}$$

from (6.115) it follows that

$$\hat{y}(p) = \hat{Q}(p,0) = \frac{p+\alpha}{p(p^2-\alpha^2)}\hat{u}(p) \tag{6.118}$$

The transfer function (6.118) relating the input and output variables is just a rational transfer function of the third order (the cancellation of the common factor in the numerator and denominator cannot be done, as it will result in a system which is not equivalent to the original one). Therefore, the state space representation of (6.118) is

$$\dot{x}_1 = x_2 \tag{6.119}$$

$$\dot{x}_2 = x_3 + u \tag{6.120}$$

$$\dot{x}_3 = \alpha^2 x_2 + \alpha u(t) \tag{6.121}$$

where $x_1 = y$. The standard sliding mode control

$$u(t) = -\mu \operatorname{sgn}(k_1 x_1 + k_2 x_2 + k_3 x_3) \tag{6.122}$$

can be used to stabilize this system. The coefficients k_1, k_2, k_3 are chosen so that the system in the sliding mode is stable. Again to obtain the values x_2 from measurements of $y = x_1$, the observer

$$\dot{\hat{x}}_1 = \hat{x}_2 + L_1(y - \hat{x}_1) \tag{6.123}$$

$$\dot{\hat{x}}_2 = \hat{x}_3 + u + L_2(y - \hat{x}_1) \tag{6.124}$$

$$\dot{\hat{x}}_2 = \alpha^2 \hat{x}_2 + \alpha u(t) + L_3(y - \hat{x}_1) \tag{6.125}$$

can be used.

6.5.4 Dispersive Wave Equation

Consider now the more general wave equation

$$\frac{\partial^2 Q(t,x)}{\partial t^2} = a(x)\frac{\partial^2 Q(t,x)}{\partial x^2} + b(x)\frac{\partial Q(t,x)}{\partial x} \tag{6.126}$$

with first order derivatives on the right hand side and spatially distributed parameters We will show that the same integral transformation approach can be used for the above class of equations. Let the boundary conditions of (6.126) be

$$\frac{\partial Q(t,0)}{\partial x} = u(t) \tag{6.127}$$
$$\frac{\partial Q(t,1)}{\partial x} = 0 \tag{6.128}$$

Applying the integral transform

$$P(t,\xi) = \int_0^1 \mathcal{D}(\xi,x)Q(t,x)\,\mathrm{d}x \tag{6.129}$$

to (6.126), and using integration by parts, we obtain

$$\begin{aligned}\frac{\partial^2 P(t,\xi)}{\partial t^2} &= \int_0^1 \mathcal{D}(\xi,x)(a(x)Q''_{xx}(t,x) + b(x)Q'_x(t,x))\,\mathrm{d}x \\ &= a(x)\mathcal{D}(\xi,x)Q'_x(t,x)|_{x=0}^1 \\ &\quad - (a(x)\mathcal{D}(\xi,x))'_x Q(t,x)|_{x=0}^1 + b(x)\mathcal{D}(\xi,x)Q(t,x)|_{x=0}^1 \\ &\quad + \int_0^1 \left[\frac{\partial^2}{\partial x^2}(a(x)\mathcal{D}(\xi,x)) - \frac{\partial}{\partial x}(b(x)\mathcal{D}(\xi,x))\right] Q(t,x)\,\mathrm{d}x\end{aligned} \tag{6.130}$$

It follows from the above expression that, if $\mathcal{D}$ satisfies the adjoint homogeneous boundary value problem

$$\frac{\partial^2\mathcal{D}(\xi,x)}{\partial\xi^2} = \frac{\partial^2}{\partial x^2}(a(x)\mathcal{D}(\xi,x)) - \frac{\partial}{\partial x}(b(x)\mathcal{D}(\xi,x)) \tag{6.131}$$

$$-\frac{\partial}{\partial x}(a(x)\mathcal{D}(\xi,x))|_{x=0} + b(0)\mathcal{D}(\xi,0) = 0 \tag{6.132}$$

$$-\frac{\partial}{\partial x}(a(x)\mathcal{D}(\xi,x))|_{x=1} + b(1)\mathcal{D}(\xi,1) = 0, \tag{6.133}$$

then $P(\xi,x)$ satisfies the equation

$$\frac{\partial^2 P(t,\xi)}{\partial t^2} = \frac{\partial^2 P(t,\xi)}{\partial\xi^2} + \varphi(\xi)u(t) \tag{6.134}$$

where

$$\varphi(\xi) = -a(0)\mathcal{D}(\xi,0) \tag{6.135}$$

The similar problem of stabilization of (6.134), (6.135) by using manifold control as was described earlier.

6.6 Diffusion Equation

In this section we consider the one dimensional diffusion equation

$$\begin{aligned}\frac{\partial Q(t,x)}{\partial t} &= \frac{\partial^2 Q(t,x)}{\partial x^2} && (6.136)\\ \frac{\partial Q(t,0)}{\partial x} &= f(t) && (6.137)\\ \frac{\partial Q(t,1)}{\partial x} &= u(t) && (6.138)\end{aligned}$$

where $f(t)$ represent a disturbance which may be the incoming heat flow and $u(t)$ a control which regulates the outgoing heat flow (cooler).

Let the variable

$$y(t) = \int_0^1 \psi(x)Q(t,x)\,\mathrm{d}x \qquad (6.139)$$

represent the measurements in the system. The function $\psi(x)$ is a characteristic of the sensor. By applying the integral transform

$$P(t,\xi) = \int_0^1 \mathcal{D}(\xi,x)Q(t,x)\,\mathrm{d}x \qquad (6.140)$$

with $\mathcal{D}$ satisfying

$$\begin{aligned}\frac{\partial \mathcal{D}(\xi,x)}{\partial \xi} &= \frac{\partial^2 \mathcal{D}(\xi,x)}{\partial x^2} && (6.141)\\ \frac{\partial \mathcal{D}(\xi,0)}{\partial x} &= 0 && (6.142)\\ \frac{\partial \mathcal{D}(\xi,1)}{\partial x} &= 0 && (6.143)\end{aligned}$$

and integrating by parts, we obtain the first order equation

$$\frac{\partial P(t,\xi)}{\partial t} = \frac{\partial P(t,\xi)}{\partial \xi} + \mathcal{D}(\xi,1)u(t) - \mathcal{D}(\xi,0)f(t) \qquad (6.144)$$

If the initial condition $\mathcal{D}(0,x)$ is such that

$$\mathcal{D}(0,x) = \psi(x) \qquad (6.145)$$

then

$$y(t) = P(t,0) \qquad (6.146)$$

For the case of an averaging uniform sensor $\psi(x) \equiv 1$ the dependence between input and output is very simple

$$\dot{y}(t) = u(t) - f(t) \qquad (6.147)$$

The following control law can be used to stabilize the output y

$$u(t) = -\operatorname{sgn} y(t) = -\operatorname{sgn}\left[\int_0^1 \psi(x)Q(t,x)\,\mathrm{d}x\right] \qquad (6.148)$$

A similar control for Dirichlet type boundary conditions has been obtained by Rebiai and Zinober (1992) using a different method.

6.7 Fourth Order Equation

6.7.1 The Euler-Bernoulli Beam

Consider now the problem of supressing normal vibrations along a flexible beam of unit length described by equations of fourth order. One end of the beam is assumed to be clamped while a control force is applied to the other end. The Euler-Bernoulli model of the beam with normalized parameters is

$$\frac{\partial^2 Q(t,x)}{\partial t^2} = -\frac{\partial^4 Q(t,x)}{\partial x^4} \tag{6.149}$$

$$Q(t,0) = 0 \tag{6.150}$$

$$Q'_x(t,0) = 0 \tag{6.151}$$

$$Q''_{xx}(t,1) = 0 \tag{6.152}$$

$$Q'''_{xxx}(t,1) = u(t) \tag{6.153}$$

The main idea behind our approach is to reduce the order of the controlled part of the system by applying an integral transformation

$$P(t,\xi) = \int_0^1 \mathcal{D}(\xi,x)Q(t,x)\,\mathrm{d}x \tag{6.154}$$

Here $P(t,\xi)$ is a new controlled variable and ξ is a new independent spatial variable $(0 \leq \xi < \infty)$. The kernel of the transformation $\mathcal{D}$ is assumed to satisfy the same type of boundary value problem as Q but with homogeneous boundary conditions

$$\frac{\partial^2 \mathcal{D}(\xi,x)}{\partial \xi^2} = -\frac{\partial^4 \mathcal{D}(\xi,x)}{\partial x^4} \tag{6.155}$$

$$\mathcal{D}(\xi,0) = 0 \tag{6.156}$$

$$\mathcal{D}'(\xi,0) = 0 \tag{6.157}$$

$$\mathcal{D}''(\xi,1) = 0 \tag{6.158}$$

$$\mathcal{D}'''(\xi,1) = 0 \tag{6.159}$$

ξ in this equation is analogous to a time variable and its value can change from zero to infinity. Let us show that under these conditions $P(t,\xi)$ satisfies an equation of the same class as (6.108), i.e. second order with control on the right hand side. ¿From (6.149) and (6.154)

$$\frac{\partial^2 P(t,\xi)}{\partial t^2} = -\int_0^1 \mathcal{D}(\xi,x)Q^{IV}_{xxxx}(t,x)\,\mathrm{d}x \tag{6.160}$$

Or using integration by parts

$$\begin{aligned} -\int_0^1 \mathcal{D}(\xi,x)Q^{IV}_{xxxx}(t,x)\,\mathrm{d}x &= -\mathcal{D}(\xi,x)Q'''_{xxx}(t,x)|_{x=0}^1 + \mathcal{D}'_x(\xi,x)Q''_{xx}(t,x)|_{x=0}^1 \\ &\quad - \mathcal{D}''_{xx}(\xi,x)Q'_x(t,x)|_{x=0}^1 + \mathcal{D}'''_{xxx}(\xi,x)Q(t,x)|_{x=0}^1 \\ &\quad - \int_0^1 \mathcal{D}^{IV}_{xxxx}(\xi,x)Q(t,x)\,\mathrm{d}x \end{aligned}$$

Taking into account equation (6.155) and the boundary conditions (6.150)-(6.153) and (6.156)-(6.159), it can be shown that P satisfies an equation of the form (6.108)

$$\frac{\partial^2 P(t,\xi)}{\partial t^2} = \frac{\partial^2 P(t,\xi)}{\partial \xi^2} + \varphi(\xi)u(t) \tag{6.161}$$

where

$$\varphi(\xi) = -\mathcal{D}(\xi, 1) \tag{6.162}$$

In (6.161) in contrast to that (6.155) with $\mathcal{D}$, ξ is a spatial variable. In order to define uniquely the solution of (6.155), two initial conditions must be assigned: $\mathcal{D}(0,x)$ and $\mathcal{D}'_\xi(0,x)$. If $\mathcal{D}'_\xi(0,x) = 0$ then from (6.154) the boundary value for (6.161) may be obtained as

$$P'_\xi(t,0) = 0 \tag{6.163}$$

The possibility to choose $\mathcal{D}(0,x)$ is an additional degree of freedom that can be used to assign the desired value of $\varphi(\xi)$. The other restriction imposed on $\mathcal{D}(0,x)$ is that the transformation (6.154) should be nonsingular in the sense that $P \equiv 0$ must imply $Q \equiv 0$. For equation (6.161) the design technique developed earlier can be used. The output variable for this case is

$$y(t) = P(t,0) = \int_0^1 \mathcal{D}(0,x)Q(t,x)\,\mathrm{d}x \tag{6.164}$$

Therefore only the values of this functional are needed for the control algorithm. We can say that the transformation (6.154) "absorbs" the dispersive properties of the equation (6.161) which describes how the waves are travelling.

6.7.2 General Fourth Order Equation

Consider the general type of fourth order equation representing a flexible beam

$$\frac{\partial^2 Q(t,x)}{\partial t^2} = a(x)\frac{\partial^4 Q(t,x)}{\partial x^4} + b(x)\frac{\partial^2 Q(t,x)}{\partial x^2} + c(x)\frac{\partial Q(t,x)}{\partial x} \tag{6.165}$$

The boundary conditions for (6.165) are

$$Q(t,0) = 0 \tag{6.166}$$

$$\frac{\partial Q(t,0)}{\partial x} = 0 \tag{6.167}$$

$$\frac{\partial^2 Q(t,1)}{\partial x^2} = u_1(t) \tag{6.168}$$

$$\frac{\partial^3 Q(t,1)}{\partial x^3} = u_2(t) \tag{6.169}$$

We consider the case when both force and torque are applied to one end of the beam. Using the integral transform of (6.165) and integrating by parts, we obtain

$$\begin{aligned}
\frac{\partial^2 P(t,\xi)}{\partial t^2} &= \int_0^1 \mathcal{D}(\xi,x)(-a(x)Q^{IV}_{xxxx}(t,x) + b(x)Q''_{xx}(t,x) + c(x)Q'_x(t,x))\,dx \\
&= -a(x)\mathcal{D}(\xi,x)Q'''_{xxx}(t,x)|_{x=0}^1 + (a(x)\mathcal{D}(\xi,x))'_x Q''_{xx}(t,x)|_{x=0}^1 \\
&\quad - (a(x)\mathcal{D})''_{xx} Q'_x(t,x)|_{x=0}^1 + (a(x)\mathcal{D}(\xi,x))'''_{xxx} Q(t,x)|_{x=0}^1 \\
&\quad + b(x)\mathcal{D}(\xi,x)Q'_x(t,x)|_{x=0}^1 + c(x)\mathcal{D}(\xi,x)Q(t,x)|_{x=0}^1 \\
&\quad + \int_0^1 \left[-\frac{\partial^4}{\partial x^4}(a(x)\mathcal{D}(\xi,x)) + \frac{\partial^2}{\partial x^2}(b(x)\mathcal{D}(\xi,x)) \right. \\
&\quad \left. - \frac{\partial}{\partial x}(c(x)\mathcal{D}(\xi,x)) \right] Q(t,x)\,dx
\end{aligned}$$

If $\mathcal{D}$ satisfies the adjoint to (6.165)-(6.169) homogeneous boundary value problem

$$\frac{\partial^2 \mathcal{D}(\xi,x)}{\partial \xi^2} = -\frac{\partial^4}{\partial x^4}(a(x)\mathcal{D}(\xi,x)) + \frac{\partial^2}{\partial x^2}(b(x)\mathcal{D}(\xi,x)) - \frac{\partial}{\partial x}(c(x)\mathcal{D}(\xi,x))$$

$$a(x)\mathcal{D}(\xi,x)|_{x=0} = 0$$

$$\frac{\partial}{\partial x}(a(x)\mathcal{D}(\xi,x))|_{x=0} = 0$$

$$-\frac{\partial^2}{\partial x^2}(a(x)\mathcal{D}(\xi,x))|_{x=1} + (b(x)\mathcal{D}(\xi,x))|_{x=1} = 0$$

$$\frac{\partial^3}{\partial x^3}(a(x)\mathcal{D}(\xi,x))|_{x=1} - \frac{\partial}{\partial x}(b(x)\mathcal{D}(\xi,x))|_{x=1} + (c(x)\mathcal{D}(\xi,x))|_{x=1} = 0$$

then $P(\xi,x)$ satisfies

$$\frac{\partial^2 P(t,\xi)}{\partial t^2} = \frac{\partial^2 P(t,\xi)}{\partial \xi^2} + \varphi_1(\xi)u_1(t) + \varphi_2(\xi)u_2(t)$$

The functions φ_1 and φ_2 are

$$\varphi_1(\xi) = \frac{\partial}{\partial x}(a(x)\mathcal{D}(\xi,x))|_{x=1} \tag{6.170}$$

$$\varphi_2(\xi) = -a(1)\mathcal{D}(\xi,1) \tag{6.171}$$

The design approach described earlier can also be used for this problem.

6.8 Conclusions

In this chapter we have introduced a new approach for stabilization of distributed parameter systems based on the sliding mode control approach. Previous use of sliding modes has been mainly accomplished for finite dimensional (approximate) models of distributed parameter systems. Here, we retain the infinite-dimensional model for the systems and investigate exact solutions using sliding mode control.

In the particular strategy introduced here, the control design is developed initially for a class of differential–difference systems. It was then demonstrated that the partial differential equation with second order spatial partial derivatives, can be transformed in some cases into the above differential–difference form. A number of different boundary value problems were analyzed.

We then considered a class of distributed parameter systems with fourth order spatial partial derivatives. An integral transform was introduced that changes the model into second order form. Thus, a two-stage design process can be utilized to eventually generate the sliding mode controllers.

A similar approach can be used for the case of multidimensional spatial variables. For that case the equations describing vibrating plates, membranes and three dimensional bodies as well as systems of flexible bodies may be considered. The integral transform (6.129) for this case uses the spatial integration over the structure configuration as the transformed variable P still has only a one-dimensional spatial variable.

The sliding mode algorithms considered make the closed-loop system highly insensitive to external disturbances and parameter variations. Furthermore, from the applications viewpoint, the models considered are particularly appropriate for the utilization of distributed actuation and may be used in "smart structures" with piezoelectric materials or temperature control systems with distributed heating.

References

Bellman, R., Cooke, K.L. 1963, Differential-difference equations. Academic Press, New York

Drakunov, S.V., Özgüner, Ü 1992, Vibration Supression in Flexible Structures via the Sliding Mode Control Approach. *Proc IEEE Conference on Decision and Control,* Tucson, Arizona, 1365–1366

Drakunov, S.V., Özgüner, Ü. 1992, Vibration Supression in Flexible Structures via the Sliding Mode Control Approach. Submitted to *IEEE Transactions on Automatic Control*

Drakunov, S.V., Utkin V.I. 1991, Sliding Mode Control Concept for Abstract Dynamic Systems. *Proc Int Workshop on Nonsmooth Control and Optimization*, Vladivostok, Russia, 121

Drakunov, S.V., Utkin V.I. 1992, Sliding Mode Control in Dynamic Systems. *International Journal of Control* **55**, 1029–1037

Drakunov, S.V., Izosimov D.B., Luk'yanov A.G., Utkin V.A., Utkin V.I. 1990, The Block Control Principle. *Automation and Remote Control*, Part 1, **51**, 601–608; Part 2, **51**, 737–746

Lions, J.L., Magenes, E. 1972, *Non-homogeneous Boundary Value Problems and Applications*, Springer-Verlag, Berlin/New York

Meirovitch, D. 1986, *Elements of vibrational analysis*, McGraw-Hill, New York

Rebiai, S. E., Zinober, A. S., 1992 Stabilization of Infinite Dimensional Systems by Nonlinear Boundary Control. *International Journal of Control* **57**, 1167–1175

Treves, F. 1975, *Basic Linear Partial Differential Equations*, Academic Press, New York

Utkin, V.I. 1978, *Sliding Modes and Their Application in Variable Structure Systems*, MIR, Moscow

Yurkovich, S., Özgüner, Ü., Al-Abbas, F. 1986, Model Reference Sliding Mode Adaptive Control for Flexible Structures. *Journal of Astronautical Sciences* **36**, 285–310

Young, K-K. D., Özgüner Ü., Jian-Xin Xu 1993 Variable Structure Control of Flexible Manipulators in *Variable Structure Control for Robotics and Aerospace Applications*, ed. Young, K-K. D., Elsevier Press, 247–277

7. Digital Variable Structure Control with Pseudo-Sliding Modes

Xinghuo Yu

7.1 Introduction

The theoretical development of variable structure control (VSC) has been mainly focussed on the study of continuous-time systems. Its digital counterpart, discrete-time variable structure control (DVSC), has received less attention.

The current trend of implementation of VSC is towards using digital rather than analog computers, due to the availability of low-cost, high-performance microprocessors. In the implementation of DVSC, the control instructions are carried out at discrete instants; noting that the switching frequency is actually equal to or lower than the sampling frequency. With such a comparatively low switching frequency, the system states move in a zigzag manner about the prescribed switching surfaces. As such, the well-known main feature of VSC, the invariance properties, may be jeopardized.

This chapter aims to investigate some of the inherent properties peculiar to DVSC, and discuss the design of DVSC systems. The chapter is organized as follows. Section 7.2 presents a simulation study which shows the sampling effect on the discretization of a continuous-time VSC system. By simply increasing the sampling period gradually, the system behaviour changes from sliding on the switching line to zigzagging, and further increase leads to chaos. This demonstrates the necessity of the study of DVSC. Section 7.3 surveys the recent development of the theory of DVSC systems. A new DVSC scheme, which enables the elimination of zigzagging as well as divergence from the switching hyperplane, is discussed in Sect. 7.4. Computer simulations are presented in Sect. 7.5 to show the effectiveness of the scheme developed. The conclusions are drawn in Sect. 7.6.

7.2 Sampling Effect on a VSC System

Consider the two-dimensional continuous-time VSC system

$$\begin{aligned} \dot{x}_1 &= x_2 \\ \dot{x}_2 &= -fx_2 + u \end{aligned} \qquad (7.1)$$

where $u = -\alpha_1 |x_1| \mathrm{sgn} s$, $\alpha_1 > 0$ is a typical VSC, in which the switching line is defined by

$$s = cx_1 + x_2 = 0, \quad c > 0 \qquad (7.2)$$

which is an asymptotically stable sliding mode. It is well-known that the necessary and sufficient condition (Utkin 1977) for $s = 0$ to be a sliding mode, characterized by $\dot{x}_1 = -cx_1$, is

$$c^2 - \alpha_1 < cf < c^2 + \alpha_1 \tag{7.3}$$

Using zero-order hold (ZOH) with a sampling period h the system is discretized as

$$x(k+1) = \Phi x(k) + \Gamma u(k) \tag{7.4}$$

where $x^T = (x_1 \quad x_2)$,

$$\Phi = \begin{bmatrix} 1 & (1-\exp(-fh))/f \\ 0 & \exp(-fh) \end{bmatrix} \tag{7.5}$$

$$\Gamma = [(h/f) + (\exp(-fh) - 1)/f^2 \quad (1-\exp(-fh))/f]^T \tag{7.6}$$

and $u(k) = -\alpha_1 |x_1(k)| \mathrm{sgn} s(k)$ with $s(k) = cx_1(k) + x_2(k)$.

Letting $\alpha_1 = 8$, $f = -5$, $c = 1$, which satisfy (7.3), and the initial state $x_1(0) = 1.0$, $x_2(0) = 0.1$, the phase plane plots of $x_1(k)$ versus $x_2(k)$ with two different sampling periods are drawn in Fig. 7.1 which illustrates two zigzagging motions: one being stable discrete-time sliding mode with $h = 0.018$, the other stable without sliding at certain instants with $h = 0.01916$. Increasing the sampling period h further may lead to instability (even with $h = 0.0192$). Note that the value of f is well within the range $(-7, 9)$ defined by (7.3).

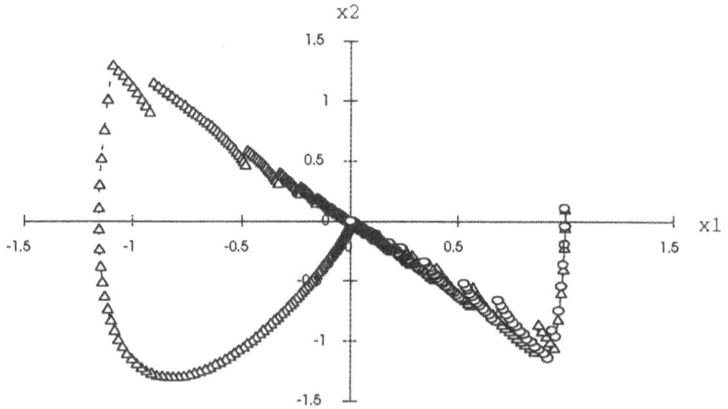

Fig. 7.1. Two motions: stable discrete-time sliding mode (o); stable without sliding (Δ)

The zigzagging behaviour can be further investigated using the Lyapunov exponents method (Grantham and Athalye 1990). The Lyapunov exponents method is often used to measure the growth rates of the distance between neighbouring trajectories of nonlinear chaotic dynamics. To present the basic idea of a Lyapunov exponent, let $\delta(t)$ denote the distance between two trajectories for a continuous-time system. If $\delta(0)$ is small and $\delta(t) \to \delta(0)\exp(\varrho t)$ as $t \to \infty$, then ϱ is called a Lyapunov exponent. The distance between trajectories grows, shrinks or remains constant depending on whether ϱ is positive, negative or zero respectively. The definition of a Lyapunov exponent for discrete-time systems is the same as for continuous-time systems except that t is replaced by kh.

In the study of zigzagging behaviour, we consider $\delta(kh)$ as a distance between a trajectory and the origin of the phase plane. Therefore we actually study the growth rates of the distance between the system trajectory and the origin. For the continuous-time system (7.1), because the system eventually exhibits an asymptotically stable sliding mode governed by $\dot{x}_1 = -x_1$, then the Lyapunov exponent for the continuous-time system is -1, indicating the trajectory shrinks with rate -1. However the Lyapunov exponent for the discrete-time system (7.4) is not obvious. Using the Gramm-Schmidt algorithm (Grantham and Athalye 1990), the Lyapunov exponent versus the sampling period h can be calculated. Figure 7.2 shows the plot of Lyapunov exponent versus the sampling period h. While h increases from 0 to about 0.019, the Lyapunov exponent slowly decreases from -1, the slope of the sliding line (7.2), indicating that the chattering becomes increasingly worse with the progressive increase of h. The chaotic phenomenon starts when h is about 0.019. The Lyapunov exponent jumps up sharply and irregularly with little oscillation with respect to the increase of h. The Lyapunov exponent becomes positive when $h > 0.02$, indicating the trajectory exponentially grows, i.e. the system is unstable.

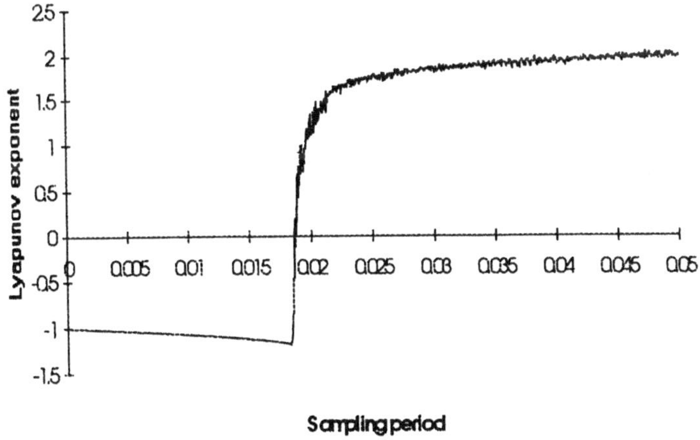

Fig. 7.2. Lyapunov exponent with respect to sampling period h

The above example demonstrates that the implementation is not simply a discretization of a VSC system with a *small enough* sampling period. The *small enough* sampling period may cause chaotic behaviour. The inherent properties of DVSC need to be investigated.

7.3 Conditions for Existence of Discrete-Time Sliding Mode

The existence of a continuous-time sliding mode implies that in a vicinity of the prescribed switching surface, the velocity vectors of the state trajectories always point towards the switching surface (DeCarlo et al 1988). An ideal sliding mode exists only when the system state satisfies the dynamic equation governing the sliding mode for all $t \geq t_0$, for some t_0. This requires infinitely fast switching.

It is obvious that the definition for continuous-time sliding modes can not be applied to the discrete-time sliding modes since the concept of velocity vectors of the system state trajectories is not available. The switching frequency is actually equal to or lower than the sampling frequency. The comparatively low switching frequency causes the discrete-time system state to move about the switching surface in a zigzag manner.

Discrete-time sliding modes were first named "quasi-sliding modes" (Milosavljevic 1985). However, the similarity between discrete-time sliding modes and continuous-time sliding modes disappears as the sampling period increases with the system trajectory appearing to zigzag within a bounded domain. Therefore "pseudo-sliding mode" is a more precise statement.

Consider the single-input discrete-time dynamic system

$$x(k+1) = f(k, x(k), u(k)) \tag{7.7}$$

where $x \in \mathbb{R}^n$, and $u(k)$ is the sliding mode control which may not necessarily be discontinuous on the switching surface defined by $s(k) = s(x(k)) = 0$.

Definition 7.1 *The pseudo-sliding mode is said to exist if in an open neighbourhood of the manifold* $\{x : \ s(k) = 0\}$*, denoted by* Ω_s*, the condition*

$$\nabla s(k) s(k) < 0 \tag{7.8}$$

holds where $\nabla s(k) = s(k+1) - s(k)$.

Definition 7.1 is actually a mild modification of the definition by Milosavljevic (1985) who first proposed a necessary condition for the existence of the pseudo-sliding mode by replacing the derivative term in the well-known condition

$$\lim_{s \to 0} \dot{s}s < 0 \tag{7.9}$$

with a forward difference such that

$$\lim_{s(k) \to 0+} \nabla s(k) < 0, \qquad \lim_{s(k) \to 0^-} \nabla s(k) > 0 \tag{7.10}$$

It is obvious that the conditions $s(k) \to 0^+$ and $s(k) \to 0^-$ are rarely satisfied in practice, since it is impossible for the system states to approach a switching surface sufficiently closely.

Definition 7.2 (Sarpturk et al 1987) *A system is said to exhibit a convergent pseudo-sliding mode, if in the neighbourhood Ω_s, the condition*

$$|s(k+1)| < |s(k)| \tag{7.11}$$

holds.

Condition (7.11) actually imposes upper and lower bounds on the DVSC. Kotta (1989) pointed out that the upper and lower bounds depend on the distance of the system state from the sliding surfaces.

Definition 7.2 can also be set up equivalently by replacing the condition (7.11) with $s^2(k+1) < s^2(k)$ (Furuta 1990) and $|s(k)s(k+1)| < s^2(k)$ (Sira-Ramírez 1991).

As shown in Yu and Potts (1992a) and Spurgeon (1992), the condition (7.11) and its equivalents are only sufficient conditions for the existence of pseudo-sliding mode. It is not necessary to satisfy the condition (7.11) and its equivalents while $s(k)s(k+1) < 0$.

The equivalent control plays an important role in the theory of VSC. When a system is sliding, its dynamics can be considered to be driven by an equivalent control. However in the discrete-time case, since the system states are rarely very close to the sliding surfaces, how to define sliding is an open question. Ideally we can find a control $u_{eq}(k)$, which can be called discrete-time equivalent control such that $s(k) = 0$ and $s(k+1) = 0$. The existence of discrete-time equivalent control in DVSC has been proved by Sira-Ramírez (1991).

Theorem 7.3 (Sira-Ramírez 1991) *Suppose that the system (7.7) has relative degree one, i.e. $\partial s(f(k, x(k), u(k)))/\partial u(k) \neq 0$ for all k, and the DVSC structure is chosen as*

$$u = \begin{cases} u^+(x) & \text{for } s > 0 \\ u^-(x) & \text{for } s < 0 \end{cases} \tag{7.12}$$

then the equivalent control u_{eq} exists and satisfies

$$u^-(x) < u_{eq} < u^+(x) \tag{7.13}$$

In contrast to the above methodologies Utkin and Drakunov (1989) proposed a different approach which uses contraction mapping to guarantee the existence of the pseudo-sliding mode. For further development, readers are referred to Utkin (1992).

We now define the existence of the pseudo-sliding mode.

Definition 7.4 *The discrete-time dynamic system (7.7) is said to exhibit a pseudo-sliding mode, if there exists an integer $K > 0$, such that for all $k > K$, $x(k) \in \Omega_s$. It follows that $f(k, x(k), u(k)) \in \Omega_s$.*

Remark. Definition 7.4 includes the case that the DVSC system state may never reside in the sliding mode. It may also not be necessary that in an open neighbourhood Ω_s the system state always approaches the sliding surface so long as it does not leave Ω_s. There are two often used definitions of the neighbourhood Ω_s: one being

$$\Omega_s = \{x : \quad |s| < \epsilon \parallel x \parallel, \quad \epsilon > 0\} \tag{7.14}$$

which is for state-feedback type DVSC, and the other

$$\Omega_s = \{x : \quad |s| < \epsilon, \quad \epsilon > 0\} \tag{7.15}$$

which is for relay type DVSC.

Spurgeon (1992) further questioned the appropriateness of the application of traditional hyperplane design philosophy to uncertain DVSC systems, and developed a linear equivalent control structure which has superior performance.

It can be argued that if the sample period is chosen to be small enough, the implementation of continuous-time VSC shall still enjoy the superior invariance properties of VSC. The question is how small the sampling period should be. There may exist an upper bound of sampling period for implementation of a continuous-time VSC such that if the sampling period to be chosen is less than the upper bound, the continuous-time VSC structure can still be used, otherwise one should adopt totally different DVSC design methodologies, such as those developed by Spurgeon (1992), Magaña and Żak (1987).

The effect of sampling on the "best" discretization of VSC has been studied and the formulae for upper bounds of sampling stepsize have been obtained (Potts and Yu 1991, Yu and Potts 1992a, Yu 1993). Note that these upper bounds are independent of the distance of the system state from the sliding surfaces. However the discretization scheme is too complicated for practical implementation.

In practice most applications are done using ZOH. It is therefore necessary to address the problems associated with sampling on those discrete-time systems discretized using ZOH. The following sections are devoted to investigation of digital VSC of linear systems using discontinuous VSC structures based on linear feedback with switched gains.

7.4 Digital VSC Systems

We consider the digital linear control system in the controllable canonical form

$$x(k+1) = \Phi x(k) + \Gamma u(k) \tag{7.16}$$

where

$$\Phi = \begin{bmatrix} 0 & 1 & 0 & \cdots & 0 \\ 0 & 0 & 1 & \cdots & 0 \\ \vdots & \vdots & \vdots & \vdots & \vdots \\ 0 & 0 & 0 & \cdots & 1 \\ -a_1 & -a_2 & -a_3 & \cdots & -a_n \end{bmatrix} \tag{7.17}$$

$$\Gamma = [0 \; 0 \; 0 \; \ldots \; 0 \; 1]^T \tag{7.18}$$

in which $x \in \mathbb{R}^n$, $u \in \mathbb{R}$, and the paramcter variations are assumed bounded. This canonical form is assumed to have been obtained by a linear transformation from the digital system (Ogata 1987)

$$z(k+1) = \exp(Ah)z(k) + \int_0^h \exp(A\tau)\mathrm{d}\tau Bu(k) \tag{7.19}$$

which is the exact discretization of the continuous-time system $\dot{z} = Az + Bu$ using ZOH, with $z \in \mathbb{R}^n$, and A, B are of appropriate dimensions.

The switching hyperplane is represented by

$$s(k) = c^T x(k) = c_1 x_1(k) + c_2 x_2(k) + \ldots + x_n(k) = 0 \tag{7.20}$$

The phenomenon of interest is the possible occurrence of $s = 0$ as an asymptotically stable sliding hyperplane, i.e. all zeros (eigenvalues) of the characteristic polynomial for (7.20) are inside the unit circle in the complex plane.

7.4.1 Control Strategy

The structure of DVSC, which will be used in the following sections, is based on linear feedback with switched gain type of discrete-time variable structure control (SDVSC) defined by

$$u(k) = -\sum_{i=1}^{l} \Psi_i x_i(k) \tag{7.21}$$

with

$$\Psi_i = \begin{cases} \alpha_i & \text{if } x_i(k)s(k) \geq 0 \\ -\alpha_i & \text{if } x_i(k)s(k) < 0 \end{cases} \tag{7.22}$$

where $\alpha_i > 0$, $i = 1, \ldots, l$, and l is the number of switched gains used. Apart from the nonlinear behaviour at the switching hyperplanes $x_i = 0$, $i = 1, \ldots, l$ and $s(x) = 0$, the system (7.16) with (7.21) is of nth-order with 2^l linear feedback control structures.

We propose two different control laws for the cases $x \in \Omega_s$ and $x \notin \Omega_s$. The open neighbourhood Ω_s is defined by (7.14) for a properly chosen ϵ since the control structure is linear feedback with switched gains. For $x \notin \Omega_s$ we use the control law SDVSC (7.21), (7.22) to force the system state to approach and/or cross Ω_s. This will be discussed in Sects. 7.4.3 and 7.4.4. For $x \in \Omega_s$, we design another control law to eliminate the zigzagging within Ω_s. Section 7.4.5 will deal with the design of such a control law.

7.4.2 Partitions in the State Space

Before further discussion of the design of DVSC, we shall partition the state space so that we can easily identify which subset in the state space uses which control structure. Recall that SDVSC (7.21), (7.22) actually represents 2^l linear feedback control structures. Each linear structure is activated in a subset in the state space. In order to identify which subset uses which control structure, we define the set

$$\Xi = \{\boldsymbol{\xi} \in \mathbb{R}^n,\ \boldsymbol{\xi} = [\Psi_1 \dots \Psi_l\ 0\ \dots 0]^T\} \tag{7.23}$$

where $\Psi_i = \pm\alpha_i$ according to (7.22). Therefore Ξ represents a set with 2^l elements representing 2^l possible linear control structures.

We define another set Θ

$$\Theta = \{\boldsymbol{\theta} \in \mathbb{R}^n, \boldsymbol{\theta} = [\theta_1 \dots \theta_l\ 0 \dots 0]^T,\ \theta_i = \pm 1, (i = 1, \dots, l)\} \tag{7.24}$$

which has 2^l elements, and will be used for partitioning the state space.

A partition in the state space is defined by

$$\begin{aligned} R(\boldsymbol{\theta}) = \{x \in \mathbb{R}^n,\ & x_i\theta_i \geq 0 \text{ if } \theta_i = 1, \\ & x_i\theta_i > 0 \text{ if } \theta_i = -1,\ (i = 1, \dots, l) \text{ for } \boldsymbol{\theta} \in \Theta\} \end{aligned} \tag{7.25}$$

Obviously

$$\mathbb{R}^n = \bigcup_{\boldsymbol{\theta} \in \Theta} R(\boldsymbol{\theta}) \tag{7.26}$$

The partition $R(\boldsymbol{\theta})$ with some $\boldsymbol{\theta} \in \Theta$ can be further partitioned into two subsets according to $s \geq 0$ and $s < 0$. For $s \geq 0$, there exists a control structure denoted by $\boldsymbol{\xi}^+ \in \Xi$, where the superscript "+" represents the subset $R(\boldsymbol{\theta}) \bigcap \{x \in \mathbb{R}^n,\ s \geq 0\}$. Correspondingly, for $s < 0$, in $R(\boldsymbol{\theta}) \bigcap \{x \in \mathbb{R}^n,\ s < 0\}$ with the same $\boldsymbol{\theta} \in \Theta$, there exists another control structure $\boldsymbol{\xi}^- \in \Xi$. In the following sections the superscripts "+" and "–" are referred to the cases of $s \geq 0$ and $s < 0$ respectively.

Note that there may exist a different $\boldsymbol{\theta}_0 = -\boldsymbol{\theta}$ such that in the corresponding $R(\boldsymbol{\theta}_0) \bigcap \{x \in \mathbb{R}^n,\ s < 0\}$ and $R(\boldsymbol{\theta}_0) \bigcap \{x \in \mathbb{R}^n,\ s \geq 0\}$ the same control structures, $\boldsymbol{\xi}^+$ and $\boldsymbol{\xi}^-$, are activated respectively.

The characteristic polynomial for the system (7.16)–(7.18) with the control (7.21), (7.22) is represented by $P(\lambda; \boldsymbol{\xi})$ which is defined by

$$\begin{aligned} P(\lambda; \boldsymbol{\xi}) = \lambda^n &+ a_n\lambda^{n-1} + \dots \\ &+ (a_l + \Psi_l)\lambda^{l-1} + \dots + (a_2 + \Psi_2)\lambda + a_1 + \Psi_1 \end{aligned} \tag{7.27}$$

for $\boldsymbol{\xi} \in \Xi$.

Using the definition of the set Ξ, the SDVSC can be written alternatively as

$$u(k, \boldsymbol{\xi}) = u(k) = -\boldsymbol{\xi}^T x(k) \quad \text{for } \boldsymbol{\xi} \in \Xi \tag{7.28}$$

Therefore the system (7.16)–(7.18) under the control (7.21), (7.22) can be represented alternatively by

$$x(k+1) = (\Phi - \Gamma \xi^T) x(k) \tag{7.29}$$

with $\xi \in \Xi$.

7.4.3 Design of SDVSC; Acquisition of Lower Bounds

For the system state to reach Ω_s from any initial state outside Ω_s, we impose the condition

$$\nabla s(k) s(k) < 0 \tag{7.30}$$

for the design of SDVSC so that the system state will approach and/or cross the switching hyperplane (7.20). In this section we discuss the design of SDVSC in the limiting case that $\epsilon = 0$, i.e. $\Omega_s = \{x \in \mathbb{R}^n,\ s = 0\}$. This will enable us to investigate the performance of the discrete-time discontinuous VSC structure SDVSC.

A sufficient condition for (7.30) to hold is deduced by taking

$$\nabla s(k) = c^T (\Phi - I_n) x(k) - \sum_{i=1}^{l} \Psi_i x_i(k) \tag{7.31}$$

where I_n is the $n \times n$ unit matrix. From (7.20)

$$x_n(k) = s(k) - \sum_{i=1}^{n-1} c_i x_i(k) \tag{7.32}$$

and substituting (7.31) and (7.32) into (7.30) yields

$$\begin{aligned} \nabla s(k) s(k) \quad = \quad & \sum_{i=1}^{l} (c_{i-1} - c_i - a_i - c_i \rho - \Psi_i) x_i(k) s(k) + \\ & \sum_{i=l+1}^{n-1-l} (c_{i-1} - c_i - a_i - c_i \rho) x_i(k) s(k) + \rho s^2(k) < 0 \end{aligned} \tag{7.33}$$

where $\rho = c_{n-1} - a_n - 1$. We immediately conclude that for the condition (7.30) to be satisfied, it is sufficient that

$$\begin{aligned} & \rho \le 0 \\ & |c_{i-1} - c_i - a_i - c_i \rho| < \alpha_i \quad \text{for} \quad i = 1, \ldots, l, \quad c_0 = 0 \\ & c_{i-1} - c_i - a_i - c_i \rho = 0 \quad \text{for} \quad i = l+1, \ldots, n-1 \end{aligned} \tag{7.34}$$

Remark. Equation (7.34) actually gives the lower bounds of α_i, denoted by $\underline{\alpha}_i$, $i = 1, \ldots, l$, since the parameter variations are assumed bounded.

7.4.4 Design of SDVSC; Acquisition of Upper Bounds

A question arises from using the condition (7.30). Can any value of α_i, which satisfies $\alpha_i > \underline{\alpha}_i (i = 1, \ldots, l)$, be used for SDVSC? The answer is no, because the existence of asymptote hyperplanes, on which the system trajectory diverges (on one side the trajectory tending towards the switching hyperplane, on the other side moving away from it), restricts the choice of the α_i. This will be fully discussed and a sufficient condition for the system to avoid such divergence will be derived. For the proof of the existence of asymptote hyperplanes, readers are referred to Appendix 7.1.

The derivation follows from the argument that the values of $\alpha_i (i = 1, \ldots, l)$ must be sufficiently close to the $\underline{\alpha}_i$ satisfying (7.34), so that a step across the switching hyperplane does not extend beyond the region which forces an immediate return towards the hyperplane. This region is bounded by the asymptote hyperplanes and the switching hyperplane.

Without loss of generality, we choose a $\boldsymbol{\theta}$, where $\theta_i = 1,\ i = 1, \ldots, l$ such that

$$R(\boldsymbol{\theta}) = \{x \in \mathrm{I\!R}^n,\ x_i \geq 0,\ i = 1, \ldots, l\} \tag{7.35}$$

Suppose that the system state is in the subset $R(\boldsymbol{\theta}) \bigcap \{x \in \mathrm{I\!R}^n,\ s \geq 0\}$, in which the corresponding control structure is $\boldsymbol{\xi}^+ = [\alpha_1\ \ldots\ \alpha_l\ 0 \ldots 0]^T$, and approaches the switching hyperplane $s = 0^+$. The limiting case occurs when a single step corresponding to a particular value of k, just carries from $s(k) = 0^+$ into the region characterized by the adjoined subset $R(\boldsymbol{\theta}) \bigcap \{x \in \mathrm{I\!R}^n,\ s < 0\}$, in which another structure $\boldsymbol{\xi}^- = [-\alpha_1\ \ldots\ -\alpha_l\ 0 \ldots 0]^T$ is employed. If the characteristic polynomial of the system with $\boldsymbol{\xi}^-$ has m real eigenvalues which are greater than one, there may exist m asymptote hyperplanes represented by $r_j(x, \boldsymbol{\xi}^-) = 0,\ (j = 1, \ldots, m)\ m < n$ (see Appendix 7.1). Any larger values of $\alpha_i (i = 1, \ldots, l)$ may yield a step into the region

$$\{x \in R(\boldsymbol{\theta}),\ s < 0,\ r_j(x, \boldsymbol{\xi}^-) \leq 0,\ j \in (1, \ldots, m)\} \tag{7.36}$$

from where the trajectory moves away from the switching hyperplane. The tendency of approaching the switching hyperplane is therefore violated. Similarly, the reasoning applies to the case $x(k) \in R(\boldsymbol{\theta}) \bigcap \{x \in \mathrm{I\!R}^n,\ s < 0\}$, where $\boldsymbol{\theta} \in \boldsymbol{\Theta}$ is the same as above. There may exist another set of upper bounds of $\alpha_i (i = 1, \ldots, l)$ such that any larger values of α_i may produce a step into the region

$$\{x \in R(\boldsymbol{\theta}),\ s > 0,\ r_j(x, \boldsymbol{\xi}^+) \geq 0,\ j \in (1, \ldots, q)\} \tag{7.37}$$

from where the trajectory moves away from the switching hyperplane. Here q is the number of the real eigenvalues (which are greater than one) of the characteristic polynomial with $\boldsymbol{\xi}^+$.

Apparently any system state driven from $s(k) = 0^+$ with smaller values of $\alpha_i (i = 1, \ldots, l)$ may drop into the regions defined by

$$\Omega^-(\boldsymbol{\theta}) = R(\boldsymbol{\theta}) \bigcap \{x \in \mathrm{I\!R}^n,\ s < 0,\ r_j(x, \boldsymbol{\xi}^-) > 0,\ j = 1, \ldots, m\} \tag{7.38}$$

from where it will go back towards the switching hyperplane. The same reasoning applies to the case $x(k) \in R(\theta)\bigcap\{x \in \mathbb{R}^n,\ s < 0\}$, and

$$\Omega^+(\theta) = R(\theta)\bigcap\{x \in \mathbb{R}^n,\quad s \geq 0,\ r_j(x,\xi^+) < 0,\ j = 1,\ldots,q\} \qquad (7.39)$$

The regions (7.38) and (7.39) are the attracting regions towards the switching hyperplane that satisfy

$$\Omega^-(\theta)\bigcap\Omega^+(\theta) = \emptyset \qquad (7.40)$$

Here the superscripts "+" and "–" represents the cases $s \geq 0$ and $s < 0$ respectively.

Note that either m or q may be zero, meaning that there is no such real eigenvalue which is greater than one. For example, the attracting region may be set to

$$\Omega^+(\theta) = R(\theta)\bigcap\{x \in \mathbb{R}^n,\ s \geq 0\} \qquad (7.41)$$

if $q = 0$.

This analogy applies to $R(\theta)$ for each $\theta \in \Theta$. The ξ^+, ξ^- can be considered as a pair (or adjoined pair) relating to the partition in the state space $R(\theta)$ with $s \geq 0$ and $s < 0$. There are actually 2^{l-1} such pairs.

There are 2^l control structures. The number of constraints (or inequalities) for calculating the upper bounds of α_i, denoted by $\overline{\alpha}_i (i = 1,\ldots,l)$, depends on how many asymptote hyperplanes there are for each control structure. Each constraint (inequality) can be obtained by applying the algorithm in Appendix 7.2. The upper bounds $\overline{\alpha}_i$ can therefore be obtained by solving the inequalities. The upper bounds may not be unique.

The above analysis is summarized in the following theorem:

Theorem 7.5 *For the digital VSC system (7.16)-(7.18) with the control (7.21)-(7.22) to approach and/or cross the switching hyperplane (7.20) without divergence from the switching hyperlane, it is sufficient that (7.34) and the following conditions hold*

$$\underline{\alpha}_i < \alpha_i < \overline{\alpha}_i, \quad \textit{for} \quad i = 1,\ldots,l \qquad (7.42)$$

where $\underline{\alpha}_i = |c_{i-1} - c_i - a_i - c_i\rho|\ (i = 1,\ldots,l),\ c_0 = 0,$ *and the* $\overline{\alpha}_i (i = 1,\ldots,l)$ *are determined by the constraints obtained using the algorithm in Appendix 7.2 for the* 2^l *control structures.*

Remark. The condition (7.42) is independent of the distance from the switching hyperplane. This means that the design methodology developed is more relaxed than (7.11).

It is easy to extend the developed algorithm for the case when the following control structure is used:

$$u(k) = -\sum_{i=1}^{l} \Psi_i x_i(k) \tag{7.43}$$

and

$$\Psi_i = \begin{cases} \alpha_i & \text{if } x_i(k)s(k) > 0 \\ \beta_i & \text{if } x_i(k)s(k) < 0 \end{cases} \tag{7.44}$$

where $|\alpha_i|$ may be different from $|\beta_i|$, $i = 1, \ldots, l$.

7.4.5 Modification of SDVSC — Elimination of Zigzagging

Theorem 7.5 guarantees the switching hyperplane to be approached and/or crossed without divergence. With (7.42) the system state will zigzag about the switching hyperplane. The zigzagging behaviour is of course not acceptable for practical implementation as the system would not exhibit the motion (7.20). This section aims to develop a scheme to reduce the zigzagging.

Apparently the parameter ϵ should be chosen so that

$$\Omega_s \subset \bigcup(\Omega^+(\theta)\bigcup\Omega^-(\theta)) \tag{7.45}$$

for all $\theta \in \Theta$. The choice of ϵ is restricted by the choice of $\alpha_i (i = 1, \ldots, l)$. Ω_s is actually a coned shaped region as depicted in Fig. 7.3 for two-dimensional systems. The reason for zigzagging is that with a particular value of k and the

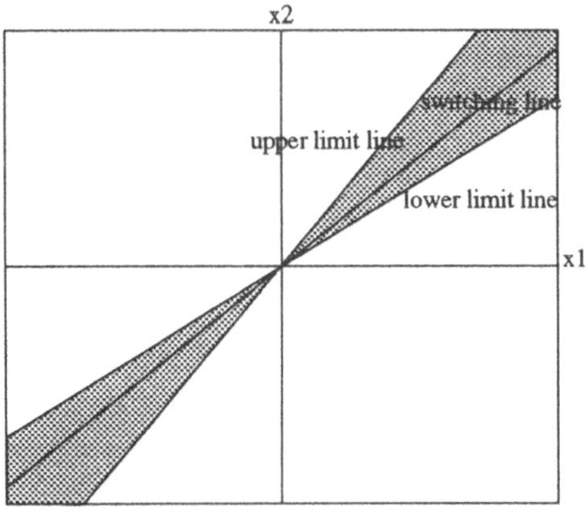

Fig. 7.3. Ω_s in a two-dimensional system

system state $x(k)$ close to $s = 0$, the control (which is constant over a sampling period) may take the system state to the next position of $x(k+1)$ which may not be close to $s = 0$. This is in contrast to the situation in continuous-time VSC in which the control changes as soon as the system state crosses the switching hyperplane $s = 0$. However, if a mechanism similar to continuous-time VSC is introduced to DVSC within Ω_s (suppose $x(k)$ is close to $s = 0$ and the control is then softened so that $x(k+1)$ is reasonably close to $s = 0$) the zigzagging will then be softened.

This can be done by means of the following scheme. For each $\boldsymbol{\theta} \in \Theta$, as discussed in Sect. 7.4.3, we can construct two adjoined subsets $\Omega^-(\boldsymbol{\theta})$ and $\Omega^+(\boldsymbol{\theta})$ in which there exist $\boldsymbol{\xi}^-, \boldsymbol{\xi}^+$ such that the controls $u(k, \boldsymbol{\xi}^-)$ and $u(k, \boldsymbol{\xi}^+)$ are respectively activated. Applying Theorem 7.3, there exists u_{eq} such that $u(k, \boldsymbol{\xi}^-) < u_{eq} < u(k, \boldsymbol{\xi}^+)$. If the system state $x(k)$ is in $\Omega_s \bigcap \Omega^-(\boldsymbol{\theta})$, since the tendency of the system trajectory is from $\Omega_s \bigcap \Omega^-(\boldsymbol{\theta})$ to $\Omega_s \bigcap \Omega^+(\boldsymbol{\theta})$, we then let

$$u_{eq}(k) = \eta^-(k, \boldsymbol{\theta})u(k, \boldsymbol{\xi}^+) + (1 - \eta^-(k, \boldsymbol{\theta}))u(k, \boldsymbol{\xi}^-) \tag{7.46}$$

According to Theorem 7.3 there exists a $\eta^-(k, \boldsymbol{\theta})$, $0 < \eta^-(k, \boldsymbol{\theta}) < 1$ such that (7.46) holds. In fact from (7.46)

$$\eta^-(k, \boldsymbol{\theta}) = \frac{u_{eq}(k) - u(k, \boldsymbol{\xi}^-)}{u(k, \boldsymbol{\xi}^+) - u(k, \boldsymbol{\xi}^-)} \tag{7.47}$$

On the other hand, the equivalent control can be obtained by letting $s(k+1) = 0$ so that $u_{eq}(k) = -(c^T \Gamma)^{-1} c^T \Phi x(k)$. To eliminate the zigzagging, we choose the following softening control

$$\hat{u}^-(k, \boldsymbol{\theta}) = \hat{\eta}^-(k, \boldsymbol{\theta})u(k, \boldsymbol{\xi}^+) + (1 - \hat{\eta}^-(k, \boldsymbol{\theta}))u(k, \boldsymbol{\xi}^-) \tag{7.48}$$

where $\hat{\eta}^-(k, \boldsymbol{\theta}) = \eta^-(k, \boldsymbol{\theta})(1 - s(k)/\hat{s}^-(k, \boldsymbol{\theta}))$ with $\hat{s}^-(k, \boldsymbol{\theta})$ defined by $s(x(k))$ along the boundaries of $\Omega_s \bigcap \Omega^-(\boldsymbol{\theta})$. Apparently $0 < s(k)/\hat{s}^-(k, \boldsymbol{\theta}) < 1$ and $0 < \hat{\eta}^-(k, \boldsymbol{\theta}) < 1$. Using control (7.48) in $\Omega_s \bigcap \Omega^-(\boldsymbol{\theta})$, the equation

$$s(k+1) = \frac{s(k)}{\hat{s}^-(k, \boldsymbol{\theta})} c^T (\Phi x(k) + \Gamma u(k, \boldsymbol{\xi}^-)) \tag{7.49}$$

can be deduced where $\Phi x(k) + \Gamma u(k, \boldsymbol{\xi}^-)$ represents the system state at $k+1$, driven by $u(k, \boldsymbol{\xi}^-)$ from $x(k)$. Since $0 < s(k)/\hat{s}^-(k, \boldsymbol{\theta}) < 1$, the equation (7.49) shows that $x(k+1)$ driven by $\hat{u}^-(k, \boldsymbol{\theta})$ is always closer to the switching hyperplane than being driven by $u(k, \boldsymbol{\xi}^-)$. When $s(k) = 0$, then $s(k+1) = 0$. The zigzagging is therefore softened.

Correspondingly if the system $x(k)$ is in the region $\Omega_s \bigcap \Omega^+(\boldsymbol{\theta})$, and since the tendency of the system trajectory is from $\Omega_s \bigcap \Omega^+(\boldsymbol{\theta})$ to $\Omega_s \bigcap \Omega^-(\boldsymbol{\theta})$, then we choose

$$\hat{u}^+(k, \boldsymbol{\theta}) = (1 - \hat{\eta}^+(k, \boldsymbol{\theta}))u(k, \boldsymbol{\xi}^+) + \hat{\eta}^+(k, \boldsymbol{\theta})u(k, \boldsymbol{\xi}^-) \tag{7.50}$$

where $\hat{\eta}^+(k, \boldsymbol{\theta}) = \eta^+(k, \boldsymbol{\theta})(1 - s(k)/\hat{s}^+(k, \boldsymbol{\theta}))$ with $\hat{s}^+(k, \boldsymbol{\theta})$ defined by $s(x(k))$ along the boundaries of $\Omega_s \bigcap \Omega^+(\boldsymbol{\theta})$ and $0 < \hat{\eta}^+(k, \boldsymbol{\theta}) < 1$. Similarly, $\eta^+(k, \boldsymbol{\theta})$ is chosen by equation

$$u_{eq}(k) = (1 - \eta^+(k, \boldsymbol{\theta}))u(k, \boldsymbol{\xi}^+) + \eta^+(k, \boldsymbol{\theta})u(k, \boldsymbol{\xi}^-) \tag{7.51}$$

i.e.

$$\eta^{+}(k,\theta) = \frac{u_{eq}(k) - u(k,\xi^{+})}{u(k,\xi^{-}) - u(k,\xi^{+})} \tag{7.52}$$

The control (7.50) has the same effect on softening the zigzagging and an equation similar to (7.49) holds.

In general this kind of control can soften the zigzagging in the sense that, when the system state $x(k)$ is close to the switching hyperplane with the control $\hat{u}^{-}(k,\theta)$ or $\hat{u}^{+}(k,\theta)$ the system state $x(k+1)$ will not be as far away as when it is driven by $u(k,\xi^{-})$ or $u(k,\xi^{+})$. The modified SDVSC (MSDVSC), denoted by $\overline{u}(k)$, which may reduce zigzagging, is obtained as

$$\overline{u}(k) = \begin{cases} \hat{u}^{+}(k,\theta) & \text{if } x \in \Omega_s \bigcap \Omega^{+}(\theta) \\ \hat{u}^{-}(k,\theta) & \text{if } x \in \Omega_s \bigcap \Omega^{-}(\theta) \\ -\sum_{i=1}^{l} \Psi_i x_i(k) & \text{if } x \notin \Omega_s \end{cases} \tag{7.53}$$

for all $\theta \in \Theta$.

Note that with (7.50) and (7.48), in order to ensure that the system state does not overshoot Ω_s, it is necessary to further reduce the upper bounds $\overline{\alpha}_i (i = 1, \ldots, l)$ such that in the limiting case, within one step of the boundaries of Ω_s, the system state may not overshoot Ω_s and the controls $\hat{u}^{+}(k,\theta)$ and $\hat{u}^{-}(k,\theta)$ can be activated.

7.5 Simulation Results

We now present simulations for a two-dimensional system to show the effectiveness of the algorithm developed.

7.5.1 Two-Dimensional System

Consider the two-dimensional system with

$$A = \begin{bmatrix} 0 & 1 \\ -f_1 & -f_2 \end{bmatrix}, \quad B = [0 \quad 1]^T \tag{7.54}$$

With the use of ZOH, we have

$$\begin{aligned} E = \exp(Ah) &= \begin{bmatrix} e_{11} & e_{12} \\ e_{21} & e_{22} \end{bmatrix} \\ &= \exp(-\kappa_1 h) \begin{bmatrix} \cos(\kappa_2 h) + \kappa_1 \kappa_2^{-1} \sin(\kappa_2 h) & \kappa_2^{-1} \sin(\kappa_2 h) \\ -f_1 \kappa_2^{-1} \sin(\kappa_2 h) & \cos(\kappa_2 h) - \kappa_1 \kappa_2^{-1} \sin(\kappa_2 h) \end{bmatrix} \end{aligned} \tag{7.55}$$

with $\kappa_1 = f_2/2$, $\kappa_2 = (1/2)\sqrt{4f_1 - f_2^2}$, where the constant κ_2 is real or pure imaginary, and

$$G = \int_0^h \exp(A\tau)d\tau B = \begin{bmatrix} g_1 \\ g_2 \end{bmatrix} \tag{7.56}$$

where

$$g_1 = \begin{cases} \frac{1}{\kappa_1^2 + \kappa_2^2}(1 - \exp(-\kappa_1 h)(\kappa_1\kappa_2^{-1}\sin(\kappa_2 h) + \cos(\kappa_2 h))) & \text{if } \kappa_1^2 + \kappa_2^2 \neq 0 \\ (h/(2\kappa_1)) + (\exp(-2\kappa_1 h) - 1)/(4\kappa_1^2) & \text{otherwise} \end{cases} \tag{7.57}$$

$$g_2 = \begin{cases} \frac{1}{\kappa_1^2 + \kappa_2^2}\exp(-\kappa_1 h)(\kappa_2 + \kappa_1^2\kappa_2^{-1})\sin(\kappa_2 h) & \text{if } \kappa_1^2 + \kappa_2^2 \neq 0 \\ \exp(-2\kappa_1 h) & \text{otherwise} \end{cases} \tag{7.58}$$

The discrete controllable canonical form is

$$\Phi = \begin{bmatrix} 0 & 1 \\ -a_1 & -a_2 \end{bmatrix}, \quad \Gamma = [0 \quad 1]^T \tag{7.59}$$

where $a_1 = -e_{12}e_{21} + e_{11}e_{22}$, $a_2 = -e_{11} - e_{22}$. This canonical form can be obtained by means of the following transformation (Ogata 1987)

$$P = [G \quad EG][\Gamma \quad \Phi\Gamma]^{-1} \tag{7.60}$$

The SDVSC is $u(k) = -\Psi_1 x_1(k)$ where Ψ_1 takes as α_1 or $-\alpha_1$ according to $x_1(k)s(k) \geq 0$ or $x_1(k)s(k) < 0$. The system is actually a linear system with two linear feedback control structures. MSDVSC is discussed in the next section.

There are only two control structures in Ξ corresponding to $\boldsymbol{\xi}_1 = [\alpha_1 \; 0]^T$ and $\boldsymbol{\xi}_2 = [-\alpha_1 \; 0]^T$. The set Θ has only two elements, $\boldsymbol{\theta}_1 = [1 \; 0]^T$ and $\boldsymbol{\theta}_2 = [-1 \; 0]^T$. The two partitions in the phase plane are $R(\boldsymbol{\theta}_1) = \{x \in \mathbb{R}^2, \; x_1 \geq 0\}$ and $R(\boldsymbol{\theta}_2) = \{x \in \mathbb{R}^2, \; x_1 < 0\}$. Obviously $\mathbb{R}^2 = R(\boldsymbol{\theta}_1) \bigcup R(\boldsymbol{\theta}_2)$.

The characteristic polynomial for SDVSC is

$$P(\lambda; \boldsymbol{\xi}) = \lambda^2 + a_2\lambda + a_1 + \Psi_1, \quad \boldsymbol{\xi} \in \Xi \tag{7.61}$$

The construction of the orbit of the system with a control structure $\boldsymbol{\xi} \in \Xi$ is as follows. In the $x_1(k), x_2(k)$ (or $x_1(k+1)$) phase plane the system state $x(k)$ follows a curve $W(k)$ satisfying

$$W(k) = x_2^2(k) + a_2 x_1(k)x_2(k) + (a_1 + \Psi_1)x_1^2(k) \tag{7.62}$$

which represents an ellipse and a hyperbola if $a_2^2 - 4(a_1 + \Psi_1)$ is less than zero and greater than zero respectively (Potts 1982). We ignore the case $a_2^2 - 4(a_1 + \Psi_1) = 0$ since we can always select Ψ_1 such that $a_2^2 - 4(a_1 + \Psi_1) \neq 0$. We obtain

$$W(k+1) = (a_1 + \Psi_1)W(k) \tag{7.63}$$

which indicates that the elliptic or hyperbolic curve $W(k)$ for increasing k, is contracting (or expanding), if $|a_1 + \Psi_1|$ is less than one (greater than one), respectively.

Figure 7.4 illustrates a typical orbit of the discrete points on an elliptic curve $W(k)$, $k = 1, 2, \ldots$. $W(k)$ (Potts 1982) from an initial point, by first projecting horizontally to the line $x_1(k) = x_2(k)$, then vertically to the next appropriate $W(k)$ curve, then horizontally and vertically in a staircase. Figure 7.5 illustrates a typical orbit of the discrete points on a hyperbolic curve $W(k)$ which proceeds in the same manner.

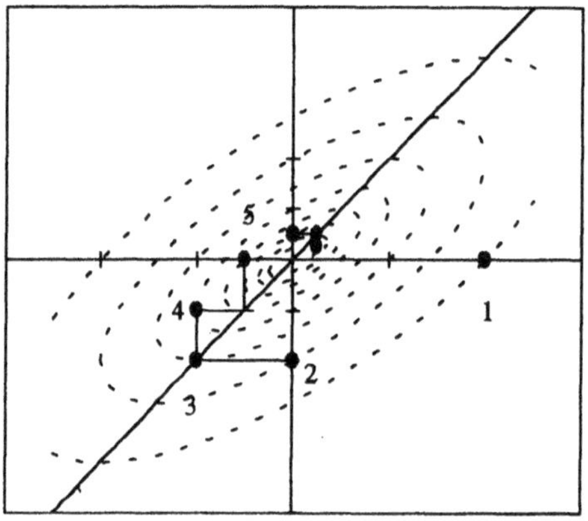

Fig. 7.4. Construction of an orbit and contracting ellipses

7.5.2 Example 1

By choosing $f_1 = 0$, $f_2 = 0.05$, $h = 0.05$ and $c_1 = -0.9975$, then $a_1 = 0.9975$, $a_2 = -1.9975$. Equation (7.34) gives $\rho = 0$, and (7.34) gives the lower bound $\underline{\alpha}_1 = 0$ which is determined by

$$\underline{\alpha}_1 = |a_1 + c_1 a_2 - c_1^2| \tag{7.64}$$

Now let us investigate the mechanism of approaching the switching line. For the given parameters, the characteristic polynomial of $\boldsymbol{\xi}_1$, $P(\lambda; \boldsymbol{\xi}_1)$, has only complex eigenvalues. So there does not exist any asymptote. However, the characteristic polynomial for $\boldsymbol{\xi}_2$, $P(\lambda; \boldsymbol{\xi}_2)$, has two real eigenvalues: one being greater than one, the other less than one; meaning that there are two asymptotes. Since in $R(\theta_1) \bigcap \{x \in \mathbb{R}^2,\ s \geq 0\}$ the control structure is $\boldsymbol{\xi}_1$, there does not exist any asymptote. Therefore $\Omega^+(\theta_1) = R(\theta_1) \bigcap \{x \in \mathbb{R}^2,\ s \geq 0\}$. Similarly in $R(\theta_2) \bigcap \{x \in \mathbb{R}^2,\ s < 0\}$, the control structure is $\boldsymbol{\xi}_1$ and

$\Omega^-(\theta_2) = R(\theta_2)\bigcap\{x \in \mathbb{R}^2,\ s < 0\}$. However in $R(\theta_1)\bigcap\{x \in \mathbb{R}^2,\ s < 0\}$ the control structure is ξ_2, meaning that there exist two asymptotes in which only the one with real eigenvalue larger than one, defined by $r_1 = 0$, is partly in $R(\theta_1)\bigcap\{x \in \mathbb{R}^2,\ s < 0\}$. Therefore $\Omega^-(\theta_1) = R(\theta_1)\bigcap\{x \in \mathbb{R}^2,\ r_1 > 0,\ s < 0\}$. Similarly in $R(\theta_2)$ and $s \geq 0$ the control structure is ξ_2, and $\Omega^+(\theta_2) = R(\theta_2)\bigcap\{x \in \mathbb{R}^2,\ r_1 < 0,\ s \geq 0\}$. These subsets are illustrated in Fig. 7.6, where $\Omega^+(\theta_1) =$ I, $\Omega^-(\theta_1) =$ IIA, $\Omega^-(\theta_2) =$ III, $\Omega^+(\theta_2) =$ IVA. In $R(\theta_1)$ the control structure pair (ξ^+, ξ^-) is (ξ_1, ξ_2), and symmetrically in $R(\theta_2)$ the control structure pair (ξ^+, ξ^-) is (ξ_2, ξ_1).

By combining two motions (see Fig. 7.6) one being the orbits of discrete points of the system with the control structure ξ_1 for $x_1 s \geq 0$ (i.e. the regions I, III), the other the orbits of discrete points of the system with the control structure ξ_2 for $x_1 s < 0$ (i.e. regions IIA and IVA), the switching line $s = 0$ can be approached and/or crossed. Figure 7.6 illustrates the mechanism of approaching the switching line $s = 0$. Divergence may happen for example, when the state $x(k)$ is in region I. The SDVSC control with a larger α_1, may drive $x(k+1)$ to the region IIB, overshooting the attraction region IIA. $x(k+2)$ will then move away from $s = 0$. This divergence may occur many times.

To avoid such divergence (or overshooting the attracting regions), apply the algorithm (see Appendix 7.2). The upper bound $\overline{\alpha}_1$ is obtained by

$$[Q(\Phi - \Gamma\xi_1^T)L]^T[v_{\xi_2}^T v_{\xi_2} cc^T - (c^T v_{\xi_2})^2 I_2][Q(\Phi - \Gamma\xi_1^T)L]^T \geq 0 \tag{7.65}$$

$$c^T v_{\xi_2} \geq 0 \tag{7.66}$$

with

$$Q = \begin{bmatrix} 0 & -1 \\ 1 & 0 \end{bmatrix}, \quad L = [1 \quad -c_1]^T \tag{7.67}$$

$$c = [c_1 \quad 1]^T, \quad v_{\xi_2} = [v_1^{\xi_2} \quad 1]^T \tag{7.68}$$

$$v_1^{\xi_2} = 0.5(a_2 + \sqrt{a_2^2 + 4(\alpha_1 - a_1)}), \quad \xi_1 = [\alpha_1 \quad 0]^T \tag{7.69}$$

The only asymptote is $r_1 = v_{\xi_2}^T x = v_1^{\xi_2} x_1 + x_2 = 0$. We use this asymptote for construction of $\Omega^-(\theta_1)$ and $\Omega^+(\theta_2)$, since $r_1 = 0$ is the only asymptote in both regions. The regions are symmetrical with respect to the origin of the coordinates. Equation (7.65) with (7.67)–(7.69) gives $\overline{\alpha}_1 = 0.946$, the upper bound, which also satisfies (7.66). Figure 7.7 illustrates the system behaviour using SDVSC with $\alpha_1 = 0.1$, which is less than the upper bound. The system trajectory exhibits zigzagging.

To eliminate the zigzagging, MSDVSC is adopted in which ϵ is chosen to be 0.5. The region Ω_s is constrained by the following two lines around the switching line $x_2 - 0.4493x_1 = 0$ and $x_2 - 2.2017x_1 = 0$, which are well within the attracting regions. To work out the new limit on α_1, we replace $v_1^{\xi_2}$ in (7.68) by -0.4493. Calculation yields $\overline{\alpha}_1 = 0.3$. Following the control scheme for $x \in \Omega_s$ in Sect. 7.4.5, we have

$$\hat{\eta}^+(k,\theta_1) = (1-\frac{s(k)}{\hat{s}^+(k,\theta_1)})\frac{(a_1+\alpha_1)x_1(k)+(a_2-c_1)x_2(k)}{2\alpha_1 x_1(k)} \quad (7.70)$$

$$\hat{\eta}^-(k,\theta_1) = (1-\frac{s(k)}{\hat{s}^-(k,\theta_1)})\frac{(a_1+\alpha_1)x_1(k)+(a_2-c_1)x_2(k)}{-2\alpha_1 x_1(k)} \quad (7.71)$$

$$\hat{s}^-(k,\theta_1) = \frac{-c_1\epsilon^2-\epsilon\sqrt{1+c_1^2-\epsilon^2}}{1-\epsilon^2}x_1(k) \quad (7.72)$$

$$\hat{s}^+(k,\theta_1) = \frac{-c_1\epsilon^2+\epsilon\sqrt{1+c_1^2-\epsilon^2}}{1-\epsilon^2}x_1(k) \quad (7.73)$$

and symmetrically $\hat{s}^+(k,\theta_2)=\hat{s}^-(k,\theta_1)$, $\hat{s}^-(k,\theta_2)=\hat{s}^+(k,\theta_1)$, $\hat{\eta}^+(k,\theta_2)=\hat{\eta}^-(k,\theta_1)$, $\hat{\eta}^-(k,\theta_2)=\hat{\eta}^+(k,\theta_1)$. Then the MSDVSC is

$$\overline{u}(k)=\begin{cases}(1-2\hat{\eta}^+(k,\theta))\alpha_1 x_1(k) & \text{if } x\in\Omega_s\bigcap\Omega^+(\theta)\\ -(1-2\hat{\eta}^-(k,\theta))\alpha_1 x_1(k) & \text{if } x\in\Omega_s\bigcap\Omega^-(\theta)\\ -\alpha_1 \mathrm{sgn}s(k)|x_1(k)| & \text{if } x\notin\Omega_s\end{cases} \quad (7.74)$$

for $\theta=\theta_1$, or θ_2. Figure 7.8 demonstrates the performance with elimination of zigzagging using MSDVSC. The system state smoothly approaches the switching line and then stays on the line when it reaches the line.

7.5.3 Example 2

If one chooses $f_1=0$, $f_2=-3.49$, $h=0.046$ and $c_1=-0.955$, then $a_1=1.1741$, $a_2=-2.1741$. Using similar analysis to that in Sect. 7.5.2, (7.64) and (7.65)–(7.69) give the lower bound 0.0099, the upper bound 0.837. $\rho=0.2$ is small and slightly violates the condition (7.34). However it only affects the system performance slightly when the system state is far away from the switching line. The tendency towards the switching line is not affected since it is characterized by the asymptote. Figure 7.9 shows the system behaviour using SDVSC with $\alpha_1=0.8$ which is just less than the upper bound. The system trajectory exhibits zigzagging about the switching line. Figure 7.10 shows the system behaviour using SDVSC with $\alpha_1=0.9$ which is just larger than the upper bound. The system state at certain instants leaves the switching line, but converges on the other side, illustrating the occurrence of divergence.

To eliminate zigzagging, we use MSDVSC with $\epsilon=0.5$. The region Ω_s is constrained by the following two lines around the switching line $x_2-0.4137x_1=0$ and $x_2-2.2138x_1=0$.

Replace $v_1^{\xi_2}$ in (7.68) by -0.4137. Calculation yields $\overline{\alpha}_1=0.44$. Using (7.74) with (7.70) – (7.73), the improved performance with elimination of zigzagging is shown in Fig. 7.11 for $\alpha_1=0.4$. The influence of the slight violation of the condition (7.34) can be seen from the small rises of $s(k)$ in the first few instants in the Figs. 7.9(b), 7.10(b) and 7.11(b).

It is interesting to note that from the various simulations we find that with SDVSC the smaller the sampling period, the bigger the value of $(\overline{\alpha}_1-\underline{\alpha}_1)$;

indicating an expansion of the upper limitation on α_1. With the same α_1, the bigger the sampling period and the more serious the zigzagging. However the MSDVSC always give smooth sliding along the switching line provided a small α_1 is chosen so that at some instants the system state may drop into Ω_s where the softening control $\hat{u}(k, \boldsymbol{\xi})$ can be activated.

7.6 Conclusions

This chapter has reviewed the recent development of the theory of DVSC and developed a DVSC scheme which enables the elimination of zigzagging as well as the divergence from the switching hyperplane. The control strategy is as follows: outside a given neighbourhood of the switching hyperplane the conventional VSC structure SDVSC is used to force the system state to approach and/or cross the switching hyperplane, and within the neighbourhood the MSDVSC is used to eliminate the zigzagging. An algorithm to calculate the upper and lower bounds for SDVSC has been proposed. The upper and lower bounds are independent of the distance of the system state from the switching hyperplane. Simulation results have been presented to show the effectiveness of the scheme developed.

The systems we have discussed are linear discrete-time systems. It is intended to extend the theory to nonlinear discrete-time systems.

Acknowledgements

The author is indebted to the Australian Research Council for a grant.

Appendix 7.1 Asymptote Hyperplane

Suppose the characteristic polynomial with a control structure $\boldsymbol{\xi} \in \Xi$ has at least one real positive eigenvalue which is greater than one. Without loss of generality, we can therefore order the eigenvalues as $\lambda_1^\xi, \lambda_2^\xi, ..., \lambda_n^\xi$, where λ_1^ξ is the real eigenvalue with value greater than one. An asymptote hyperplane

$$r_1(x, \boldsymbol{\xi}) = (v_\xi^1)^T x = 0 \tag{7.75}$$

can be constructed in which v_ξ^1 is a left eigenvector of $\Phi - \Gamma \boldsymbol{\xi}^T$ corresponding to the positive eigenvalue λ_1^ξ. In terms of the eigenvalues (assumed distinct), the components of the vector v_ξ^1 are given by

$$v_\xi^1 = [v_{1,1}^\xi \quad v_{1,2}^\xi \quad \dots \quad v_{1,n-1}^\xi \quad v_{1,n}^\xi]^T \tag{7.76}$$

where

$$v^{\xi}_{1,n} = 1 \tag{7.77}$$

$$v^{\xi}_{1,n-1} = -\sum_{i=2}^{n} \lambda^{\xi}_i \tag{7.78}$$

$$v^{\xi}_{1,n-2} = \sum_{i=2}^{n-1} \sum_{j=i+1}^{n} \lambda^{\xi}_i \lambda^{\xi}_j \tag{7.79}$$

$$\vdots \qquad \vdots$$

$$v^{\xi}_{1,1} = (-1)^{n-1} \lambda^{\xi}_2 \lambda^{\xi}_3 \ldots\ldots \lambda^{\xi}_n \tag{7.80}$$

The eigenvector is in general only defined to within a constant multiplier. It is easy to deduce that

$$\begin{aligned} \nabla r_1(x(k), \boldsymbol{\xi}) &= r_1(x(k+1), \boldsymbol{\xi}) - r_1(x(k), \boldsymbol{\xi}) \\ &= (\lambda^{\xi}_1 - 1) r_1(x(k), \boldsymbol{\xi}) \end{aligned} \tag{7.81}$$

Since $\lambda^{\xi}_1 > 1$, $\nabla r_1(x(k), \boldsymbol{\xi})$ is negative, zero or positive accordingly as $r_1(x(k), \boldsymbol{\xi})$ is negative, zero or positive. This characterizes the property of an asymptote hyperplane.

Appendix 7.2 Algorithm

An algorithm for calculating the upper bounds $\overline{\alpha}_i (i = 1, \ldots, l)$ by using one of the asymptote hyperplanes is as follows. For some $\boldsymbol{\theta} \in \Theta$, and in its partition $R(\boldsymbol{\theta})$, there exist the control structures $\boldsymbol{\xi}^+, \boldsymbol{\xi}^-$ which are to be used in the subsets $R(\boldsymbol{\theta}) \bigcap \{x \in \mathbb{R}^n,\ s \geq 0\}$ and $R(\boldsymbol{\theta}) \bigcap \{x \in \mathbb{R}^n,\ s < 0\}$ respectively. Assume that in the limiting case with a particular value of k, $x(k) \in \Omega^+(\boldsymbol{\theta})$ and satisfies

$$s(k) = c^T x(k) = 0^+ \tag{7.82}$$

which implies that $x(k)$ is just above $s = 0$. Using the control $u(k, \boldsymbol{\xi}^+)$ with certain α_i $(\alpha_i > \underline{\alpha}_i)$, $i = 1, \ldots, l$, the system state at $k+1$ just hits one of the asymptote hyperplanes $r(x, \boldsymbol{\xi}^-) = 0$ in $\Omega^-(\boldsymbol{\theta})$, i.e.

$$r(x(k+1), \boldsymbol{\xi}^-) = v^T_{\xi^-} x(k+1) = v^T_{\xi^-} (\Phi - \Gamma(\boldsymbol{\xi}^+)^T) x(k) = 0 \tag{7.83}$$

Here the normal directions of the hyperplanes $s = 0$ and $r = 0$ are c and v_{ξ^-} respectively. With smaller values of α_i $(i = 1 \ldots, l)$ $(\alpha_i > \underline{\alpha}_i)$, the control $u(k, \boldsymbol{\xi}^+)$ should bring the system state $x(k+1)$ to the region below the switching hyperplane $s = 0$ and above the asymptote hyperplane $r(x, \boldsymbol{\xi}^-) = 0^+$ (i.e. $\Omega^-(\boldsymbol{\theta})$), so that $x(k+2)$ with the control $u(k+1, \boldsymbol{\xi}^-)$ immediately returns towards the switching hyperplane $s = 0$. An algorithm for calculation of the upper bounds is now proposed. Construct a $(n-1)$-dimensional hyperplane by using a state $x(k+1)$, which is between $s = 0^-$ and $r(x, \boldsymbol{\xi}^-) = 0^+$, and a $(n-2)$-dimensional hyperplane which is the intersection of $s = 0$ and $r = 0$, through the origin of the coordinates. This $(n-1)$-dimensional hyperplane can be generated by noting

$$w(\overline{x}) = p^T\overline{x} = \begin{vmatrix} \overline{x}^T & 1 \\ x^T(k+1) & 1 \\ D^T & \underline{1} \\ \underline{0}^T & 1 \end{vmatrix} = 0 \tag{7.84}$$

where $\overline{x} \in \mathbb{R}^n$ is an arbitrary variable vector, and

$$D = \begin{bmatrix} I_{n-2} \\ D_1^{-1}D_2 \end{bmatrix} \tag{7.85}$$

$$D_1^{-1} = 1/(c_{n-1} - v_{n-1}^{\xi^-}) \begin{bmatrix} 1 & -1 \\ -v_{n-1}^{\xi^-} & c_{n-1} \end{bmatrix} \tag{7.86}$$

$$D_2 = \begin{bmatrix} -c_1 & -c_2 & \dots & -c_{n-2} \\ -v_1^{\xi^-} & -v_2^{\xi^-} & \dots & -v_{n-2}^{\xi^-} \end{bmatrix} \tag{7.87}$$

$$\underline{1} = [1 \;\; 1 \;\; \dots \;\; 1]^T \in \mathbb{R}^{n-2} \tag{7.88}$$

$$\underline{0} = [0 \;\; 0 \;\; \dots \;\; 0]^T \in \mathbb{R}^{n} \tag{7.89}$$

and I_{n-2} is a $(n-2) \times (n-2)$ unit matrix. Define

$$p = [p_1 \; p_2 \; \dots \; p_n]^T \tag{7.90}$$

where $p_i(i = 1, \dots, n)$ is the cofactor of $\overline{x}_i$ in (7.84), $i = 1, \dots, n$; i.e. p is the normal direction of this hyperplane. Thus p_i is a linear function of $x(k+1)$, and can be expressed as

$$p = Qx(k+1) \tag{7.91}$$

The tendency of the system trajectory is from $s = 0^+$ to $s < 0$ (i.e. from $\Omega^+(\boldsymbol{\theta})$ to $\Omega^-(\boldsymbol{\theta})$). We define φ_{sr}, the positive angle from $s = 0^-$ to $r = 0$, and φ_{sw}, the positive angle from $s = 0^-$ to $w = 0$. The sufficient condition to avoid divergence from $s = 0$ is

$$\varphi_{sw} \le \varphi_{sr} \tag{7.92}$$

Using the normal directions of these three hyperplanes,

$$\cos\varphi_{sw} = \frac{c^T p}{\sqrt{c^T c}\sqrt{p^T p}} \tag{7.93}$$

$$\cos\varphi_{sr} = \frac{c^T v_{\xi^-}}{\sqrt{c^T c}\sqrt{(v_{\xi^-})^T v_{\xi^-}}} \tag{7.94}$$

Note that (7.94) may be negative. We then impose the condition

$$c^T v_{\xi^-} \ge 0 \tag{7.95}$$

such that $0 \le \varphi_{sw} \le \varphi_{sr} \le \pi/2$. Therefore, from (7.95) the necessary and sufficient condition for (7.92) to hold, is that

$$\cos^2 \varphi_{sw} \geq \cos^2 \varphi_{sr} \tag{7.96}$$

which is equivalent to

$$(v_{\xi^-})^T v_{\xi^-} (c^T p)^2 \geq (c^T v_{\xi^-})^2 p^T p \tag{7.97}$$

Since $x(k) \in R(\theta) \bigcap \{x \in \mathrm{IR}^n,\ s \geq 0\}$ where ξ^+ is used, then

$$x(k+1) = (\Phi - \Gamma(\xi^+)^T)x(k) \tag{7.98}$$

and the limiting case $s(k) = 0$ implies

$$x_n(k) = -c_1 x_1(k) - c_2 x_2(k) - \ldots - c_{n-1} x_{n-1}(k) \tag{7.99}$$

Define

$$x(k) = L\underline{x}(k) \tag{7.100}$$

$$L = \begin{bmatrix} I_{n-1} \\ -\underline{c}^T \end{bmatrix} \tag{7.101}$$

$$\underline{c} = [c_1\ c_2\ \ldots\ c_{n-1}]^T \tag{7.102}$$

$$\underline{x}(k) = [x_1(k)\ x_2(k)\ \ldots\ x_{n-1}(k)]^T \tag{7.103}$$

Then substituting (7.91), (7.98) and (7.100)–(7.102) into (7.97) yields

$$\begin{aligned} &\underline{x}^T(k)(Q(\Phi - \Gamma(\xi^+)^T)L)^T[(v_{\xi^-})^T v_{\xi^-} cc^T \\ &\quad -(c^T v_{\xi^-})^2 I_n] Q(\Phi - \Gamma(\xi^+)^T)L\underline{x}(k) \geq 0 \end{aligned} \tag{7.104}$$

where I_n and I_{n-1} are $n \times n$ and $(n-1) \times (n-1)$ unit matrices respectively. The inequalities (7.104) and (7.95) give the constraint for obtaining the upper bounds $\overline{\alpha}_i$ $(i = 1, \ldots, l)$ in $\Omega^-(\theta)$ that prevent the system states from diverging from the asymptote hyperplane $r = 0$.

The upper bounds can be the values making the symmetric matrix associated with the quadratic polynomial (7.104) semi-positive definite, and also satisfying (7.95).

References

DeCarlo, R.A., Żak, S.H., Matthews, G.P. 1988, Variable structure control of nonlinear multivariable systems: a tutorial. *Proceedings of IEEE* **76**, 212–232

Furuta, K. 1990, Sliding mode control of a discrete system. *Systems and Control Letters* **14**, 145–152

Grantham, W.J., Athalye, A.M. 1990, Discretization chaos: feedback control and transition to chaos. *Control and Dynamic Systems* **34**, 205–277

Kotta, U. 1989, Comments on the stability of discrete-time sliding mode control systems. *IEEE Transactions on Automatic Control* **AC-34**, 1021–1022

Magaña, M.E., Żak, S.H. 1987, The control of discrete-time uncertain dynamical systems. *Research Report TR-EE 87-32*, School of Electrical Engineering, Purdue University, West Lafayette, Indiana, USA

Milosavljevic, C. 1985, General conditions for the existence of a quasi-sliding mode on the switching hyperplane in discrete variable structure systems. *Automat. Remote Control* **46**, 307–314

Ogata, K. 1987, *Discrete-Time Control Systems*, Prentice-Hall, Englewood Cliffs, N.J.

Potts, R.B. 1982, Differential and difference equations. *Am. Math. Monthly* **89**, 402–407

Potts, R.B., Yu, X. 1991, Discrete structure system with pseudo-sliding mode, *Journal of Australian Mathematical Society, Ser. B* **32**, 365–376

Sarpturk, S.Z., Istefanopulos, Y., Kaynak, O. 1987, The stability of discrete-time sliding mode control systems. *IEEE Transactions on Automatic Control* **AC-32**, 930–932

Sira-Ramírez, H. 1991, Nonlinear discrete variable structure systems in quasi-sliding mode. *International Journal of Control* **54**, 1171–1187

Spurgeon, S.K. 1992, Hyperplane design techniques for discrete-time variable structure control systems. *International Journal of Control* **55**, 445–456

Utkin, V.I. 1987, Variable structure systems with sliding modes. *IEEE Transactions on Automatic Control* **AC-22**, 212–222

Utkin, V.I., Drakunov, S.V. 1989, On discrete-time sliding mode control. *Proceedings of IFAC Symposium on Nonlinear Control Systems (NOLCOS)*, Capri, Italy, 484–489

Utkin, V.I. 1992, Sliding mode control in dynamic systems. *Proceedings of Second IEEE Workshop on Variable Structure and Lyapunov Control of Uncertain Dynamical Systems*, Sheffield, UK, 170–181

Yu, X., Potts, R.B. 1992a, Analysis of discrete variable structure systems with pseudo-sliding modes. *International Journal of Systems Science* **23**, 503–516

Yu, X., Potts, R.B. 1992b, Computer-controlled variable structure system. *Journal of Australian Mathematical Society, Ser. B* **34**, 1–17

Yu, X. 1992, Chaos in discrete variable structure systems. *Proc IEEE Conference on Decision and Control*, Tucson, USA, **2**, 1862–1863

Yu, X. 1993, Discrete variable structure control systems. *International Journal of Systems Science* **24**, 373–386

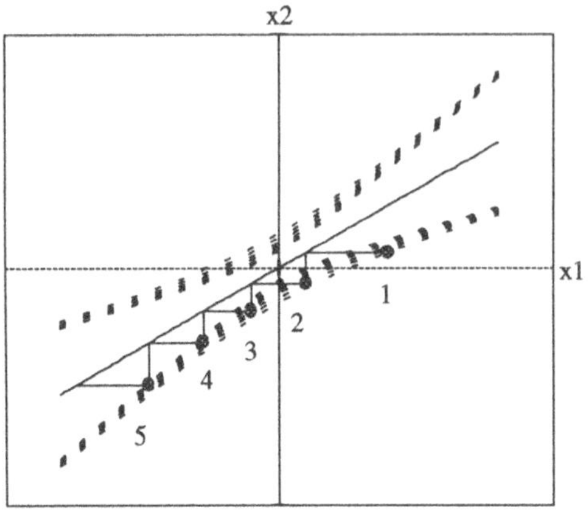

Fig. 7.5. Construction of an orbit and expanding hyperbolas

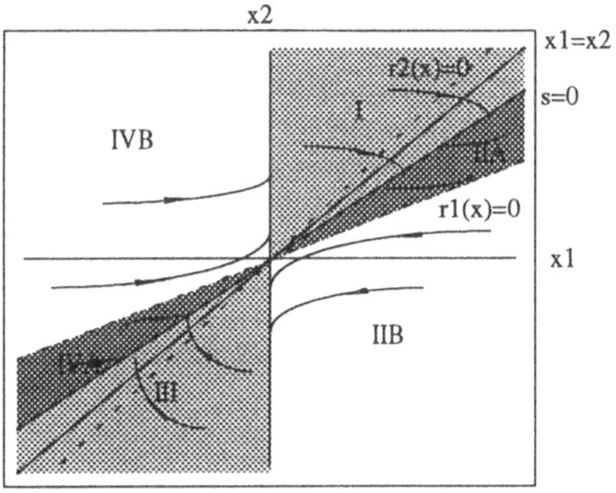

Fig. 7.6. The $x_1(k)$, $x_2(k)$ phase plane is divided into regions corresponding to the sign of $x_1(k)s(k)$. The shaded regions I and III and the heavily shaded regions IIA and IVA are the attracting regions. $r_1(x)$, $r_2(x)$ are the two asymptotes.

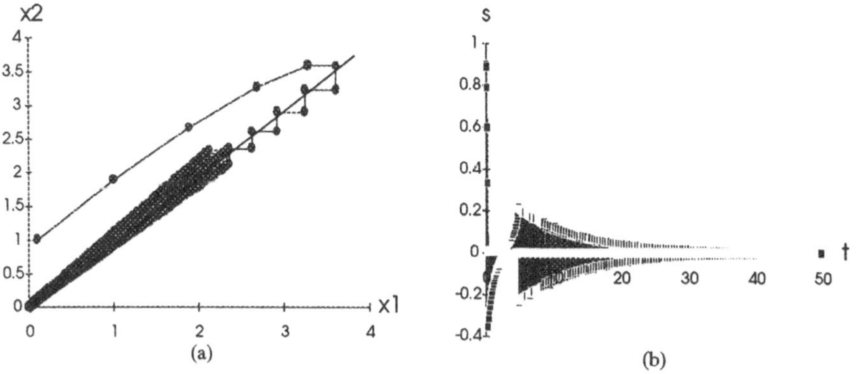

Fig. 7.7. SDVSC with $\alpha_1 = 0.1$. (a) Phase plane portrait. (b) Switching variable

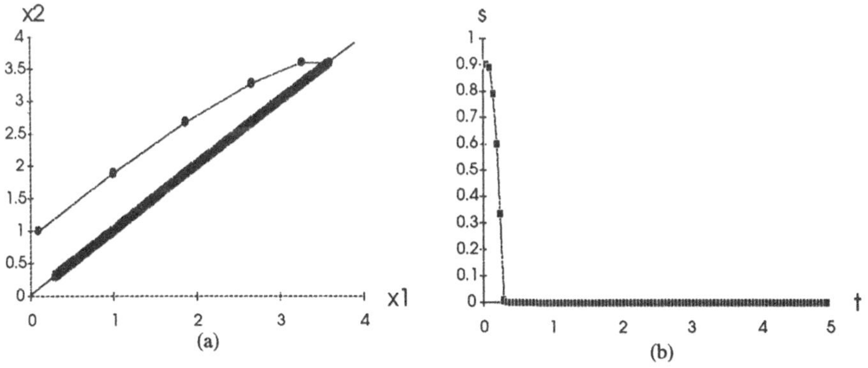

Fig. 7.8. MSDVSC with $\alpha_1 = 0.1$. (a) Phase plane portrait. (b) Switching variable

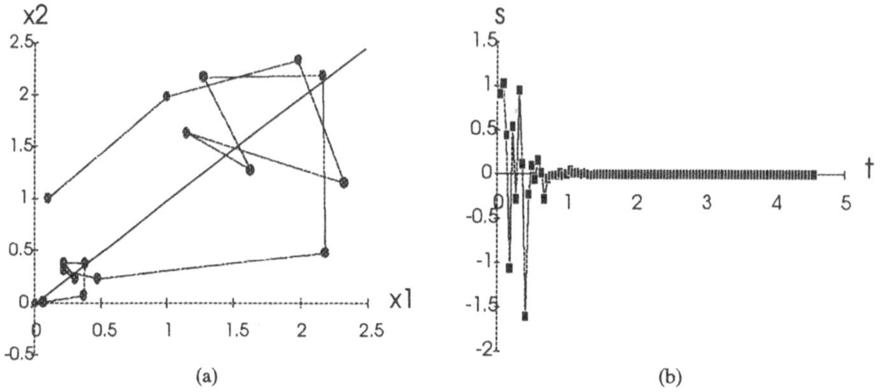

Fig. 7.9. SDVSC with $\alpha_1 = 0.8$. (a) Phase plane portrait. (b) Switching variable

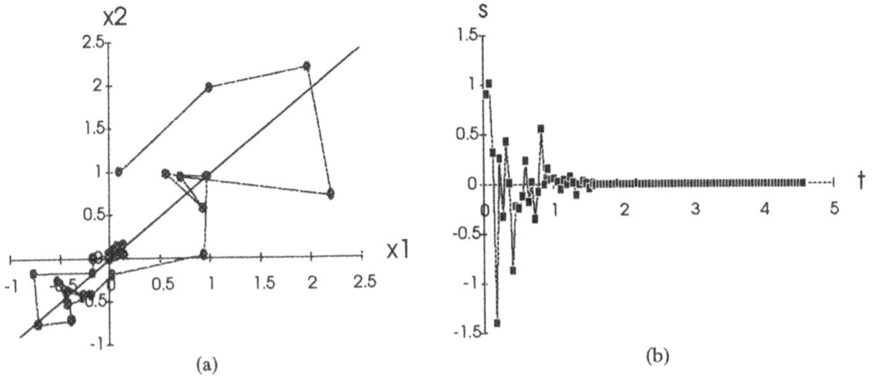

Fig. 7.10. SDVSC with $\alpha_1 = 0.9$. (a) Phase plane portrait. (b) Switching variable

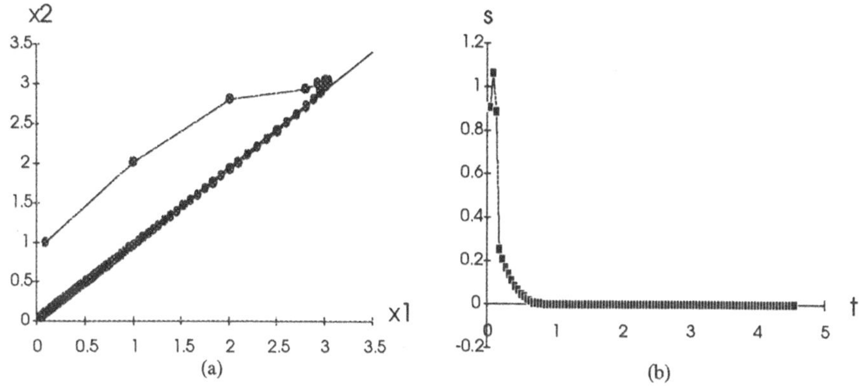

Fig. 7.11. MSDVSC with $\alpha_1 = 0.4$. (a) Phase plane portrait. (b) Switching variable

8. Robust Observer-Controller Design for Linear Systems

Hebertt Sira-Ramírez, Sarah K. Spurgeon and Alan S.I. Zinober

8.1 Introduction

Sliding mode observation and control schemes for both linear and nonlinear systems have been of considerable interest in recent times. Discontinuous nonlinear control and observation schemes, based on sliding modes, exhibit fundamental robustness and insensitivity properties of great practical value (see Utkin (1992), and also Canudas de Wit and Slotine (1991)). A fundamental limitation found in the sliding mode control of linear perturbed systems and in sliding mode feedforward regulation of observers for linear perturbed systems, is the necessity to satisfy some structural conditions of the "matching" type. These conditions have been recognized in the work of Utkin (1992), Walcott and Żak (1988) and Dorling and Zinober (1983). Such structural constraints on the system and the observer have also been linked to *strictly positive real* conditions in Walcott and Żak (1988) and in the work of Watanabe et al (1992). More recently a complete Lyapunov stability approach for the design of sliding observers, where the above-mentioned limitations are also apparent, was presented by Edwards and Spurgeon (1993).

Here a different approach to the problem of output feedback control for any controllable and observable, perturbed linear system is taken. For the sake of simplicity, single-input single-output perturbed plants are considered, but the results can be easily generalized to multivarable linear systems.

Using a *Matched Generalized Observer Canonical Form* (MGOCF), similar to those developed by Fliess (1990a), it is found that for the sliding mode state observation problem in observable systems, the structural conditions of the matching type are largely irrelevant. This statement is justified by the fact that a perturbation input "rechannelling" procedure *always* allows one to obtain a matched realization for the given system. Such rechannelling is never carried out in practice and its only purpose is to obtain a reasonable estimate (bound) of the influence of the perturbation inputs on the state equations of the proposed canonical form. It is shown that the chosen matched output reconstruction error feedforward map, which is a design quantity, uniquely determines the stability features of the reduced order sliding state estimation error dynamics. The state vector of the proposed realization is, hence, robustly asymptotically estimated, independently of whether or not the matching conditions are satisfied by the original system.

The sliding mode output regulation problem for controllable and observable minimum phase systems is then addressed, using a combination of a sliding mode observer and a sliding mode controller. For this, a suitable modification of the MGOCF is proposed. The resulting matched canonical form turns out quite surprisingly to be in a traditional Kalman state space representation form. The obtained Matched Output Regulator Canonical Form (MORCF) is constructed in such a way that it is always matched with respect to the "re-channelled" perturbation inputs. The output signal of the system, expressed now in canonical form, is shown to be controlled by a suitable dynamical "pre-compensator" input, which is physically realizable. For the class of systems treated, the combined state estimation and control problem (i.e. output regulation problem) is therefore always robustly solvable by means of a sliding mode scheme, independently of any matching conditions.

In Sect. 8.2 the role of the matching conditions in sliding mode controller, sliding mode observer and sliding mode output regulation designs, is examined from a classical state space representation viewpoint. This section addresses the rather restrictive nature of the structural conditions that guarantee the robust reconstruction and robust regulation of the system state vector components. In essence, these conditions imply that the feedforward output error injection map of the observer must be in the range space of the perturbation input distribution map of the system. For guaranteeing robustness in a sliding mode control problem, the matching conditions demand that the perturbation input channel map must be in the range space of the control input channel map. For the observer design in particular, these matching conditions imply that the freedom in choosing the stability features of the reduced order ideal sliding reconstruction error dynamics, is severely curtailed and the structure of the system must, by itself, guarantee asymptotic stability of the reduced order observation error dynamics. If the matching condition is not satisfied, then the observation error is dependent upon the external perturbations, and accurate state reconstruction is not feasible.

In Sect. 8.3 the MGOCF, based on the input-output description of the given system, is proposed and it is shown that the matching conditions can always be satisfied while placing no restrictions on the stabilizability of the feedforward regulated error dynamics. This result constitutes the "dual", in a certain sense, to that recently published by Fliess and Messager (1991), involving sliding mode controllers for linear time-invariant controllable systems. Sect. 8.4 presents the MORCF for minimum phase controllable and observable systems. The proposed canonical form is shown to be suitable for the simultaneous design of a robust sliding mode observer/sliding mode controller scheme, independently of any matching conditions. A tutorial design example which considers the design of a sliding mode controller for a power converter demonstrates the theoretical results of this chapter in Sect. 8.5. In Sect. 8.6 conclusions are drawn and further research is suggested.

8.2 Matching Conditions in Sliding Mode State Reconstruction and Control of Linear Systems

Here the classical approaches to sliding mode controller and observer design using the traditional Kalman state variable representation of linear time-invariant systems are presented. Within this constrained formulation, robust observation and control schemes are feasible if, and only if, certain structural conditions are satisfied. The structural conditions for the sliding mode controller design restrict the system's input disturbance distribution map to the range of the control input distribution map. Similar conditions for the sliding mode observer design demand that the observer's feedforward output error injection map be in the range of the system's input disturbance distribution map.

Consider a controllable and observable n-dimensional linear system of the form

$$\begin{aligned} \dot{x} &= Ax + bu + \gamma\xi \\ y &= cx \end{aligned} \tag{8.1}$$

where u and ξ are, respectively, the scalar control input signal and the (bounded) scalar external perturbation input signal. The output y is also assumed to be a scalar quantity. All matrices have the appropriate dimensions. The column vector γ is referred to as the *perturbation input distribution map*, while b is called the *control input distribution map*. The system (8.1) is assumed to be *relative degree* one, i.e. the scalar product $cb \neq 0$. It is assumed, without loss of generality, that $cb > 0$. Furthermore, it is assumed that the underlying input-output system is *minimum phase*.

8.2.1 Matching Conditions in Sliding Mode Controller Design

Suppose it is desired by means of state feedback to zero the output y of the given system. It is well known that if the system (8.1) is unperturbed (i.e. $\xi = 0$), then a variable structure feedback control law of the form

$$u = -\frac{1}{cb}(cAx + K \text{ sign } y) \tag{8.2}$$

where $K > 0$ is a constant design gain, accomplishes the desired control objective in finite time. The output signal y satisfies then the following dynamics

$$\dot{y} = -K \text{ sign } y \tag{8.3}$$

It can be shown under rather mild assumptions that the regulated output variable y of the perturbed system (8.1) still converges to zero in finite time, when the controller (8.2) is used. Indeed the resulting controlled behaviour of the output signal when the controller (8.2) is used in the system (8.1) is given by

$$\dot{y} = c\gamma\xi - K \operatorname{sign} y \tag{8.4}$$

Let the absolute value of the perturbation input ξ be bounded by a constant $M > 0$. Then, for $K > M|c\gamma|$, the feedback control policy (8.2) is seen to create in finite time a *sliding regime* on the hyperplane represented by $y = 0$, irrespective of the particular values adopted by ξ.

The *ideal sliding dynamics* satsified by the controlled state vector x are obtained from the following invariance conditions (Utkin 1992)

$$y = 0 \ , \qquad \dot{y} = 0 \tag{8.5}$$

These conditions imply the existence of a "virtual" perturbation-dependent value of the regulating input u, known as the *equivalent control*, and denoted by u_{eq} (see Utkin (1992)), which replaces the discontinuous feedback control action on the sliding hyperplane $y = 0$ and helps in describing, in an average sense, the dynamical behaviour of the constrained system. From (8.1) and $\dot{y} = 0$ in (8.5) one obtains

$$u_{eq} = -\frac{cAx}{cb} - \frac{c\gamma}{cb}\xi \tag{8.6}$$

Substituting (8.6) into (8.1) yields

$$\dot{x} = (I - \frac{bc}{cb})Ax + (I - \frac{bc}{cb})\gamma\xi \tag{8.7}$$

which represents a *redundant* dynamics taking place on any of the linear varieties y = constant. In particular, when the initial conditions are such that $y = cx = 0$, then (8.7) in combination with $y = 0$ is called the *reduced order ideal sliding dynamics.*

Note that the matrix $P = [I - (bc)/(cb)]$ is a *projection operator* along the range space of b onto the null space of c (El-Ghezawi et al 1983), i.e.

$$Pb = 0 \ , \qquad Px = x \quad \forall\ x \ \text{ s.t. } \ cx = 0$$

Thus, in general, the reduced order ideal sliding dynamics will be dependent upon the perturbation signal ξ. However, under structural constraints on the distribution maps b and γ, known as the *matching conditions*, it is possible to obtain a reduced order ideal sliding dynamics (8.7) which is free of the influence of the perturbation signal ξ. One may establish that the ideal sliding dynamics (8.7) are independent of ξ if, and only if,

$$\gamma = \rho b \tag{8.8}$$

for some constant scalar ρ. In other words, the ideal sliding dynamics are independent of ξ if, and only if, the range spaces of the maps γ and b coincide. The proof is as follows. If the matrix feeding the perturbations ξ into the (average) sliding dynamics equation (8.7) is identically zero, then no perturbations are present in the average system behaviour. This would require the following identity to hold

$$(I - \frac{bc}{cb})\gamma = 0 \tag{8.9}$$

which simply means that γ may be expressed as $\gamma = \rho b$ where $\rho = (c\gamma)/(cb)$. On the other hand if γ is a column vector of the form $\gamma = \rho b$, then

$$(I - \frac{bc}{cb})\gamma = (b - \frac{bc}{cb}b)\rho = (b - b)\rho = 0$$

If the matching condition (8.8) is satisfied, the ideal sliding dynamics is specified by the following constrained dynamics

$$\begin{aligned} \dot{x} &= (I - \frac{bc}{cb})Ax \\ y &= cx = 0 \end{aligned} \tag{8.10}$$

The robust sliding mode controller design problem, for systems satisfying the matching condition (8.8), consists of specifying an output vector c (i.e. a *sliding surface* $y = cx = 0$) and a discontinuous state feedback control policy u of the form (8.2), such that the reduced order ideal sliding dynamics (8.10) is guaranteed to exhibit asymptotically stable behaviour to zero. As may easily be seen, such a stability property is a *structural property* associated with the particular form of the maps A, c and γ. It can be shown that the asymptotic stability of (8.10) can be guaranteed if a *strictly positive real condition*, associated with the constrained system, is satisfied (see Utkin (1992)).

8.2.2 Matching Conditions in Sliding Mode Observer Design

An asymptotic observer for the system (8.1), including an external feedforward compensation signal v, may be proposed as follows

$$\begin{aligned} \dot{\hat{x}} &= A\hat{x} + bu + h(y - \hat{y}) + \lambda v \\ \hat{y} &= c\hat{x} \end{aligned} \tag{8.11}$$

The vector h is called the *vector of observer gains* and the column vector λ is the *feedforward injection map*.

The *state reconstruction error*, defined as $e = x - \hat{x}$, obeys the following dynamical behaviour, from (8.1) and (8.11)

$$\begin{aligned} \dot{e} &= (A - hc)e + \gamma\xi - \lambda v \\ e_y &= ce \end{aligned} \tag{8.12}$$

The signal $e_y = y - \hat{y}$ is called the *output reconstruction error*.

Because of the observability assumption on the system (8.1), there always exists a vector of observer gains h which assigns any arbitrarily prespecified set of n eigenvalues (with complex conjugate pairs) to the matrix $(A - hc)$.

The robust sliding mode observer design problem consists of specifying a vector of observer gains h, a feedforward injection map λ and a discontinuous feedforward injection policy v, based solely on output reconstruction error measurements e_y, such that the reconstruction error dynamics (8.12) is guaranteed to exhibit asymptotically stable behaviour to zero, in spite of all possible bounded values of the external perturbation input signal ξ.

Consider the time derivative of the output reconstruction error signal

$$\begin{aligned} \dot{e}_y &= c(A - hc)e + c\gamma\xi - c\lambda v \\ &= cAe - che_y + c\gamma\xi - c\lambda v \end{aligned} \tag{8.13}$$

We assume, without loss of generality, that the quantity $c\lambda$ is nonzero and positive (i.e. $c\lambda > 0$). As before, let the absolute value of the the perturbation input ξ be bounded by a constant $M > 0$. Also let W be a sufficiently large positive scalar constant. Then, a discontinuous feedforward input v of the form

$$v = W \text{sign } e_y \tag{8.14}$$

is seen to create a *sliding regime* on a bounded region of the reconstruction error space. Such a region would necessarily be contained in the hyperplane $e_y = 0$.

As may be easily verified, from (8.13) and (8.14), in the region characterized by $e_y = 0$ and $|cAe| + |c\gamma\xi| \leq Wc\lambda$, the above choice of the feedforward signal v results in the *sliding condition* $e_y\dot{e}_y < 0$ (see Utkin (1992)) being satisfied. Using the known bound M on the signal ξ, such a region can be expressed as

$$|cAe| \leq Wc\lambda - M|c\gamma|$$

Thus, the discontinuous feedforward policy (8.14) drives the output observation error e_y to zero in finite time, irrespective of both the initial conditions of e and the values of the perturbation input ξ, provided $c\lambda W > |c\gamma|M$.

The *ideal reduced order sliding behaviour* of the state reconstruction error signal e is obtained from the following version of the *invariance conditions*

$$e_y = 0 \ , \qquad \dot{e}_y = 0 \tag{8.15}$$

The conditions (8.15) imply a "virtual" perturbation-dependent value of the output error feedforward injection signal v, which constitutes the *equivalent feedforward* signal, denoted by v_{eq}. This "virtual" feedforward signal is useful in describing the average behaviour of the error system (8.12) when regulated by the feedforward signal v. Using (8.13) and (8.15) one readily obtains

$$v_{eq} = \frac{cAe}{c\lambda} + \frac{c\gamma}{c\lambda}\xi \tag{8.16}$$

Substitution of the equivalent feedforward signal expression (8.16) in the state observation error equation (8.12), leads to the following (redundant) *ideal sliding error dynamics*, taking place on a bounded region of $e_y = 0$

$$\dot{e} = (I - \frac{\lambda c}{c\lambda})Ae + (I - \frac{\lambda c}{c\lambda})\gamma\xi \tag{8.17}$$

Note that the matrix $S = [I - (\lambda c)/(c\lambda)]$ is a *projection operator* along the range space of λ onto the null space of c, i.e.

$$S\lambda = 0 \ , \qquad Sx = x \qquad \forall x \qquad \text{s.t. } cx = 0$$

The reduced order ideal sliding error dynamics will, in general, be dependent upon the perturbation signal ξ. However, under a structural constraint on the distributions maps γ and λ, known as the *matching condition*, it is possible to obtain an ideal sliding error dynamics (8.17) which is free of the influence of the perturbation signal ξ. One may establish that the ideal sliding error dynamics (8.17) is independent of ξ if, and only if,

$$\gamma = \mu\lambda \tag{8.18}$$

for some constant scalar μ. In other words, the sliding error dynamics is independent of ξ if, and only if, the range spaces of the maps γ and λ coincide. The proof of this result is similar to the one carried out for the sliding mode controller case in Sect. 8.2.2 and is omitted.

If the matching condition (8.18) is satisfied, then the reconstruction error dynamics is specified by the following constrained dynamics

$$\begin{aligned} \dot{e} &= (I - \frac{\gamma c}{c\gamma})Ae \\ e_y &= ce \ = \ 0 \end{aligned} \tag{8.19}$$

The resulting reduced order unforced error dynamics obtained from (8.19), must be asymptotically stable. As can be seen, such a stability property is a structural property linked to the particular form of the maps A, c and γ. It can be shown that the asymptotic stability of (8.19) can be guaranteed if a *strictly positive real condition*, associated with the constrained system, is satisfied (see also Walcott and Żak (1988)).

8.2.3 The Matching Conditions for Robust Output Regulation

If the state variables x of the system are not available for measurement, then the variable structure feedback control law (8.2) must be modified to include the dynamical observer states, instead of those of the given system. The *estimated* variable structure feedback control law is now

$$\hat{u} = -\frac{1}{cb}(cA\hat{x} + K \text{ sign } y) \tag{8.20}$$

The regulated state variables x now obey the following variable structure controlled dynamics

$$\begin{aligned}\dot{x} &= Ax - \frac{b}{cb}(cA\hat{x} + K \text{ sign } y) \qquad (8.21)\\ &= (I - \frac{bc}{cb})Ax + \frac{b}{cb}cAe - \frac{bK}{cb}\text{ sign } y\end{aligned}$$

where e is the state reconstruction error dynamics.

The output signal evolution is therefore governed by the dynamical system

$$\dot{y} = cAe - K \text{ sign } y \qquad (8.22)$$

Since the observation error e is guaranteed to converge asymptotically to zero, the output signal y is clearly seen to converge to zero in finite time, provided a sufficiently large value of K is chosen.

It is clear that the ideal sliding dynamics simultaneously taking place on $y = 0$ and $e_y = 0$, will be independent of the perturbation input ξ if, and only if, the matching conditions (8.8) and (8.18) are satisfied, i.e. if the maps γ and λ are both in the range space of the control input distribution channel map b.

8.3 A Generalized Matched Observer Canonical Form for State Estimation in Perturbed Linear Systems

Suppose a linear system of the form (8.1) is given such that the matching condition discussed in Sect. 8.2.3 does not yield an asymptotically stable reduced observation error system (8.19). By resorting to an input-output description of the perturbed system, one can find a canonical state space realization, in generalized state coordinates, which *always* satisfies the matching condition of the form (8.18) while producing a prespecified asymptotically stable constrained error dynamics. The state of the matched canonical realization can therefore always be estimated robustly.

By means of straightforward state vector elimination, the input-output representation of the linear time-invariant perturbed system (8.1) is assumed to be in the form

$$\begin{aligned}y^{(n)} + k_n y^{(n-1)} + \cdots + k_2\dot{y} + k_1 y &= \beta_0 u + \beta_1 \dot{u} + \cdots + \beta_{n-1}u^{(n-1)}\\ &\quad + \gamma_0\xi + \gamma_1\dot{\xi} + \cdots + \gamma_q\xi^{(q)} \qquad (8.23)\end{aligned}$$

where ξ represents the bounded external perturbation signal and the integer q satisfies, without loss of generality, $q \leq n-1$.

The *Generalized Matched Observer Canonical Form* (GMOCF) of the above system is given by the following generalized state representation model (see Fliess (1990a) for a similar canonical form)

$$\begin{aligned}\dot{\chi}_1 &= -k_1\chi_n + \beta_0 u + \beta_1\dot{u} + \cdots + \beta_{n-1}u^{(n-1)} + \lambda_1\eta\\ \dot{\chi}_2 &= \chi_1 - k_2\chi_n + \lambda_2\eta\end{aligned}$$

$$
\begin{aligned}
&\vdots \\
\dot{\chi}_{n-1} &= \chi_{n-2} - k_{n-1}\chi_n + \lambda_{n-1}\eta \\
\dot{\chi}_n &= \chi_{n-1} - k_n\chi_n + \eta \\
y &= \chi_n
\end{aligned}
\tag{8.24}
$$

where η is an "auxiliary" perturbation signal, modelling the influence of the external signal ξ on every equation of the proposed system realization.

The relation existing between the signal η and its generating signal ξ, is obtained by computing the input-output description of system (8.24) in terms of the perturbation input η. The input-output description of the hypothesized model (8.24) is then compared with that of the original system (8.23). This procedure results in a scalar linear time-invariant differential equation for η which accepts the signal ξ as an input.

The models presented below constitute realizations of such an input-output description, according to the order q of the differential polynomial for ξ in (8.23).

For $q < n-1$, the perturbation input η is obtained as the output of the following dynamical system

$$
\begin{aligned}
\dot{z}_1 &= z_2 \\
\dot{z}_2 &= z_3 \\
&\vdots \\
\dot{z}_{n-1} &= -\lambda_1 z_1 - \lambda_2 z_2 - \cdots - \lambda_{n-1} z_{n-1} + \xi \\
\eta &= \gamma_0 z_1 + \gamma_1 z_2 + \cdots + \gamma_{q-1} z_q
\end{aligned}
\tag{8.25}
$$

For $q = n-1$ the state space realization corresponding to (8.25) is simply

$$
\begin{aligned}
\dot{z}_1 &= z_2 \\
\dot{z}_2 &= z_3 \\
&\vdots \\
\dot{z}_{n-1} &= -\lambda_1 z_1 - \lambda_2 z_2 - \cdots - \lambda_{n-1} z_{n-1} + \xi \\
\eta &= (\gamma_0 - \gamma_{n-1}\lambda_1) z_1 + (\gamma_1 - \gamma_{n-1}\lambda_2) z_2 + \cdots \\
&\quad + (\gamma_{n-2} - \gamma_{n-1}\lambda_{n-1}) z_{n-1} + \gamma_{n-1}\xi
\end{aligned}
\tag{8.26}
$$

Assumption 8.1 *Suppose the components of the auxiliary perturbation distribution channel map $\lambda_1, \ldots, \lambda_{n-1}$ in (8.24) are such that the following polynomial, in the complex variable s, is Hurwitz*

$$
p_r(s) = s^n + \lambda_{n-1} s^{n-2} + \cdots + \lambda_2 s + \lambda_1 \tag{8.27}
$$

Equivalently, Assumption 8.1 implies that the output η of the system (8.25) (or that of system (8.26)), generating the auxiliary perturbation η, is a *bounded signal* for every bounded external perturbation signal ξ. If, for instance, ξ satisfies $|\xi| < N$, then, given N, the signal η satisfies $|\eta| \leq M$ for some positive constant M. An easily computable, although conservative, estimate for M is given by $M = \sup_{\omega \in [0,\infty)} | N\, G(j\omega)|$ where $G(s)$ is the Laplace transfer function relating η to ξ in the complex frequency domain.

Remark. It should be stressed that the purpose of having a state space model for the auxiliary perturbation signal η, accepting as a forcing input the signal ξ, is to be able to estimate a bound for the influence of ξ on the proposed state realization (8.24) of the original system (8.1).

An observer for the system realization (8.24) is proposed as follows

$$\begin{aligned}
\dot{\hat{\chi}}_1 &= -k_1\hat{\chi}_n + \beta_0 u + \beta_1 \dot{u} + \cdots + \beta_{n-1}u^{(n-1)} \\
&\qquad + h_1(y - \hat{y}) + \lambda_1 v \\
\dot{\hat{\chi}}_2 &= -k_2\hat{\chi}_n + \hat{\chi}_1 + h_2(y - \hat{y}) + \lambda_2 v \\
&\ \vdots \\
\dot{\hat{\chi}}_{n-1} &= -k_{n-1}\hat{\chi}_n + \hat{\chi}_{n-2} + h_{n-1}(y - \hat{y}) + \lambda_{n-1} v \\
\dot{\hat{\chi}}_n &= -k_n\hat{\chi}_n + \hat{\chi}_{n-1} + h_n(y - \hat{y}) + v \\
\hat{y} &= \hat{\chi}_n
\end{aligned} \tag{8.28}$$

Note that exactly the same output error feedforward distribution map for the signal v has been chosen as the one corresponding to the auxiliary perturbation input signal η in (8.24). Consequently, the proposed canonical form (8.24) for the system always satisfies the matching condition (8.8). The crucial point is that the matched error feedforward distribution map can always be conveniently chosen to guarantee asymptotic stability of the ideal sliding error dynamics.

Use of (8.28) results in the following feedforward regulated reconstruction error dynamics

$$\begin{aligned}
\dot{\epsilon}_1 &= -(k_1 + h_1)\epsilon_n + \lambda_1(\eta - v) \\
\dot{\epsilon}_2 &= \epsilon_1 - (k_2 + h_2)\epsilon_n + \lambda_2(\eta - v) \\
&\ \vdots \\
\dot{\epsilon}_{n-1} &= \epsilon_{n-2} - (k_{n-1} + h_{n-1})\epsilon_n + \lambda_{n-1}(\eta - v) \\
\dot{\epsilon}_n &= \epsilon_{n-1} - (k_n + h_n)\epsilon_n + (\eta - v) \\
\epsilon_y &= \epsilon_n
\end{aligned} \tag{8.29}$$

where ϵ_i represents the state estimation error components $\chi_i - \hat{\chi}_i$, for $i = 1, \ldots, n$.

In order to have a reconstruction error transient response associated with a preselected nth order characteristic polynomial, such as

$$p(s) = s^n + \alpha_n s^{n-1} + \cdots + \alpha_2 s + \alpha_1 \ , \tag{8.30}$$

the gains h_i $(i = 1, \ldots, n)$ should be appropriately chosen as $h_i = \alpha_i - k_i$ $(i = 1, \ldots, n)$.

The feedforward output error injection signal v is chosen to be the discontinuous regulation policy

$$v = W\text{sign } \epsilon_y = W\text{sign } \epsilon_n \tag{8.31}$$

where W is a positive constant. From the final equation in (8.29) it is seen that, for a sufficiently large gain W, the proposed choice of the feedforward signal v results in a sliding regime on a region properly contained in the set expressed by

$$\epsilon_n = 0 \ , \qquad |\epsilon_{n-1}| \leq W - M \tag{8.32}$$

The equivalent feedforward signal, v_{eq}, is obtained from the *invariance conditions* (see also Canudas de Wit and Slotine (1991))

$$\epsilon_n = 0 \ , \qquad \dot{\epsilon}_n = 0 \tag{8.33}$$

One obtains from (8.33) and the last of (8.29)

$$v_{eq} = \eta + \epsilon_{n-1} \tag{8.34}$$

The equivalent feedforward signal is, generally speaking, dependent upon the perturbation signal η. It should be remembered that the equivalent feedforward signal v_{eq} is a *virtual* feedforward action that needs not be synthesized in practice, but one which helps to establish the salient features of the *average* behaviour of the sliding mode regulated observer. The resulting dynamics governing the evolution of the error system in the sliding region are then ideally described by

$$\begin{aligned} \dot{\epsilon}_1 &= -\lambda_1 \epsilon_{n-1} \\ \dot{\epsilon}_2 &= \epsilon_1 - \lambda_2 \epsilon_{n-1} \\ &\vdots \\ \dot{\epsilon}_{n-1} &= \epsilon_{n-2} - \lambda_{n-1}\epsilon_{n-1} \\ \epsilon_y &= \epsilon_n = 0 \end{aligned} \tag{8.35}$$

and exhibits, in a natural manner, a feedforward error injection structure of the "auxiliary output error" signal ϵ_{n-1}, through the design gains $\lambda_1, \ldots, \lambda_{n-1}$. As a result, the roots of the characteristic polynomial in (8.27) determining the behaviour of the homogeneous reduced order system (8.35), are completely determined by a suitable choice of the components of the feedforward vector, $\lambda_1, \ldots, \lambda_{n-1}$.

An asymptotically stable behaviour to zero of the estimation error components $\epsilon_1, \ldots, \epsilon_{n-1}$ is therefore achievable since the output observation error ϵ_n undergoes a sliding regime on the relevant portion of the "sliding surface" $\epsilon_n = 0$. The states of the estimator (8.28) are then seen to converge asymptotically towards the corresponding components of the state vector of the system realization (8.24).

The characteristic polynomial (8.27) of the reduced order observation error dynamics (8.35) coincides entirely with that of the transfer function relating the auxiliary perturbation model signal η to the actual perturbation input ξ. Hence, appropriate choice of the design parameters $\lambda_1, \ldots, \lambda_{n-1}$ not only guarantees asymptotic stability of the sliding error dynamics, but also ensures boundedness of the auxiliary perturbation input signal η, for any given bounded external perturbation ξ.

Remark. In general, the observed states of the matched generalized state space realization are different from the states of the particular realization (8.1). The state χ in (8.24) may even be devoid of any physical meaning. A linear relationship can always be established between the originally given state vector x of system (8.1) and the state χ, reconstructed from the canonical form (8.24). However, generally speaking, such a relationship allows a *perturbation dependent* state coordinate transformation and cannot be used in practice. Nevertheless, it will be shown that a suitable modification of the proposed matched canonical form is effective in implementing a combined observer-controller output feedback sliding mode regulator.

8.4 A Matched Canonical Realization for Sliding Mode Output Feedback Regulation of Perturbed Linear Systems

Consider a linear system of the form (8.1). It will be shown that by resorting to an input-output description of the perturbed system, one can find a canonical state space realization which *always* satisfies the matching conditions of the form (8.8) and (8.18), while producing a prespecified asymptotically stable reduced order state and observation error sliding dynamics. The state of the matched canonical realization can therefore always be robustly estimated and controlled.

By means of straightforward state vector elimination, the input-output representation of the linear time-invariant perturbed system (8.1) is assumed to be of the form given by (8.23). The *Matched Output Regulator Canonical Form* (MORCF) of the above system is given by the following state representation model

$$
\begin{aligned}
\dot{\chi}_1 &= -k_1\chi_n + \lambda_1(\eta + \vartheta) \\
\dot{\chi}_2 &= \chi_1 - k_2\chi_n + \lambda_2(\eta + \vartheta) \\
&\vdots \\
\dot{\chi}_{n-1} &= \chi_{n-2} - k_{n-1}\chi_n + \lambda_{n-1}(\eta + \vartheta) \\
\dot{\chi}_n &= \chi_{n-1} - k_n\chi_n + (\eta + \vartheta) \\
y &= \chi_n
\end{aligned}
\tag{8.36}
$$

where ϑ is an "auxiliary" input interpreted as a *precompensator* input. Note that the auxiliary input distribution map of the proposed canonical form is chosen to match precisely that of the auxiliary (rechannelled) perturbation input η. This guarantees that the realization is matched and that the sliding mode controller will be robust with respect to such perturbations. It is easy to see by computing the input-output representation of the matched realization (8.36), that the auxiliary input ϑ is related to the original control input u by means of the following proper transfer function

$$\frac{\hat{u}(s)}{\hat{\vartheta}(s)} = \frac{s^{n-1} + \lambda_{n-1}s^{n-2} + \cdots + \lambda_1}{b_{n-1}s^{n-1} + \cdots + b_1 s + b_0} \tag{8.37}$$

We refer to (8.37) as the *precompensator* transfer function. Alternatively, a state space realization of the dynamical precompensator is given by

$$\begin{aligned}
\dot{\zeta}_1 &= \zeta_2 \\
\dot{\zeta}_2 &= \zeta_3 \\
&\vdots \\
\dot{\zeta}_{n-2} &= \zeta_{n-1} \\
\dot{\zeta}_{n-1} &= -\frac{b_0}{b_{n-1}}\zeta_1 - \frac{b_1}{b_{n-1}}\zeta_2 - \cdots - \frac{b_{n-3}}{b_{n-1}}\zeta_{n-2} - \frac{b_{n-2}}{b_{n-1}}\zeta_{n-1} + \frac{1}{b_{n-1}}\vartheta \\
u &= (\lambda_1 - \frac{b_0}{b_{n-1}})\zeta_1 + (\lambda_2 - \frac{b_1}{b_{n-1}})\zeta_2 + \cdots \\
&\quad + (\lambda_{n-1} - \frac{b_{n-2}}{b_{n-1}})\zeta_{n-1} + \frac{1}{b_{n-1}}\vartheta
\end{aligned} \tag{8.38}$$

The perturbation input η in (8.36) is, as before, an "auxiliary" perturbation signal, modelling the influence of the external signal ξ on every equation of the proposed system realization. It is straightforward to verify that the signal η in (8.36) is obtained from the signal ξ in the same manner as it was obtained in (8.25) (or (8.26)).

The components of the auxiliary perturbation distribution channel map $\lambda_1, ..., \lambda_{n-1}$ in (8.36), are such that the characteristic polynomial in the complex variable s is Hurwitz. This, in turn, guarantees a truly minimum phase dynamical precompensator (8.37) (or (8.38)). The minimum phase condition on the zeroes of the precompensator transfer function also guarantees simultaneously that the the output η of system (8.25) or (8.26), generating the auxiliary perturbation η, is a ***bounded signal*** for every bounded external perturbation signal ξ.

8.4.1 Observer Design

An observer for the system realization (8.36) is proposed as follows

$$\begin{aligned}
\dot{\hat{\chi}}_1 &= -k_1\hat{\chi}_n + h_1(y-\hat{y}) + \lambda_1(v+\vartheta) \\
\dot{\hat{\chi}}_2 &= -k_2\hat{\chi}_n + \hat{\chi}_1 + h_2(y-\hat{y}) + \lambda_2(v+\vartheta) \\
&\vdots \\
\dot{\hat{\chi}}_{n-1} &= -k_{n-1}\hat{\chi}_n + \hat{\chi}_{n-2} + h_{n-1}(y-\hat{y}) + \lambda_{n-1}(v+\vartheta) \\
\dot{\hat{\chi}}_n &= -k_n\hat{\chi}_n + \hat{\chi}_{n-1} + h_n(y-\hat{y}) + (v+\vartheta) \\
\hat{y} &= \hat{\chi}_n
\end{aligned} \tag{8.39}$$

Note that exactly the same output error feedforward distribution map λ for the signal v has been chosen as that corresponding to the auxiliary perturbation input signal η and to the control input distribution map in (8.36). As a consequence, the matching conditions (8.8) and (8.18) are satisfied by the proposed matched canonical realization (8.36). Use of the observer (8.39) results exactly in the same sliding mode feedforward regulated reconstruction error dynamics already given in (8.29).

A reconstruction error transient response may be chosen which is associated with a preselected n th order characteristic polynomial, such as (8.30), by means of the appropriate choice of the observer gains h_i, $i = 1, \ldots, n$.

The feedforward output error injection signal v is chosen, as before, as a discontinuous regulation policy of the variable structure type

$$v = W\text{sign } \epsilon_y = W\text{sign } \epsilon_n \tag{8.40}$$

with W being a positive constant. For a sufficiently large gain W, the proposed choice of the feedforward signal v results in a sliding regime on a region properly contained in the set

$$\epsilon_n = 0 \ , \qquad |\epsilon_{n-1}| \leq W - M \tag{8.41}$$

The resulting reduced order dynamics governing the evolution of the sliding mode regulated error system in the computed sliding region of the error space, is then ideally described by the same asymptotically stable unforced differential equation as in (8.35).

8.4.2 Sliding Mode Controller Design

We first show that the proposed matched canonical form (8.24) also facilitates the design of a sliding mode controller when all states of the realization are directly measurable. Once the sliding mode controller based on full state feedback information has been obtained, a similar sliding mode controller in which all the required state variables are derived from the observer, will be developed.

8.4.2.1 Sliding Mode Controller Based on Full State Information A sliding mode controller may be obtained by considering the unperturbed version of the final equation in the canonical form (8.36), (i.e. from the differential equation governing the behaviour of the output $y = \chi_n$ with $\eta = 0$), and the discontinuous regulated policy proposed in (8.3). Such a sliding mode control policy is given by

$$\vartheta = k_n \chi_n - \chi_{n-1} - W \text{ sign } \chi_n \tag{8.42}$$

Using the above controller in the perturbed output equation, results in the following controlled output dynamics

$$\dot{\chi}_n = \eta - W \text{ sign } y \tag{8.43}$$

Therefore a sliding mode controller gain W, which is assumed to satisfy $W > M$, guarantees the convergence of y to zero in finite time, irrespectively of the bounded values of the computed perturbation effect η.

The invariance conditions $\chi_n = 0$, $\dot{\chi}_n = 0$ result in the following perturbation dependent equivalent auxiliary control input

$$\vartheta_{eq} = -\chi_{n-1} - \eta \tag{8.44}$$

The ideal sliding dynamics, obtained from substitution of (8.44) in the canonical realization (8.36), is

$$\begin{aligned} \dot{\chi}_1 &= -\lambda_1 \chi_{n-1} \\ \dot{\chi}_2 &= \chi_1 - \lambda_2 \chi_{n-1} \\ &\vdots \\ \dot{\chi}_{n-1} &= \chi_{n-2} - \lambda_{n-1} \chi_{n-1} \\ y &= \chi_n = 0 \end{aligned} \tag{8.45}$$

The characteristic polynomial of the constrained dynamics is given again by the Hurwitz polynomial (8.27), and the ideal sliding dynamics (8.45) is asymptotically stable to zero.

8.4.2.2 Sliding Mode Controller Based on Observer State Information If the state χ_{n-1} is not directly available for measurement, the feedback control (8.42) should be modified to employ the estimated state obtained from the sliding observer (8.39) as

$$\hat{\vartheta} = k_n y - \hat{\chi}_{n-1} - W \text{ sign } y \tag{8.46}$$

where the fact that the output y is clearly available for measurement, has been used. This control policy still results in finite time convergence of y to zero as can be seen from the closed-loop output dynamical equation

$$\begin{aligned} \dot{y} &= (\chi_{n-1} - \hat{\chi}_{n-1}) + \eta - W \text{ sign } y \\ &= \epsilon_{n-1} + \eta - W \text{ sign } y \end{aligned} \tag{8.47}$$

Since ϵ_{n-1} is decreasing asymptotically to zero, the output y is seen to go to zero in finite time for sufficiently large values of $W > M$.

The output observation error signal e_y, and the output signal y itself, are seen to converge to zero in finite time. The combined reduced order ideal sliding/ideal observer dynamics is obtained from the same invariance conditions $\chi_n = 0$, $\dot{\chi}_n = 0$ as before. This results in precisely the same equivalent control input and the same equivalent feedforward signals. The resulting reduced order ideal sliding/ideal observation error dynamics is still given by (8.35) and (8.45). The overall scheme is therefore asymptotically stable.

8.5 Design Example: The Boost Converter

Consider the average Boost converter model derived by Sira-Ramírez and Lischinsky-Arenas (1991)

$$\begin{aligned} \dot{z}_1 &= -\omega_0 z_2 + \mu\omega_0 z_2 + b \\ \dot{z}_2 &= \omega_0 z_1 - \omega_1 z_2 - \mu\omega_0 z_1 \end{aligned} \tag{8.48}$$

where $z_i, i = 1,2$ denote the corresponding "averaged components" of the state vector x where $x_1 = I\sqrt{L}$, $x_2 = V\sqrt{C}$ represent the normalized input current and output voltage variables respectively. The quantity $b = E/\sqrt{L}$ is the normalised external input voltage. The LC (input) circuit natural oscillating frequency and the RC output circuit time constant are denoted by $\omega_0 = 1/\sqrt{LC}$ and $\omega_1 = 1/(RC)$ respectively. The variable μ is the control input. The equilibrium points of the average model (8.48) are obtained as

$$\mu = U \;\; ; \quad Z_1(U) = \frac{b\omega_1}{\omega_0{}^2(1-U)^2} \;\; ; \quad Z_2(U) = \frac{b}{\omega_0(1-U)} \tag{8.49}$$

where U denotes a particular constant value for the duty ratio function. The linearisation of the average PWM model (8.48) about the constant operating points (8.49) is given by

$$\begin{aligned} \dot{z}_{1\delta} &= -(1-U)\omega_0 z_{2\delta} + \frac{b}{1-U}\mu_\delta \\ \dot{z}_{2\delta} &= (1-U)\omega_0 z_{1\delta} - \omega_1 z_{2\delta} - \frac{b\omega_1}{(1-U)^2\omega_0}\mu_\delta \end{aligned} \tag{8.50}$$

with

$$\mu_\delta(t) = \mu(t) - U \;\; ; \quad z_{i\delta}(t) = z_i(t) - Z_i(U) \;\; , \quad i = 1,2 \tag{8.51}$$

Taking the averaged normalised input inductor current z_1 as the system output in order to meet the relative degree 1 and minimum phase assumptions, the following input/output relationship is obtained

$$\frac{z_{1\delta}(s)}{\mu_\delta(s)} = \omega_0 Z_2(U)\frac{s + 2\omega_0}{s^2 + \omega_1 s + (1-U)^2\omega_0{}^2} \tag{8.52}$$

The controller/observer pair (8.46), (8.39) is now implemented on the average boost converter model. For simulation purposes nominal parameter values of $R = 30\Omega$, $C = 20\mu$F, $L = 20$mH and $E = 15$V are assumed. The desirable set point for the average normalized input inductor current is $z_1 = 0.4419$ which corresponds to a constant value $U = 0.6$. In order to demonstrate the robustness of the approach, the effects of noise on both the input current and output voltage dynamics will be considered. The system representation then becomes, from (8.52),

$$\begin{aligned} \dot{z}_{1\delta} &= -632.46z_{2\delta} + 265.17\mu_\delta + \alpha\xi \\ \dot{z}_{2\delta} &= 632.46z_{1\delta} - 1666.67z_{2\delta} - 698.77\mu_\delta + \beta\xi \end{aligned} \tag{8.53}$$

Here α and β define the noise distribution channel which is not necessarily matched. The polynomial (8.27) which defines the auxiliary perturbation distribution map is chosen to be

$$p_r(s) = s + 3000 \tag{8.54}$$

The rate of decay of the reconstruction error dynamics (8.30) is determined by the roots of the following characteristic polynomial

$$p(s) = s^2 + 8500s + 18000000 \tag{8.55}$$

Using (8.54) and (8.55) an observer (8.39) for the system is given by

$$\begin{aligned} \dot{\hat{\chi}}_1 &= 400000\hat{\chi}_2 + 17600000(y - \hat{y}) + 3000(v + \vartheta) \qquad (8.56) \\ \dot{\hat{\chi}}_2 &= -1666.67\hat{\chi}_2 + \hat{\chi}_1 + 6833.33(y - \hat{y}) + (v + \vartheta) \\ \hat{y} &= \hat{\chi}_2 \\ v &= W_{obs}\,\mathrm{sign}\,(y - \hat{y}) \qquad (8.57) \end{aligned}$$

The following state-space realisation may be used to determine the plant input μ_δ

$$\begin{aligned} \dot{w} &= -3333.33w + 0.0038\vartheta \qquad (8.58) \\ \mu_\delta &= -333.33z + 0.0038\vartheta \\ \vartheta &= -W_{con}\,\mathrm{sign}\,y - \hat{\chi}_1 + 1666.67y \end{aligned}$$

The magnitude of the discontinuous gain elements W_{con} and W_{obs} were chosen to be 120 and 220 respectively. These were tailored to provide the required speeds of response as well as appropriate disturbance rejection capabilities. Using a disturbance distribution map defined by $\alpha = 0.01$ and $\beta = -0.02$, which is clearly unmatched with respect to the input and output distributions of the system realisation (8.53), and a high frequency cosine representing the system noise, the following simulation results were obtained. Fig. 8.1 shows the convergence of the estimated inductor current to the actual inductor current. A sliding mode is reached whereby $z_1(t) - Z_i(t) = 0$. The required set point is thus attained and maintained despite the disturbance which is acting upon the system. Fig. 8.2 shows the control effort μ. The discontinuous nature of this signal supports the assertion that a sliding mode has been attained.

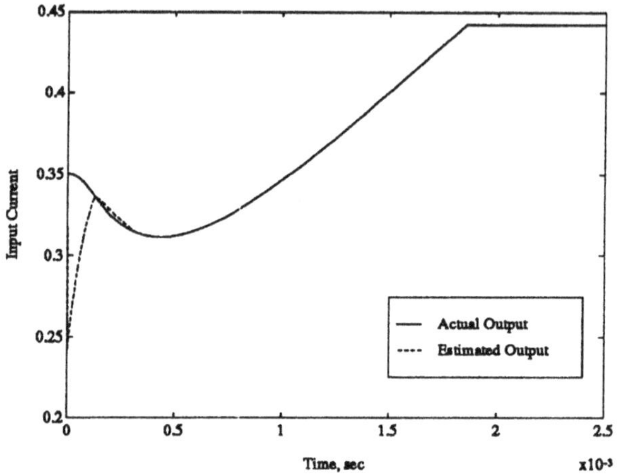

Fig. 8.1. Response of the actual and estimated average normalized inductor current

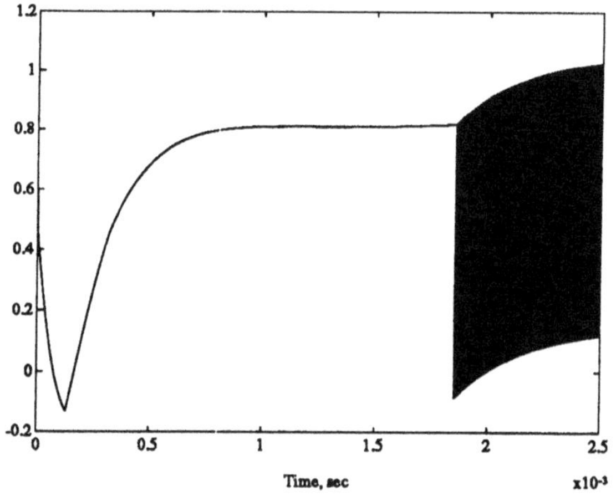

Fig. 8.2. Response of the control effort μ

8.6 Conclusions

It has been shown that, when using a sliding mode approach, structural conditions of the *matching type*, are largely irrelevant for robust state reconstruction and regulation of linear perturbed systems. The class of linear systems for which robust sliding mode output feedback regulation can be obtained, independently of any *matching conditions*, comprises the entire class of controllable (stabilizable) and observable (reconstructible) linear systems with the appropriate relative degree and minimum phase condition.

This result, first postulated by Sira-Ramírez and Spurgeon (1993b), is of particular practical interest when the designer has the freedom to propose a convenient state space representation for a given unmatched system. This is in total accord with the corresponding results found in Fliess and Messager (1991), and in Sira-Ramírez and Spurgeon (1993b) regarding, respectively, the robustness of the sliding mode control of perturbed controllable linear systems, expressed in the *Generalized Observability Canonical Form*, and the dual result for the sliding mode observation schemes based on the *Generalized Observer Canonical Form.*

Sliding mode output regulator theory (i.e. addressing an observer-controller combination) for linear systems may also be examined from an algebraic viewpoint using *Module Theory* (see Fliess (1990b)). The conceptual advantages of using a module theoretic approach to sliding mode control were recently addressed by Fliess and Sira-Ramírez (1993) and Sira-Ramírez in Chapter 2. The module theoretic approach can also provide further generalizations and insights related to the results presented.

8.7 Acknowledgments

Professor Sira-Ramírez is grateful to Professor Michel Fliess of the Laboratoire des Signaux et Systèmes, CNRS (France), for many interesting discussions relating to the results in this chapter.

References

Canudas de Wit, C., Slotine, J.J.E. 1991, Sliding Observers for Robot Manipulators. *Automatica* **27** , 859–864

Dorling, C.M., Zinober, A.S.I. 1983, A Comparative Study of the Sensitivity of Observers. *Proceedings IASTED Symposium on Applied Control and Identification*, Copenhagen, 6.32–6.38

El-Ghezawi, O.M.E., Zinober, A.S.I., Billings, S.A. 1983, Analysis and design of variable structure systems using a geometric approach. *International Journal of Control* **38**, 657–671

Edwards, C., Spurgeon, S.K. 1993, On the Development of Discontinuous Observers. *International Journal of Control*, to appear

Fliess, M. 1990a, Generalized Controller Canonical Forms for Linear and Nonlinear Dynamics. *IEEE Transactions on Automatic Control* **AC-35**, 994–1001

Fliess, M. 1990b, Some basic structural properties of generalized linear systems. *Systems and Control Letters* **15**, 391–396

Fliess, M., Messager, F. 1991, Sur la Commande en Régime Glissant. *C.R. Acad. Sci. Paris* **313 Series I**, 951–956

Fliess, M., Sira-Ramírez, H. 1993, Regimes glissants, structure variables lineaires et modules. *C.R. Acad. Sci. Paris* Series I, submitted for publication

Sira-Ramírez, H., Lischinsky-Arenas, P. 1991, Differential Algebraic Approach in Nonlinear Dynamical Compensator Design for d.c.-d.c. Power Converters. *International Journal of Control* **54**, 111–133

Sira-Ramírez, H., Spurgeon, S.K. 1993a, On the robust design of sliding observers for linear systems. *Systems and Control Letters* , to appear

Sira-Ramírez, H., Spurgeon, S.K. 1993b, Robust Sliding Mode Control using Measured Outputs. *IEEE Transactions on Automatic Control* , submitted for publication

Utkin, V.I. 1992, *Sliding Modes in Control Optimization*, Springer-Verlag, New York

Walcott, B.L., Żak, S.H. 1988, Combined Observer-Controller Synthesis for Uncertain Dynamical Systems with Applications, *IEEE Transactions on Systems, Man and Cybernetics* **18**, 88–104

Watanabe, K., Fukuda, T., Tzafestas, S.G. 1992, Sliding Mode Control and a Variable Structure System Observer as a Dual Problem for Systems with Nonlinear Uncertainties. *International Journal of Systems Science* **23**, 1991–2001

9. Robust Stability Analysis and Controller Design with Quadratic Lyapunov Functions

Martin Corless

9.1 Introduction

The use of Lyapunov functions to guarantee the stability of an equilibrium state of a dynamical system dates back to the original work of Lyapunov himself (Lyapunov 1907). Roughly speaking, an equilibrium state is stable if one can find a scalar valued function (called a *Lyapunov function*) of the system state which has a strict minimum at the equilibrium state and whose value decreases along every trajectory of the system (except, of course the equilibrium trajectory.) The significance of this result is that it allows one to guarantee stability without having to solve the differential equations describing the system. This is particularly important for nonlinear systems where explicit solutions are generally not available. The major difficulty in applying Lyapunov theory is in finding an appropriate Lyapunov function.

Consider a dynamical system described by

$$\dot{x} = f(t, x) \tag{9.1}$$

where $t \in \mathbb{R}$ is the time variable and $x(t) \in \mathbb{R}^n$ is the state vector. Suppose $x = 0$ is an equilibrium state of (9.1) and one is interested in the stability of this equilibrium state.

We say that a function V is a *quadratic Lyapunov function* for (9.1) if there exist real, positive-definite, symmetric matrices P and Q such that for all t and x

$$V(x) = x^T P x = \sum_{i=1}^{n} \sum_{j=1}^{n} P_{ij} x_i x_j \tag{9.2}$$

and

$$x^T P f(t, x) \leq -x^T Q x \tag{9.3}$$

It follows from (9.3) that along any solution $x(\cdot)$ of (9.1),

$$\mathrm{d}V(x(t))/\mathrm{d}t \leq -2x(t)^T Q x(t) \ ; \tag{9.4}$$

hence $V(x(t))$ decreases along every non-zero trajectory. From (9.4) one can show that every solution $x(\cdot)$ of (9.1) satisfies

$$\|x(t)\| \leq \beta \|x(t_0)\| \exp[-\alpha(t - t_0)] \qquad \forall \quad t \geq t_0 \tag{9.5}$$

where

$$\alpha = \lambda_{min}(P^{-1}Q), \qquad \beta = [\lambda_{max}(P)/\lambda_{min}(P)]^{1/2} \ ;$$

i.e., (9.1) is ***globally uniformly exponentially stable (GUES)*** with rate of convergence α.

For exponentially stable linear time-invariant systems, one can obtain quadratic Lyapunov functions by solving a linear matrix equation. To be more specific, consider a linear time-invariant system described by

$$\dot{x} = Ax \tag{9.6}$$

where A is a real matrix. The main Lyapunov result for linear systems is that system (9.6) is exponentially stable iff for each positive-definite symmetric matrix $Q \in \mathbb{R}^{n\times n}$ the *Lyapunov matrix equation (LME)*

$$PA + A^T P + 2Q = 0 \tag{9.7}$$

has an unique solution for P and this solution is positive-definite symmetric; see, e.g. Kalman and Bertram (1960). A quadratic Lyapunov function for (9.6) is then given by

$$V(x) = x^T P x \ ; \tag{9.8}$$

using LME (9.7), it can be readily seen that

$$x^T PAx = -x^T Qx$$

In recent years, there has been considerable research activity in the use of quadratic Lyapunov functions to guarantee *robust stability* of uncertain/nonlinear systems; by robust stability we mean stability in the presence of any allowable uncertainty or nonlinearity. Consider an uncertain/nonlinear system described by

$$\begin{aligned} \dot{x} &= Ax + g(\delta, x) \\ \delta &\in \Delta \end{aligned}$$

where δ is a vector or matrix of uncertain parameters and g is a known continuous function. All the uncertainty and nonlinearity in the system is characterized by the term $g(\delta, \cdot)$. If the nominal linear portion, $\dot{x} = Ax$, of the system is exponentially stable, one could could choose a quadratic Lyapunov function V for this system and attempt to guarantee stability of the original system with V as a *Lyapunov function candidate*. An advantage of this approach is that one only requires knowledge of the bounding set Δ; also it guarantees stability in the presence of time-varying and/or state dependent parameters.

The constructive use of Lyapunov functions for control design dates back to at least Kalman and Bertram (1960). Much of the early work on the design of stabilizing controllers for uncertain systems was based on the constructive use of Lyapunov functions; see, for example, Gutman (1979), Gutman and Leitmann (1976), Leitmann (1978, 1979a, 1979b). In recent years there has been considerable activity in the use of quadratic Lyapunov functions for robust control design of uncertain systems.

9.2 Quadratic Stability

Consider an *uncertain system* described by

$$\dot{x} = f(x, \delta) \tag{9.9a}$$
$$\delta \in \Delta \tag{9.9b}$$

where $t \in \mathbb{R}$ is time and $x(t) \in \mathbb{R}^n$ is the state. All the uncertainty in the system is modelled by the lumped uncertain term δ. The only information assumed on δ is the bounding set Δ to which it belongs.

Definition 9.1 *System (9.9) is* quadratically stable *if there exist positive-definite symmetric matrices $P, Q \in \mathbb{R}^{n \times n}$ such that for all $x \in \mathbb{R}^n$*

$$x^T P f(x, \delta) \leq -x^T Q x \qquad \forall \ \ \delta \in \Delta \tag{9.10}$$

If uncertain system (9.9) is quadratically stable, then *any* system of the form

$$\dot{x} = f(x, \delta(t, x)) \tag{9.11a}$$
$$\delta(t, x) \in \Delta \tag{9.11b}$$

is GUES with rate $\alpha = \lambda_{min}(P^{-1}Q)$ and this stability is guaranteed by the quadratic Lyapunov function given by $V(x) = x^T P x$. Hence, without loss of generality, we will consider δ constant. Note that the Lyapunov function is independent of the uncertainty. In what follows we call P a *common Lyapunov matrix (CLM)* for (9.9).

In the initial research (Becker and Grimm 1988, Corless and Da 1988, Corless et al 1989, Eslami and Russel 1980, Patel and Toda 1986, Yedavalli 1985 and 1989, Yedavalli et al 1985, Zhou and Khargonekar 1987) on using quadratic Lyapunov functions to guarantee stability of an uncertain system, the approach was to consider a nominal linear portion of system (9.9), choose a quadratic Lyapunov function for this nominal part and then consider this a Lyapunov function candidate for the uncertain system (9.9). In general, this approach produces sufficient conditions for quadratic stability. Subsequent research produced readily verifiable conditions which are both necessary and sufficient for quadratic stability of specific classes of uncertain systems; some of these results are presented in the next two sections.

9.2.1 Systems Containing Uncertain Scalar Parameters

Consider first a general uncertain linear system described by

$$\dot{x} = A(\delta) x \tag{9.12a}$$
$$\delta \in \Delta \tag{9.12b}$$

where Δ *is compact* and the matrix-valued function $A(\cdot)$ *is continuous.* One can readily show that this system is quadratically stable with common Lyapunov matrix P iff

$$PA(\delta) + A(\delta)^T P < 0 \qquad \forall\, \delta \in \Delta \tag{9.13}$$

The set of positive-definite symmetric matrices P satisfying this requirement is clearly a convex set.

Consider now a linear system,

$$\dot{x} = [A_0 + \delta_1 A_1 + \ldots + \delta_r A_r]x \tag{9.14a}$$
$$|\delta_i| \leq 1 \tag{9.14b}$$

with several uncertain scalar parameters, $\delta_i \in \mathbb{R}, i = 1, 2, \ldots, r$.

Example 9.2 The second order system

$$\dot{x}_1 = x_2 \tag{9.15a}$$
$$\dot{x}_2 = (\delta - 2)x_1 - x_2 \tag{9.15b}$$
$$|\delta| \leq 1\ , \tag{9.15c}$$

with uncertain scalar parameter δ, can be described by (9.14) with

$$A_0 = \begin{bmatrix} 0 & 1 \\ -2 & -1 \end{bmatrix}, \quad A_1 = \begin{bmatrix} 0 & 0 \\ 1 & 0 \end{bmatrix} \tag{9.16}$$

Utilizing (9.13) one may readily deduce the following result; see Horisberger and Belanger (1976) which contains a more general result.

Theorem 9.3

A positive-definite symmetric matrix P is a common Lyapunov matrix for (9.14) iff it satisfies the following linear matrix inequalities:

$$PA_0 + A_0{}^T P + \sum_{i=1}^{r} \delta_i \left[PA_i + A_i{}^T P\right] < 0 \tag{9.17a}$$
$$for \qquad \delta_i = -1, 1, \quad i = 1, 2, \ldots, r \tag{9.17b}$$

Thus the determination of quadratic stability for system (9.18) can be reduced to the problem of finding a positive-definite matrix P which satisfies a *finite* number of linear inequality constraints (9.17). Such problems can be solved numerically in a finite number of steps; see, e.g., Bernussou et al (1989), Boyd et al (1993), and Boyd and Yang (1989). These papers also contains many other results on solving quadratic stability problems via convex programming methods.

9.2.2 Systems Containing a Single Uncertain Matrix

Here we consider uncertain linear systems in which all the uncertainty is characterized by a single uncertain matrix $\delta \in \mathbb{R}^{p \times q}$:

$$\dot{x} = [A + D\delta E]x \tag{9.18a}$$
$$\|\delta\| \leq 1 \tag{9.18b}$$

Example 9.4 The system of Example 9.2 can be described by (9.18) with

$$A = \begin{bmatrix} 0 & 1 \\ -2 & -1 \end{bmatrix}, \quad D = \begin{bmatrix} 0 \\ 1 \end{bmatrix}, \quad E = \begin{bmatrix} 1 & 0 \end{bmatrix} \tag{9.19}$$

Remark. One may readily show that if the uncertain system (9.18) is quadratically stable then *any nonlinear/nonautonomous* system of the form

$$\dot{x} = Ax + D\delta(t,x) \tag{9.20a}$$
$$\|\delta(t,x)\| \leq \|Ex\| \tag{9.20b}$$

is GUES with Lyapunov function given by $V(x) = x^T P x$.

Example 9.5 Consider an inverted pendulum under linear control described by

$$\begin{aligned} \dot{x}_1 &= x_2 \\ \dot{x}_2 &= -2x_1 - x_2 + \sin x_1 \end{aligned}$$

Letting

$$\delta(t,x) := \sin x_1$$

this system can be described by (9.20) with A, D, E given by (9.19). Hence quadratic stability of the linear system (9.15) guarantees GUES of this nonlinear system.

The following result can be established using results in Khargonekar et al (1990) and Rotea and Khargonekar (1989).

Theorem 9.6 *A positive-definite symmetric matrix P is a common Lyapunov matrix for (9.18) iff there is a real scalar $\mu > 0$ such that the following* quadratic matrix inequality (QMI) *is satisfied:*

$$PA + A^T P + \mu P D D^T P + \mu^{-1} E^T E < 0 \tag{9.21}$$

Note that if $\tilde{P}$ is a common Lyapunov matrix for (9.18), then $P := \mu\tilde{P}$ is also a common Lyapunov matrix for this system and it satisfies the following quadratic matrix inequality:

$$PA + A^TP + PDD^TP + E^TE < 0 \tag{9.22}$$

This leads to the following result.

Corollary 9.7 *System (9.18) is quadratically stable iff there exists a positive-definite symmetric matrix $P \in \mathbb{R}^{n\times n}$ which satisfies the quadratic matrix inequality (9.22).*

Using properties of QMI (9.22) (see Ran and Vreugdenhil, 1988), one can readily deduce the following corollary from Corollary 9.7.

Corollary 9.8 *The uncertain system (9.18) is quadratically stable iff for* any *positive-definite symmetric matrix Q there is an $\bar{\epsilon} > 0$ such that for all $\epsilon \in (0, \bar{\epsilon}]$ the following* Riccati equation *has a positive-definite symmetric solution for P*

$$PA + A^TP + PDD^TP + E^TE + \epsilon Q = 0 \tag{9.23}$$

Using this corollary, the search for a common Lyapunov matrix is reduced to a one parameter search.

Remark. Satisfaction of the quadratic matrix inequality (9.22) is equivalent to satisfaction of the following linear affine matrix inequality :

$$\begin{bmatrix} -PA - A^TP - E^TE & PD \\ D^TP & I \end{bmatrix} > 0 \tag{9.24}$$

The set of positive-definite symmetric matrices satisfying this last inequality is clearly a convex set. This reduces the determination of quadratic stability for system (9.18) to the problem of finding a positive-definite matrix P which satisfies a linear affine inequality constraint.

9.2.3 Quadratic Stability and H_∞

In many situations, one is only interested in whether a given uncertain system is quadratically stable or not; one may not actually care what the common Lyapunov matrix is. To this end, the following frequency domain characterization of quadratic stability is useful.

First, define the transfer matrix $H(s)$ by

$$H(s) = E(sI - A)^{-1}D \tag{9.25}$$

and let

$$\|H\|_\infty := \sup_{\omega \in \mathbb{R}} \|H(j\omega)\| \tag{9.26}$$

Then we have the following result from Khargonekar et al (1990).

Theorem 9.9 *The uncertain system (9.18) is quadratically stable iff*

(i) A is asymptotically stable and

(ii)

$$\|H\|_\infty < 1 \tag{9.27}$$

Remark. Hinrichsen and Pritchard (1986a, 1986b) also demonstrate that satisfaction of conditions (i) and (ii) above is also *necessary and sufficient* for system (9.18) to be stable for all *constant complex* δ with $\|\delta\| \leq 1$.

Example 9.10 Consider Example 9.5 again. Here the matrix A is asymptotically stable and

$$H(s) = 1/(s^2 + s + 2)$$

One may readily compute that

$$\|H\|_\infty = 2/\sqrt{7} < 1$$

Hence, this nonlinear system is exponentially stable.

9.2.4 Systems Containing Several Uncertain Matrices

Here we consider linear systems whose uncertainty is characterized by several uncertain matrices, $\delta_i \in \mathbb{R}^{p_i \times q_i}$, $i = 1, 2, \ldots, r$:

$$\dot{x} = [A + D_1\delta_1 E_1 + \ldots + D_r\delta_r E_r]x \tag{9.28a}$$

$$\|\delta_i\| \leq 1. \tag{9.28b}$$

This system can also be described by

$$\dot{x} = [A + D\delta E]x \tag{9.29a}$$

$$\delta \in \Delta \tag{9.29b}$$

where

$$\Delta = \{\text{block-diag}\,(\delta_1, \delta_2, \ldots, \delta_r) : \quad \delta_i \in \mathbb{R}^{p_i \times q_i}, \ \|\delta_i\| \leq 1\} \tag{9.30}$$

and

$$D := [D_1\ D_2\ \ldots D_r], \qquad E := [E_1'\ E_2'\ \ldots E_r']' \tag{9.31}$$

Consider now any r positive scalars $\mu_1, \mu_2, \ldots, \mu_r$ and let

$$\Lambda_i := \text{block-diag}\,(\mu_1 I_{p_1}, \mu_2 I_{p_2}, \ldots, \mu_r I_{p_r})$$
$$\Lambda_o := \text{block-diag}\,(\mu_1 I_{q_1}, \mu_2 I_{q_2}, \ldots, \mu_r I_{q_r})$$

Then, as a consequence of the structure of Δ, each $\delta \in \Delta$ satisfies

$$\delta = \Lambda_i \delta \Lambda_o{}^{-1};$$

hence this system can also be described by

$$\dot{x} = [A + \tilde{D}\delta\tilde{E}]x \quad (9.32a)$$

$$\delta \in \Delta \quad (9.32b)$$

with

$$\tilde{D} := D\Lambda_i, \qquad \tilde{E} := \Lambda_o^{-1}E \quad (9.33)$$

Using this observation and the sufficiency part of Theorem 9.6 one can readily obtain the following result:

Theorem 9.11 *A positive-definite symmetric matrix $P \in \mathbb{R}^{n \times n}$ is a common Lyapunov matrix for system (9.28) if there exist r positive scalars $\mu_1, \mu_2, \ldots, \mu_r$ which satisfy the following quadratic matrix inequality:*

$$PA + A^T P + \sum_{i=1}^{r} [\mu_i P D_i D_i{}^T P + \mu_i^{-1} E_i{}^T E_i] < 0 \quad (9.34)$$

Proof. Note that (9.34) can be written as

$$PA + A^T P + P\tilde{D}\tilde{D}^T P + \tilde{E}^T \tilde{E} < 0$$

Since $||\delta|| \leq 1$, it follows from representation (9.32) and Theorem 9.6 that P is a CLM.

We immediately have the following corollary.

Corollary 9.12 *System (9.28) is quadratically stable if there exist a positive-definite symmetric matrix $P \in \mathbb{R}^{n \times n}$ and r positive scalars $\mu_1, \mu_2, \ldots, \mu_r$ such that (9.34) is satisfied.*

Remark. Note that Corollary 9.12 provides only a *sufficient condition* for quadratic stability of system (9.28). This condition is not necessary for quadratic stability; Rotea et al (1993) contains an example which is quadratically stable but for which the above condition is not satisfied.

It should be clear that one may also obtain a sufficient condition involving a Riccati equation with scaling parameters μ_i using Corollary 9.8 and a H_∞ sufficient condition using Theorem 9.9.

9.3 Quadratic Stabilizability

Consider now an uncertain control system described by

$$\dot{x} = F(x, u, \delta) \tag{9.35a}$$
$$\delta \in \Delta \tag{9.35b}$$

where t, x, δ are as previously defined and $u(t) \in \mathbb{R}^m$ is the control input. Suppose (9.35) is subject to a memoryless state feedback controller $k(\cdot)$, i.e.

$$u(t) = k(x(t)) \tag{9.36}$$

Then the resulting closed-loop system is described by

$$\dot{x} = F(x, k(x), \delta) \tag{9.37a}$$
$$\delta \in \Delta \tag{9.37b}$$

and we have the following definition.

Definition 9.13 *System (9.35) is* quadratically stabilizable *iff there exists a controller $k : \mathbb{R}^n \to \mathbb{R}^m$ such that the corresponding closed-loop system (9.37) is quadratically stable.*

Remark. We do not lose any generality in the above definition by considering only *time-invariant memoryless controllers.* This is because, in the context of quadratic stabilizability of (9.35) via full state feedback, *time-varying dynamic implies time-invariant memoryless*; i.e. if there exists a time-varying dynamic controller which yields a quadratically stable closed-loop system, then there also exists a quadratically stabilizing time-invariant memoryless controller (Corless et al 1993). This was first noted in Petersen (1988) and Rotea (1990) for linear systems.

9.3.1 Linear vs. Nonlinear Control

Much of the literature on quadratic stabilizability is concerned with linear uncertain systems described by

$$\dot{x} = A(\delta)x + B(\delta)u \tag{9.38a}$$
$$\delta \in \Delta \tag{9.38b}$$

It is reasonable to conjecture that if such a system is quadratically stabilizable then quadratic stability can be achieved with a linear controller

$$u = Kx \tag{9.39}$$

This conjecture is *false* in general; see Petersen (1985b) for a counterexample.

However, it can be shown that a linear uncertain system (9.38) is quadratically stabilizable via a linear controller if it satisfies one of the following conditions:

(i) It is quadratically stabilizable via a continuously differentiable controller (Barmish 1983).

(ii) It is quadratically stabilizable, $B(\delta)$ is independent of δ, $A(\cdot)$ is continuous, and Δ is compact (Hollot and Barmish 1980).

(iii) It belongs to the class considered in section 9.3.4 (Rotea and Khargonekar 1989).

(iv) It satisfies a matching or generalized matching condition; see next section.

9.3.2 Matching, Generalized Matching and Other Structural Conditions

Consider a linear uncertain system described by (9.38). In the early literature a common assumption was the following matching condition

A matching condition. There are matrices $A_0 \in \mathbb{R}^{n\times n}$ and $B_0 \in \mathbb{R}^{n\times m}$ such that for all $\delta \in \Delta$

(i)

$$A(\delta) = A_0 + B_0 E(\delta), \qquad B(\delta) = B_0 G(\delta) \tag{9.40}$$

(ii)

$$G(\delta) + G(\delta)^T > 0 \tag{9.41}$$

Thus a matched uncertain system can be described by

$$\begin{aligned} \dot{x} &= A_0 x + B_0[E(\delta)x + G(\delta)u] && (9.42a) \\ \delta &\in \Delta && (9.42b) \end{aligned}$$

Assuming (A_0, B_0) is stabilizable, $E(\cdot), G(\cdot)$ are continuous functions and the uncertainty set Δ is compact, then, regardless of the "size" of Δ, this system can be quadratically stabilized by nonlinear (Leitmann 1978, 1979a, 1979b, 1981) or linear controllers (Barmish et al 1983, Swei and Corless 1989).

Initial research aimed at eliminating the matching condition introduced a notion of "measure of mismatch" (Barmish and Leitmann 1982, Chen and Leitmann 1987, Yedavalli and Liang 1987). Thorp and Barmish (1981) introduced *generalized matching conditions*. These are structural conditions on the uncertainty which are less restrictive than matching conditions and permit quadratic stabilization via linear control, regardless of the size of most of the uncertain elements. These conditions were further generalized in Swei (1993). Other structural conditions were introduced in Wei (1990).

9.3.3 A Convex Parameterization of Linear Quadratically Stabilizing Controllers

Consider a linear uncertain system described by (9.38) where Δ *is compact* and the functions $A(\cdot), B(\cdot)$ *are continuous*. Then this system is quadratically stabilizable by linear controller (9.39) iff, there is a positive-definite, symmetric matrix $P \in \mathbb{R}^{n\times n}$ satisfying

$$P[A(\delta)+B(\delta)K]+[A(\delta)+B(\delta)K]^T P<0 \qquad \forall\, \delta \in \Delta \tag{9.43}$$

Following Bernussou et al (1989), let

$$S := P^{-1}, \quad L := KP^{-1} \tag{9.44}$$

Then S is positive-definite symmetric and condition (9.43) can be written as

$$A(\delta)S + SA(\delta)^T + B(\delta)L + L^T B(\delta)^T < 0 \qquad \forall\, \delta \in \Delta \tag{9.45}$$

The set of matrix pairs (S, L) which satisfy (9.45) with S positive-definite symmetric is a clearly a convex set. Also the set of linear quadratically stabilizing controllers is given by (9.39) where

$$K = LS^{-1} \tag{9.46}$$

and (S, L) satisfy (9.45) with S positive-definite symmetric.

Systems containing uncertain scalar parameters. As a particular application of the above result, consider an uncertain system described by (9.38) with

$$A(\delta) = A_0 + \delta_1 A_1 + \ldots + \delta_r A_r \tag{9.47a}$$

$$B(\delta) = B_0 + \delta_1 B_1 + \ldots + \delta_r B_r \tag{9.47b}$$

$$\Delta = \{\delta \in \mathbb{R}^r : \quad |\delta_i| \le 1, \quad i = 1, 2, \ldots, r\} \tag{9.47c}$$

Then, we can readily deduce the following result.

Theorem 9.14 *The uncertain system (9.38)-(9.47) is quadratically stabilizable via linear control (9.39) iff there exist matrices S, L, with S positive-definite symmetric, which satisfy*

$$A_0S + SA_0{}^T + B_0L + L^T B_0{}^T + \sum_{i=1}^{r} \delta_i \left[A_iS + SA_i{}^T + B_iL + L^T B_i{}^T\right] < 0 \tag{9.48a}$$

$$for \qquad \delta_i = -1, 1 \qquad i = 1, 2, \ldots, r \tag{9.48b}$$

If (9.38)-(9.47) is quadratically stabilizable via linear control, the set of linear quadratically controllers is given by (9.39) where

$$K = LS^{-1}$$

and S, L satisfy (9.48) with S positive-definite symmetric.

9.3.4 Systems Containing Uncertain Matrices

Consider first a uncertain linear system in which all the uncertainty is characterized by a single matrix $\delta \in \mathbb{R}^{p\times q}$:

$$\dot{x} = Ax + Bu + D\delta[Ex + Gu] \tag{9.49a}$$
$$||\delta|| \leq 1 \tag{9.49b}$$

Rotea and Khargonekar (1989) show that if (9.49) is quadratically stabilizable then it is quadratically stabilizable via linear control. Using this result, Corollary 9.7, parameterization (9.44), and the remark following Corollary 9.8, one can obtain the following result.

Theorem 9.15 *The uncertain system (9.49) is quadratically stabilizable (9.39) iff there exist matrices S, L, with S positive-definite symmetric, which satisfy the following linear affine matrix inequality:*

$$\begin{bmatrix} -AS - SA^T - BL - L^TB^T - DD^T & SE^T + L^TG^T \\ ES + GL & I \end{bmatrix} > 0 \tag{9.50}$$

If (9.49) is quadratically stabilizable, the set of linear quadratically stabilizing controllers is given by (9.39) where

$$K = LS^{-1}$$

and S, L satisfy (9.50) with S positive-definite symmetric.

One can also solve the quadratic stabilization problem for (9.49) by solving a parameterized Riccati equation. To this end, we suppose, without loss of generality, that if G is non-zero, it is partitioned as

$$G = [G_1 \ 0]$$

where G_1 is a matrix of full column rank. Let

$$u = \begin{bmatrix} u_1 \\ u_2 \end{bmatrix}, \qquad B = [B_1 \ B_2]$$

be the corresponding partitions of u and B, respectively. Hence, (9.49) can be rewritten as

$$\dot{x} = Ax + B_1u_1 + B_2u_2 + D\delta[Ex + G_1u_1]$$

where G_1 is either zero or a full column rank matrix. Define

$$\Theta := \begin{cases} 0 & \text{if } G_1 = 0 \\ (G_1^TG_1)^{-1} & \text{if } G_1 \neq 0 \end{cases} \tag{9.51}$$

and let

$$\tilde{A} := A - B\Theta G_1^TE\ , \qquad \tilde{E} := [I - G_1\Theta G_1^T]E$$

The following theorem is an outcome of Khargonekar et al (1990) and Rotea and Khargonekar (1989).

Theorem 9.16 *for any positive-definite symmetric matrix Q there exists $\bar{\epsilon} > 0$ such that for all $\epsilon \in (0, \bar{\epsilon}]$, the following Riccati equation has a positive-definite symmetric solution for P*

$$P\tilde{A} + \tilde{A}^T P + P[DD^T - B_1 \Theta B_1^T - \epsilon^{-1} B_2 B_2^T]P + \tilde{E}^T \tilde{E} + \epsilon Q = 0 \tag{9.52}$$

If (9.49) is quadratically stabilizable, a stabilizing controller is given by (9.39) with

$$K = -\Theta G_1^T E - [(1/2\epsilon)B_2^T + \Theta B_1^T]P \tag{9.53}$$

where P is a positive-definite symmetric solution to (9.52).

Consider now a system

$$\dot{x} = Ax + Bu + \sum_{i=1}^{r} D_i \delta_i [E_i x + G_i u] \tag{9.54a}$$

$$||\delta_i|| \leq 1 \tag{9.54b}$$

containing several uncertain matrices $\delta_i, i = 1, 2, \ldots, r$. Considering *any* r positive scalar parameters $\mu_1, \mu_2, \ldots, \mu_r$ and letting

$$D := \begin{bmatrix} \mu_1 D_1 & \mu_2 D_2 & \ldots & \mu_r D_r \end{bmatrix}, \qquad E := \begin{bmatrix} {\mu_1}^{-1} E_1 \\ {\mu_2}^{-1} E_2 \\ \vdots \\ {\mu_r}^{-1} E_r \end{bmatrix}$$

$$G := \begin{bmatrix} {\mu_1}^{-1} G_1 \\ {\mu_2}^{-1} G_2 \\ \vdots \\ {\mu_r}^{-1} G_r \end{bmatrix}$$

this system can be described by

$$\dot{x} = Ax + Bu + D\delta[Ex + Gu] \tag{9.55a}$$

where δ has block diagonal structure (9.30) and satisfies $||\delta|| \leq 1$. Hence, using Theorems 9.15 and 9.16, one can obtain sufficient conditions for the quadratic stabilizability of (9.54); recall Section 9.2.4. For extended fun along these lines, see Petersen (1985a), Petersen and Hollot (1986), and Schmitendorf (1988).

9.4 Controllers Yielding Robustness in the Presence of Persistently Acting Disturbances

Consider an uncertain system described by

$$\begin{aligned} \dot{x} &= f(x,\delta) + B(\delta)u + w(\delta) & \text{(9.56a)} \\ \delta(t,x) &\in \Delta & \text{(9.56b)} \end{aligned}$$

where w is to be regarded as a persistently acting disturbance. We assume the following.

Assumption 9.17 *The "undisturbed" uncertain system*

$$\begin{aligned} \dot{x} &= f(x,\delta) + B(\delta)u & \text{(9.57a)} \\ \delta &\in \Delta & \text{(9.57b)} \end{aligned}$$

is quadratically stabilizable.

One can use the results of the previous section to attempt to satisfy this assumption and obtain a stabilizing controller $k(\cdot)$ and a common Lyapunov matrix P.

Assumption 9.18 *There exists a matrix $B_0 \in \mathbb{R}^{n\times m}$ and positive scalars λ, κ such that for all $\delta \in \Delta$*

$$\begin{aligned} B(\delta) &= B_0 G(\delta) & \text{(9.58a)} \\ w(\delta) &= B_0 d(\delta) & \text{(9.58b)} \\ \lambda_{min}[G(\delta) + G(\delta)^T] &\geq \lambda & \text{(9.58c)} \\ \|d(\delta)\| &\leq \kappa & \text{(9.58d)} \end{aligned}$$

9.4.1 Discontinuous Controllers

The earliest controllers proposed in the literature for the class of systems considered here were discontinuous; see Gutman (1979) and Gutman and Leitmann (1976). Let $k(\cdot)$ be a controller which guarantees quadratic stability of (9.57) with common Lyapunov matrix P. Choose any scalar $\rho \geq 0$ which satisfies

$$\|d(\delta)\| \leq \frac{1}{2}\lambda_{min}[G(\delta) + G(\delta)^T]\rho \tag{9.59}$$

A quadratically stabilizing controller is given by:

$$u = k(x) - \rho \operatorname{sgn}(B_0{}^T Px) \tag{9.60}$$

where sgn(·) is the signum function defined by

$$\operatorname{sgn}(y) := \begin{cases} -\|y\|^{-1}y & \text{if} \quad y \neq 0 \\ 0 & \text{if} \quad y = 0 \end{cases}$$

These controllers are sometimes referred to as *Lyapunov Min-Max Controllers*; see Gutman (1979) and Gutman and Palmor (1982).

However, a feedback controller which is a discontinuous function of the state is undesirable for both practical and theoretical reasons. From a theoretical viewpoint, the closed-loop system does not satisfy the usual requirements which guarantee existence of solutions. Thus a solution may not exist in the usual sense. To illustrate this, consider a simple scalar closed-loop system described by

$$\begin{aligned} \dot{x} &= -x - \operatorname{sgn}(x) + w(t) \\ |w(t)| &\leq 1 \end{aligned}$$

and suppose one has defined $\operatorname{sgn}(0) = 0$. Since this system is GUES (let $V(x) = x^2$), the only possible solution for initial condition $x(0) = 0$ is $x(t) \equiv 0$. Substitution of $x = \dot{x} = 0$ in the differential equation yields $\operatorname{sgn}(0) := w(t)$. Hence, unless $w(t) \equiv 0$, this differential equation has no solution in the usual sense. Also, since $w(t)$ is unknown, one cannot let $\operatorname{sgn}(0) = w(t)$. For differential equations with right-hand sides which are discontinuous functions of the state, one must resort to generalized dynamical systems or the theory of differential inclusions to attempt to guarantee solutions; see Aubin and Cellina (1984) and Filippov (1960). This requires much heavier mathematical machinery than that normally required in studying ordinary differential equations.

If one attempts to practically implement a discontinuous controller, the result is "chattering" of the control when the state reaches the discontinuity region; i.e, the control oscillates at a very high frequency between its limits. Control actuators in many practical systems usually do not like this; they die (fail) at an early age from fatigue.

The above controllers are very similar to *Variable Structure Controllers*; see DeCarlo et al (1988), Slotine and Sastry (1983), Utkin (1977), Young (1978), Zinober (1990).

9.4.2 Continuous Controllers

Let $k(\cdot)$, P and ρ be as defined in the previous section and consider any $\epsilon > 0$. The following controller can be regarded as a continuous approximation of the discontinuous controller presented above.

$$u = k(x) - \rho s(\epsilon^{-1} {B_0}^T P x) \tag{9.61}$$

where $s(\cdot)$ is the saturation function which satisfies

$$s(y) := \begin{cases} -\|y\|^{-1}y & \text{if} \quad \|y\| > 1 \\ y & \text{if} \quad \|y\| \leq 1 \end{cases}$$

This controller does not guarantee exponential stability; however it guarantees that all solutions approach (in an exponential fashion) a neighborhood of the origin and this neighborhood can be made arbitrarily small by choosing $\epsilon > 0$ sufficiently small (Corless 1993, Corless and Leitmann 1981, 1988, 1990). If ϵ is sufficiently small, then from a practical viewpoint, the behavior of the closed loop system is the same as that of GUES.

Remark. Uncertain systems whose uncertainty bounds λ and κ depend on x can also be readily "stabilized" using a modification of the controllers presented in this section; see Corless (1993), Corless and Leitmann (1981), (1988), (1990). For an alternative approach using Lyapunov functions, see Barmish et al (1983).

9.5 Conclusions

In recent years, considerable progress has been achieved in the use of quadratic Lyapunov functions for the robust analysis and stabilization of uncertain systems. This paper presents a subjective account of some of the main results in this area.

Some of the topics not discussed here include

- Numerical techniques for solving quadratic stability and quadratic stabilizability problems
- Quadratic stabilization using only a measured output instead of the full state
- System order reduction in quadratic stabilization problems
- Discrete-time systems
- Adaptive control design with quadratic Lyapunov functions
- Quadratic stabilization of singularly perturbed systems
- Finite time quadratic stabilizability
- Applications

9.6 Acknowledgements

The author is grateful to Professor Mario Rotea of Purdue University and Professor George Leitmann of the University of California-Berkeley for useful, illuminating, and informative discussions on the topics of this paper.

References

Abdul-Wahab, A. A. 1990, Robustness measure bounds for Lyapunov-type state-feedback systems. *Proc IEE* **137D**, 321–337

Aubin, J.P., Cellina, A. 1984, *Differential inclusions*, Springer, New York

Badr, R. I., Hassan, F., Bernussou, J., Bilal, A. Y. 1988, Time-domain robustness and performance optimisation of linear quadratic control problems. *Proc IEE* **135D**, 223–231

Bahnasawi, A. A., Al-Fuhaid, A. S., Mahmoud, M. S. 1989, Linear feedback approach to the stabilization of uncertain discrete systems. *Proc IEE* **136D**, 47–52

Balakrishnan, V., Boyd, S., Balemi, S. 1991, Branch and bound algorithm for computing the minimum stability degree of parameter-dependent linear systems. *International Journal of Control* **1**, 295–317

Barmish, B.R. 1983, Stabilization of uncertain systems via linear control. *IEEE Transactions on Automatic Control* **28**, 848–850

Barmish, B. R. 1985, Necessary and sufficient conditions for quadratic stabilizability of an uncertain system. *Journal of Optimization Theory and Applications* **46**, 399–408

Barmish, B.R., Corless, M., Leitmann, G. 1983, A new class of stabilizing controllers for uncertain dynamical systems. *SIAM Journal on Control and Optimization* **21**, 246–255

Barmish, B.R., Galimidi, A.R. 1986, Robustness of Luenberger observers: linear systems stabilized via nonlinear control. *Automatica* **22**, 413–423

Barmish, B.R., Leitmann, G. 1982, On ultimate boundedness control of uncertain systems in the absence of matching assumptions. *IEEE Transactions on Automatic Control* **27**, 153–158

Barmish, B.R., Petersen, I.R., Feuer, A. 1983, Linear ultimate boundedness control of uncertain dynamic systems. *Automatica* **19**, 523–532

Basar, T., Bernhard, P. 1991, *H_∞ optimal control and related minimax design problems: a dynamic game approach*, Birkhäuser, Boston

Basile, G., Marro, G. 1987, On the robust controlled invariant. *Systems and Control Letters* **9**, 191–195

Becker, N., Grimm, W. 1988, Comments on 'reduced conservatism in stability bounds by state transformation.' *IEEE Transactions on Automatic Control* **AC-33**, 223–224

Bernstein, D. S. 1987, Robust static and dnamic output-feedback stabilization: deterministic and stochastic perspectives. *IEEE Transactions on Automatic Control* **AC-32**, 1076–1084

Bernstein, D. S., Haddad, W. M. 1988, The optimal projection equations with Petersen-Hollot bounds: robust stability and performance via fixed-order dynamic compensation for systems with structured real-valued parameter uncertainty. *IEEE Transactions on Automatic Control* **AC-33**, 578–582

Bernstein, D. S., Haddad, W. M. 1989, Robust stability and performance analysis for linear dynamic systems. *IEEE Transactions on Automatic Control* **AC-34**, 751–758

Bernstein, D. S., Hollot, C. V. 1989, Robust stability for sampled data control systems. *Systems and Control Letters* **13**, 217–226

Bernussou, J., Peres, P. L. D., Geromel, J. C. 1989, A linear programming oriented procedure for quadratic stabilization of uncertain systems. *Systems and Control Letters* **13**, 65–72

Boyd, S., El Ghaoui, L., Feron, E., Balakrishan, V. 1993, *Linear matrix inequalities in system and control theory*, monograph in preparation

Boyd, S., Yang, Q. 1989, Structured and simulantaneous Lyapunov functions for system stability problems. *International Journal of Control* **49**, 2215–2240

Breinl, V.W., Leitmann, G. 1987, State feedback for uncertain dynamical systems. *Appl. Math. Comput.* **22**, 65–87

Chen, Y. H. 1987, Robust output feedback controller: direct design. *International Journal of Control* **46**, 1083–1091

Chen, Y. H. 1987, Robust output feedback controller: indirect design. *International Journal of Control* **46**, 1093–1103

Chen, Y. H. 1988, Design of robust controllers for uncertain dynamical systems. *IEEE Transactions on Automatic Control* **33**, 487–491

Chen, Y.H., Leitmann, G. 1987, Robustness of uncertain systems in the absence of matching assumptions. *International Journal of Control* **45**, 1527–1542

Corless, M. 1993, Control of uncertain nonlinear systems. *ASME Journal of Dynamic Systems, Measurements and Control* **115**, to appear

Corless, M., Da, D. 1988, New criteria for robust stability. *International Workshop on Robustness in Identification and Control*, Turin, Italy

Corless, M., Leitmann, G. 1981, Continuous state feedback guaranteeing uniform ultimate boundedness for uncertain dynamical systems. *IEEE Transactions on Automatic Control* **26**, 1139–1144

Corless, M., Leitmann, G. 1988, Controller design for uncertain systems via Lyapunov functions. *Proc American Control Conference,* Atlanta, Georgia

Corless, M., Leitmann, G. 1990, Deterministic control of uncertain systems: a Lyapunov theory approach. in *Deterministic Control of Uncertain Systems*, ed. Zinober, A.S.I., Peter Peregrinus Ltd., London

Corless, M., Ryan, E.P. 1991, Robust feedback control of singularly perturbed uncertain dynamical systems. *Dynamics and Stability of Systems* **6**, 107–121

Corless, M., Rotea, M.A., Swei, S. M. 1993, System order reduction in quadratic stabilization problems. *12th IFAC World Congress*, Sydney, Australia

Corless, M., Zhu, G., Skelton, R. 1989, Improved robustness bounds using covariance matrices. *Proc IEEE Conference on Decision and Control,* Tampa, Florida

DeCarlo, R.A., Zak, S.H., Matthews, G. P. 1988, Variable structure control of nonlinear multivariable systems: a tutorial. *Proc IEEE* **76**, 212–232

Dorato, P. 1987, *Robust control*, IEEE Press, New York

Dorato, P., Yedavalli, R.K. 1990, *Recent advances in robust control,* IEEE Press, New York

Doyle, J. 1982, Analysis of feedback systems with structured uncertainties. *Proc IEE* **129D**, 242–250

Doyle, J., Packard, A. 1987, Uncertain multivariable systems from a state space perspective. *Proc American Control Conference,* Minneapolis, Minnesota

Eslami, M., Russel, D.L. 1980, On stability with large parameter variations: stemming from the direct method of Lyapunov. *IEEE Transactions on Automatic Control* **AC-25**, 1231-1234

Galimidi, A. R., Barmish, B.R. 1986, The constrained Lyapunov problem and its application to robust output feedback design. *IEEE Transactions on Automatic Control* **AC-31**, 410–419

Garofalo, F., Celentano, G., Glielmo, L. 1993, Stability robustness of interval matrices via Lyapunov quadratic forms. *IEEE Transactions on Automatic Control* **AC-38**, 281–284

Garofalo, F., Leitmann, G. 1989, Guaranteeing ultimate boundedness and exponential rate of convergence for a class of nominally linear uncertain systems. *ASME Journal of Dynamic Systems, Measurements and Control* **111**, 584–588

Garofalo, F., Leitmann, G. 1990, A composite controller ensuring ultimate boundedness for a class of singularly perturbed uncertain systems. *Dynamics and Stability of Systems* **3**, 135–145

Geromel, J.C., Peres, P.L.D., Bernussou, J. 1991, On a convex parameter space method for linear control design of uncertain systems. *SIAM Journal on Control and Optimization* **29**, 381–402

Gibbens, P.W., Fu, M. 1991, Output feedback control for output tracking of nonlinear uncertain systems. *Technical Report EE9121,* University of Newcastle, Newcastle, Australia

Gu, K., Chen, Y. H., Zohdy, M. A., Loh, N. K. 1991, Quadratic stabilizability of uncertain systems: a two level optimization setup. *Automatica* **27**, 161–165

Gu, K., Zohdy, M. A., Loh, N. K. 1990, Necessary and sufficient conditions of quadratic stability of uncertain linear systems. *IEEE Transactions on Automatic Control* **AC-35**, 601–604

Gutman, S. 1979, Uncertain dynamical systems-Lyapunov min-max approach. *IEEE Transactions on Automatic Control* **AC-24**, 437-443

Gutman, S., Leitmann, G. 1976, Stabilizing feedback control for dynamical systems with bounded uncertainty. *IEEE Transactions on Automatic Control* Clearwater, Florida

Gutman, S., Palmor, Z. 1982, Properties of min-max controllers in uncertain dynamical systems. *SIAM Journal on Control and Optimization* **20**, 850–861

Hinrichsen, D., Pritchard, A.J. 1986a, Stability radii of linear systems. *Systems and Control Letters* **7**, 1–10

Hinrichsen, D., Pritchard, A.J. 1986b, Stability radius for structured perturbations and the algebraic Riccati equation. *Systems and Control Letters* **8**, 105-113

Hollot, C.V. 1987, Bound invariant Lyapunov functions: a means for enlarging the class of stabilizable uncertain systems. *International Journal of Control* **46**, 161–184

Hollot, C. V., Barmish, B.R. 1980, Optimal quadratic stabilizability of uncertain linear systems. *18th Allerton Conference on Communications, Control, and Computing*, University of Illinois, Monticello, Illinois

Hopp, T.H., Schmitendorf, W.E. 1990, Design of a linear controller for robust tracking and model following. *ASME Journal of Dynamic Systems, Measurements and Control* **112**, 552–558

Horisberger, H. P., and Belanger, P. R. 1976, Regulators for linear, time invariant plants with uncertain parameters. *IEEE Transactions on Automatic Control* **AC-21**, pp. 705–708

Hyland, D. C., Bernstein, D. S. 1987, The majorant Lyapunov equation: a nonnegative matrix equation for robust stability and performance of large scale systems. *IEEE Transactions on Automatic Control* **AC-32**, 1005–1013

Hyland, D. C., Collins, E. G. 1989, An M-matrix and majorant approach to robust stability and performance analysis for systems with structured uncertainty. *IEEE Transactions on Automatic Control* **AC-34**, 699–710

Hyland, D. C., Collins, E. G. 1991, Some majorant robustness results for discrete-time systems. *Automatica* **27**, 167–172

Jabbari, F., Benson, R.W. 1992, Observers for stabilization of systems with matched uncertainty. *Dynamics and Control* **2**, 303–323

Jabbari, F., Schmitendorf, W.E. 1991, Robust linear controllers using observers. *IEEE Transactions on Automatic Control* **AC-36**, 1509–1511

Jabbari, F., Schmitendorf, W.E. 1993, Effects of using observers on stabilization of uncertain linear systems. *IEEE Transactions on Automatic Control* **AC-38**, 266–271

Kalman, R.E., Bertram, J.E. 1960, Control system analysis and design via the "second method" of Lyapunov, I: continuous-time systems. *Journal of Basic Engineering* **32**, 317–393.

Khargonekar, P.P., Petersen, I.R., Zhou, K. 1990, Robust stabilization of uncertain linear systems: quadratic stabilizability and H^∞ control theory. *IEEE Transactions on Automatic Control* **35**, 356–361

Kolla, S. P., Yedavalli, R. K., Farison, J. B. 1989, Robust stability bounds on time-varying perturbations for state-space models of linear discrete-time systems. *International Journal of Control* **50**, 151–159

Leitmann, G. 1978, Guaranteed ultimate boundedness for a class of uncertain linear dynamical systems. *IEEE Transactions on Automatic Control* **AC-23**, 1109–1110

Leitmann, G. 1979a, Guaranteed asymptotic stability for some linear systems with bounded uncertainties. *ASME Journal of Dynamic Systems, Measurements and Control* **101**, 212–216

Leitmann, G. 1979b, Guaranteed asymptotic stability for a class of uncertain linear dynamical systems. *Journal of Optimization Theory and Applications* **27**, pp. 99–106

Leitmann, G. 1981, On the efficacy of nonlinear control in uncertain linear systems. *ASME Journal of Dynamic Systems, Measurements and Control* **102**, 5–102

Leitmann, G., Ryan, E.P., Steinberg, A. 1986, Feedback control of uncertain systems: robustness with respect to neglected actuator and sensor dynamics. *International Journal of Control* **43**, 1243–1256

Lyapunov, A.M. 1907, Problème general de la stabilite du mouvement (in French). *Ann. Fac. Sci. Toulouse* **9**, 203–474. Reprinted in *Ann. Math. Study* No. 17, 1949, Princeton University Press

Michael, G. J., Merriam, C. W., III 1969, Stability of parametrically disturbed linear optimal control systems. *Journal of Mathematical Analysis and Applications* **28**, pp. 294-302

Olas, A., Optimal quadratic Lyapunov functions for robust stability problems. *Dynamics and Control* , to appear.

Patel, R.V., Toda, M. 1980, Quantitative measures of robustness for uncertain systems. *Joint Automatic Control Conference,* San Francisco, California

Petersen, I.R. 1985a, A Riccati equation approach to the design of stabilizing controllers and observers for a class of uncertain linear systems. *IEEE Transactions on Automatic Control* **AC-30**, 904–907

Petersen, I.R. 1985b, Quadratic stabilizability of uncertain linear systems: existence of a nonlinear stabilizing control does not imply existence of a linear stabilizing control. *IEEE Transactions on Automatic Control* **30**, 291–293

Petersen, I. R. 1987, Notions of stabilizability and controllability for a class of uncertain linear systems. *International Journal of Control* **46**, 409–422

Petersen, I.R. 1988, Quadratic stabilizability of uncertain linear systems containing both constant and time-varying uncertain parameters. *Journal of Optimization Theory and Applications* **57**, 439–461

Petersen, I.R., Corless, M., Ryan, E.P. 1992, A necessary and sufficient condition for quadratic finite time feedback controllability. *International Workshop on Robust Control*, Ascona, Switzerland

Petersen, I.R., Hollot, C.V. 1986, A Riccati equation approach to the stabilization of uncertain linear systems. *Automatica* **22**, 397–411

Petersen, I.R., Hollot, C.V. 1988, High gain observers applied to problems in stabilization of uncertain linear systems, disturbance attenuation, and H_∞ optimization. *International Journal of Adaptive Control and Signal Processing* **2**, 347–369

Petersen, I.R., Pickering, M. R. 1992, An algorithm for the quadratic stabilization of uncertain systems with structured uncertainty of the one-block type. *Proc American Control Conference,* Chicago, Illinois

Qu, Z., Dorsey, J. 1991, Robust control of generalized dynamic systems without the matching conditions. *ASME Journal of Dynamic Systems, Measurements and Control* **113**, 582–589

Ran, A.C.M., R. Vreugdenhil, R. 1988, Existence and comparison theorems for algebraic Riccati equations for continuous- and discrete-time systems. *Linear Alg. Appl.* **99**, 63–83

Rotea, M. A. 1990, *Multiple objective and robust control for linear systems*, Ph.D. Thesis, University of Minnesota, Minneapolis

Rotea, M. A., Corless, M., Da, D., Petersen, I.R. 1993, Systems with structured uncertainty: relations between quadratic and robust stability. *IEEE Transactions on Automatic Control* **AC-38**, to appear

Rotea, M. A., Khargonekar, P.P. 1989, Stabilization of uncertain systems with norm bounded uncertainty - a control Lyapunov approach. *SIAM Journal on Control and Optimization* **27**, 1462–1476

Ryan, E.P., Corless, M. 1984, Ultimate boundedness and asymptotic stability of a class of uncertain dynamical systems via continuous and discontinuous feedback control. *IMA Journal of Mathematical Control and Information* **1**, 223–242

Schmitendorf, W.E. 1988, Designing stabilizing controllers or uncertain systems using the Riccati equation approach. *IEEE Transactions on Automatic Control* **33**, 376–379

Slotine, J.J., Sastry, S.S. 1983, Tracking control of nonlinear systems using sliding surfaces, with application to robot manipulator manipulators. *International Journal of Control* **48**, 465–492

Sobel, K. M., Banda, S. S., Yeh, H. H. 1989, Robust control for linear systems with structured state space uncertainty. *International Journal of Control* **50**, 1991–2004

Soldatos, A.G., Corless, M. 1991, Stabilizing uncertain systems with bounded control. *Dynamics and Control* **3**, 227–238

Stalford, H. 1987, Robust control of uncertain systems in the absence of matching conditions:scalar input. *Proc IEEE Conference on Decision and Control,* Los Angeles, California

Swei, S. M. 1993, *Quadratic stabilization of uncertain systems: reduced gain controllers, order reduction, and quadratic controllability*, Ph.D. Thesis, Purdue University, West Lafayette, Indiana

Swei, S. M., Corless, M. 1989, Reduced gain controllers for a class of uncertain dynamical systems. *IEEE International Conference on Systems Engineering*, Dayton, Ohio

Swei, S. M., Corless, M. 1991, On the necessity of the matching condition in robust stabilization. *Proc IEEE Conference on Decision and Control,* Brighton, U.K.

Thorp, J. S., Barmish, B. R. 1981, On guaranteed stability of uncertain linear systems via linear control. *Journal of Optimization Theory and Applications* **35**, 559–579

Utkin, V.I. 1977, Variable structure systems with sliding modes. *IEEE Transactions on Automatic Control* **AC-22**, 212–222

Wei, K. 1990, Quadratic stabilizability of linear systems with structural independent time-varying uncertainties. *IEEE Transactions on Automatic Control* **35**, 268–277

Yedavalli, R.K. 1985, Improved measures of stability robustness for linear state space models. *IEEE Transactions on Automatic Control* **AC-30**, 557–559

Yedavalli, R.K. 1989, On Measures of stability robustness for linear state space systems with real parameter perturbations: a perspective. *Proc American Control Conference,* Pittsburgh, Pennsylvania

Yedavalli, R.K., Banda, S.S., Ridgely, D. B. 1985, Time domain stability robustness measures for linear regulators. *Journal of Guidance, Control and Dynamics* **4**, 520–525

Yedavalli, R.K., Liang, Z. 1987, Reduced conservatism in the ultimate boundedness control of mismatched uncertain systems. *ASME Journal of Dynamic Systems, Measurements and Control* **109**, 1–6

Young, K.-K. D 1978, Design of variable structure model-following control systems. *IEEE Transactions on Automatic Control* **AC-23**, 1079–1085

Zak. S. H. 1990, On the stabilization and the observation of non-linear uncertain dynamic systems. *IEEE Transactions on Automatic Control* **AC-35**, 604–607

Zhou, K., Khargonekar, P.P. 1987, Stability robustness bounds for linear state-space models with structured uncertainty. *IEEE Transactions on Automatic Control* **AC-32**, 621–623

Zhou, K., Khargonekar, P. P. 1988, On the stabilization of uncertain linear systems via bound invariant Lyapunov functions. *SIAM Journal on Control and Optimization* **26**, 1265–1273

Zinober, A.S.I. 1990, *Deterministic control of uncertain systems*, Peter Peregrinus Ltd., London

10. Universal Controllers: Nonlinear Feedback and Adaptation

Eugene P. Ryan

10.1 Introduction

A priori information sufficient for (adaptive) stabilization and questions of existence and construction of universal controllers for various classes of dynamical systems have been the subject of many recent studies with a variety of viewpoints: see, for example, Byrnes and Willems (1984), Carera and Furuta (1989), Corless (1991), Corless and Leitmann (1984), Logemann and Owens (1988), Logemann and Zwart (1991), Mårtensson (1985), Miller and Davison (1989); a comprehensive bibliography is given in the survey by Ilchmann (1991). Linear single-input systems, possibly with high-frequency gain of unknown sign (Helmke and Prätzel-Wolters 1988; Helmke, Prätzel-Wolters and Schmid 1990; Ilchmann and Logemann 1992; Morse 1984, 1985; Willems and Byrnes 1984) and possibly subject to 'mild' nonlinear perturbations, feature prominently: 'strongly' nonlinear single-input systems, such as those considered in Mårtensson (1991), Ryan (1990, 1991a, 1991b), have received less attention. One possible source of this disparity is that the linear "L^p-type" stability arguments, prevalent in the former, may fail to have counterparts in the latter context: analysis (and synthesis) of universal stabilizers for nonlinear systems can differ in an essential way from the linear case.

A second body of work (Byrnes and Willems 1984; Ilchmann and Logemann 1992; Ilchmann and Owens 1990; Ilchmann, Owens and Prätzel-Wolters 1987; Mårtensson 1985, 1986, 1987, 1991; Townley and Owens 1991) is concerned with linear multi-input systems, again possibly subject to 'mild' nonlinear perturbations. Even in the linear case (with high-frequency gain of unknown sign), the transition from single to multiple inputs is not straightforward: the existence of "finite spectrum-unmixing sets", conjectured by Byrnes and Willems (1984) and proved by Mårtensson (1986, 1987, 1991) (see Lemma 10.4 below), plays a central role. In Sects. 10.2, 10.3 and 10.4, some extensions of the latter investigations to more general multi-input nonlinear systems are described. We focus on three particular (but not mutually exclusive) classes:

Class I: systems modelled by pth-order controlled differential inclusions on $\mathbb{R}^m$. We assume that the full state is available for feedback purposes. Weak a priori assumptions (Assumptions 10.1, 10.2 and 10.3 below) on the operators and set-valued maps of the model determine the class. In Sect.10.2, we describe an adaptive discontinuous feedback controller (as developed in Ryan (1993)) which is shown to be a universal stabilizer for this class.

Class II: nonlinearly perturbed linear systems with output constraints.
In Section 10.3, we address the problem of *output feedback*: under a minimum-phase assumption on the unperturbed system and for a specific class of nonlinear perturbations, the above strategy is one of output feedback (which, in the unperturbed case, reduces to a linear adaptive strategy coincident with those of Ilchmann and Logemann (1992), Mårtensson (1987)). Here, we extend the investigation to a problem (akin to that of Ryan (1992)) of tracking $\mathbb{R}^m$-valued reference signals of Sobolev class $W^{1,\infty}$.
Class III: two-input systems.
Finally we specialize to two-input systems of Class I, with one additional a priori assumption, namely, that the determinant of a particular invertible operator is of known sign. Under this extra structural assumption, in Sect.4, we show that a universal stabilizer, of less complexity than that of the general Class I, is feasible.

By way of motivation for Class I, consider an m-controlled-degree-of-freedom mechanical system of the form

$$M^{\#}\ddot{q}(t) + B^{\#}u(t) = g(t, q(t), \dot{q}(t))$$

where $q(t) \in \mathbb{R}^m$ is a vector of generalized coordinates, $M^{\#}$ is an inertia matrix, $B^{\#}$ is a control interface matrix, and the function g (assumed measurable in t and continuous in its other arguments) represents elastic, damping, friction, Coriolis and other forces as well as extraneous disturbances.
We assume that $M^{\#}$ and $B^{\#}$ are unknown but invertible and that g is unknown but bounded, modulo an unknown scalar multiplier $\mu > 0$, by a known continuous function of the state in the sense that, for almost all $t \in \mathbb{R}$,

$$\|g(t, q, v)\| \leq \mu\gamma(q, v) \qquad \forall\ (q, v) \in \mathbb{R}^m \times \mathbb{R}^m$$

For example, if g is polynomial in (q, v) of *unknown* degree with *unknown* t-dependent coefficients, $a_i(t)$ say, and $a_i(\cdot) \in L^\infty(\mathbb{R})$, then the above assumption holds with $\gamma(q, v) = \exp(\|(q, v)\|)$.
Writing

$$M = \mu^{-1}M^{\#}, \quad B = \mu^{-1}B^{\#}$$

and defining the set-valued map

$$\mathcal{Z} : (q, v) \mapsto \gamma(q, v)\overline{\mathcal{B}}$$

where $\overline{\mathcal{B}}$ denotes the closed unit ball centred at the origin in $\mathbb{R}^m$, the non-autonomous system can be embedded in the following autonomous differential inclusion

$$M\ddot{q}(t) + Bu(t) \in \mathcal{Z}(q(t), \dot{q}(t))$$

Systems of this nature fall within the main category to be studied.

10.2 Class I: Universal Adaptive Stabilizer

We first consider Class I of uncertain dynamical systems modelled by a pth-order controlled differential inclusion on $\mathbb{R}^m$ of the form

$$Mz^{(p)}(t) + Bu(t) \in \mathcal{Z}(z(t), \dot{z}(t), ..., z^{(p-1)}(t)) \qquad (10.1)$$
$$(z(0), \dot{z}(0), ..., z^{(p-1)}(0)) = \zeta^0$$

with $z(t), u(t) \in \mathbb{R}^m$.
Only the following a priori information is required:

Assumption 10.1 *M, B are invertible.*

Assumption 10.2 *There exists a* known *finite set*

$$\mathcal{K} = \{K_1, K_2, ..., K_r\} \subset Gl(m; \mathbb{R})$$

such that, for some $K_j \in \mathcal{K}$, $\sigma(M^{-1}BK_j) \subset \mathbb{C}^+$ (where $\sigma(\cdot)$ denotes spectrum and $\mathbb{C}^+$ is the open right half complex plane).

Assumption 10.3 *$\mathcal{Z}$ is a known continuous set-valued map from $\mathbb{R}^{pm}$ to the non-empty, convex and compact subsets of $\mathbb{R}^m$.*

Remark. M and B need not be known, but are non-singular by Assumption 10.1. Any finite set $\mathcal{K}$ satisfying Assumption 10.2 is referred to as *a spectrum-unmixing set* for $M^{-1}B$. Such sets have been the subject of recent study (see Mårtensson (1987, 1991), Zhu (1989)); for completeness, we reiterate Mårtensson's fundamental result, namely, that, for each $m \in \mathbb{N}$, there exists a finite set $\mathcal{K}$ of *orthogonal* matrices which is an unmixing set for every element of $Gl(m; \mathbb{R})$.

Lemma 10.4 (Mårtensson) *There exists a finite set*

$$\mathcal{O} = \{O_1, O_2, ..., O_r\} \subset \mathbb{R}^{m \times m}$$

of orthogonal *matrices with the property that, for every invertible $L \in \mathbb{R}^{m\times m}$, there exists $O_j \in \mathcal{O}$ such that $\sigma(LO_j) \subset \mathbb{C}^+$.*

Thus, the essence of Assumption 10.2 is not the *existence* of a spectrum-unmixing set but rather that one such set is *known* to the controller. Assumption 10.3 is a regularity condition on the known set-valued map $\mathcal{Z}$, which, in conjunction with the class of feedback strategies to be studied, ensures existence of solutions of the feedback-controlled initial-value problem (10.1). Recall that a compact-set-valued map $x \mapsto F(x) \subset \mathbb{R}^N$, defined on an open set $G \subset \mathbb{R}^P$, is *continuous* if it is both upper and lower semicontinuous at each $\bar{x} \in G$: F is *upper semicontinuous* at $\bar{x}$ if, and only if, for each $\varepsilon > 0$, there exists $\delta > 0$ such that $F(\bar{x} + \delta\mathcal{B}) \subset F(\bar{x}) + \varepsilon\mathcal{B}$, where $\mathcal{B}$ denotes the open unit ball centred

at zero in the appropriate space; F is *lower semicontinuous* at $\bar{x}$ if, and only if, for every sequence $(x_k) \subset G$ converging to $\bar{x}$ and for every $\phi \in F(\bar{x})$, there exists a sequence $(\phi_k \in F(x_k))$ converging to ϕ.
We will show, by construction, that Assumptions 10.1, 10.2 and 10.3 are sufficient for the existence of a *Class I universal stabilizer*, that is, a feedback control strategy which does not depend on the unknown parameters M and B and which ensures that, for each $\zeta^0 \in \mathbb{R}^{mp}$, every state solution of (10.1) under feedback control approaches the zero state.
The analytical framework is that of differential inclusions (Aubin and Cellina 1984, Filippov 1988, Roxin 1965, Ryan 1990, 1991a,b).
In Sects. 10.3 and 10.4, the construction is modified to provide controllers which are universal for Classes II and III.

10.2.1 Coordinate transformation

Let $C_i \in \mathbb{R}^{m\times m}$, $i = 1, 2, ..., p-1$, be such that all poles of the linear system

$$\mathcal{S}: \qquad z^{(p-1)} + C_{p-1}z^{(p-2)} + \cdots + C_2\dot{z}(t) + C_1 z(t) = 0$$

lie in the open left half complex plane $\mathbb{C}^-$. Let T denote the coordinate transformation

$$(z(t), \dot{z}(t), ..., z^{(p-1)}(t)) \mapsto (w(t), y(t))$$

where

$$w(t) = (w_1(t), w_2(t), \cdots, w_{p-1}(t)) = (z(t), \dot{z}(t), ..., z^{(p-2)}) \in \mathbb{R}^{(p-1)m}$$

$$y(t) = C_1 z(t) + C_2\dot{z}(t) + \cdots + C_{p-1}z^{(p-2)} + z^{(p-1)} \in \mathbb{R}^m$$

This transformation takes (10.1) into the form

$$\left.\begin{aligned} &\dot{w}(t) = L_1 w(t) + L_2 y(t) \\ &M[\dot{y}(t) + L_3 w(t) - C_{p-1}y(t)] + Bu(t) \in \mathcal{F}(w(t), y(t)) \\ &(w(0), y(0)) = T\zeta^0 \end{aligned}\right\} \qquad (10.2)$$

where $\mathcal{F} := \mathcal{Z} \circ T^{-1}$ and L_1, L_2, L_3 are linear with

$$L_1: \ (w_1, w_2, \cdots, w_{p-2}, w_{p-1}) \mapsto \left(w_2, w_3, \cdots, w_{p-1}, -\sum_{i=1}^{p-1} C_i w_i\right) \qquad (10.3)$$

The spectrum of L_1 is precisely that of the linear system $\mathcal{S}$ and so $\sigma(L_1) \subset \mathbb{C}^-$. Therefore, the Lyapunov equation

$$PL_1 + L_1^T P + I = 0$$

has unique symmetric positive-definite solution P.

10.2.2 Adaptive Feedback Strategy

Let $\mathcal{K} = \{K_1, K_2, ..., K_r\}$ be a finite unmixing set for $M^{-1}B$. We may assume that $\|K_i\| = 1$ for all $i \in \{1, 2, ..., r\}$.
Let $(\tau_n)_{n=0}^{\infty}$ be any strictly increasing sequence such that

$$\tau_n \to \infty \quad \text{and} \quad \frac{\tau_{n-1}}{\tau_n} \to 0 \quad \text{as } n \to \infty.$$

For example, with $\tau_0 > 1$ and $\rho > 1$, one such sequence is generated by the recursion $\tau_n = \tau_{n-1}^{\rho}$.
Let $(\sigma_n)_{n\in\mathbb{N}}$ be any sequence such that

$$\tau_{n-1} < \sigma_n < \tau_n \quad \text{and} \quad \frac{\sigma_n}{\tau_n} \to 0 \quad \text{as } n \to \infty.$$

For example, $\sigma_n = \frac{1}{2n}(\tau_n + (2n-1)\tau_{n-1})$ suffices.
Let $\kappa \mapsto s(\kappa) \in \mathbb{R}$ be any continuous function with the properties

$$s(\kappa) \in \{1, 2, ..., r\} \quad \forall\, \kappa \in (-\infty, \tau_0] \bigcup_{n\in\mathbb{N}} [\sigma_n, \tau_n]$$

$$s([k, \infty)) = [1, r] \quad \forall k \in \mathbb{R}.$$

In the case $r = 3$, Fig. 10.1 depicts the graph of one such function. Finally, let

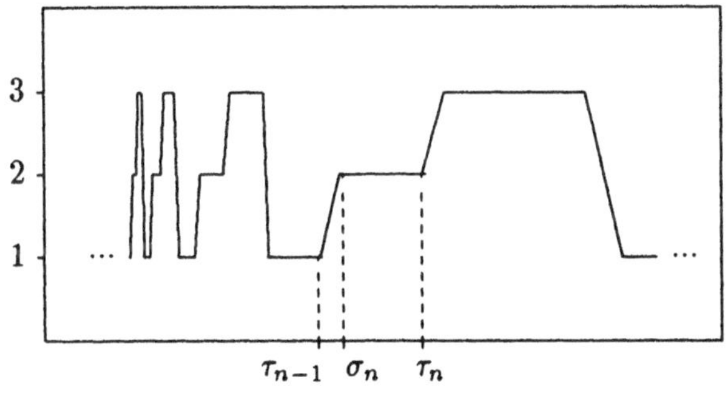

Fig. 10.1. Graph of typical function $s(\cdot)$

$s \mapsto K(s) \in \mathbb{R}^{m \times m}$ be any continuous map with the properties

$$K(s) \in \operatorname{conv} \mathcal{K} \quad \forall\, s \in [1, r]$$

$$K(s) = K_s \quad \forall\, s \in \{1, 2, ..., r\}.$$

Thus, for each s in the interval $[1, r]$, $K(s)$ is a convex combination of the elements of the unmixing set $\mathcal{K}$ and, whenever s belongs to the index set $\{1, 2, ..., r\}$, $K(s)$ coincides with the corresponding element $K_s \in \mathcal{K}$.
Our proposed adaptive strategy is given formally as

$$u(t) = \kappa(t)\,[f(w(t), y(t)) + \|y(t)\|]\,\|y(t)\|^{-1}K(s(\kappa(t)))y(t)$$

$$\dot{\kappa}(t) = f(w(t), y(t))\|y(t)\| + \|y(t)\|^2, \qquad \kappa(0) = \kappa^0$$

where $f : (w, y) \mapsto \max\{\|\phi\| \mid \phi \in \mathcal{F}(w, y)\}$. Continuity of the set-valued map $\mathcal{F}$, together with compactness of its values, ensures that f is a well-defined *continuous* map $\mathbb{R}^{pm} \to [0, \infty)$.

Remark. The function $K(s(k(\cdot))$ is central to the strategy; it provides a facility for cycling through the elements of the unmixing set $\mathcal{K}$ and dwelling at each element of the set for progressively longer time intervals.

Noting the discontinuous nature of the feedback and writing

$$x(t) = (w(t), y(t), \kappa(t))$$

we interpret the control in the following generalized sense

$$u(t) \in \Psi(x(t))$$

where

$$\Psi(x) := \kappa\,[f(w, y) + \|y\|]\,K(s(\kappa))\psi(y)$$

with

$$\psi(y) := \begin{cases} \{\|y\|^{-1}y\}, & y \neq 0 \\ \overline{\mathcal{B}}, & y = 0 \end{cases}$$

where, as before, $\overline{\mathcal{B}}$ denotes the closed unit ball centred at the origin in $\mathbb{R}^m$. The overall adaptively controlled system may now be embedded in the following initial-value problem in $\mathbb{R}^N$, $N := pm + 1$,

$$\dot{x}(t) \in F(x(t)), \quad x(0) = x^0 = (T\zeta^0, \kappa^0) \tag{10.4}$$

with the set-valued map $x \mapsto F(x) \subset \mathbb{R}^N$ defined by

$$F(x) := F_1(x) \times F_2(x) \times F_3(x)$$

$$F_1(x) := \{L_1 w + L_2 y\}$$

$$F_2(x) := \{M^{-1}[\phi - Bu] - L_3 w + C_{p-1} y \mid \phi \in \mathcal{F}(w, y),\ u \in \Psi(x)\}$$

$$F_3(x) := \{\|y\|^2 + f(w, y)\|y\|\}$$

for all $x = (w, y, \kappa) \in \mathbb{R}^{(p-1)m} \times \mathbb{R}^m \times \mathbb{R} = \mathbb{R}^N$.
It is clear that F takes convex and compact values. Continuity of the function f, together with upper semicontinuity of the set-valued map ψ, implies upper semicontinuity of Ψ. Upper semicontinuity of F follows immediately. Therefore, for each $x^0 \in \mathbb{R}^N$, (10.4) admits a solution (see, for example, Aubin and Cellina (1984), p.98, Theorem 3) and every solution has a maximal extension (see, for example, Ryan (1990)). Moreover, if a maximally extended solution is bounded, then its (maximal) interval of existence is the half line $[0, \infty)$.

10.2.3 Stability Analysis

We now arrive at the first result which, in the context of the original system representation (10.1), may be paraphrased as follows: under the adaptive strategy with arbitrary κ^0, for each ζ^0, (i) every solution of (10.1) can be extended indefinitely (finite escape times do not occur), (ii) the adaptive gain $\kappa(t)$ tends to a finite limit, and (iii) every solution of (10.1) tends to the zero state.

Theorem 10.5 *Let* $x(\cdot) = (w(\cdot), y(\cdot), \kappa(\cdot)) : [0, \omega) \to \mathbb{R}^N$ *be a maximal solution of (10.4). Then*
(i) $\omega = \infty$,
(ii) $\lim_{t\to\infty} \kappa(t)$ *exists and is finite,*
(iii) $\lim_{t\to\infty} \|(w(t), y(t))\| = 0$.

Proof. For convenience, we write $D = M^{-1}B$. Since $\mathcal{K} = \{K_1, K_2, ..., K_r\}$ is a finite unmixing set for D, there exists $j \in \{1, 2, ..., r\}$ such that $\sigma(DK_j) \subset \mathbb{C}^+$. Therefore, the Lyapunov equation

$$Q(DK_j) + (DK_j)^T Q - I = 0$$

has unique symmetric positive-definite solution Q. Write $\alpha = 2\|QD\|$ and define a map $\nu : \mathbb{R} \to \mathbb{R}$ as follows

$$\nu : \kappa \mapsto \begin{cases} \frac{1}{2}, & s(\kappa) = j \\ -\frac{1}{2}\alpha, & s(\kappa) \neq j \end{cases}$$

Let

$$W_1 : w \mapsto \tfrac{1}{2}\langle w, Pw\rangle \quad \text{and} \quad W_2 : y \mapsto \tfrac{1}{2}\langle y, Qy\rangle$$

and

$$W : x = (w, y, \kappa) \mapsto W_1(w) + W_2(y)$$

Since $\dot{w}(t) = L_1 w(t) + L_2 y(t)$ and $\sigma(L_1) \subset \mathbb{C}^-$, there exist constants c_0 and c_1 such that, for all $t_0, t \in [0, \omega)$ with $t \geq t_0$,

$$\int_{t_0}^{t} |\langle Qy(s), L_3 w(s)\rangle| \, ds \;\leq\; c_0\|w(t_0)\|^2 + c_1 \int_{t_0}^{t} \|y(s)\|^2 \, ds$$

$$\leq c_0\|w(t_0)\|^2 + c_1(\kappa(t) - \kappa(t_0)) \tag{10.5}$$

Writing $c_2 = \|QC_{p-1}\| + \|QM^{-1}\|$, we have

$$\langle \nabla W_2(y), \eta\rangle \leq -\langle Qy, L_3x\rangle + (c_2 - \kappa\nu(\kappa))[f(w,y)\|y\| + \|y\|^2]$$

for all $\eta \in F_2(x)$.
Therefore, for the maximal solution $x(\cdot) = (w(\cdot), y(\cdot), \kappa(\cdot))$, we have

$$\frac{\mathrm{d}}{\mathrm{d}t} W_2(y(t)) \leq -\langle Q(y(t)), L_3 w(t)\rangle + (c_2 - \kappa(t)\nu(\kappa(t)))\dot{\kappa}(t)$$

for almost all t, which, on integration and using (10.5), yields

$$0 \leq W_2(y(t)) \leq W_2(y(t_0)) + c_0\|w(t_0)\|^2 + (c_1 + c_2)(\kappa(t) - \kappa(t_0)) - \int_{\kappa(t_0)}^{\kappa(t)} \theta\nu(\theta)\,\mathrm{d}\theta$$

valid for all $t, t_0 \in [0,\omega)$ with $t \geq t_0$.
Seeking a contradiction, suppose that the monotone increasing function $\kappa(\cdot)$ is unbounded. Then there exists $t_0 \in [0,\omega)$ such that $\kappa(t) \geq 1$ for all $t \in [t_0, \omega)$. Therefore,

$$0 \leq \liminf_{t\uparrow\omega} \frac{W_2(y(t))}{\kappa(t)} \leq [W_2(y(t_0)) + c_0\|w(t_0)\|^2 - (c_1 + c_2)\kappa(t_0)] + c_1 + c_2$$

$$- \limsup_{\xi\to\infty} \frac{1}{\xi}\int_{\kappa(t_0)}^{\xi} \theta\nu(\theta)\,\mathrm{d}\theta \tag{10.6}$$

By supposition, $\kappa(\cdot)$ is unbounded and so, by definition of s, there exists an increasing sequence $(t_k)_{k\in\mathbb{N}}$, with $t_k \to \omega$ as $k \to \infty$, and associated increasing subsequences $(\sigma_{n_k})_{k\in\mathbb{N}}$ and $(\tau_{n_k})_{k=0}^{\infty}$ of (σ_n) and (τ_n) such that $\kappa(t_0) \leq \tau_{n_0}$ and, for all $k \in \mathbb{N}$,

$$\kappa(t_k) = \tau_{n_k} \quad \text{and} \quad s(\lambda) = j \quad \forall\, \lambda \in [\sigma_{n_k}, \tau_{n_k}]$$

Now,

$$\limsup_{\xi\to\infty} \frac{1}{\xi}\int_{\kappa(t_0)}^{\xi} \theta\nu(\theta)\,\mathrm{d}\theta \;\geq\; \limsup_{k\to\infty} \frac{1}{\tau_{n_k}}\int_{\kappa(t_0)}^{\tau_{n_k}} \theta\nu(\theta)\,\mathrm{d}\theta$$

and

$$\frac{1}{\tau_{n_k}}\int_{\kappa(t_0)}^{\tau_{n_k}} \theta\nu(\theta)\,\mathrm{d}\theta = \text{constant} + \frac{1}{2\tau_{n_k}}\sum_{i=1}^{k}\left(\int_{\sigma_{n_i}}^{\tau_{n_i}} \theta\,\mathrm{d}\theta - \alpha\int_{\tau_{n_{i-1}}}^{\sigma_{n_i}} \theta\,\mathrm{d}\theta\right)$$

$$= \text{constant} + \frac{1}{4\tau_{n_k}}\sum_{i=1}^{k-1}\left(\tau_{n_i}^2 - (1+\alpha)\sigma_{n_i}^2 + \alpha\tau_{n_{i-1}}^2\right)$$

$$+ \frac{1}{4\tau_{n_k}}\left(\tau_{n_k}^2 - (1+\alpha)\sigma_{n_k}^2 + \alpha\tau_{n_{k-1}}^2\right)$$

Recalling that $\sigma_n/\tau_n \to 0$ as $n \to \infty$, we see that the second term (summation) on the right hand side of the latter equation is bounded from below uniformly in k; furthermore, since $\tau_n \to \infty$ and $\tau_{n-1}/\tau_n \to 0$ as $n \to \infty$, we may conclude that

$$\frac{1}{\tau_{n_k}}\left(\tau_{n_k}^2 - (1+\alpha)\sigma_{n_k}^2 + \alpha\tau_{n_{k-1}}^2\right) \;\to\; \infty \quad \text{as } k \to \infty$$

Therefore

$$\limsup_{\xi\to\infty} \frac{1}{\xi}\int_{\kappa(t_0)}^{\xi} \theta\nu(\theta)\,\mathrm{d}\theta \;=\; \infty$$

This contradicts (10.6), and so $\kappa(\cdot)$ is bounded.
Define

$$\kappa^* := \tfrac{1}{4} + \|QM^{-1}\| + \|QC_{p-1}\| + (\|PL_2\| + \|QL_3\|)^2$$

Then, for all $\phi \in F(x)$,

$$\begin{aligned}\langle \nabla W(x), \phi\rangle \;&\le \langle Pw, L_1 w\rangle + \langle Pw, L_2 y\rangle - \langle Qy, L_3 w\rangle + \langle Qy, C_{p-1} y\rangle \\ &\quad + f(w,y)\|QM^{-1}\|\|y\| - \kappa\nu(\kappa)[f(w,y)\|y\| + \|y\|^2] \\ &\le -\tfrac{1}{4}(\|w\|^2 + \|y\|^2) + (\kappa^* - \kappa\nu(\kappa))[f(w,y)\|y\| + \|y\|^2]\end{aligned}$$

which is valid for all $x = (w, y, \kappa) \in \mathbb{R}^N$. Therefore, for the maximal solution $x(\cdot)$, we may conclude that

$$\frac{\mathrm{d}}{\mathrm{d}t}W(x(t)) \le -\tfrac{1}{4}(\|w(t)\|^2 + \|y(t)\|^2) + (\kappa^* - \kappa(t)\nu(\kappa(t))\dot{\kappa}(t) \qquad (10.7)$$

for almost all t.
Integrating (10.7) and using boundedness of $\kappa(\cdot)$, we see that $W(x(\cdot))$ is bounded, whence boundedness of $w(\cdot)$ and $y(\cdot)$. We have now shown that the solution $x(\cdot) = (w(\cdot), y(\cdot), \kappa(\cdot))$ is bounded and so $\omega = \infty$. This establishes assertion (i). Assertion (ii) follows by boundedness and monotonicity of $\kappa(\cdot)$. It remains to prove assertion (iii).
Boundedness of the solution $x(\cdot)$ ensures that it has non-empty ω-limit set Ω. Since the solution approaches its ω-limit set, we will prove assertion (iii) by showing that Ω is contained in the set

$$\Pi := \{x = (w, y, \kappa) \in \mathbb{R}^N \mid \|w\|^2 + \|y\|^2 = 0\}$$

Define

$$V:\; x = (w, y, \kappa) \mapsto W(x) - \int_0^{\kappa} (\kappa^* - \theta\nu(\theta))\,\mathrm{d}\theta$$

For all $x = (w, y, \kappa) \in \mathbb{R}^N$, we have

$$\langle \nabla V(x), \phi\rangle \le -\tfrac{1}{4}(\|w\|^2 + \|y\|^2) \quad \forall\, \phi \in F(x) \qquad (10.8)$$

Seeking a contradiction, suppose $\Omega \not\subset \Pi$. Then there exists $\bar{x} = (\bar{w}, \bar{y}, \bar{\kappa}) \in \Omega$ and $\epsilon > 0$ such that $\|\bar{w}\|^2 + \|\bar{y}\|^2 \geq 2\epsilon$. By continuity, there exists $\delta_1 > 0$ such that

$$\|(w,y) - (\bar{w},\bar{y})\| < \delta_1 \implies \|\bar{w}\|^2 + \|\bar{y}\|^2 > \epsilon.$$

By upper semicontinuity of F and compactness of its values, there exist $\delta > 0$ and $r > 0$ such that $F(z) \subset F(\bar{x}) + \epsilon B \subset rB$ for all $z \in \bar{x} + \delta B$. We may assume that $\delta \leq \delta_1$. Since $\bar{x} \in \Omega(x)$, there exists an increasing sequence $\{t_n\} \subset \mathbb{R}^+$ with $t_n \to \infty$ and $x(t_n) \to \bar{x}$ as $n \to \infty$. By continuity of V we have

$$V(x(t_n)) - V(\bar{x}) < \frac{\delta\epsilon}{4r} \tag{10.9}$$

for all n sufficiently large. Let n^* be such that $x(t_n) \in \bar{x} + \frac{1}{2}\delta B$ for all $n > n^*$. Since $F(z) \subset rB$ for all $z \in \bar{x} + \delta B$, it follows that, for all $n > n^*$, $x(t) \in \bar{x} + \delta B$ for all $t \in [t_n, t_n + (\delta/3r)]$. Hence, using (10.8), we may conclude that

$$V(x(t_n)) - V(\bar{x}) \geq \int_{t_n}^{t_n+(\delta/3r)} [\|w(s)\|^2 + \|y(s)\|^2]\, ds \geq \frac{\delta\epsilon}{3r}$$

for all $n > n^*$. This contradicts (10.9). Therefore, $\Omega \subset \Pi$ and so $(w(t), y(t)) \to (0,0)$ as $t \to \infty$.

10.3 Class II: Nonlinearly Perturbed Linear Systems and Tracking by Output Feedback

In this section, we indicate how the above control strategy (and attendant stability analysis) may be carried over to a tracking problem for a class $\mathcal{N}$ of nonlinearly perturbed m-input, m-output linear systems of the form:

$$\begin{aligned} \dot{\tilde{x}}(t) &= \tilde{A}\tilde{x}(t) + \tilde{B}\left[u(t) + g(t, \tilde{x}(t))\right], \quad \tilde{x}(0) = \tilde{x}^0, \\ \tilde{y}(t) &= \tilde{C}\tilde{x}(t) \end{aligned} \tag{10.10}$$

with state $\tilde{x}(t) \in \mathbb{R}^n$, control $u(t) \in \mathbb{R}^m$ and output $\tilde{y}(t) \in \mathbb{R}^m$. The following assumptions (counterparts of Assumptions 10.1,10.2.10.3) determine the class $\mathcal{N}$.

Assumption 10.6 *The triple $(\tilde{C}, \tilde{A}, \tilde{B})$ defines a minimum phase linear system $\mathcal{L}$ of relative degree one, that is,*

$$\operatorname{rank}\begin{bmatrix} sI - \tilde{A} & \tilde{B} \\ \tilde{C} & 0 \end{bmatrix} = n + m \qquad \forall\, s \in \bar{\mathbb{C}}^+$$

(where $\bar{\mathbb{C}}^+$ denotes the closed right half complex plane), and

$$B := \tilde{C}\tilde{B} \in Gl(m; \mathbb{R})$$

Assumption 10.7 *A finite spectrum unmixing set $\mathcal{K}$ is known for B.*

Assumption 10.8 *For each $\tilde{x} \in \mathbb{R}^n$, the function $g(\cdot, \tilde{x})$ is measurable; for almost all $t \in \mathbb{R}$, the function $g(t, \cdot)$ is continuous; there exist scalar $\mu > 0$ and continuous function γ such that, for almost all $t \in \mathbb{R}$,*

$$\|g(t, \tilde{x})\| \leq \mu\gamma(\tilde{C}\tilde{x}) \quad \forall\ \tilde{x} \in \mathbb{R}^n.$$

(In words, g is a Carathéodory function which is bounded, modulo an unknown scalar multiplier, by a known continuous function of the output.*)*

The problem to be addressed is that of determining an adaptive *output feedback* strategy that guarantees $(\mathcal{R}, \mathcal{N})$-universal tracking in the following sense: for each reference signal r of some given class $\mathcal{R}$, for each system Σ of class $\mathcal{N}$, and for each initial state $\tilde{x}^0 \in \mathbb{R}^n$, the state $\tilde{x}(\cdot)$ is bounded on $[0, \infty)$ and the output $\tilde{y}(\cdot)$ asymptotically tracks $r(\cdot)$, that is, the tracking error $e(t) = \tilde{y}(t) - r(t)$ tends to zero as $t \to \infty$.

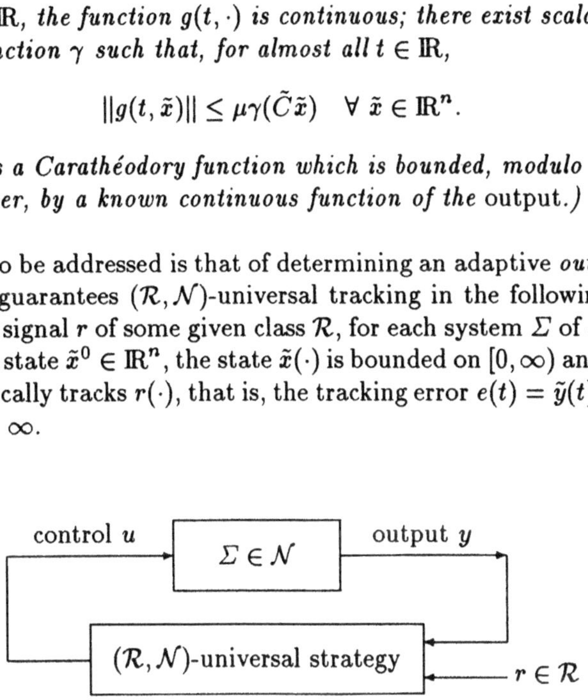

Fig. 10.2. The tracking framework

10.3.1 Class of Reference Signals

As the class $\mathcal{R}$ of reference signals to be tracked, we take the space of functions $r : \mathbb{R} \to \mathbb{R}^m$ that are absolutely continuous on compact intervals and that are bounded with essentially bounded derivative. Equipped with the norm

$$\|r\|_{1,\infty} = \|r\|_\infty + \|\dot{r}\|_\infty$$

this class can be identified as the Sobolev space

$$\mathcal{R} = W^{1,\infty}(\mathbb{R}, \mathbb{R}^m)$$

This class includes, for example, outputs from stable linear systems driven by L^∞ inputs. However, we stress that the control strategy developed below need not have recourse to dynamical systems (linear or otherwise) which may replicate the reference signals: in this sense, an *internal model principle* is not invoked in the controller construction.

10.3.2 Coordinate Transformation

Let $T_1 : \mathbb{R}^n \to \mathbb{R}^{n-m}$ be any linear map such that $\ker T_1 = \operatorname{im} \tilde{B}$. Then the coordinate transformation

$$T : x \mapsto (w, y) := (T_1\tilde{x}, \tilde{C}\tilde{x})$$

takes system (10.10) into the form

$$\dot{w}(t) = L_1 w(t) + L_2 y(t)$$

$$\dot{y}(t) + L_3 w(t) + L_4 y(t) + Bu \in \mu\mathcal{F}(w, y)$$

where $\mathcal{F}(w, y) = f(w, y)\overline{\mathcal{B}}$ and

$$f(w, y) := \gamma(\tilde{C}T^{-1}(w, y)) \ (= \gamma(\tilde{C}\tilde{x}) = \gamma(\tilde{y}))$$

Note that the spectrum of L_1 coincides with the set of zeros of the linear system $\mathcal{L}$ and so, by Assumption 1′, $\sigma(L_1) \subset \mathbb{C}^-$.

10.3.3 Adaptive Output Feedback Strategy

Let sequences (σ_n), (τ_n) and functions $s(\cdot)$, $K(\cdot)$ be as in Section 10.2.2. Then, with $r \in \mathcal{R}$, the adaptive *output* feedback strategy is given by

$$u(t) \in \Psi(\tilde{y}(t), r(t), \kappa(t)),$$

$$\dot{\kappa}(t) = (1 + \gamma(\tilde{y}(t)))\|\tilde{y}(t) - r(t)\| + \|\tilde{y}(t) - r(t)\|^2, \quad \kappa(0) = \kappa^0,$$

where

$$\Psi(\tilde{y}, r, \kappa) := \kappa\,[1 + \gamma(\tilde{y}) + \|\tilde{y} - r\|]\, K(s(\kappa))\psi(\tilde{y} - r)$$

with ψ defined as in Section 10.2.2.
Writing $e(t) = \tilde{y}(t) - r(t)$, $x(t) = (w(t), e(t), \kappa(t))$ and defining

$$\rho_r := \|r\|_{1,\infty}$$

the overall adaptively-controlled system may be embedded in the following initial-value problem in $\mathbb{R}^N$, $N := n + 1$,

$$\dot{x}(t) \in F_r(t, x(t)), \quad x(0) = x^0 = (T_1\tilde{x}^0, \tilde{C}\tilde{x}^0 - r(0), \kappa^0) \tag{10.11}$$

with the *upper semicontinuous* set-valued map $(t, x) \mapsto F_r(t, x)$ defined by

$$F_r(t,x) := F_{r1}(x) \times F_{r2}(t,x) \times F_{r3}(t,x)$$

$$F_{r1}(x) := \{L_1 w + L_2[e+v] \mid \|v\| \le \rho_r\}$$

$$F_{r2}(t,x) := \{[\phi - Bu] - L_3 w - L_4[e+v] \mid \|\phi\| \le \mu\gamma(e+r(t)),\ u \in \Psi(e+r(t), r(t), \kappa),\ \|v\| \le \rho_r\}$$

$$F_{r3}(t,x) := \{\|e\|^2 + (1+\gamma(e+r(t)))\|e\|\}$$

for all $(t, x) = (t, w, e, \kappa) \in \mathbb{R} \times \mathbb{R}^{n-m} \times \mathbb{R}^m \times \mathbb{R} = \mathbb{R}^{N+1}$.

Remark. The time-dependence of F_r in (10.11) arises solely through the reference signal $r(\cdot)$. In the trivial case $r(\cdot) \equiv 0$, the problem is autonomous and the stability analysis of Sect. 10.2.3 may be applied (with minor modifications – mainly of a notational nature) to conclude that the above output feedback strategy is an $\mathcal{N}$-universal stabilizer in the same sense as Theorem 10.5. The non-trivial case $r(\cdot) \not\equiv 0$ will now be pursued. Note that upper semicontinuity of F_r, together with convexity and compactness of its values, ensures that the initial-value problem (10.11) has a solution: moreover, every solution can be maximally extended.

10.3.4 Stability Analysis

In the context of the system (10.10), the next result may be paraphrased as follows: let $r \in \mathcal{R}$, then under the proposed adaptive output-feedback strategy with arbitrary κ^0, for each $\tilde{x}^0$, every solution of (10.10) is bounded and so can be extended indefinitely, the adaptive gain $\kappa(t)$ tends to a finite limit, and the tracking error tends to zero.

Theorem 10.9 *Let $r \in \mathcal{R}$ and $(\tilde{x}^0, \kappa^0) \in \mathbb{R}^N$ be arbitrary. Let $x(\cdot) = (w(\cdot), e(\cdot), \kappa(\cdot)) : [0, \omega) \to \mathbb{R}^N$ be a maximal solution of (10.11). Then*
(i) $\omega = \infty$,
(ii) $x(\cdot)$ is bounded,
(iii) $\lim_{t\to\infty} \kappa(t)$ exists and is finite,
(iii) $\lim_{t\to\infty} \|e(t)\| = 0$.

Proof. By Assumption 10.6, $\sigma(L_1) \subset \mathbb{C}^-$ and so the Lyapunov equation

$$PL_1 + L_1^T P + I = 0$$

has unique symmetric positive-definite solution P. By Assumption 10.7, there exists $K_j \in \mathcal{K}$ such that $\sigma(BK_j) \subset \mathbb{C}^+$ and so the Lyapunov equation

$$Q(BK_j) + (BK_j)^T Q - I = 0$$

has unique symmetric positive-definite solution Q. Write $\alpha = 2\|QB\|$ and let the map ν be precisely as in the proof of Theorem 10.5. Define

$$W : x = (w, e, \kappa) \mapsto W_1(w) + W_2(e) = \tfrac{1}{2}\langle w, Pw\rangle + \tfrac{1}{2}\langle e, Qe\rangle$$

Since $\sigma(L_1) \subset \mathbb{C}^-$ and

$$\dot{w}(t) - L_1 w(t) - L_2 e(t) \in \{L_2 v \mid \|v\| \le \rho_r\}$$

for almost all t, there exist positive constants c_0 and c_1 such that, for all $t_0, t \in [0, \omega)$ with $t \ge t_0$,

$$\int_{t_0}^{t} |\langle Qe(s), L_3 w(s)\rangle| \, ds \le c_0\|w(t_0)\|^2 + c_1 \int_{t_0}^{t} \left[\|e(s)\| + \|e(s)\|^2\right] \, ds$$

Analogous to the proof of Theorem 10.5, for some positive constant c_2 we have

$$\langle \nabla W_2(e), \eta\rangle \le -\langle Qe, L_3 w\rangle + (c_2 - \kappa\nu(\kappa)\,[1 + \gamma(e + r(t)) + \|e\|]\,\|e\|$$

for all $\eta \in F_{r2}(t, x)$. Therefore, for the maximal solution $x(\cdot) = (w(\cdot), e(\cdot), \kappa(\cdot))$, we find

$$0 \le W_2(e(t)) \le W_2(e(t_0)) + c_0\|w(t_0)\|^2 + (c_1 + c_2)(\kappa(t) - \kappa(t_0)) - \int_{\kappa(t_0)}^{\kappa(t)} \theta\nu(\theta)\, d\theta$$

valid for all $t, t_0 \in [0, \omega)$ with $t \ge t_0$. By the same argument as that in the proof of Theorem 10.5, we may now conclude boundedness of $\kappa(\cdot)$. Boundedness of $e(\cdot)$ follows immediately by the last inequality. For almost all $t \in [0, \omega)$, we have

$$\frac{d}{dt} W_1(w(t)) \le -\tfrac{1}{2}\|w(t)\|^2 + \|L_2^T P w(t)\|[\|e(t)\| + \rho_r]$$

and so, by boundedness of $e(\cdot)$, we may conclude boundedness of $w(\cdot)$. We have now shown boundedness of the solution $x(\cdot)$ and so $\omega = \infty$. This establishes assertions (i) and (ii) of the theorem. Assertion (iii) follows by monotonicity and boundedness of $\kappa(\cdot)$. It remains to prove assertion (iv). This we do by the following argument. First observe that $\dot{\kappa}(t) \ge \|e(t)\|^2$ and so, by boundedness of $\kappa(\cdot)$, we may conclude that $e(\cdot) \in L^2([0, \infty); \mathbb{R}^m)$. Now observe that, by boundedness of $r(\cdot)$ and $x(\cdot)$, there exists a compact set K such that $F_r(t, x(t)) \subset K$ for all $t \in [0, \infty)$. Therefore, $\dot{e}(\cdot) \in L^\infty([0, \infty); \mathbb{R}^m)$ and so we may conclude that $\|e(t)\| \to 0$ as $t \to \infty$.

10.4 Class III: Two-Input Systems

In this final section, we consider the following special case of (10.1) or, equivalently, (10.2): $m = 2$ and the sign of $\det(M^{-1}B)$ known. We write $D = M^{-1}B$

(as before) and, without loss of generality, we assume that its determinant is *positive*. (If $\det D < 0$, then simply replace u by Ju, where $J = \mathrm{diag}\{1, -1\}$.) Since the determinant of D is the product of its eigenvalues, the condition $\det D > 0$ is equivalent to knowing a priori that the eigenvalues of D are non-zero and lie either in the *closed* right half or the *closed* left half complex plane. Note, in particular, that D may have spectrum on the imaginary axis.
We remark, in passing, that the set of three orthogonal matrices

$$\mathcal{K} = \{I, -I, K\}, \quad \text{where} \quad K := \frac{1}{\sqrt{2}} \begin{bmatrix} 1 & -1 \\ 1 & 1 \end{bmatrix}$$

is an unmixing set for each D with positive determinant.
Our objective is to show that the following alternative (and simpler) strategy is a universal stabilizer for this particular class of systems:

$$u(t) \in \Psi^o(x(t)), \quad \dot{\kappa}(t) = f(w(t), y(t))\|y(t)\| + \|y(t)\|^2, \quad \kappa(0) = \kappa^0$$

where

$$\Psi^o : x = (w, y, \kappa) \mapsto \kappa^2[f(w, y) + \|y\|]O(\kappa)\psi(y)$$

with $O : \mathbb{R} \to SO(2; \mathbb{R})$ given by

$$O(\kappa) := \begin{bmatrix} \cos\kappa & \sin\kappa \\ -\sin\kappa & \cos\kappa \end{bmatrix}$$

Remark. This simplified strategy dispenses with the explicit reliance of the earlier strategy on the unmixing set $\mathcal{K}$ and the associated sequences (τ_n) and (σ_n) governing the cycling therethrough: loosely speaking, these features are implicit in the "rotation" of the control direction induced by the orthogonal-valued term $O(\kappa(t))$ in the present strategy.

Let the set-valued maps F_1 and F_2 be as in Section 10.2.2 and define $x \mapsto F_2^o(x) \subset \mathbb{R}^2$ by

$$F_2^o(x) := \left\{ M^{-1}[\phi - Bu] - L_3 w + C_{p-1} y \mid \phi \in \mathcal{F}(w, y),\ u \in \Psi^o(x) \right\}.$$

Then the feedback-controlled system can be embedded in the initial-value problem in $\mathbb{R}^N$, $N = 2p + 1$:

$$\dot{x}(t) \in F^o(x(t)), \quad x(0) = x^0 = (T\zeta^0, \kappa^0) \tag{10.12}$$

where $F^o : x \mapsto F_1(x) \times F_2^o(x) \times F_3(x) \subset \mathbb{R}^N$.

Theorem 10.10 *Let $x(\cdot) = (w(\cdot), y(\cdot), \kappa(\cdot)) : [0, \omega) \to \mathbb{R}^N$ be a maximal solution of (10.12). Then*
(i) $\omega = \infty$,
(ii) $\lim_{t\to\infty} \kappa(t)$ *exists and is finite,*
(iii) $\lim_{t\to\infty} \|(w(t), y(t))\| = 0$.

Proof. By assumption, $D = M^{-1}B$ has positive determinant. Therefore, by polar decomposition together with surjectivity of the map $O : \mathbb{R} \to SO(2;\mathbb{R})$, there exists symmetric $R > 0$ and $\kappa^{\#}$ such that

$$D = RO(\kappa^{\#})$$

Let symmetric $P > 0$ be as in Sect.10.2.1, define $Q := R^{-1}$ and let $W : x = (w, y, \kappa) \mapsto W_1(w) + W_2(y)$, with

$$W_1(w) := \tfrac{1}{2}\langle w, Pw\rangle, \quad \text{and} \quad W_2(y) := \tfrac{1}{2}\langle y, Qy\rangle$$

Define

$$\kappa^* := \tfrac{1}{4} + \|QM^{-1}\| + \|QC_{p-1}\| + (\|PL_2\| + \|QL_3\|)^2$$

Then, for all $\phi \in F^o(x)$,

$$\begin{aligned} \langle \nabla W(x), \phi\rangle &\le \langle Pw, L_1 w\rangle + \langle Pw, L_2 y\rangle - \langle Qy, L_3 w\rangle + \langle Qy, C_{p-1} y\rangle \\ &\quad + f(w,y)\|QM^{-1}\|\|y\| + \kappa^2 \cos(\kappa^{\#} + \kappa)[f(w,y)\|y\| + \|y\|^2] \\ &\le -\tfrac{1}{4}(\|w\|^2 + \|y\|^2) \\ &\quad + (\kappa^* + \kappa^2 \cos(\kappa^{\#} + \kappa))[f(w,y)\|y\| + \|y\|^2] \end{aligned}$$

which is valid for all $x = (w, y, \kappa) \in \mathbb{R}^N$. Therefore, for the maximal solution $x(\cdot)$, we may conclude that, for almost all t,

$$\frac{\mathrm{d}}{\mathrm{d}t} W(x(t)) \le [\kappa^* + \kappa^2(t)\cos(\kappa^{\#} + \kappa(t))]\dot{\kappa}(t) \tag{10.13}$$

We first show that the monotone function $\kappa(\cdot)$ is bounded. Suppose that $\kappa(\cdot)$ is unbounded. Let $t_0 \in [0, \omega)$ be such that $\kappa(t) \ge 1$ for all $t \in [t_0, \omega)$. From (10.13), we have

$$0 \le W(x(t)) \le W(x(t_0)) + \kappa^*[\kappa(t) - \kappa(t_0)] + \int_{\kappa(t_0)}^{\kappa(t)} \theta^2 \cos(\kappa^* + \theta)\, \mathrm{d}\theta \tag{10.14}$$

for all $t \in [\tau, \omega)$.
Dividing by $\kappa(t) \ge 1$ and taking limit inferior as $t \uparrow \omega$ ($\kappa(t) \to \infty$), yields the contradiction:

$$0 \le W(x(\tau)) + \kappa^* + \liminf_{\xi \to \infty} \frac{1}{\xi} \int_{\kappa(t_0)}^{\xi} \theta^2 \cos(\kappa^* + \theta)\, \mathrm{d}\theta = -\infty$$

Therefore $\kappa(\cdot)$ is bounded. From (10.14), it immediately follows that $W(x(\cdot))$ is bounded, whence boundedness of $w(\cdot)$ and $y(\cdot)$. We have now shown that $x(\cdot) = (w(\cdot), y(\cdot), \kappa(\cdot))$ is bounded and so $\omega = \infty$. This establishes assertion (i). Assertion (ii) is evident by boundedness and monotonicity of $\kappa(\cdot)$.

Finally, define

$$V : \ x = (w, y, \kappa) \mapsto W(x) - \int_0^\kappa [\kappa^* + \theta^2 \cos(\kappa^* + \theta)] \, d\theta$$

Then, for almost all $x = (w, y, \kappa) \in \mathbb{R}^N$, we have

$$\langle \nabla V(x), \phi \rangle \leq -\tfrac{1}{4}(\|w\|^2 + \|y\|^2) \qquad \forall \, \phi \in F^o(x).$$

By the same argument as that employed in the proof of Theorem 10.5, assertion (iii) follows.

10.4.1 Example

Consider the problem of simultaneous stabilization of two uncertain systems that are coupled through the scalar control inputs u_1 and u_2:

$$\ddot{z}_1(t) + b_{11}u_1(t) + b_{12}u_2(t) = g_1(t, z_1(t), \dot{z}_1(t))$$

$$\ddot{z}_2(t) + b_{21}u_1(t) + b_{22}u_2(t) = g_2(t, z_2(t), \dot{z}_2(t))$$

We assume that the unknown continuous functions g_1 and g_2 are bounded in the following sense: for some (unknown) scalar $\mu > 0$,

$$|g_i(t, v_1, v_2)| \leq \mu[1 + \|v\|^3], \quad i = 1, 2,$$

for all $(t, v_1, v_2) = (t, v) \in \mathbb{R} \times \mathbb{R}^2$. For example, g_i can exhibit polynomial state dependence of degree not exceeding three with continuous bounded t-dependent coefficients. Two particular system realizations could correspond to oscillators of van der Pol and Duffing type, respectively, with periodic forcing: these systems, in the absence of control, can exhibit highly irregular dynamic behaviour (Guckenheimer and Holmes 1983, Thompson and Stewart 1986):

$$g_1(t, z_1, \dot{z}_1) = a_1 z_1 + a_2(z_1^2 - a_3)(\dot{z}_1 + a_4 \sin \omega_1 t) + a_5 \cos \omega_1 t$$

$$g_2(t, z_2, \dot{z}_2) = a_6 z_2 + a_7 \dot{z}_2 + a_8 z_2^3 + a_9 \sin \omega_2 t$$

with unknown parameters $a_i, \omega_i \in \mathbb{R}$. For example, Fig. 10.3 depicts the evolution of the van der Pol and Duffing system variables $z_1(t)$ and $z_2(t)$, respectively, in the absence of control and for the particular parameter values

$$a_1 = -0.7, \ a_2 = -10, \ a_3 = -0.1, \ a_4 = 0.25, \ a_5 = 0.4, \omega_1 = \pi/2$$

$$a_6 = 0, \ a_7 = -0.05, \ a_8 = -1, \ a_9 = 7.5, \omega_2 = 1$$

The coupling parameters b_{ij} are unknown, but are assumed to satisfy:

$$b_{11}b_{22} - b_{12}b_{21} > 0.$$

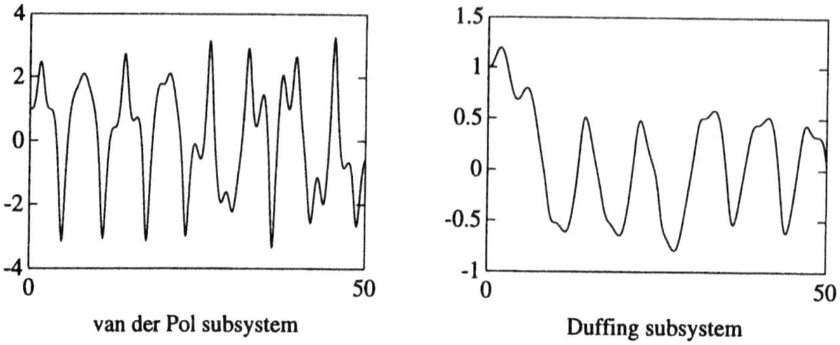

Fig. 10.3. Uncontrolled evolution of variables $z_1(t)$ and $z_2(t)$

It is readily verified that all systems of this class can be embedded in a differential inclusion of form (10.1), with $z(t) = (z_1(t), z_2(t)) \in \mathbb{R}^2$ and $u(t) = (u_1(t), u_2(t)) \in \mathbb{R}^2$. Writing $x(t) = (z_1(t), \dot{z}_1(t), z_2(t), \dot{z}_2(t))$ and $y(t) = (y_1(t), y_2(t))$, where

$$y_1(t) = \dot{z}_1 + c_1 z_1(t), \quad y_2 = \dot{z}_2(t) + c_2 z_2(t)$$

with $c_1, c_2 > 0$, then, by Theorem 10.10, the following is a universal controller for the class of systems under consideration:

$$u(t) \in \kappa^2(t)[1 + \|y(t)\| + \|x(t)\|^3] O(\kappa(t)) \psi(y(t))$$

$$\dot{\kappa}(t) = [1 + \|y(t)\| + \|x(t)\|^3] \|y(t)\|, \quad \kappa(0) = \kappa^0$$

Typical dynamic behaviour is depicted in Fig. 10.4 wherein, for purposes of illustration, (i) the controller parameter values $c_1 = 2 = c_2$ are adopted, (ii) the functions g_i (unknown to the controller) are taken to be those of the above-cited van der Pol and Duffing systems (with parameter values – unknown to the controller – as above) and input parameters (again unknown to the controller)

$$b_{11} = 1 = b_{22}, \quad b_{12} = 0.5 = b_{21}$$

Note that the scalings of the time axes Figs. 10.3 and 10.4 differ by a factor of 10.

References

Aubin, J-P., Cellina, A. 1984, *Differential Inclusions*, Springer-Verlag, Berlin-New York

Byrnes, C.I., Willems, J.C. 1984, Adaptive stabilization of multivariable linear systems. *Proc IEEE Conference on Decision and Control,* 1574–1577

Cabera, J.B.D., Furuta, K. 1989, Improving the robustness of Nussbaum type regulators by the use of σ-modification - local results, *Systems and Control Letters* **12**, 421–429

Corless, M. 1991, Simple adaptive controllers for systems which are stabilizable via high gain feedback, *IMA Journal of Mathematical Control and Information* **8**, 379–387

Corless, M., Leitmann, G. 1984, Adaptive control for uncertain dynamical systems, in *Dynamical Systems and Microphysics: Control Theory and Mechanics* (A. Blaquière and G. Leitmann, eds), Academic Press, New York

Filippov, A.F. 1988, *Differential Equations with Discontinuous Righthand Sides*, Kluwer Academic Publishers, Dordrecht

Guckenheimer, J., Holmes, P. 1983, *Nonlinear Oscillations, Dynamical Systems, and Bifurcations of Vector Fields*, Springer-Verlag, Berlin-NewYork

Helmke, U., Prätzel-Wolters, D. 1988, Stability and robustness properties of universal adaptive controllers for first order linear systems, *International Journal of Control* **48**, 1153–1183

Helmke, U., Prätzel-Wolters, D., Schmid, S. 1990, Adaptive tracking for scalar minimum phase systems, in *Control of Uncertain Systems* (D. Hinrichsen and B. Märtensson, eds), Birkhäuser, Boston

Ilchmann, A. 1991, Non-identifier-based adaptive control of dynamical systems: a survey, *IMA Journal of Mathematical Control and Information* **8**, 321–366

Ilchmann, A., Logemann, H. 1992, High-gain adaptive stabilization of multivariable linear systems - revisited, *Systems and Control Letters* **18**, 355–364

Ilchmann, A., Owens, D.H. 1990, Adaptive stabilization with exponential decay, *Systems and Control Letters* **14**, 437–443

Ilchmann, A., Owens, D.H., Prätzel-Wolters, D. 1987, High-gain robust adaptive controllers for multivariable systems, *Systems and Control Letters* **8**, 397–404

Logemann, H., Owens, D.H. 1988, Input-output theory of high-gain adaptive stabilization of infinite-dimensional systems with non-linearities, *International Journal of Adaptive Control and Signal Processing* **2**, 193–216

Logemann, H., Zwart, H. 1991, Some remarks on adaptive stabilization of infinite dimensional systems, *Systems and Control Letters* **16**, 199–207

Mårtensson, B. 1985, The order of any stabilizing regulator is sufficient a priori information for adaptive stabilization, *Systems and Control Letters* **6**, 87–91

Mårtensson, B. 1986, *Adaptive Stabilization*, PhD Thesis, Lund Institute of Technology, Sweden

Mårtensson, B. 1987, Adaptive stabilization of multivariable linear systems, *Contemporary Mathematics* **68**, 191–225

Mårtensson, B. 1990, Remarks on adaptive stabilization of first order non-linear systems, *Systems and Control Letters* **14**, 1–7

Mårtensson, B. 1991, The unmixing problem, *IMA Journal of Mathematical Control and Information* **8**, 367–377

Miller, D.E., Davison, E.J. 1989, An adaptive controller which provides Lyapunov stability, *IEEE Transactions on Automatic Control* **AC-34**, 599–609

Morse, A.S. 1984, New directions in parameter adaptive control, *Proc IEEE Conference on Decision and Control,* 1566–1568

Morse, A.S. 1985, A Three Dimensional Universal Controller for the Adaptive Stabilization of Any Strictly Proper Minimum Phase System with Relative Degree Not Exceeding Two, *IEEE Transactions on Automatic Control* **AC-30**, 1188–1191

Nussbaum, R.D. 1983, Some remarks on a conjecture in parameter adaptive control, *Systems and Control Letters* **3**, 243–246

Roxin, E. 1965, On generalized dynamical systems defined by contingent equations, *Journal of Differential Equations* **1**, 188–205

Ryan, E.P. 1990, Discontinuous feedback and universal adaptive stabilization, in *Control of Uncertain Systems* (D. Hinrichsen and B. Mårtensson, eds), Birkhäuser, Basel-Boston

Ryan, E.P. 1991a, A universal adaptive stabilizer for a class of nonlinear systems, *Systems and Control Letters* **16**, 209–218

Ryan, E.P. 1991b, Finite-time stabilization of uncertain nonlinear planar systems, *Dynamics and Control* **1**, 83–94

Ryan, E.P. 1992, Universal $W^{1,\infty}$-tracking for a class of nonlinear systems, *Systems and Control Letters* **18**, 201–210

Ryan, E.P. 1993, Adaptive stabilization of multi-input nonlinear systems, *International Journal of Robust and Nonlinear Control* to appear

Tao, G., Ioannou, P.A. 1991, Robust adaptive control of plants with unknown order and high frequency gain, *International Journal of Control* **53**, 559–578

Thompson, J.M.T., Stewart, H.B. 1986, *Nonlinear Dynamics and Chaos*, Wiley, New York

Townley, S., Owens, D.H. 1991, A note on the problem of multivariable adaptive tracking, *IMA Journal of Mathematical Control and Information* **8**, 389–395

Willems, J.C., Byrnes, C.I. 1984, Global adaptive stabilization in the absence of information on the sign of the high frequency gain, in *Lecture Notes in Control and Information Sciences*, Vol. 62, Springer-Verlag, Berlin-New York, 49–57

Xin-jie Zhu 1989, A finite spectrum unmixing set for $GL(3,\mathbb{R})$, in *Computation and Control* (K. Bowers and J. Lund, eds), Birkhäuser, Basel-Boston

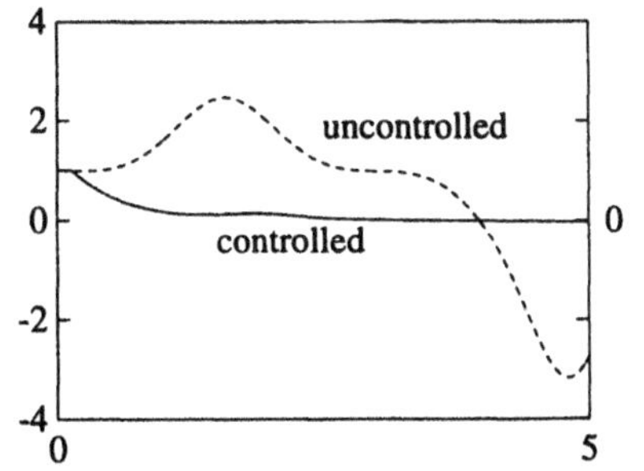

Controlled and uncontrolled evolution of $z_1(t)$

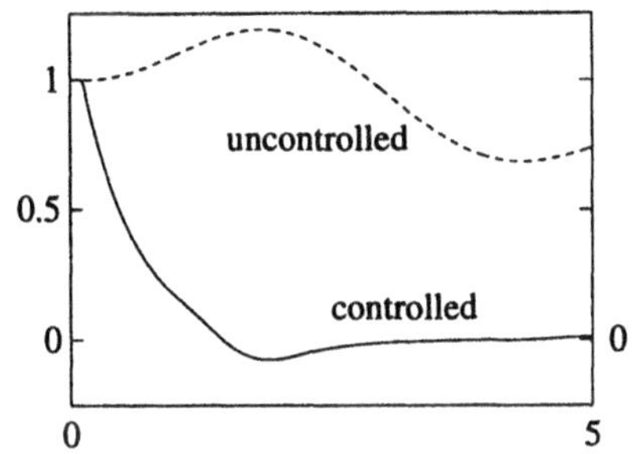

Controlled and uncontrolled evolution of $z_2(t)$

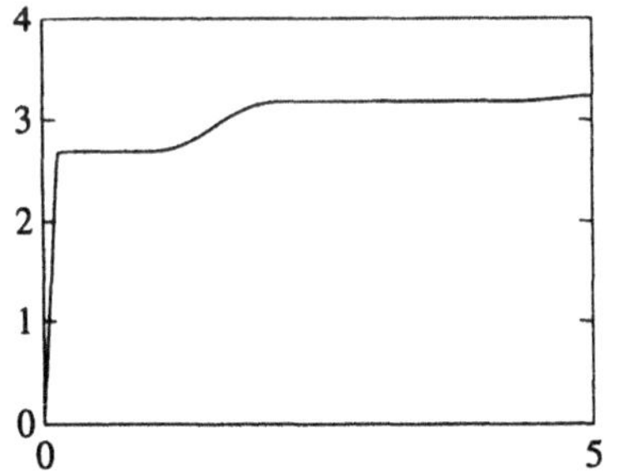

Evolution of adapting parameter $\kappa(t)$

Fig. 10.4. Example: typical controlled and uncontrolled behaviour

11. Lyapunov Stabilization of a Class of Uncertain Affine Control Systems

David P. Goodall

11.1 Introduction

In this chapter we consider the problem of global feedback stabilization of controlled dynamical systems in the presence of uncertainty. Some of the early work on deterministic feedback stabilization of uncertain systems, using Lyapunov theory, was developed by Leitmann (1979), Gutman (1979), and Corless and Leitmann (1981), amongst others. The structure of the mathematical model representing the system, essentially a perturbation of a linear system, has the form

$$\dot{x}(t) = Ax(t) + Bu(t) + h(t, x(t), u(t)) \tag{11.1}$$

where $x(t) \in \mathbb{R}^n$ is the state vector, $u(t) \in \mathbb{R}^m$ is the control (input) vector, A, B are known matrices of appropriate dimension, and $(t, x, u) \mapsto h(t, x, u)$, modelling the uncertainty in the system, is assumed unknown. Later, Ryan and Corless (1984) utilized the concept of an invariant manifold, adopted from Variable Structure Control theory, to establish the required stability property and, in addition, obtain ultimate attainment of prescribed dynamic behaviour. Since discontinuous feedback is a natural candidate in many stabilization problems, the right hand side of (11.1) may be discontinuous, which gives rise to many analytical difficulties (Filippov 1988). However, using a differential inclusion formulation, in which the right hand side of (11.1) is replaced by a known set-valued map, such difficulties can be overcome (see, for example, Leitmann (1979), Gutman (1979), Gutman and Palmor (1982), Aubin and Cellina (1984), Goodall and Ryan (1988)). For stabilization of uncertain systems using continuous feedback controls subject to constraints, see Soldatos and Corless (1991). More recently, attention has been focused on systems characterized in terms of additive perturbations to a known nonlinear system with specific structure (Elmali and Olgac 1992; Goodall 1992a). Here the known nonlinear system is assumed to be affine in the control variable with structure

$$\dot{x}(t) = f(x(t)) + G(x(t))u(t), \qquad x(t) \in \mathbb{R}^n, \ u(t) \in \mathbb{R}^m \tag{11.2}$$

There have been a number of approaches to the global stabilization problem for nonlinear systems, affine in the control (see, for example, Andreini et al. 1988, Byrnes and Isidori 1991, Kokotovic and Sussman 1989, Seibert and Suarez 1991, and Tsinias 1990). One of the main techniques used in this paper is the same as that considered in Seibert and Suarez (1991), viz. transformation, by feedback, of affine control systems into the "regular" form

$$\dot{y}^1(t) = f_1(y^1(t), y^2(t)) \tag{11.3}$$
$$\dot{y}^2(t) = f_2(y^1(t), y^2(t)) + G(y^1(t), y^2(t))u(t) \tag{11.4}$$

where $y^1(t) \in \mathbb{R}^{n-m}$, $y^2(t)$, $u(t) \in \mathbb{R}^m$, f_1 and f_2 are vector fields defined on $\mathbb{R}^{n-m} \times \mathbb{R}^m$ and G is a $m \times m$ matrix. Systems transformable to the "regular" form (11.3)–(11.4) have been investigated by Hunt et al (1983) and Luk'yanov and Utkin (1981), amongst others. In Luk'yanov and Utkin (1981), transformations are constructed with the purpose of using variable structure feedback controls to stabilize uncertainty in the system.

The class of uncertain systems, to be stabilized, comprises systems which are nonlinear perturbations to a known class of affine control systems. This work extends that of Goodall and Ryan (1988) in which the class of uncertain systems consisted of perturbed known linear systems. The asymptotic stability property is established by employing Lyapunov techniques and utilizing the concept of an invariant, attractive manifold $\mathcal{M} \subset \mathbb{R}^n$, adopted from variable structure control theory. Many aspects of variable structure control theory are considered in Utkin (1992). The underlying approach, in this chapter, is based on the deterministic theory of feedback control in the presence of uncertainty (for more details, see Zinober (1990)).

A differential inclusion formulation provides a framework for modelling the uncertainty in the system by set-valued maps. For a differential inclusion approach to adaptive stabilization of a class of uncertain systems, see Ryan (1988). The proposed feedback controls are embedded in set-valued maps, henceforth, referred to as *generalized feedbacks*. A class of generalized feedbacks is chosen such that the solutions to the controlled differential inclusion system are attracted to some specified manifold, $\mathcal{M}$, attaining $\mathcal{M}$ in finite time, and each solution being ultimately constrained to $\mathcal{M}$. The manifold $\mathcal{M}$ is not restricted to be linear. Nonlinear manifolds can arise naturally in some problems when stability is being investigated. One possible advantage of using a nonlinear manifold is that the time taken for a solution to reach its 'stable state' can be minimized. A class of discontinuous feedback controls is presented, which renders the zero state of a class of uncertain affine control systems globally uniformly asymptotically stable.

11.2 Decomposition into Controlled and Uncontrolled Subsystems

The following notation is adopted. Let $\langle \cdot\, , \cdot \rangle$ and $\|\cdot\|$ denote the Euclidean inner product and induced norm, respectively. $\mathcal{L}(\mathbb{R}^p, \mathbb{R}^q)$ denotes the set of all continuous linear maps from $\mathbb{R}^p$ into $\mathbb{R}^q$.
For $\alpha \in \mathbb{R}$, $x \in \mathbb{R}^p$ and S_1, $S_2 \subset \mathbb{R}^p$,

$$\alpha S_1 := \{\alpha x : x \in S_1\}, \qquad x + S_1 := \{x + s_1 : s_1 \in S_1\}$$
$$S_1 + S_2 := \{s_1 + s_2 : s_1 \in S_1,\ s_2 \in S_2\}, \quad \langle x,\ S_1\rangle := \{\langle x,\ s_1\rangle : s_1 \in S_1\} \subset \mathbb{R}$$

Let $\mathbb{B}_p$ denote the open unit ball centred at the origin in $\mathbb{R}^p$, with closure $\overline{\mathbb{B}}_p$. Finally, let Π_K denote the orthogonal projector onto K, where K is a linear subspace of $\mathbb{R}^p$.

Consider the nonlinear control system (11.2), affine in the control input, where f is a C^∞ vector field on $\mathbb{R}^n$ satisfying $f(0) = 0$, and $G(x) \in \mathcal{L}(\mathbb{R}^n, \mathbb{R}^m)$ has a $m \times m$ invertible minor which is full rank for all x. Here, it is assumed that f and G are known. Without loss of generality, $G(x)$ can be partitioned as

$$G(x) = \begin{bmatrix} \hat{G}_1(x^1, x^2) \\ \hat{G}_2(x^1, x^2) \end{bmatrix}$$

where $x = [x^1\ x^2]^T$, $x^1 \in \mathbb{R}^{n-m}$, $x^2 \in \mathbb{R}^m$, $\hat{G}_1(x^1, x^2) \in \mathcal{L}(\mathbb{R}^{n-m}, \mathbb{R}^m)$ and $\hat{G}_2(x^1, x^2) \in \mathcal{L}(\mathbb{R}^m, \mathbb{R}^m)$ is nonsingular for all (x^1, x^2). Hence, system (11.2) may be expressed as

$$\dot{x}^1(t) = \hat{f}_1(x^1(t), x^2(t)) + \hat{G}_1(x^1(t), x^2(t))u(t) \tag{11.5}$$
$$\dot{x}^2(t) = \hat{f}_2(x^1(t), x^2(t)) + \hat{G}_2(x^1(t), x^2(t))u(t) \tag{11.6}$$

where $\hat{f}_1$ and $\hat{f}_2$ are C^∞ vector fields. It is assumed that there exists a diffeomorphic map $x \mapsto \phi(x) : \mathbb{R}^n \to \mathbb{R}^{n-m}$, with $\phi(0) = 0$, which satisfies the (Pfaffian) system of $(n-m)m$ partial differential equations

$$(\mathrm{D}\phi)(x)G(x) = 0 \tag{11.7}$$

for all x, where $(\mathrm{D}\phi)(x)$ denotes the Fréchet derivative of ϕ at x, i.e. the Jacobian matrix of ϕ. For this case, (11.5)–(11.6) can be reduced to a regular form, where the control only appears in the second subsystem.

Remark. The partial differential equations (11.7) may be restated in terms of the Pfaffian form (i.e. a linear differential 1-form)

$$\omega = W(x)dx = 0 \tag{11.8}$$

where W satisfies $W(x)G(x) = 0$. Locally, the Frobenius Theorem for 1-forms provides necessary and sufficient conditions for (11.8) to be completely integrable (see Choquet-Bruhat et al (1982) and Luk'yanov and Utkin (1981)).

The nonlinear transformation

$$y^1 = \phi(x), \qquad y^2 = x^2$$

transforms system (11.5)–(11.6) into the form

$$\dot{y}^1(t) = \tilde{f}_1(y^1(t), y^2(t)) \tag{11.9}$$
$$\dot{y}^2(t) = \tilde{f}_2(y^1(t), y^2(t)) + \tilde{G}_2(y^1(t), y^2(t))u(t) \tag{11.10}$$

Subsystem (11.9) is now independent of the control influence.

11.3 The Class of Uncertain Systems

Severe conditions on the structure of subsystem (11.9) are now imposed.

Hypothesis 11.1 *The map $\tilde{f}_1 : \mathbb{R}^{n-m} \times \mathbb{R}^m \to \mathbb{R}^{n-m}$ is affine in its second argument, having the form*

$$\tilde{f}_1(y^1, y^2) = f_1(y^1) + F_1(y^1)h(y^2)$$

where $f_1 : \mathbb{R}^{n-m} \to \mathbb{R}^{n-m}$, $F_1(y^1) \in \mathcal{L}(\mathbb{R}^{n-m}, \mathbb{R}^m)$, $h : \mathbb{R}^m \to \mathbb{R}^m$ is bijective and $[(\mathrm{D}h)(y^2)]^{-1}$ exists for all $y^2 \in \mathbb{R}^m$.

Remark. Alternative hypotheses on the nonlinear coupling term h for system (11.9)–(11.10) are considered in Goodall (1992b).

The class of controlled uncertain dynamical systems, to be considered, is essentially a nonlinear perturbation to system (11.9)–(11.10). With respect to system (11.9)–(11.10), the system is assumed to be subject to uncertainty modelled by augmenting the nominal differential equation by an unknown function $(t, y^1, y^2) \mapsto g_1(t, y^1, y^2)$ for subsystem (11.9) and a set-valued map $(t, y^1, y^2, u) \mapsto \mathcal{G}_2(t, y^1, y^2, u)$ in the case of subsystem (11.10). The uncertainty in the system is characterized by the following hypotheses :

Hypothesis 11.2

(a) There exist C^∞ vector fields $g_{\mathrm{m}} : \mathbb{R}^{n-m} \times \mathbb{R}^m \to \mathbb{R}^{n-m}$, $g_{\mathrm{r}} : \mathbb{R} \times \mathbb{R}^{n-m} \times \mathbb{R}^m \to \mathbb{R}^{n-m}$ such that

(i) $\Pi_{im(F_1)}\, g_1(t, y^1, y^2) = F_1(y^1) g_{\mathrm{m}}(y^1, y^2)$

(ii) $\Pi_{ker(F_1^T)}\, g_1(t, y^1, y^2) = g_{\mathrm{r}}(t, y^1, y^2)$

(b) There exists a known upper semicontinuous map $\mathcal{H} : \mathbb{R} \times \mathbb{R}^{n-m} \times \mathbb{R}^m \to 2^{\mathbb{R}^m}$ with nonempty, convex and compact values, a real positive constant κ, and a known continuous function $h : \mathbb{R} \to [0, \kappa]$, $\kappa < 1$, such that

$$\mathcal{G}_2(t, y^1, y^2, u) = \tilde{G}_2(y^1, y^2)\left[\mathcal{H}(t, y^1, y^2) + h(t)\|u\|\overline{\mathbb{B}}_m\right]$$

Remarks.

(*i*) Here, $2^{\mathbb{R}^m}$ denotes the subsets of $\mathbb{R}^m$.

(*ii*) A set-valued map $\mathcal{A} : \mathbb{R}^p \to 2^{\mathbb{R}^q}$, with compact values, is upper semicontinuous at $\alpha \in \mathbb{R}^p$ iff, for each $\varepsilon > 0$, there exists $\delta > 0$ such that $\mathcal{A}(\bar{\alpha}) \subset \mathcal{A}(\alpha) + \varepsilon\mathbb{B}_q$, for all $\bar{\alpha} \in \alpha + \delta\mathbb{B}_p$.

Analogous to the terminology used in Barmish et al (1983) and Goodall and Ryan (1988), the vector field g_{m} is said to model the *matched* uncertainty in the system, while g_{r} models *residual* uncertainty. In this chapter, only matched uncertainty is considered, i.e. it is assumed that

$$g_{\mathrm{r}}(t, y^1, y^2) = 0, \qquad \forall\ (t, y^1, y^2)$$

To reiterate, the class of uncertain systems, to be investigated here, is comprised of systems with the following structure :

$$\begin{aligned} \dot{y}^1(t) &= f_1(y^1(t)) + F_1(y^1(t))[h(y^2(t)) + g(y^1(t), y^2(t))] \quad (11.11) \\ \dot{y}^2(t) &\in f_2(y^1(t), y^2(t)) + G_2(y^1(t), y^2(t))u(t) \\ &\quad + \mathcal{G}_2(t, y^1(t), y^2(t), u(t)) \quad (11.12) \end{aligned}$$

The vector field g and the set-valued map $\mathcal{G}_2$ model the uncertainty in the system as nonlinear perturbations to the 'known' system (11.9)–(11.10).

11.4 Subsystem Stabilization

For a vector field $f \in C^\infty(\mathbb{R}^p)$ and a scalar function $z : \mathbb{R}^p \to \mathbb{R}$, let $L_f z$ denote the *Lie derivative* of z along f which is defined by

$$(L_f z)(x) := \langle \nabla z(x),\ f(x) \rangle$$

where $\nabla z(x)$ denotes the gradient vector of z. For two vector fields $f,\ g \in C^\infty$, the notation $ad^j(f, g)(x)$ is defined recursively by

$$\begin{aligned} ad^0(f, g)(x) &:= g(x) \\ ad^{j+1}(f, g)(x) &:= [f,\ ad^j(f, g)](x), \quad j = 0, 1, \ldots \end{aligned}$$

where $[\ \cdot\ , \cdot\ ]$ denotes the *Lie bracket* and is defined by $[f,\ g] := (\mathrm{D}g)f - (\mathrm{D}f)g$. In terms of the above notation the following proposition holds.

Proposition 11.3 *Let $f,\ g$ be C^∞ vector fields and ψ a C^∞ scalar field defined on $\mathbb{R}^p$. If, for the system*

$$\dot{x}(t) = f(x(t)), \qquad x(t) \in \mathbb{R}^p$$

the set

$$\Psi := \{x \in \mathbb{R}^p \ :\ \nabla(L_f \psi)(x) = 0, \quad (L_g \psi)(x) = 0\}$$

is invariant under f, then for all $x \in \Psi$

$$(L_{ad^j(f,g)} \psi)(x) = 0, \qquad \forall\ j = 0,\ 1, \ldots$$

Proof. This is easily proved using an inductive argument and the identity

$$L_{[f,g]}\psi = L_f(L_g\psi) - L_g(L_f\psi)$$

which is given in appendix A6, Isidori (1989) (also, Nijmeijer and van der Schaft 1990). Let $a(t) = (L_g\psi)(x(t))$ along all solutions to $\dot{x}(t) = f(x(t))$. Assume

$$\frac{\mathrm{d}^j}{\mathrm{d}t^j}a(t) = (L_{ad^j(f,g)}\psi)(x(t))$$

for all $x(t) \in \Psi$, then

$$\begin{aligned}\frac{\mathrm{d}^{j+1}}{\mathrm{d}t^{j+1}}a(t) &= L_f(L_{ad^j(f,g)}\psi)(x(t)) \\ &= (L_{[f,ad^j(f,g)]}\psi)(x(t)) + L_{ad^j(f,g)}(L_f\psi)(x(t)) \\ &= (L_{ad^{j+1}(f,g)}\psi)(x(t)) + L_{ad^j(f,g)}(L_f\psi)(x(t)) \\ &= (L_{ad^{j+1}(f,g)}\psi)(x(t))\end{aligned}$$

since $x(t) \in \Psi$. Since $a(t)$ vanishes identically for all $x(t) \in \Psi$ along solutions to $\dot{x}(t) = f(x(t))$,

$$(L_{ad^{j+1}(f,g)}\psi)(x) = 0$$

□

Initially subsystem (11.11) is regarded as a isolated system with input y^2 and a smooth feedback function $w : y^1 \mapsto y^2 = w(y^1)$, $\mathbb{R}^{n-m} \to \mathbb{R}^m$, is sought to stabilize this system. The approach is similar to that used by Goodall and Ryan (1991) in which Lyapunov theory and the invariance principle of LaSalle are invoked.

For global uniform asymptotic stability of the zero state, system (11.11) must exhibit the properties :

(*i*) *Existence and continuation of solutions.* For each $y_0^1 \in \mathbb{R}^{n-m}$, there exists a local solution $y^1 : [0, t_1) \to \mathbb{R}^{n-m}$ (i.e. an absolutely continuous function satisfying (11.11) a.e. and $y^1(0) = y_0^1$) and every such solution can be extended into a solution on $[0, \infty)$.

(*ii*) *Uniform boundedness of solutions.* For each $\varrho > 0$, there exists $r(\varrho) > 0$ such that $y^1(t) \in r(\varrho)\mathbb{B}_{n-m}$, for all $t \geq 0$ on every solution $y^1 : [0,\infty) \to \mathbb{R}^{n-m}$ with $y_0^1 \in \varrho\mathbb{B}_{n-m}$.

(*iii*) *Uniform stability of the state origin.* For each $\delta > 0$, there exists $d(\delta) > 0$ such that $y^1(t) \in \delta\mathbb{B}_{n-m}$ for all $t \geq 0$ on every solution $y^1 : [0,\infty) \to \mathbb{R}^{n-m}$ with $y_0^1 \in d(\delta)\mathbb{B}_{n-m}$.

(*iv*) *Global uniform attractivity of the state origin.* For each $\varrho > 0$ and $\varepsilon > 0$, there exists $T(\varrho, \varepsilon) \geq 0$ such that $y^1(t) \in \varepsilon\mathbb{B}_{n-m}$ for all $t \geq T(\varrho,\varepsilon)$ on every solution $y^1 : [0,\infty) \to \mathbb{R}^{n-m}$ with $y_0^1 \in \varrho\mathbb{B}_{n-m}$.

To achieve global asymptotic stability of the zero state of system (11.11) additional hypotheses are assumed to hold :

Hypothesis 11.4

(a) There exist a C^∞ function $v_1 : \mathbb{R}^{n-m} \to \mathbb{R}$ and a continuous function $\lambda : \mathbb{R}^{n-m} \to [0,\infty)$, satisfying $\lambda(0) = 0$ and $\lambda(y^1) \geq 0 \ \ \forall\, y^1 \neq \mathbf{0}$, such that

(i) $v_1(\mathbf{0}) = 0$ *and* $v_1(y^1) > 0$ *for* $y^1 \neq \mathbf{0}$

(ii) $(L_{f_1} v_1)(y^1) \leq -\lambda(y^1)$

(iii) $v_1(y^1) \to \infty$ *as* $\|y^1\| \to \infty$.

(b) There exist known real constants $\alpha \in [0,1)$, $\beta > 0$ and a known C^1 function q which satisfies

$$\beta\nu^2 \leq \nu q(\nu) \qquad \forall\, \nu \in \mathbb{R}$$

such that

$$|g_i(y^1, y^2)| \leq |q((L_{f_i^*} v_1)(y^1))| + \alpha |h_i(y^2)|$$

where $f_i^* = f_i^*(y^1) \in C^\infty(\mathbb{R}^{n-m})$ denotes the i th column of the matrix $F_1(y^1)$, h_i and g_i are the i th components of h and g, respectively.

(c) There exists a nonempty set $\Omega \subset \mathbb{R}^{n-m} \setminus \{0\}$ such that

(i) for each $y^1 \in \Omega$,

$$\mathrm{span}\{f_1(y^1),\ ad^k(f_1, f_i^*)(y^1);\ i = 1, \ldots, m;\ k = 0,\ 1, \ldots\} = \mathbb{R}^{n-m}$$

(ii) $\{0\}$ is the unique proper subset of $\Omega^c \cap \Gamma$ which is invariant with respect to f_1, where Ω^c denotes the complement of Ω,

$$\Gamma := \Lambda \cap \{y^1 \in \mathbb{R}^{n-m} : \ (L_{f_i^*} v_1)(y^1) = 0, \qquad i = 1, \ldots, m\}$$

and $\Lambda := \{y^1 \in \mathbb{R}^{n-m} : \lambda(y^1) = 0\}$.

Remarks.

(*i*) Essentially, the suppositions of Hypothesis 11.4(a) imply the existence of a Lyapunov function for f_1 when $y^1 \notin \Lambda$. A critical case arises when $y^1 \in \Lambda \neq \emptyset$. In this case, the state origin of the system

$$\dot{y}^1 = f_1(y^1(t))$$

is stable but not necessarily asymptotically stable.

(*ii*) Conditions in Hypothesis 11.4(c) originated from the work of Jurdjevic and Quinn (1978), Slemrod (1978) for bilinear systems, and subsequently modified in Ryan and Buckingham (1983). The condition $\Omega^c = \{0\}$ would suffice for (c)(*ii*), however, this condition is unnecessarily strong. Consider

$$\dot{x} = f(x) + G(x)u, \qquad x \in \mathbb{R}^3,\ u \in \mathbb{R}$$

$$f(x) = Ax, \quad G(x) = Bx, \quad A = \begin{bmatrix} 0 & 1 & 0 \\ -1 & 0 & 0 \\ 0 & 0 & 0 \end{bmatrix}, \quad B = \begin{bmatrix} 0 & 0 & 0 \\ 0 & -1 & 0 \\ 0 & 0 & 1 \end{bmatrix}$$

For this example

$$\text{span}\{Ax,\ ad^k(A,B)(x);\ k = 0, 1, \ldots\} = \mathbb{R}^3$$

iff

$$x \in \Omega := \{x \in \mathbb{R}^3 \ : \ x_1^2 + x_2^2 \neq 0,\ x_3 \neq 0\}$$

In this case, $\{0\}$ is not the only subset of Ω^c invariant under f (i.e. $\exp(At)$, $-\infty < t < \infty$). Therefore, no conclusion can be made concerning the stability of the system. However, with $v(x) = ||x||^2$,

$$\Gamma = \{x \in \mathbb{R}^3 \ : \ (L_g v)(x) = 0\} = \{x \in \mathbb{R}^3 \ : \ -x_2^2 + x_3^2 = 0\}$$

where $g(x) = Bx$. Hence, $\Omega^c \cap \Gamma = \{x \in \mathbb{R}^3 \ : \ x_2 = x_3 = 0\}$, of which $\{0\}$ is the only subset invariant under $\exp(At)$, $-\infty < t < \infty$. Thus, a statement concerning the stability of the system can be made.

In order to show indefinite continuation of solutions for subsystem (11.11), the concept of a *maximal solution* is required.

Definition 11.5 *A* **maximal solution** *is any function* $t \mapsto y^1(t) : [0, \tau) \to \mathbb{R}^m$ *which is absolutely continuous on compact subintervals, satisfying* (11.11) *a.e. with prescribed initial condition, and does not have a proper extension which is also a solution.*

A smooth stabilizing feedback for subsystem (11.11) is presented in the following lemma.

Lemma 11.6 *Under the conditions stated in* Hypothesis 11.4, *the feedback function*

$$w : y^1 \mapsto y^2 := (h^{-1} \circ s)(y^1) \tag{11.13}$$

where $s = [s_1,\ s_2, \ldots,\ s_m]^{\mathrm{T}}$,

$$s_i(y^1) := -\gamma(1-\alpha)^{-1} q((L_{f_i^*} v_1)(y^1)) \tag{11.14}$$

and $\gamma > 1$, *renders the zero state of subsystem* (11.11), *with* $y^1(0) = y_0^1$, *globally asymptotically stable.*

Proof. For each $y_0^1 \in \mathbb{R}^{n-m}$, Hypothesis 11.4(b) guarantees that the feedback controlled system (11.11), with $y^1(0) = y_0^1$, has at least one maximal solution $y^1(\cdot) : [0, \tau) \to \mathbb{R}^{n-m}$. Along each maximal solution of (11.11), for almost all $t \in [0, \tau)$,

$$\begin{aligned}\dot v_1(y^1(t)) &= (L_{f_1}v_1)(y^1(t))\\ &\quad+\langle\nabla v_1(y^1(t)),\ F(y^1(t))[s(y^1(t))+g(y^1(t),w(y^1(t)))]\rangle\\ &= (L_{f_1}v_1)(y^1(t))\\ &\quad+\sum_{i=1}^{m}(L_{f_i^\bullet}v_1)(y^1(t))[s_i(y^1(t))+g_i(y^1(t),w(y^1(t)))]\end{aligned}$$

Whence, in view of Hypothesis 11.4(a) and (b),

$$\begin{aligned}\dot v_1(y^1(t)) &\le -\lambda(y^1(t))-\beta(\gamma-1)\sum_{i=1}^{m}|(L_{f_i^\bullet}v_1)(y^1(t))|^2\\ &\le 0\end{aligned}$$

holds for almost all $t\in[0,\tau)$, along every maximal solution of (11.11), which implies that all solutions can be continued indefinitely. Also, properties of boundedness of solutions and stability clearly hold. Let Θ be the largest invariant subset of Γ (defined in Hypothesis 11.4(c)); invariant in the sense that, under Hypotheses 11.4(a) and (b), any solution of (11.11) starting in Θ remains in Θ for all t. By LaSalle's invariance principle, all solutions of (11.11) approach Θ as $t\to\infty$, where, for each $y_0^1\in\Theta$, every solution $y^1(t)$ of (11.11) satisfies

$$\lambda(y^1(t))=0,\qquad (L_{f_i^\bullet}v_1)(y^1(t))=0\qquad \text{for } i=1,\dots,m$$

and, hence,

$$\dot y^1(t)=f_1(y^1(t))$$

for almost all $t\in\mathbb{R}$. As a consequence of assumption Hypothesis 11.4(a)(*ii*), the Lie derivative of v_1 along f_1 is non-positive for all $y^1\in\mathbb{R}^{n-m}$. However, for every $y^1\in\Lambda$, where Λ is defined in Hypothesis 11.4(c)(*ii*), $(L_{f_1}v_1)(y^1)$ has a maximum and therefore

$$\nabla((L_{f_1}v_1)(y^1))=\mathbf{0}$$

Hence, as a consequence of Proposition 11.3, Θ can be characterized as

$$\begin{aligned}\Theta := \{y^1\in\mathbb{R}^{n-m}\ :\ &\lambda(y^1)=0,\qquad (L_{f_1}v_1)(y^1)=0\\ &(L_{ad^k(f_1,f_i^\bullet)}v_1)(y^1)=0;\qquad k=0,\ 1,\dots,\qquad i=1,\dots,\ m\}\end{aligned}$$

Clearly, $\Theta\subset\Omega^c\cap\Gamma$ and thus Hypothesis 11.4(c) ensures that $\Theta=\{\mathbf{0}\}$. Finally, as a consequence of Hypothesis 11.4(a)(*iii*), one can conclude that the feedback control $y^2=w(y^1)$ (defined by (11.13)–(11.14)) renders the state origin of (11.11) globally asymptotically stable.

□

11.5 Proposed Class of Generalized Feedback Controls

The desired feedback controls are embedded in a class $\mathcal{U}$ of *generalized feedbacks.*

Definition 11.7 *A* $\mathcal{F} : \mathbb{R} \times \mathbb{R}^{n-m} \times \mathbb{R}^m \rightarrow 2^{\mathbb{R}^m}$ *is a* **generalized feedback** *if*

(i) $\mathcal{F}$ *is upper semicontinuous with nonempty, convex and compact values;*

(ii) $\mathcal{F}$ *is singleton-valued except on a set* $\Sigma_{\mathcal{F}}$ *of Lebesgue measure zero in* $\mathbb{R} \times \mathbb{R}^{n-m} \times \mathbb{R}^m$.

Generalized feedbacks are set-valued maps associating a subset of controls to each state of the system. A simple example of a generalized feedback, defined on $\mathbb{R}$, is the map

$$x \mapsto \mathcal{A}(x) := \begin{cases} \{-1\}, & x < 0 \\ [-1, 1], & x = 0 \\ \{1\}, & x > 0 \end{cases}$$

of which any selection is an example of a relay-type control function.

The primary objective is to determine a generalized feedback strategy $\mathcal{F}$ such that any selection, $u(t) \in \mathcal{F}(t, y^1(t), y^2(t))$, renders the zero state of the feedback controlled differential inclusion system

$$\dot{y}^1(t) = f_1(y^1(t)) + F_1(y^1(t))[h(y^2(t)) + g(y^1(t), y^2(t))] \tag{11.15}$$

$$\dot{y}^2(t) \in f_2(y^1(t), y^2(t)) + G_2(y^1(t), y^2(t)) \left[u(t) + \mathcal{H}(t, y^1(t), y^2(t)) + h(t)\|u(t)\|\overline{\mathbb{B}}_m\right] \tag{11.16}$$

with initial condition $y^1(0) = y^1_0$, $y^2(0) = y^2_0$, globally uniformly asymptotically stable.

Choose $A, Q \in \mathcal{L}(\mathbb{R}^m, \mathbb{R}^m)$ such that $\sigma(A) \subset \mathbb{C}^-$ and $Q > 0$. Let $P > 0$ denote the unique symmetric solution of the Lyapunov equation

$$PA + A^{\mathrm{T}}P + Q = 0 \tag{11.17}$$

Define the set-valued maps $\mathcal{P} : \mathbb{R}^{n-m} \times \mathbb{R}^m \rightarrow 2^{\mathbb{R}^m}$ and $\mathcal{D} : \mathbb{R}^m \rightarrow 2^{\mathbb{R}^m}$ by

$$(y^1, y^2) \mapsto \mathcal{P}(y^1, y^2) := \{\nu \in \mathbb{R}^m : \langle (\mathrm{D}h)(y^2)G_2(y^1, y^2)\nu,\ P(h(y^2) - s(y^1))\rangle \geq 0\}$$

$$u \mapsto \mathcal{D}(u) := \begin{cases} \{\|u\|^{-1}u\}, & u \neq 0 \\ \overline{\mathbb{B}}_m, & u = 0 \end{cases}$$

Introducing the notation :

$$\xi(\Xi) := \max\{\|\mu\| : \mu \in \Xi\}$$

for a compact set $\Xi \neq \emptyset$, with $\xi(\emptyset) := 0$, and a design parameter $\delta > 0$, the proposed generalized feedback is

$$(t, y^1, y^2) \mapsto \mathcal{F}(t, y^1, y^2) = k(y^1, y^2) + \mathcal{N}(t, y^1, y^2) \tag{11.18}$$

where

$$k(y^1, y^2) := G_2^{-1}(y^1, y^2)\Big[-f_2(y^1, y^2) + [(\mathrm{D}h)(y^2)]^{-1}[A(h(y^2) - s(y^1)) + (\mathrm{D}s)(y^1)(f_1(y^1) + F_1(y^1)h(y^2))]\Big]$$

and

$$\mathcal{N}(t, y^1, y^2) := -\rho(t, y^1, y^2)\mathcal{D}\left([(\mathrm{D}h)(y^2)G_2(y^1, y^2)]^{\mathrm{T}} P(h(y^2) - s(y^1))\right)$$

where ρ is any continuous functional satisfying

$$\begin{aligned}\rho(t, y^1, y^2) \geq (1 - h(t))^{-1}\Big[& h(t)\|k(y^1, y^2)\| + \xi(\mathcal{H}(t, y^1, y^2) \cap \mathcal{P}(y^1, y^2)) \\ &+ \left(\sum_{i=1}^{m} (\alpha|h_i(y^2)| + \gamma^{-1}(1-\alpha)|s_i(y^1)|)\ \|(\mathrm{D}s)(y^1)f_i^*(y^1)\| + \delta\right) \\ &\times \left\| [(\mathrm{D}h)(y^2)G_2(y^1, y^2)]^{-1}\right\| \Big]\end{aligned} \tag{11.19}$$

and $\gamma > 1$. Loosely speaking, the function k is designed to stabilize system (11.11)–(11.12) in the absence of uncertainty, whilst the set-valued map $\mathcal{N}$ is constructed to counteract the uncertainty in the system.

Clearly, $\mathcal{F}$ has convex and compact values; moreover, the continuity of ρ and the upper semicontinuity of $\mathcal{D}$ ensure, by Proposition 11.8 (Goodall 1989), that $\mathcal{F}$ is upper semicontinuous. Hence, $\mathcal{F}$ qualifies as a generalized feedback.

Proposition 11.8 *Let Y_1, Y_2 be a real Banach spaces. If $f : Y_1 \to \mathbb{R}$ is continuous and $\mathcal{Y} : Y_1 \to 2^{Y_2}$ is upper semicontinuous with compact values, then $f\mathcal{Y} : Y_1 \to 2^{Y_2}$, $y \mapsto f(y)\mathcal{Y}(y)$ is upper semicontinuous with compact values.*

Remarks.

(*i*) The proposed design strategy for $\mathcal{F}$ ensures that any selection is discontinuous in nature. Thus, the set $\Sigma_{\mathcal{F}}$ (introduced in Defn. 11.7) may be interpreted, in the ensuing analysis, as a switching surface of control discontinuities.

(*ii*) The intersection $\mathcal{H}(t, y^1, y^2) \cap \mathcal{P}(y^1, y^2)$ is adopted in (11.19) in order to economize on the gain ρ by exploiting the possible occurrence of "stability enhancing" uncertainties.

11.6 Global Attractive Manifold $\mathcal{M}$

The feedback function $y^1 \mapsto w(y^1)$, specified explicitly in Lemma 11.6 by (11.13)–(11.14), defines a smooth nonlinear manifold $\mathcal{M} \subset \mathbb{R}^{n-m} \times \mathbb{R}^m$, where

$$\mathcal{M} := \{(y^1, y^2) \ : \ y^2 = w(y^1)\}$$

With respect to the uncertain system (11.15)–(11.16), the generalized feedback strategy $\mathcal{F}$ is designed to render $\mathcal{M}$ invariant and globally finite-time attractive. Consider the function $t \mapsto e(t) := ((h \circ y^2) - (s \circ y^1))(t)$ which satisfies the differential inclusion :

$$\dot{e}(t) \in \mathcal{E}(t, y^1(t), e(t))) \tag{11.20}$$

with initial condition :

$$e(0) = e_0 := h(y_0^2) - s(y_0^1) \tag{11.21}$$

where the set-valued map $\mathcal{E} : \ \mathbb{R} \times \mathbb{R}^{n-m} \times \mathbb{R}^m \to 2^{\mathbb{R}^m}$ is defined by

$$\begin{aligned}
\mathcal{E}(t, y^1, e) := \Big\{ &(\mathrm{D}h)(h^{-1} \circ (e + s(y^1))) \Big[f_2(y^1, h^{-1} \circ (e + s(y^1))) \\
&+ G_2(y^1, h^{-1} \circ (e + s(y^1))) \, [u(t) \\
&+ \mathcal{H}(t, y^1, h^{-1} \circ (e + s(y^1))) + \ h(t) \|u(t)\| \overline{\mathbb{B}}_m \ ] \Big] \\
- (\mathrm{D}s)(y^1) &\Big[f_1(y^1) + F_1(y^1)[e + s(y^1) + g(y^1, h^{-1} \circ (e + s(y^1)))] \Big] \\
&: \ u(t) \in \mathcal{F}(t, y^1, h^{-1} \circ (e + s(y^1))) \ \Big\}
\end{aligned}$$

The following proposition (see Aubin and Cellina (1984) and Ryan (1990)) is required to show existence of a solution, with respect to differential inclusion systems, and that all solutions can be continued indefinitely.

Proposition 11.9 *If the set-valued map $(t, y^1, e) \mapsto \mathcal{E}(t, y^1, e)$ is upper semi-continuous with nonempty, convex and compact values then, for each $e_0 \in \mathbb{R}^m$, there exists a local solution of* (11.20)–(11.21) *which can be extended into a maximal solution $e : [0, \tau) \to \mathbb{R}^m$ and if $e(\cdot)$ is bounded then $\tau = \infty$.*

Before Proposition 11.9 can be invoked, $\mathcal{E}$ must be shown to be a upper semi-continuous map with nonempty, convex and compact values. This can be achieved using the following propositions (Aubin and Cellina 1984). Let Y_1, Y_2, Y_3 denote real Banach spaces.

Proposition 11.10 *Let $K \subset Y_1$ be compact and let $\mathcal{Y} : Y_1 \to 2^{Y_2}$ be upper semicontinuous with compact values. Then $\mathcal{Y}(K) \subset Y_2$ is compact.*

Proposition 11.11 *Let $\mathcal{Y} : Y_1 \to 2^{Y_2}$ and $\mathcal{Z} : Y_2 \to 2^{Y_3}$ have non-empty values. If $\mathcal{Y}$ and $\mathcal{Z}$ are upper semicontinuous, then $\mathcal{Z} \circ \mathcal{Y}$ is upper semicontinuous, where the composition $\mathcal{Z} \circ \mathcal{Y} : Y_1 \to 2^{Y_3}$ is defined by*

$$y \mapsto (\mathcal{Z} \circ \mathcal{Y})(y) := \bigcup_{z \in \mathcal{Y}(y)} \mathcal{Z}(z) \ .$$

Let

$$\mathcal{E}^*(t,y^1,y^2,\mathcal{F}(t,y^1,y^2)) := \bigcup_{u\in\mathcal{F}(t,y^1,y^2)} \mathcal{E}^*(t,y^1,y^2,u)$$

where $\mathcal{E}^*(t,y^1,y^2,u) := u+\mathcal{H}(t,y^1,y^2)+h(t)\|u\|\overline{\mathbb{B}}_m$ is the sum of upper semicontinuous, compact-valued set-valued maps. It follows from Propositions 11.10 and 11.11 that $(t,y^1,y^2)\mapsto(\mathcal{E}^*\circ\mathcal{F})(t,y^1,y^2)$, and, hence, $\mathcal{E}$, is upper semicontinuous with compact values. Also, convexity of values of $\mathcal{E}^*$ and $\mathcal{F}$ imply that $\mathcal{E}$ has convex values. Hence, invoking Proposition 11.9, for each e_0, the initial value problem (11.20)–(11.21) admits a maximal solution $e:\ [0,\tau)\to\mathbb{R}^m$.

Consider the behaviour of the function $v_2\ :\ \mathbb{R}^m\to[0,\infty),\ e\mapsto v_2(e):=\frac{1}{2}\langle e,\ Pe\rangle$ along solutions of (11.20). Along each maximal solution $e:\ [0,\tau)\to\mathbb{R}^m$,

$$\dot v_2(e(t))\in\mathcal{V}(t,y^1(t),e(t)) := \big\langle\mathcal{E}(t,y^1(t),e(t)),\ Pe(t)\big\rangle$$

As a consequence of (11.17) and (11.18),

$$\begin{aligned}\max\mathcal{V}(t,y^1(t),e(t)) \ \le\ & -\frac{1}{2}\langle e(t),\ Qe(t)\rangle\\ & -\big\langle(\mathrm{D}s)(y^1(t))F_1(y^1(t))g(y^1(t),h^{-1}\circ(e+s(y^1))(t)),\ Pe(t)\big\rangle\\ & -\rho(t,y^1(t),h^{-1}\circ(e+s(y^1))(t))\langle\mathcal{D}(p(t)),\ p(t)\rangle\\ & +\big\langle\mathcal{H}(t,y^1(t),h^{-1}\circ(e+s(y^1))(t)),\ p(t)\big\rangle\\ & +h(t)\xi(\mathcal{F}(t,y^1(t),h^{-1}\circ(e+s(y^1))(t)))\|p(t)\| \end{aligned}\qquad (11.22)$$

where

$$p(t) := \big[(\mathrm{D}h)(h^{-1}\circ(e+s(y^1))(t))G(y^1(t),h^{-1}\circ(e+s(y^1))(t)\big]^{\mathrm{T}}Pe(t)$$

Using Hypothesis 11.4(b) and (11.14),

$$\begin{aligned}&\big\|[(\mathrm{D}h)(y^2)G(y^1,y^2)]^{-1}(\mathrm{D}s)(y^1)F_1(y^1)g(y^1,y^2)\big\| \le\\ &\qquad\sum_{i=1}^{m}\{\alpha|h_i(y^2)|+\gamma^{-1}(1-\alpha)|s_i(y^1)|\}\\ &\qquad\times\big\|[(\mathrm{D}h)(y^2)G(y^1,y^2)]^{-1}(\mathrm{D}s)(y^1)f_i^*(y^1)\big\|\end{aligned}\qquad (11.23)$$

From (11.18),

$$\xi(\mathcal{F}(t,y^1,y^2))\le\big\|k(y^1,y^2)\big\|+\rho(t,y^1,y^2)\qquad (11.24)$$

Substituting (11.23)–(11.24) in (11.22) and noting that, for $p(t)\neq\mathbf{0}$,

$$\langle\mathcal{D}(p(t)),\ p(t)\rangle=\|p(t)\|$$

$$\begin{aligned}\max\mathcal{V}(t,y^1(t),e(t)) \ \le\ & -\frac{1}{2}\langle e(t),\ Qe(t)\rangle\\ & -\rho(t,y^1(t),h^{-1}\circ(e+s(y^1))(t))[1-h(t)]\,\|p(t)\|\end{aligned}$$

$$
\begin{aligned}
&+ \|p(t)\| \Bigg\{ \sum_{i=1}^{m} [\alpha |e_i(t) + s_i(y^1(t))| + \gamma^{-1}(1-\alpha)|s_i(y^1(t))|] \times \\
&\big\| [(\mathrm{D}h)(h^{-1} \circ (e + s(y^1))(t)) G_2(y^1, h^{-1} \circ (e + s(y^1))(t)]^{-1} \\
&\times (\mathrm{D}s)(y^1(t)) f_i^*(y^1(t)) \big\| \\
&+ h(t) \big\| k(y^1(t), h^{-1} \circ (e + s(y^1))(t)) \big\| \\
&+ \xi(\mathcal{H}(t, y^1(t), h^{-1} \circ (e + s(y^1))(t)) \\
&\qquad\qquad \cap \mathcal{P}(y^1(t), h^{-1} \circ (e + s(y^1))(t))) \Bigg\} \qquad (11.25)
\end{aligned}
$$

with

$$
p(t) := \left[(\mathrm{D}h)(h^{-1} \circ (e + s(y^1))(t)) G(y^1(t), h^{-1} \circ (e + s(y^1))(t) \right]^{\mathrm{T}} Pe(t)
$$

However, from (11.19),

$$
\begin{aligned}
(1 - h(t))\rho(t, y^1, y^2) \ \geq \ & \Big[h(t) \big\| k(y^1, y^2) \big\| + \xi(\mathcal{H}(t, y^1, y^2) \cap \mathcal{P}(y^1, y^2)) \\
& + \left(\sum_{i=1}^{m} [\alpha |h_i(y^2)| + \gamma^{-1}(1-\alpha)|s_i(y^1)|] \, \big\| (\mathrm{D}s)(y^1) f_i^*(y^1) \big\| + \delta \right) \\
& \times \big\| [(\mathrm{D}h)(y^2) G(y^1, y^2)]^{-1} \big\| \ \Big]
\end{aligned}
$$

Hence, (11.25) becomes

$$
\begin{aligned}
\max \mathcal{V}(t, y^1(t), e(t)) \ \leq \ & -\frac{1}{2} \langle e(t), \ Qe(t) \rangle \\
- \delta \big\| [(\mathrm{D}h)(h^{-1} \circ (e + s(y^1))(t)) & G(y^1, h^{-1} \circ (e + s(y^1))(t))]^{-1} \big\| \, \|p(t)\|
\end{aligned}
$$

But

$$
\begin{aligned}
\|p(t)\| \ \geq \ & \big\| [(\mathrm{D}h)(h^{-1} \circ (e + s(y^1))(t)) G(y^1, h^{-1} \circ (e + s(y^1))(t))]^{-1} \big\|^{-1} \\
& \times \|Pe(t)\|
\end{aligned}
$$

and so

$$
\begin{aligned}
\max \mathcal{V}(t, y^1(t), e(t)) \ & \leq \ -\frac{1}{2} \langle e(t), \ Qe(t) \rangle - \delta \|Pe(t)\| \qquad (11.26) \\
& \leq \ 0
\end{aligned}
$$

a.e.. Therefore, $e(\cdot)$ is bounded and hence every maximal solution $e : [0, \tau) \to \mathbb{R}^m$ can be continued indefinitely (see Proposition 11.9). Using (11.26), it can be shown (see, for example, Goodall and Ryan (1991)) that the manifold $\mathcal{M}$ is finite-time attractive and invariant.

Lemma 11.12 *For each* (y_0^1, y_0^2), $(y^1(t), y^2(t)) \in \mathcal{M}$ *for all* $t > T$ *where* T *satisfies*

$$
T < \delta^{-1} \{ 2 \big\| P^{-1} \big\| v_2(e_0) \}^{\frac{1}{2}}
$$

Proof. From (11.26), it follows for $Pe \neq \mathbf{0}$ that

$$\begin{aligned} \dot{v}_2(e(t)) &< -\delta\|Pe(t)\| \\ &\leq -\delta\sqrt{2}\|P^{-1}\|^{-\frac{1}{2}}[v_2(e(t))]^{\frac{1}{2}} \end{aligned}$$

a.e. along solutions of (11.20). On integration

$$\sqrt{v_2(e(t))} - \sqrt{v_2(e(0))} < -\delta\|2P^{-1}\|^{-\frac{1}{2}}t$$

If the manifold $\mathcal{M}$ is attained in finite time τ, then $v_2(e(\tau)) = 0$, (i.e. $(y^1(\tau), y^2(\tau)) \in \mathcal{M}$). Clearly, τ satisfies

$$\tau < \delta^{-1}[2\|P^{-1}\|v_2(e(0))]^{\frac{1}{2}}$$

Moreover, for all $t > \tau$, $v_2(e(t)) = 0 \iff h(y^2) - s(y^1) = \mathbf{0}$, i.e. $(y^1(t), y^2(t)) \in \mathcal{M}$ and so $\mathcal{M}$ is (positively) invariant.

□

It is noted that the upper bound on the time required to attain $\mathcal{M}$ is inversely proportional to the controller design parameter δ and so the time taken to reach $\mathcal{M}$ can be controlled through δ.

11.7 Lyapunov Stabilization

In this final stage the generalized feedback $\mathcal{F}$, defined by (11.18)–(11.19), is shown to render the zero state of the differential inclusion system (11.15)–(11.16) globally uniformly attractive. Consider the differential inclusion system

$$\begin{bmatrix} \dot{y}^1(t) \\ \dot{e}(t) \end{bmatrix} \in \tilde{\mathcal{G}}(t, y^1(t), e(t)) \tag{11.27}$$

with initial condition

$$\begin{bmatrix} y^1(0) \\ e(0) \end{bmatrix} = \begin{bmatrix} y_0^1 \\ e_0 \end{bmatrix} \tag{11.28}$$

where $\tilde{\mathcal{G}} : \mathbb{R} \times \mathbb{R}^{n-m} \times \mathbb{R}^m \to 2^{\mathbb{R}^n}$ is defined by

$$\begin{aligned} \tilde{\mathcal{G}}(t, y^1, e) \quad := \quad & \left\{ \begin{bmatrix} f_1(y^1) + F_1(y^1)(h(e + w(y^1)) \\ \eta \end{bmatrix} \right. \\ & \left. + \begin{bmatrix} g(y^1, h^{-1} \circ (e + s(y^1))) \\ 0 \end{bmatrix} : \eta \in \mathcal{E}(t, y^1, e) \right\} \end{aligned}$$

The set-valued map $(t, y^1, e) \mapsto \tilde{\mathcal{G}}(t, y^1, e)$ is upper semicontinuous with nonempty, convex and compact values and, hence, for each $\begin{bmatrix} y_0^1 \\ e_0 \end{bmatrix}$, the initial value problem (11.27)–(11.28) admits a maximal solution $\begin{bmatrix} y^1 \\ e \end{bmatrix} : [0, \tau) \to \mathbb{R}^n$. A Lyapunov function candidate, $v : \mathbb{R}^n \to [0, \infty)$, is defined by

$$v\left(\begin{bmatrix} y^1 \\ e \end{bmatrix}\right) := v_1(y^1) + \zeta v_2(e)$$

where $\zeta \in \mathbb{R}^+$, a real constant, is to be specified. With reference to inequality (11.26), it is seen that, along solutions to (11.27) and for almost all t,

$$\begin{aligned} \dot{v}\left(\begin{bmatrix} y^1(t) \\ e(t) \end{bmatrix}\right) &= \dot{v}_1(y^1(t)) + \zeta \dot{v}_2(e(t)) \\ &\leq \dot{v}_1(y^1(t)) - \frac{\zeta}{2}\sigma_{\min}(Q)\|e(t)\|^2 \end{aligned}$$

where $\sigma_{\min}(\cdot)$ denotes the minimum eigenvalue of a real, symmetric matrix. Under the assumptions in Hypothesis 11.4(b) and along solutions to (11.27),

$$\begin{aligned} \dot{v}_1(y^1(t)) \leq & -\lambda(y^1(t)) - \beta(\gamma - 1)\sum_{i=1}^{m} |(L_{f_i^*}v_1)(y^1(t))|^2 \\ & + (1+\alpha)\sum_{i=1}^{m} |(L_{f_i^*}v_1)(y^1(t))|\,|e_i(t)| \end{aligned}$$

a.e.. Therefore, along each maximal solution of (11.27),

$$\begin{aligned} \dot{v}\left(\begin{bmatrix} y^1(t) \\ e(t) \end{bmatrix}\right) \leq & -\lambda(y^1(t)) \\ & -\frac{1}{2}\sum_{i=1}^{m}\left\langle \begin{bmatrix} |(L_{f_i^*}v_1)(y^1(t))| \\ e_i(t) \end{bmatrix}, E_\zeta \begin{bmatrix} |(L_{f_i^*}v_1)(y^1(t))| \\ e_i(t) \end{bmatrix}\right\rangle \end{aligned}$$

for almost all $t \in [0, \tau)$, where

$$E_\zeta = \begin{bmatrix} 2\beta(\gamma-1) & -(1+\alpha) \\ -(1+\alpha) & \zeta\mu \end{bmatrix}$$

and $\mu := \sigma_{\min}(Q)$. Choosing

$$\zeta > (1+\alpha)^2[2\beta(\gamma-1)\mu]^{-1}$$

ensures that, along each maximal solution $\begin{bmatrix} y^1 \\ e \end{bmatrix} : [0, \tau) \to \mathbb{R}^n$,

$$\dot{v}\begin{bmatrix} y^1(t) \\ e(t) \end{bmatrix} \leq 0 \qquad (11.29)$$

a.e.. With reference to Proposition 11.9, it follows that every maximal solution $\begin{bmatrix} y^1 \\ e \end{bmatrix} : [0, \tau) \to \mathbb{R}^n$ can be continued indefinitely. Hence, as a consequence of (11.29) and a similar analysis to that in Lemma 11.6, the following theorem can be concluded.

Theorem 11.13 *Under assumptions* Hypotheses 11.1–11.4, *the generalized feedback* $\mathcal{F} \in \mathcal{U}$, *having the form* (11.18)–(11.19), *renders the zero state of the differential inclusion system* (11.15)–(11.16) *globally uniformly asymptotically stable.*

Finally, the following corollary is an immediate consequence of Lemmas 11.6 and 11.12.

Corollary 11.14 *With conditions specified in* Hypotheses 11.1–11.4, *the manifold*

$$\mathcal{M} = \{(y^1, y^2) \in \mathbb{R}^{n-m} \times \mathbb{R}^m \ : \ y^2 = w(y^1)\}$$

is a global attractive manifold, (positively) invariant, for the uncertain system (11.15)–(11.16).

11.8 Example of Uncertain System Stabilization via Discontinuous Feedback

The prototype model is to be based on Rayleigh's equation

$$\begin{aligned} \dot{x}_1(t) &= x_2(t) \\ \dot{x}_2(t) &= -x_1(t) + \varepsilon \left(x_2(t) - \frac{1}{3} x_2^3(t) \right) \end{aligned}$$

for which existence of limit cycles, for certain range of values of the parameter ε, is well-known (see, for example, Birkhoff and Rota (1989)). It is supposed that ε is a time-varying parameter which is governed by the function $t \mapsto y(t) \ : \mathbb{R} \to \mathbb{R}$. It is assumed that ε satisfies

$$\varepsilon(t) := y^3(t) + y(t)$$

$$\dot{y}(t) = \ell(y(t)) + u(t)$$

where $\ell \ : \ \mathbb{R} \to \mathbb{R}$ is known, and is controlled through the function $t \mapsto u(t)$. In this case, the prototype model takes the form

$$\begin{aligned} \dot{x}_1(t) &= x_2(t) \\ \dot{x}_2(t) &= -x_1(t) + \left(y^3(t) + y(t)\right) \left(x_2(t) - \frac{1}{3} x_2^3(t) \right) \\ \dot{y}(t) &= \ell(y(t)) + u(t) \end{aligned}$$

Suppose the system is subject to uncertainty and the uncertain dynamics are modelled by

$$\dot{x}_1(t) = x_2(t) \tag{11.30}$$

$$\dot{x}_2(t) = -x_1(t) + (1+k)(y^3(t) + y(t))\left(x_2(t) - \frac{1}{3}x_2^3(t)\right) \tag{11.31}$$

$$\dot{y}(t) = \ell(y(t)) + \phi(t, x_1(t), x_2(t), y(t)) + [1 + \psi(t, x_1(t), x_2(t), y(t))]u(t) \tag{11.32}$$

where k is unknown but satisfies $|k| < 1$, ϕ is unknown but is assumed to be bounded by some known continuous function $\tilde{\phi}$ and satisfies

$$|\phi(t, x_1, x_2, y)| \leq \tilde{\phi}(x_1, x_2, y)$$

while ψ is also unknown and is assumed to be uniformly bounded by a constant κ such that $\kappa \in (0,1)$ is known. Defining $(x, y) \mapsto \mathcal{H}(x, y) := \tilde{\phi}(x_1, x_2, y)\overline{\mathbb{B}}_{\mathbb{R}}$, where $x = [x_1 \; x_2]^{\mathrm{T}}$, the system (11.30–11.32) can be expressed as the controlled differential inclusion system

$$\dot{x}_1(t) = x_2(t) \tag{11.33}$$

$$\dot{x}_2(t) = -x_1(t) + (1+k)(y^3(t) + y(t))\left(x_2(t) - \frac{1}{3}x_2^3(t)\right) \tag{11.34}$$

$$\dot{y}(t) \in \ell(y(t)) + u(t) + \mathcal{H}(x(t), y(t)) + \kappa|u(t)|\overline{\mathbb{B}}_{\mathbb{R}} \tag{11.35}$$

Identifying system (11.33–11.35) with system (11.11)–(11.12),

$$y^1 = x, \quad y^2 = y, \quad f_1(x) = \begin{bmatrix} x_2 \\ -x_1 \end{bmatrix}, \quad F_1(x) = \begin{bmatrix} 0 \\ x_2 - \frac{1}{3}x_2^3 \end{bmatrix}$$

$$h(y) = y^3 + y, \quad g(x, y) = k(y^3 + y), \quad f_2(x, y) = \ell(y)$$

$G_2(x, y)$ is the identity matrix, and $\mathcal{G}_2(t, x, y, u) = \mathcal{H}(x, y) + \kappa|u|\overline{\mathbb{B}}_{\mathbb{R}}$. Clearly, the conditions in Hypothesis 11.1 hold. Since $\kappa < 1$, Hypothesis 11.2(b) is satisfied. Also, Hypothesis 11.4(b) is satisfied with $q(\nu) := \nu$; in which case, $\beta = 1$ and $\alpha = |k|$. Choosing $v_1(x) = \|x\|^2$, $(L_{f_1}v_1)(x) = 2\langle x, \; f_1(x)\rangle = 0$ for all x and, hence, Hypothesis 11.4(a) is satisfied. However, this corresponds to the critical case for which $\Lambda = \mathbb{R}^2$.

Since $f_1^*(x) = \begin{bmatrix} 0 \\ x_2 - \frac{1}{3}x_2^3 \end{bmatrix}$,

$$ad^0(f_1, f_1^*)(x) = f_1^*(x) = \begin{bmatrix} 0 \\ x_2 - \frac{1}{3}x_2^3 \end{bmatrix}$$

Hence,

$$\det([f_1(x) \; ad^0(f_1, f_1^*)(x)]) = \frac{1}{3}x_2^2(3 - x_2^2)$$

Thus, defining

$$\Omega := \left\{x \in \mathbb{R}^2 \; : \; x_2 \neq 0, \; x_2 \neq \pm\sqrt{3}\right\}$$

Hypothesis 11.4(c)(i) is satisfied. Now, consider Hypothesis 11.4(c)(ii).

$$(L_{f_1^*}v_1)(x) = \frac{2}{3}x_2^2(3 - x_2^2)$$

and therefore

$$\{x \in \mathbb{R}^2 \ : \ (L_{f_1^*}v_1)(x) = 0\} = \{x \in \mathbb{R}^2 \ : \ x_2 = 0, \ x_2 = \pm\sqrt{3}\}$$

Since $\Lambda = \mathbb{R}^2$,

$$\Gamma = \{x \in \mathbb{R}^2 \ : \ x_2 = 0, \ x_2 = \pm\sqrt{3}\} = \Omega^c$$

and so

$$\Omega^c \cap \Gamma = \{x \in \mathbb{R}^2 \ : \ x_2 = 0, \ x_2 = \pm\sqrt{3}\}$$

A solution to $\dot{x} = f_1(x)$ has the form

$$\begin{bmatrix} x_1(t) \\ x_2(t) \end{bmatrix} = \begin{bmatrix} A\sin(t) - B\cos(t) \\ A\cos(t) + B\sin(t) \end{bmatrix}$$

where A, B are real constants. Suppose

$$x(0) = x^0 \in \{x \in \mathbb{R}^2 : \ x_2 = \sqrt{3}\} \subset \Omega^c \cap \Gamma$$

i.e. $x^0 = [r \ \sqrt{3}]^{\mathrm{T}}$, $r \in \mathbb{R}$, then

$$x(t) = \begin{bmatrix} \sqrt{3}\sin(t) + r\cos(t) \\ \sqrt{3}\cos(t) - r\sin(t) \end{bmatrix}$$

Thus, in this case, $x(t) \notin \Omega^c \cap \Gamma$. Similarly, if

$$x^0 \in \{x \in \mathbb{R}^2 : \ x_2 = -\sqrt{3}\} \subset \Omega^c \cap \Gamma$$

then $x(t) \notin \Omega^c \cap \Gamma$. However, if

$$x^0 \in \{x \in \mathbb{R}^2 : \ x_2 = 0\} \subset \Omega^c \cap \Gamma$$

i.e $x^0 = [r \ 0]^{\mathrm{T}}$, $r \in \mathbb{R}$,

$$x(t) = \begin{bmatrix} r\cos(t) \\ -r\sin(t) \end{bmatrix}$$

and, clearly, $x(t) \in \Omega^c \cap \Gamma$ if and only if $r = 0$. Hence, $\left\{\begin{bmatrix} 0 \\ 0 \end{bmatrix} \in \mathbb{R}^2\right\}$ is the unique proper subset of $\Omega^c \cap \Gamma$, invariant with respect to f_1.
Selecting $Q = [1]$ and choosing $A = [a]$, where a is a negative real constant, the solution of the Lyapunov equation (11.17) is

$$P = \left[-\frac{1}{2a}\right]$$

The global attractive manifold for the system (11.33–11.35) is given by

$$\mathcal{M} = \{(x,y) \in \mathbb{R}^2 \times \mathbb{R} \;:\; y^3 + y = \frac{2}{3}\gamma(1-|k|)^{-1}x_2^2(x_2^2-3), \quad \gamma > 1\}$$

Introducing $s(x) := \frac{2}{3}\gamma(1-|k|)^{-1}x_2^2(x_2^2-3)$ and $z(x,y) = h(y) - s(x)$, the stabilizing generalized feedback is

$$\begin{aligned}\mathcal{F}(x,y) \;=\; & -\ell(y) + \frac{1}{3y^2+1}\Big[az(x,y) \\ & + \frac{4}{3}\gamma(1-|k|)^{-1}x_2(2x_2^2-3)\left[-x_1 + h(y)\left(x_2 - \frac{1}{3}x_2^3\right)\right] \\ & + \rho(x,y)\mathcal{D}\left(\frac{(3y^2+1)z(x,y)}{2a}\right)\Big]\end{aligned}$$

where

$$\mathcal{D}\left(\frac{(3y^2+1)z(x,y)}{2a}\right) = (3y^2+1)\begin{cases}\left\{\dfrac{\operatorname{sgn}(z(x,y))}{2a}\right\}, & \text{if} \quad z(x,y) \neq 0 \\ [-1,\; 1], & \text{if} \quad z(x,y) = 0\end{cases}$$

and

$$\begin{aligned}\rho(x,y) \;\geq\; & \frac{(1-\kappa)^{-1}}{(3y^2+1)}\Big[\kappa\Big| -(3y^2+1)\ell(y) + az(x,y) \\ & + \frac{4}{3}\gamma(1-|k|)^{-1}x_2(2x_2^2-3)\left[-x_1 + h(y)\left(x_2 - \frac{1}{3}x_2^3\right)\right]\Big| \\ & + \frac{2}{3}|s(x)|\;|2x_2^2-3|(|k|\;|h(y)| + \gamma^{-1}(1-|k|)|s(x)|) + \delta \\ & + (3y^2+1)\xi(\tilde{\phi}(x_1,x_2,y)\overline{\mathbb{B}}_{\mathbb{R}} \cap \mathcal{P}(x,y))\;\Big], \qquad \gamma > 1, \quad \delta > 0\end{aligned}$$

The set-valued map $\mathcal{P}$ has the form :

$$\begin{aligned}(x,y) \mapsto \mathcal{P}(x,y) \;&=\; \{r \in \mathbb{R} \;:\; r(h(y) - s(x)) \geq 0\} \\ &=\; \begin{cases}[0,\infty) & \text{if} \quad h(y) > s(x) \\ \mathbb{R} & \text{if} \quad h(y) = s(x) \\ (-\infty, 0] & \text{if} \quad h(y) < s(x)\end{cases}\end{aligned}$$

If, for example,

$$\mathcal{H}(x,y) := \{\delta d(x,y) : \; \delta \in [-\delta_1,\; \delta_2], \quad \delta_1,\; \delta_2 > 0\}$$

where $d : \mathbb{R}^2 \times \mathbb{R} \to \mathbb{R}$ is known but the parameter δ is unknown, the function $(x,y) \mapsto \xi(\mathcal{H}(x,y) \cap \mathcal{P}(x,y))$ reduces to the function

$$(x,y) \mapsto \begin{cases}\delta_2 d^+(x,y) + \delta_1|d^-(x,y)| & h(y) > s(x) \\ (\delta_1+\delta_2)|d(x,y)| & h(y) = s(x) \\ \delta_1 d^+(x,y) + \delta_2|d^-(x,y)| & h(y) < s(x)\end{cases}$$

where d^+ and d^- denote the positive and negative parts of the function $(x, y) \mapsto d(x, y)$.

References

Andreini, A., Bacciotti, A., Stefani, G. 1988, Global stabilizability of homogeneous vector fields of odd degree. *Systems and Control Letters* **10**, 251–256

Aubin, J-P., Cellina, A. 1984, *Differential inclusions*, Springer-Verlag, New York, pp 342

Barmish, B.R., Corless, M., Leitmann, G. 1983, A new class of stabilizing controllers for uncertain dynamical systems. *SIAM Journal on Control and Optimization* **21**, 246–255

Birkhoff, G., Rota, G.-C. 1989, *Ordinary differential equations*, Wiley, New York, pp 342

Byrnes, C.I., Isidori, A. 1991, Asymptotic stabilization of minimum phase nonlinear systems. *IEEE Transactions on Automatic Control* **36**, 1122–1137

Choquet-Bruhat, Y., DeWitt-Morette, C., Dillard-Bleick, M. 1982, *Analysis, manifolds and physics*, North-Holland, Amsterdam, pp 630

Corless, M., Leitmann, G. 1981, Continuous state feedback guaranteeing uniform ultimate boundedness for uncertain dynamic systems. *IEEE Transactions on Automatic Control* **AC-26**, 1139–1144

Elmali, H., Olgac, N. 1992, Robust output tracking control of nonlinear MIMO systems via sliding mode technique. *Automatica* **28**, 145–151

Filippov, A.F. 1988, *Differential equations with discontinuous righthand sides*, Reidel, Dordrecht, pp 304

Goodall, D.P. 1989, *Deterministic feedback stabilization of uncertain dynamical systems*, Ph.D. Thesis, University of Bath, Bath

Goodall, D.P. 1992a, Lyapunov stabilization of uncertain systems, affine in the control. *Proc. IEEE Int. Workshop on Variable Structure and Lyapunov Control of Uncertain Dynamical Systems*, Sheffield, 89–94

Goodall, D.P. 1992b, Lyapunov stabilization of a class of uncertain composite systems with nonlinear coupling. *Systems Science*, to appear

Goodall, D.P., Ryan, E.P. 1988, Feedback controlled differential inclusions and stabilization of uncertain dynamical systems. *SIAM Journal on Control and Optimization* **26**, 1431–1441

Goodall, D.P., Ryan, E.P. 1991, Feedback stabilization of a class of nonlinearly coupled uncertain dynamical systems. *IMA Journal of Mathematical Control and Information* **8**, 81–92

Gutman, S. 1979, Uncertain dynamical systems – a Lyapunov min-max approach. *IEEE Transactions on Automatic Control* **AC-24**, 437–443

Gutman, S., Palmor, Z. 1982, Properties of min-max controllers in uncertain dynamical systems. *SIAM Journal on Control and Optimization* **20**, 850–861

Hunt, L.R., Su, R., Meyer, G. 1983, Global transformations of nonlinear systems. *IEEE Transactions on Automatic Control* **AC-28**, 24–30

Isidori, A. 1989, *Nonlinear control systems*, Springer-Verlag, Berlin, pp 479

Jurdjevic, V., Quinn, J.P. 1978, Controllability and stability. *Journal of Differential Equations* **28**, 381–389

Kokotovic P.V., Sussmann, H.J. 1989, A positive real condition for global stabilization of nonlinear systems. *Systems and Control Letters* **13**, 125–133

Leitmann, G. 1979, Guaranteed asymptotic stability for some linear systems with bounded uncertainties. *ASME Journal of Dynamic Systems, Measurements and Control* **101**, 212–216

Luk'yanov, A.G., Utkin, V.I. 1981, Methods of reducing equations for dynamic systems to a regular form. *Automation and Remote Control* **42**, 413–420

Nijmeijer, H., van der Schaft, A. 1990, *Nonlinear dynamical control systems*, Springer-Verlag, New York, pp 467

Ryan, E.P. 1990, Discontinuous feedback and universal adaptive stabilization. in *Control of uncertain dynamical systems*, eds. Hinrichsen, D., Mårtensson, B., Birkhauser, Boston, 245–258

Ryan, E.P. 1988, Adaptive stabilization of a class of uncertain nonlinear systems: A differential inclusion approach. *Systems and Control Letters* **10**, 95–101

Ryan, E.P., Buckingham, N.J. 1983, On asymptotically stabilizing feedback control of bilinear systems. *IEEE Transactions on Automatic Control* **AC-28**, 863–864

Seibert, P., Suarez, R. 1991, Global stabilization of a certain class of nonlinear systems. *Systems and Control Letters* **16**, 17–23

Slemrod, M. 1978, Stabilization of bilinear control systems with applications to nonconservative problems in elasticity. *SIAM Journal on Control and Optimization* **16**, 131–141

Soldatos, A.G., Corless, M. 1991, Stabilizing uncertain systems with bounded control. *Dynamics and Control* **1**, 227–238

Tsinias, J. 1990, Optimal controllers and output feedback stabilization. *Systems and Control Letters* **15**, 277–284

Utkin, V.I. 1992, *Sliding modes in control and optimization*, Springer-Verlag, Berlin, pp 286

Zinober, A.S.I. 1990, *Deterministic control of uncertain systems*, Peter Peregrinus, London, pp 362

12. The Role of Morse-Lyapunov Functions in the Design of Nonlinear Global Feedback Dynamics

Efthimios Kappos

12.1 Introduction

The purpose of this chapter is to present a global control design methodology that is based on a consideration of classes of Lyapunov functions. The control aims are first translated into an equivalence class of dynamics. The crucial point is that the description of the dynamics is done not through vector fields but through Lyapunov functions for them. We then try to find some feedback law that yields dynamics in that class. This is achieved if we can find a member of that class that we can make into a Lyapunov function for the controlled dynamics. The approach presented here is, in a sense, the natural Lyapunov control design approach: it deals with the existence problem of feedback controls to achieve specific dynamical behaviour (a controllability problem), where the dynamics are determined by Lyapunov functions. It will be seen that it generalizes the fundamental philosophy of some basic linear control methodologies to the case of nonlinear systems; it also generalizes the Lyapunov stability theory to the extent that the functions considered (the 'Lyapunov function candidates') are more complex than the ones in Lyapunov's second method (they do not have to be positive definite, for example). To the extent that this chapter deals with a general framework for Lyapunov control design, it goes further than the more specialized, but not as general work of, for example, Corless (see Chapter 9). On the other hand, the results presented are preliminary and point the way to further work.

The aim of this presentation is twofold: the primary aim is to formulate the Morse-Lyapunov approach in its full generality, since many of its concepts and methods are quite unfamiliar to control theorists. This will be outlined in the first three sections. The remainder of the chapter addresses some particular cases. The problem of stabilization is one such case. More generally, we treat the problem of achieving dynamics of saddle type (of a given index) and of designing arbitrary gradient-like dynamics.

A large part of nonlinear systems theory consists of applications of the second method of Lyapunov. In fact, the Lyapunov function method is one of the very few aids available to the nonlinear control designer. The way the method works is to first select a function in an appropriate class (namely with a local minimum at the chosen point) and then prove it is a Lyapunov function for some (possibly controlled) dynamics. This then proves that the chosen dynamics are stable. Here we generalize Lyapunov's method to wider classes of dynamics by considering equivalence classes of so-called Morse-Lyapunov functions that have dynamics more complicated than a single attractor. In particular, we consider gradient-like dynamics. Now it is well known that it is very difficult, in general, to come up with good 'Lyapunov function candidates'. The nonlinear controllability problem can be considered to be the search for conditions that guarantee the existence of controls to accomplish some control task, for example stabilization. By interpreting controllability in this general sense, we define the *smooth controllability problem relative to a Morse-Lyapunov function* to be the search for conditions that guarantee the existence of a smooth feedback control that yields dynamics that have a Morse-Lyapunov function in a specific equivalence class as a Lyapunov function.

The only aspect of nonlinear feedback control design that has received substantial attention so far is the problem of stabilization. It has recently been an increasingly active research area (Dayawansa 1992, Coron 1990, and the survey book Bacciotti, 1991). Its relation to the traditional control concept of controllability has been recognised in Kappos (1992b). Early work on the generalization of the familiar linear controllability conditions has led to a consideration of the Lie bracket of vector fields. This differential-geometric approach has come up against the problem that it is not, in general, true that the negative $-X$ of a vector field X is available if X is available (the set of control vector fields is not *'symmetric'* (see Banks (1988), p. 78). Thus, except when one considers the control vector fields alone (the state vector field is assumed zero), it is not possible to use the full Lie algebra generated by all the control vector fields. This has led to weaker forms of controllability, such as local accessibility. In any case, no satisfactory general theory is available.

More recently, the tendency has been to examine in detail systems of low dimension, often ones that are not (smoothly) stabilizable (e.g. Kawski (1989)). What has been realised is that the stabilization problem is very complex and that some times only *ad hoc* methods of solution succeed. A set of necessary conditions for stabilizability have also been obtained that can be used to prove that even some simple systems are not stabilizable, at least using smooth controls. This has led to the search for alternative methods (for example using periodic controls (Coron, 1990)) that can be used to stabilize systems for which smooth feedback controls fail.

In all of this research, however, the fundamental question of when a control system *is* smoothly stabilizable has not been answered at all and has been relatively ignored (primarily because it is thought to be too difficult). This chapter is an attempt to address this question using global topological methods. Earlier work (Kappos 1992a, 1992b) has given some answers for the case of convex stabilization. This approach generalized the linear controllability and stabilizability conditions in a natural way that does not involve Lie conditions. We go further here, in that we examine conditions for achieving dynamics more complex than that of an asymptotically stable attractor, namely quite arbitrary gradient-like dynamics.

A classical example of optimal stabilization is the linear quadratic regulator. By choosing a quadratic cost functional which penalizes both the use of control action and the deviation from the equilibrium position, we end up (assuming a stabilizability condition) with a solution of the algebraic Ricatti equation that provides a symmetric, positive definite matrix P that yields the (global) Lyapunov function

$$V(x) = \frac{1}{2}x^T Px \tag{12.1}$$

The important part of the cost functional, as far as the resulting dynamics are concerned, is the quadratic term in the state. The result is a linear feedback law which gives control dynamics that are globally asymptotically stable, with the origin the unique attractor.

It is important to remark that a precise Lyapunov function comes out only as a result of this procedure: it is not possible to choose, a priori, a quadratic Lyapunov function that will work. However, and this is the crucial point, we are assured that the resulting Lyapunov function will be of a *particular topological type*, namely a Lyapunov function for a system with a unique, global (asymptotic) attractor. The stabilizability assumption is equivalent to the topological condition that there are no obstructions to the existence of this function (see Kappos (1992a)). The method presented here follows essentially the same steps.

In order to generalize the above method to a wider class of dynamics, we first need to have a way of describing global dynamics for the purpose of control. We give a first outline of some of the issues involved, referring the reader to the following sections for more details. The dynamical description of any flow falls into two parts, which we may roughly describe as the transient and asymptotic parts. (The general flow decomposition theorem of Conley (1978) separates the chain-recurrent from the strongly gradient part, see Sect. 12.2 for definitions.) We shall be concerned in this paper only with systems that are *gradient-like*. This means that they allow Lyapunov functions that are strict everywhere except at a finite number of hyperbolic equilibrium points. For this class, the asymptotic part is trivial (it is composed of equilibrium points) and hence it is only the transient part that is of interest. Moreover, this transient part is completely described, qualitatively, by any Lyapunov function for the given flow.

To understand what this means we can, for example, look at the *orbit diagram* (or Smale diagram) of the flow. This is a diagram with $n+1$ levels, corresponding to the possible index of each equilibrium point (the dimension of its unstable manifold) and with a pointed arrow connecting two equilibria if there is an orbit of the flow whose alpha limit point is one of the equilibria and the omega limit point the other. In a sense, the orbit diagram captures all the essential global dynamical features of the flow. Now any Lyapunov function can be used to obtain an orbit diagram by studying its level sets, for example. A more detailed presentation of the relation between flows and their Lyapunov functions is given by the Conley index theory (see Conley (1978) and Franzosa (1989)).

We make these notions more precise in the sections that follow. We begin in Sect. 12.2 by setting up the problem of design of feedback dynamics. In the presentation that follows, we opt for informal definitions to make the content of the chapter more readable; simple examples are given to motivate some of the definitions. Our aim throughout is to convince the reader that this novel approach is worth considering, even though some of the technical background may be unfamiliar to control theorists.

12.2 Nonlinear Systems and Control Dynamics

We shall consider control systems that are affine in the control. This means that, in the traditional control description, we have systems of the type

$$\dot{x} = f(x) + \sum_{i=1}^{m} u_i g_i(x) \tag{12.2}$$

with each $u_i \in \mathbb{R}$.

We give a coordinate-free description using the concept of a *distribution*. The control system state space will be the manifold M^n. (We shall use the term 'locally' to mean 'in a neighbourhood of the given point'.) The state dynamics are then given by a smooth vector field $X_0 \in \mathcal{X}(M^n)$ (the space of smooth vector fields on M^n). In this chapter, 'smooth' means C^∞. A (control) distribution D is an assignment of a subspace $D(p)$ of the tangent space T_pM^n to every point p of M^n that is smooth, in the following sense. It is possible to pick, locally, a basis $X_1(p), \ldots, X_m(p)$ of $D(p)$ for p in an open set U of M^n such that the (locally defined) vector fields X_i, $i = 1, \ldots, m$ are smooth. The rank of the distribution is then m and is constant in each component of the manifold M^n. It is also possible—and relevant to control theory—to define distributions of non-constant rank. Starting with a distribution D, we can consider, at each point p, the Lie algebra generated by the vector fields X_i. The subspace spanned by the resulting vectors based at p yield another subspace of T_pM^n, call it $\tilde{D}(p)$. The distribution obtained, $\tilde{D}$, is not, in general, of constant rank.

A distribution is involutive if, for any vector fields X, Y, locally in D, their Lie bracket $[X, Y]$ is also in D. A rank m distribution is integrable if, for any $p \in M^n$, there exists locally near p a submanifold $F(p)$ of dimension m such that $D(q) = T_qF(p) \subset T_pM^n$ for q near p. An involutive distribution is integrable. If the rank of a distribution D is locally constant, the Frobenius Theorem (see Abraham, Marsden and Ratiu (1983), p.260) gives the converse result: an integrable distribution must be involutive. A foliation $\mathcal{F}$ of dimension m of the manifold M^n is a decomposition of M^n into submanifolds of dimension m (the leaves of the foliation) that is smooth. This means that, locally, it is possible to choose an integrable distribution D of rank m such that the leaves of $\mathcal{F}$ are the integral manifolds of D. Conversely, an integrable distribution D defines a foliation $\mathcal{F}_D$.

Generalizations to the case of distributions of non-constant rank have been obtained by Sussmann (1977). In this case, the dimension of the leaves may vary and the manifold is stratified by the leaves of different dimensions.

Foliations of dimension one are the easiest to understand. They correspond to smooth direction fields or, equivalently, to everywhere nonzero vector fields. The fundamental existence theorem for solutions to smooth ordinary differential equations guarantees the global existence of a foliation for any direction field.

At the other extreme, we have foliations of codimension one (i.e. of dimension $n-1$). These correspond to non-zero, closed one-forms on M^n. In local coordinates, the foliation is defined by the Pfaffian equation

$$\omega = \omega_i(x)\mathrm{d}x_i = 0 \tag{12.3}$$

where ω is a 1-form such that $\mathrm{d}\omega = 0$ (we are using the convention of summing over repeated indices). This is because, locally (in a simply connected open set), ω closed means that we can write $\omega = \mathrm{d}h$, with h a smooth function on M^n. The leaves of the foliation are then the level sets of h. It is of course not possible to define the function h globally, unless $H^1(M^n)$ is trivial; in other words every closed 1-form is exact.

The relation between foliations of dimension one and of codimension one is crucial in dynamical systems theory, in general, and in control theory in particular. It lies effectively at the core of the geometrical approach presented here. Roughly speaking, this relation relies on the fact that a function in the state space (a 'Lyapunov' function) captures all the essential topological aspects of a dynamical system. Thus, the study of control dynamics will be reduced, in our approach, to a study of a class of functions, the Morse-Lyapunov functions. We start by giving some fundamental results on dynamical systems defined on a manifold M^n.

A smooth dynamical system will in this chapter be considered to be equivalent to a complete vector field X on M^n, which in turn is equivalent to the globally defined flow $\phi : M^n \times \mathbb{R} \to M^n$, $(p,t) \mapsto \phi(p,t)$ (these terms will be used interchangeably). With E representing the set of equilibrium points of X, consider the foliation of $M^n \setminus E$ by the orbits of X. A Lyapunov function V defined in an open subset $S \subset M^n$ is a smooth function such that $\mathrm{d}V(X)(p) < 0$ for all $p \in S$. In the traditional control terminology V is a strict Lyapunov function in S. It is obviously not reasonable to expect an arbitrary dynamical system X to admit a Lyapunov function V in the set $S = M^n \setminus E$ (V must be constant on a limit cycle of X, for example). What is indeed remarkable is that, provided we exclude a set containing in some sense all the recurring behaviour of the flow (generalizing the concept of limit cycles), Lyapunov functions exist for all dynamical systems. This is a theorem of Conley (1978) and it asserts the existence of a Lyapunov function on the quotient flow obtained by collapsing (topologically) each connected component of the chain-recurrent set of X to a point. Roughly speaking, this theorem (whose precise technical meaning is not important for our purposes) means that Lyapunov functions exist, at least in the part of the state space where the flow is transient. (The theorem can in fact be used to give a definition of the transient and asymptotic, or chain-recurrent parts of a flow.) The manifold M^n is assumed compact. For the definition of chain-recurrence, see for example Guckenheimer and Holmes (1983), p. 236. The chaotic behaviour of dissipative systems, to give an illustration, takes place on a strange attractor, which provides an example of a connected component of the chain-recurrent set.

The class of dynamics that will be considered here is simpler. It is precicely the class of vector fields that admit global Lyapunov functions in the set $M^n \setminus E$. These vector fields are called *gradient-like*. They are still a very useful category for control purposes. The asymptotic behaviour of a gradient-like system consists of asymptotic attractors, repellers and saddle points. Because a (strict) Lyapunov function is assumed to exist, this implies there cannot be any homoclinic connections. More generally, it implies that the set of equilibrium points is partially ordered by the Lyapunov function: if there is an orbit having the equilibrium point e_i as its α-limit set and the equilibrium point e_j as its ω-limit set, then the index of e_i must be less than the index of e_j (since the Lyapunov function decreases along the orbit). We write $e_i \prec e_j$. The orbit diagram of the gradient flow is the graph of this partial order; in other words, its vertices are the equilibrium points of the flow and a directed edge connects e_i to e_j iff $e_i \prec e_j$. Thus, heteroclinic connections are also precluded, unless they are consistent with the above dimensional argument. Gradient-like systems therefore have structural stability built-in, as it were.

We turn next to some topological remarks pertaining to gradient-like systems. Let us start, for simplicity, with the case $M^n = \mathbb{R}^n$ (with M^n compact). If the set $E = \{e_1, \ldots, e_N\}$, then $M^n \setminus E$ has the homotopy type of a union of N copies of the sphere S^{n-1} attached together by smooth maps obtained by the gradient flow. We write

$$M^n \setminus E = S_1^{n-1} \cup_{f_1} S_2^{n-1} \cup_{f_2} \cdots \cup_{f_{N-1}} S_N^{n-1} \tag{12.4}$$

where the attaching maps $f_i : S_i^{n-1} \to \cup_{j=1}^N S_j^{n-1}$ map the boundaries of disjoint isolating blocks for the equilibrium points and are given by the (gradient) flow, except for points of 'external tangency' that are to be mapped to their positive time intersection with the latter set. (An isolating block is an isolating neighbourhood that has no 'internal tangencies', see Conley (1978).) For the case of the flow on the two-sphere with one attractor and one repeller, one easily computes that $S^2 \setminus E \simeq S^1$.

A more familiar and useful decomposition of M^n is provided by Morse theory. Assuming that the Lyapunov function yielding the gradient-like flow is in fact a Morse function for the manifold M^n, its main result is that M^n has the homotopy of a cell complex, with one cell of dimension k for each equilibrium point of index k. Moreover, the manifold can be constructed by gluing a cell of dimension k for every equilibrium of index k, starting with an n-disk (a zero cell, from the point of view of homotopy type) for each attractor. The attaching maps are obtained using the gradient flow of the Morse function. It is the Lyapunov function itself that provides this information. It was in fact the realization that the gradient flow of the Morse function can be used to yield a cobordism between the level sets of the Morse function that led Smale to the proof of the h-cobordism theorem and the proof of the Poincaré conjecture in dimension five or more. (How this rather simpe idea yielded this and so many other new results is explained in the fascinating article by Bott (1988), which reviews the various striking developments in Morse theory.)

For a gradient-like system the open set $M^n \setminus E$ is foliated in two ways: the one-dimensional foliation by the orbits of the flow and the codimension one foliation by the level sets of the Lyapunov function, which we shall call Lyapunov surfaces. These foliations are in some sense dual to each other; they are transverse, in that at every point of $M^n \setminus E$, the orbit through it is transverse to the level set of the Lyapunov function. Consider the two flows on M^n: the first being the given flow and the second the gradient flow of a Lyapunov function for this flow. It is easily checked that these flows are topologically equivalent. Thus, the Lyapunov function captures the qualitative behaviour of the dynamics completely. One aspect which is missed by this description of the dynamics through a single function on the state space is the fact that the linearization of the gradient-like system at an equilibrium point may have eigenvalues with non-zero imaginary parts, whereas the linearization of the gradient flow is forced to have real eigenvalues (since it is a symmetric matrix).

A gradient-like system can be thought of as a memory system (the different stable equilibria representing the possible memory states), a context that is relevant to neural networks, for example. A large class of dynamics met in areas as diverse as nonlinear circuit theory, robotics and power systems are gradient-like, at least for a large subset of some parameter space. Besides, this class is in some sense the simplest category of systems that are truly nonlinear (linear systems have unique equilibria, if we demand that equilibria be isolated). We have thus chosen to study this class here, the generalization to other classes being perhaps possible, but cumbersome. We leave it for later study.

In the next section we describe a way of translating a control specification into dynamical terms using Lyapunov functions.

12.3 Morse Specifications

The aim of control is to alter the dynamical behaviour of a given system in some desirable way through the choice of appropriate control action. The most prevalent form of control action in practice is feedback control. We shall deal mainly with smooth state feedback controls.

The requirement that the system dynamics be altered in a desirable way can take one of two forms, which we shall label topological and geometrical; they correspond roughly to the distinction between qualitative and quantitative. An example of a topological specification is that we require the controlled system to be stable, or even that it follows asymptotically a given reference trajectory. A geometrical specification, on the other hand, deals with the local geometry of the controlled dynamics, for example when we want to have some control over the speed of response or the placement of eigenvalues. The above distinction, of course, only mirrors the distinction between differential geometry and differential topology. Both kinds are important, although it is perhaps fair to say that topological specifications are more fundamental; after all, well established linear techniques, such as the optimal LQ method yield only a topological category of system, namely a stable one. The fine tuning to obtain desirable geometric behaviour (loop shaping) comes later and is more ad hoc. We shall deal in this work only with the topological aspects of control system design.

For us, then, a control problem takes the following form: A control system (X_0, D) is given, where X_0 is the vector field of the state dynamics and D is the control distribution, assumed to be of constant rank m. A smooth feedback control X_u on the set $S \subset M^n$ is a smooth vector field defined in D over the set S, i.e. a section of the distribution D, $X_u : S \to D$ such that $\pi \circ X_u = id$, the identity on S, where $\pi : D \to S$ is the natural projection map. Thus, for all p, $X_u(p) \in D(p)$. If the vector fields $X_1, X_2, \ldots, X_m$ are a local basis for D in some subset of S, then a smooth feedback control corresponds to a choice of smooth functions $u_i : S \to \mathbb{R}$, $i = 1, \ldots, m$ such that $X_u(p) = \sum_{i=1}^{m} u_i(p) X_i(p)$. If the distribution admits a global basis (i.e. is trivial as a vector bundle), then the functions u_i are also globally defined.

The purpose of control is to find a smooth feedback control X_u such that the controlled dynamics $X_0 + X_u$ have desirable topological properties. Examples of control specifications are that we require the controlled dynamics to be globally asymptotically stable or that we require the system to be bistable, in other words that there are two asymptotic attractors whose regions of attraction are separated by the stable manifold of a saddle equilibrium of index one.

A more general specification may involve several equilibrium points of specified stability type arranged in state space in some way that makes sense from the control point of view. This vague description needs to be made precise; this we now proceed to do through the concept of a Morse specification. In the paragraphs that follow, we consider the manifold M^n ignoring the fact that there is a special vector field and a special distribution defined on it.

Definition 12.1 *A* **Morse specification** $\mathcal{M}$ *on the manifold* M^n *consists of:*

(i) a finite set of distinct points $E = \{e_1, e_2, \ldots, e_N\}$ *of* M^n *(the 'equilibrium points' of the desirable dynamics)*

(ii) integers k_i *such that* $0 \le k_i \le n$ *for* $i = 1, \ldots, N$ *(the 'indices' of the 'equilibrium points')*

(iii) a Morse function F *on* M^n *such that the points* e_i *of* E *are the only critical points of* F *and the Morse index of each* e_i *is exactly* k_i*.*

The Morse function F may be called the 'validating function' of the Morse specification. It ensures that the configuration of 'equilibrium points' specified in parts (i) and (ii) is realizable on the manifold M^n. It is of course possible to avoid requiring the existence of such an F and to give alternative, more direct topological conditions guaranteeing the consistency of the Morse specification. To give an idea of the constraints imposed by the topology of M^n on the Morse specification, a (gradient-like) vector field on a sphere S^2 must have at least two equilibrium points. This result is a special case of the Morse inequalities, which relate for any vector field the number of equilibrium points of given index to the dimension of certain free modules associated with the state space manifold (which is assumed to be compact, without boundary). Explicitly, if c_i is the number of nondegenerate equilibrium points of index i and $b_i = \dim H_i(M^n, \mathbb{R})$ is the ith Betti number (the dimension of the ith homology group), then the following inequalities hold (Bott 1988)

$$c_0 \geq b_0 \tag{12.5}$$

$$c_1 - c_0 \geq b_1 - b_0 \tag{12.6}$$

$$\dots\dots \tag{12.7}$$

$$c_{n-1} - c_{n-2} + \ldots + (-1)^{n-2} c_0 \geq b_{n-1} - b_{n-2} + \ldots + (-1)^{n-2} b_0 \tag{12.8}$$

and

$$\chi(M^n) = c_0 - c_1 + \ldots + (-1)^n c_n = b_0 - b_1 + \ldots + (-1)^n b_n \tag{12.9}$$

where $\chi(M^n)$ is the Euler characteristic of M^n. These imply in particular that for each i, the number of equilibrium points of index i is at least equal to the corresponding Betti number, i.e. $c_i \geq b_i$.

The origin of the name Morse specification comes from the Conley index theory. In this theory (Conley 1978), one meets the concept of a Morse decomposition of an isolated invariant set of a flow. In the case of a gradient-like system on a compact manifold, an example of a Morse decomposition is the set of equilibrium points. The defining property is that every other orbit must tend, in both positive and negative time, to a member of the decomposition. A less trivial Morse decomposition is obtained if we take as elements two or more equilibria together with the connecting (unstable) manifolds; for example, one component may consist of an attractor, a saddle of index one and the component of the unstable manifold of the saddle that is part of the stable manifold of the attractor. Now a Morse specification exists without reference to any vector field; it is a blueprint for constructing dynamics that share certain characteristics. The validating Lyapunov function serves the purpose of fixing an orbit diagram that relates the 'equilibrium points' of parts (i) and (ii) of the definition.

If more detail is required in the specification of the desirable dynamics, the following condition can be added to conditions (i) and (ii) of the definition of a Morse specification

(iv) *optional* for each e_i such that $k_i \neq 0, n$, we are given subspaces $E^u(e_i)$ and $E^s(e_i)$ of the tangent space $T_{e_i}M^n$ of dimension k_i and $n - k_i$ respectively (the 'unstable and stable eigenspaces' of the 'saddle equilibrium' at e_i).

We want to consider the gradient vector field of a function h defined on the manifold M^n. For this, we need to fix a Riemannian metric g on the tangent space TM^n. It establishes an isomorphism between the tangent and cotangent spaces of M^n, TM^n and T^*M^n, given by $T_pM^n \ni X \mapsto g_p(X, .) \in T_p^*M^n$. The one-form $\mathrm{d}h$, the 'derivative' of h, is then mapped to a vector field, ∇h, called the gradient vector field of h.

Definition 12.2 *Given a Morse specification $\mathcal{M}$, the class of* **Morse-Lyapunov functions** *$\mathcal{F}(\mathcal{M})$ is the class of all Morse functions on M^n such that the corresponding gradient flows are equivalent to the gradient flow of the validating Morse function F of part (iii) of the definition of $\mathcal{M}$ and such that their equilibrium sets are precicely as in parts (i) and (ii) of the definition.*

The notion of equivalence used in the above definition is of course that of topological orbital equivalence, see for example (Arnol'd 1983), Chapter 2. By definition, $\mathcal{F}(\mathcal{M})$ is nonempty. In fact, this class is very rich; for example, given a single member h of $\mathcal{F}(\mathcal{M})$, one obtains a collection of other members of the class, one for each function $T : M^n \setminus E \to \mathbf{R}$ as follows: $h_T(p) = h(\phi(p, T(p)))$, where ϕ is the gradient flow of h (see Kappos (1986) for a discussion). There exist other, more topological ways of obtaining members of the Morse-Lyapunov class, but we shall not study them here.

The fundamental question for control theory (from our topological viewpoint) is: is it possible to select a feedback control X_u such that the controlled vector field $X_0 + X_u$ has a member of the class $\mathcal{F}(\mathcal{M})$ as a Lyapunov function? We call this the problem of **smooth controllability relative to the Morse specification** $\mathcal{M}$. If the answer is yes, then, by the above discussion, we have achieved dynamics in the specified class. We shall mainly use smooth feedback controls, even though we shall have an occasion to mention piecewise smooth ones.

In the following section, we examine this control problem and study some of the fundamental geometrical questions that arise.

12.4 Obstructions to Smooth Controllability

The question we posed in the last section can be examined at two levels, the local and the global. The *local level* is concerned with the control directions available near any given point to accomplish smooth control through the level sets of some member of $\mathcal{F}(\mathcal{M})$. The *global level* turns to the question of the existence of a Morse-Lyapunov function for which the smooth controllability problem has a solution. The local level thus essentially boils down to the local geometry of the control directions, while at the global level, the question is essentially a topological one.

We examine here the local question first, by giving the directions that are not available for control on the unit sphere. We then turn to the global problem, by examining obstructions to the existence of an appropriate Morse-Lyapunov function. We fix a control system (X_0, D) and a Morse specification $\mathcal{M}$, together with its class of Morse-Lyapunov functions $\mathcal{F}(\mathcal{M})$. We also assume we have fixed a Riemannian metric g.

12.4.1 Local Smooth Controllability

We consider the *(unit) sphere bundle* SM^n obtained from the tangent bundle TM^n by taking at each point $p \in M^n$ the subset of T_pM^n consisting of unit vectors (for the given metric g).

Definition 12.3 *For any $h \in \mathcal{F}(\mathcal{M})$, the* **Gauss map** *$G_h : M^n \setminus E \to SM^n$ is the map $p \mapsto \nabla h(p)/\|\nabla h(p)\|$ (where E is the set of critical points of h). For a closed, orientable hypersurface Σ of M^n, the* **Gauss map** *$G_\Sigma : \Sigma \to SM^n$ is defined by $p \mapsto n(p)$, where $n(p)$ is the unit outward normal to Σ at p (by 'outward', we mean the positive direction in a chosen orientation for Σ).*

Note that if $\Sigma = h^{-1}(c)$, the inverse image of a regular value c of the function h, then Σ is an orientable hypersurface of M^n.

Let us for the time being suppose that $X_0(p) \notin D(p)$ and let us call the affine subset $\mathcal{I}(p) = X_0(p) + D(p)$ of T_pM^n the **indicatrix** of the control system at p. If a smooth feedback X_u has been selected, the vector field $X_0 + X_u$ at p is then an element of $I(p)$. If h in $\mathcal{F}(\mathcal{M})$ is such that we have achieved smooth controllability, then $dh(X_0(p) + X_u(p)) < 0$, or $G_h(p) \cdot (X_0(p) + X_u(p)) < 0$. We would therefore like to study the set of directions (in S_pM^n) that are not available to the control action. The indicatrix $\mathcal{I}(p)$ is of dimension m and its image on S_pM^n is an open hemisphere $S^m_+(p)$; its boundary is the set of points 'at infinity'. To describe this hemisphere more explicitly, let $\pi_{D(p)}$ be the orthogonal projection onto $D(p)$ in T_pM^n. The vector $(I - \pi_{D(p)})X_0(p)$ is then orthogonal to $D(p)$. Let x_0 be the corresponding unit vector. Choose an orthonormal basis $\{x_1, \ldots, x_m\}$ for $D(p)$. Then if S^m is the unit sphere in the span of $x_0, x_1, \ldots, x_m$, S^m_+ is the hemisphere $\{u \in S^m \ ; \ x_0 > 0\}$.

Now any direction u in S_pM^n defines an open hemisphere $S^{n-1}_-(u)$ of dimension $n-1$ as the set of all directions v such that $u \cdot v < 0$. A direction u on the unit sphere is unavailable if $S^{n-1}_-(u)$ does not intersect S^m_+. It is now not difficult to see that the following lemma is true.

Lemma 12.4 *If $X_0(p) \notin D(p)$, the set of unavailable directions in S_pM^n at the point p is the $(n-m-1)$-dimensional closed hemisphere $S^{n-m-1}_+(p)$, where $S^{n-m-1}(p)$ is orthogonal to $D(p)$ and we take the hemisphere $\{u \in S^{n-m-1} \ ; \ u \cdot x_0 \geq 0\}$.*

If $X_0(p) \in D(p)$, then the set of unavailable directions is simply the sphere $S^{n-m-1}(p)$.

The local aspect of smooth controllability is now clear. We call the set $S_+^{n-m-1}(p)$ (or the set $S^{n-m-1}(p)$, if $X_0 \in D(p)$) the **local obstruction** to controllability. Remembering that we have assumed that the control distribution comes with a foliation (whose leaves are stacked, locally, like the leaves $Y_c = \{y \in \mathbb{R}^n \; ; \; y_1 = c\}$), we see that the local obstruction varies smoothly with p. The local obstruction can be bypassed so long as there is a member h of $\mathcal{F}(\mathcal{M})$ with $G_h(p) \notin S_+^{n-m-1}(p)$. This, even for $m = 1$, is generic (is true for the 'general' h), since the local obstruction is a 'thin set'. Thus, locally, we can bypass the obstruction by perturbing h slightly.

12.4.2 Global Obstructions

Consider the sphere bundle SM^n, together with a smooth assignment of a local obstruction (an $(n-m-1)$-dimensional (hemi)sphere on the sphere fibre), to every point p of M^n. The global obstruction bypassing problem is the following:

When does there exist a Morse-Lyapunov function h in the class $\mathcal{F}(\mathcal{M})$ such that G_h (considered as a section of the sphere bundle) bypasses the obstruction everywhere in $M^n \setminus E$?

This is equivalent to the smooth controllability problem posed at the end of Sect. 12.3. The data—the obstruction sets—are fixed by the control system (X_0, D). The Morse-Lyapunov class is the translation of the control aims into topological terms. Of its many elements, we just need to find one that succeeds in bypassing the obstruction to achieve the desirable dynmaical behaviour. This is then an existence problem that is far from easy. One of the complicating factors, from the topological point of view, is that the obstruction is, for points where $X_0(p) \notin D(p)$, a hemisphere and not a sphere.

It is of course easier to give necessary conditions for the existence of such sections rather than sufficient ones. Examples of necessary conditions for stabilization are given in Bacciotti (1991). We shall give a generalization of these results.

We first generalize our notion of a Morse specification to include attractors more general than equilibrium points. A compact, connected subset $K \subset M^n$ is an **attractor** for the vector field X if:

(i) it is an isolated invariant set for the flow of X. (An invariant set is one consisting of a union of complete orbits; it is isolated if it is the maximal invariant set in a neighborhood of itself)

(ii) it is stable (for any neighbourhood U of K, we can find a neighbourhood V such that all trajectories starting in V stay in U for all positive time)

(iii) nearby trajectories tend to K as $t \to \infty$.

Using a result of Conley, one can find Lyapunov functions for X that are constant on K. Let us now reverse our point of view: let us start with the set K and require that we find dynamics for which it is an attractor. We shall take $M^n = \mathbb{R}^n$ here. By the above, we know the resulting dynamics will admit Lyapunov functions. By property (iii), we know that there must be a Lyapunov surface (a level set of the Lyapunov function) contained in a small neighbourhood of K. This is a compact, orientable submanifold of $\mathbb{R}^n$ of codimension one (it is closed, orientable and of dimension $n-1$ because it is the inverse image of a real function on $\mathbb{R}^n$; by the above remark, it is bounded). The vector field we are seeking must be transverse to K. Just as in the simple index theory for planar systems (see, for example, Vidyasagar (1980)), this imposes limitations on the vector field. Explicitly, we have the following

Theorem 12.5 *Let K be a compact, connected set in $\mathbb{R}^n$. Suppose K is required to be an attractor of some vector field X. Let Σ be a compact Lyapunov surface as above, with h the corresponding Lyapunov function. Then the Gauss map $G_\Sigma : \Sigma \to S^{n-1}$ is onto.*

Proof. Choose a direction $u \in S^{n-1}$. We shall use the identification of $\mathbb{R}^n$ with $T_p\mathbb{R}^n$, for any p. Pick vectors to complete an orthonormal basis $\{u, e_2, \ldots, e_n\}$ of $\mathbb{R}^n$. The submanifold Σ is embedded in $\mathbb{R}^n$. Consider $\pi_u : \mathbb{R}^n \to \mathbb{R}$, $p \mapsto u \cdot p$, the function giving the first coordinate in the above basis. It is a smooth function.

Lemma 12.6 *The function $\pi_u|_\Sigma$ is smooth.*

Proof. Since Σ is a submanifold, let $\iota : \Sigma \to \mathbb{R}^n$ be the injection map for Σ (it is smooth). Then $\pi_u|_\Sigma = \pi_u \circ \iota$ and the lemma follows. □

Since Σ is compact, the function $\pi_u|_\Sigma$ achieves its maximum and minimum at the points p_M and p_m of Σ. Since the function is smooth, its gradient vanishes at these two points. Thus

$$\nabla(\pi_u \circ \iota)(p_M) = \nabla(\pi_u \circ \iota)(p_m) = 0. \tag{12.10}$$

But

$$\nabla(\pi_u \circ \iota)(p) = \nabla\pi_u(\iota(p))\nabla\iota(p) \tag{12.11}$$

and $\nabla\pi_u(p) = u$, and so u is orthogonal to every vector of $T_{p_M}\Sigma$ (and of $T_{p_m}\Sigma$), since these vectors are given by $\nabla\iota(p_M)v$, $v \in \mathbb{R}^{n-1}$. Thus u is orthogonal to $\Sigma(p_M)$ and hence u is the unit outward normal at p_M. Thus $u = G_\Sigma(p_M)$. □

Theorem 12.7 *Let the control system (X_0, D) be given in $\mathbb{R}^n$. If the set of indicatrices $\mathcal{I}(p)$, $p \in \mathbb{R}^n$, mapped by the Gauss map to the unit sphere is not onto, then the control system is not smoothly stabilizable.*

There are some important special cases where sufficient conditions can be obtained. A major simplification is when the state space is actually $\mathbb{R}^n$ and the control distribution is constant. It is then possible to give sufficient conditions for achieving gradient-like dynamics. We do this in Sect. 12.5.1.

In the remainder of this section we give some definitions and present some genericity resuts that provide us with more detail about the difficulties in achieving controllability.

We first examine the controllability problem for the control distribution on its own.

Definition 12.8 *Let a control system* (X_0, D) *and a Morse function* h *be given. The* **singular set** $M^D(h)$ *is defined as the set of points* $p \in M^n$ *where* $dh(D(p)) = 0$.

Since we are free to choose any h in $\mathcal{F}(\mathcal{M})$, we assume from now on, when needed, that the chosen h is generic (i.e in some residual subset of $\mathcal{F}(\mathcal{M})$).

Theorem 12.9 *There is a residual set* K *in the Whitney product space* $\mathcal{X}(M^n) \times \cdots \times \mathcal{X}(M^n)$ *such that for a generic* h *and* $g \in K$, $M^D(h)$ *is a closed submanifold of* M^n *of dimension* $n - m$.

Proof. This is contained essentially in Kappos (1992b).

The above theorem is central to the search for appropriate nonlinear controllability conditions. The traditional approach, for example, has concentrated on the dimension of the Lie algebra generated by the control and state vector fields at all points of some subset of the state space. Theorem 12.9 tells us that, in general, at almost all points except a 'thin' (measure zero) subset, the controllability problem is trivial, since we can find at least one control vector field transverse to the level set of some Morse function. Thus, once we have specified our control aim and we have translated it into topological terms by selecting the class $\mathcal{F}(\mathcal{M})$, controllability need only be examined on the thin singular set. The problem, of course, is that h is not fixed but is only taken to be a member of the above class, so the question becomes one of finding conditions for the existence of an h that works. This is a topological question.

We first formalize this discussion in the form of a theorem.

Theorem 12.10 *Let* $\mathcal{M}$ *be a Morse specification. Let* E *be the set of equilibrium points of the Morse set and let* $\mathcal{F}(\mathcal{M})$ *be the corresponding class of Morse-Lyapunov functions. If there exists a function* h *in* $\mathcal{F}(\mathcal{M})$ *such that for all points* $p \in M^D(h)$, $dh(X_0(p)) < 0$, *then the control system* (X_0, D) *is smoothly controllable and there exists a smooth feedback control* $X_u : M^n \setminus E \to D$ *such that* h *is a strict Lyapunov for the dynamics* $X_0 + X_u$ *in the set* $M^n \setminus E$.

The content of this theorem is the assertion that smooth feedback controls exist, provided an appropriate h can be found.

Proof. The desired control can be obtained as in linearization theory (see Kappos (1992b)), using X_0 near the singular set and patching smoothly together.

In local coordinates $X_1, \dots, X_m$ for D, we have, away from the singular set, $(\mathcal{L}_{X_1}h, \dots, \mathcal{L}_{X_m}h) \neq 0$. We can use the smooth linearizing control

$$u_i = -\frac{1}{\mathcal{L}_{X_i}h}(\mathcal{L}_{X_0} + \alpha h) \tag{12.12}$$

whenever $\mathcal{L}_{X_i}h \neq 0$ and, using a partition of unity, patch these together to get a smooth feedback control in $M^n \setminus (E \cup N_\epsilon(M^D(h)))$, where$N_\epsilon(M^D(h)$ is a small tubular neighbourhood of the singular set.

Inside the neighborhood $N_{\epsilon'}(M^D(h))$, with $\epsilon' > \epsilon > 0$, we set the control equal to zero. Since, by assumption, $dh(X_0(p)) < 0$ on $M^D(h)$, this will be true in a sufficiently small neighborhood of $M^D(h)$ and thus X_0 is enough to give controllability with respect to h. Finally patch the vector fields in the two sets $M^n \setminus (E \cup N_\epsilon(M^D(h)))$ and $N_{\epsilon'}(M^D(h))$ together smoothly to get the desired vector field. □

The difficult problem is, of course, to determine when such an h exists. When the control system is linear and the control aim is the stabilization of the origin, the existence of an h is equivalent to the classical linear stabilizability condition on the matrices A and B (see Kappos (1992a)).

12.5 Some Special Cases

12.5.1 Constant Control Distribution

In the next three subsections, we work in $\mathbb{R}^n$ and we derive conditions for the existence of a function h satisfying the requirements of Theorem 12.10 assuming that the control distribution is constant.

12.5.1.1 Stabilization We shall consider the subset $C\mathcal{F}(\mathcal{M})$ of $\mathcal{F}(\mathcal{M})$ consisting of functions h that are convex in a neighborhood of the chosen 'attractor' e of $\mathcal{M}$. In a fixed neighborhood U of the attractor e, define the sets

$$\mathcal{O}_- = \{p \in U \ ; \ \pi_{D^\perp}(p) \cdot X_0(p) < 0\} \tag{12.13}$$

$$\mathcal{O}_0 = \{p \in U \ ; \ \pi_{D^\perp}(x) \cdot X_0(p) = 0\} \tag{12.14}$$

$$\mathcal{O}_+ = \{p \in U \ ; \ \pi_{D^\perp}(x) \cdot X_0(p) > 0\} \tag{12.15}$$

where $\pi_{D^\perp}$ is the projection onto the subspace orthogonal to D in $T_p\mathbb{R}^n$.

Theorem 12.11 *Suppose there is a function ϕ from a neighborhood of 0 in $D^\perp$ to $\mathbb{R}^n$ with $\phi(0) = e$ whose graph is contained in the set $\mathcal{O}_- \cup \{0\}$. Then* *there is a convex function h^* in $C\mathcal{F}(\mathcal{M})$ defined in a neighborhood of e relative to which the control system (X_0, D) is smoothly controllable.*

This theorem is found in Kappos (1992a), p. 427. It can be generalized to more general projections, giving us very much the same freedom as we have in the linear case in the choice of the state weighting matrix Q. Simple linear and nonlinear examples are given in the above reference. For the uncontrollable and non-stabilizable system

$$A = \begin{pmatrix} 1 & 0 \\ 1 & 1 \end{pmatrix}, \; b = \begin{pmatrix} 0 \\ 1 \end{pmatrix}, \tag{12.16}$$

for example, the set $\mathcal{O}_0$ is the p_2-axis and the set $\mathcal{O}_-$ is empty (since $\pi_{D^\perp}(p) \cdot X_0(p) = p_1^2 \geq 0$). Notice that if the sets $\mathcal{O}_0$ and $\mathcal{O}_-$ are both nonempty, $\mathcal{O}_0$ is on the boundary of $\mathcal{O}_-$ and it includes the sets where p or $f(x)$ are in the span of D. The strength of this result is, of course, that the condition of existence is independent of any h. The proof relies on the analysis of the local obstructions on the unit sphere studied before. Since we consider only convex Morse functions, the image by G^{-1}, the inverse Gauss map, of this sphere is a disk of dimension $n - m$ through e transverse to D. Note that the case of a 'repeller' can be handled similarly by looking at the set $\mathcal{O}_+$, instead of the set $\mathcal{O}_-$.

12.5.1.2 Achieving Saddle Dynamics Let s be a point of the Morse specification of index k, $0 < k < n$ (a 'k-saddle'). We shall also assume that the specification includes the stable and unstable eigenspaces of s, E^s and E^u. As for the case of index zero, we look at the set of directions on the Gauss sphere that are not available using the controls D and then map back with the Gauss map. Note that the Gauss map is onto near s, by degree considerations, for any h. We shall consider functions h in $\mathcal{F}(\mathcal{M})$ that, near s and after a change of coordinates, are quadratic of the form

$$h(p) = \frac{1}{2} p^T \begin{pmatrix} -P_1(p) & 0 \\ 0 & P_2(p) \end{pmatrix} p \tag{12.17}$$

where the square, symmetric matrices P_1 and P_2 are positive definite for all p of dimension $k \times k$ and $(n-k) \times (n-k)$, respectively (we have assumed $s = 0$). After a further change of coordinates, we can take each P_i diagonal, with positive entries. The set $G_h^{-1}(S^{n-m-1})$, where G_h is the Gauss map for the Morse function h, is an $(n-m)$-dimensional submanifold through s. The positioning of the eigenspaces E^s and E^u means that the Gauss sphere is divided into regions bounded by $G(E^s)$ and $G(E^u)$ (which we assume vary smoothly further away from s, as we move along the stable and unstable manifolds of the gradient flow of the selected h). When $n = 2$, and $m = 1$, it is a line L that we write as

$$L = L_+ \cup \{s\} \cup L_- \tag{12.18}$$

where L_+ is the component that corresponds to the point γ_+ and L_- to γ_-, where $\gamma_\pm$ are the two points in the orthogonal 0-sphere to D. A two-dimensional results will now be given ($k = 1$).

Theorem 12.12 *If it is possible to choose a line L as above, consistent with the spaces E^s and E^u such that L_+ lies in the set $\{p \ ; \ X_0(p) \cdot D_+^\perp > 0\}$ and similarly for L_- (with the inequality reversed), then the saddle controllability problem has a solution h^* in a neighborhood of s.*

12.5.1.3 Achieving Gradient Dynamics Let $\mathcal{M}$ be a Morse specification and $\mathcal{F}(\mathcal{M})$ the corresponding set of Morse-Lyapunov functions.

Let $S^{n-m-1} \subset S^{n-1}$ be as before, orthogonal to D. For $h \in \mathcal{F}(\mathcal{M})$, let G_h be the Gauss map in $\mathbb{R}^n$.

Theorem 12.13 *Suppose for all $p \in G_h^{-1}(S^{n-m-1})$ we have that $X_0(p) \cdot G(x) < 0$. Then the control system (X_0, D) is smoothly stabilizable relative to h.*

This result makes more precise the conditions for controllability for a given Morse function h. Except in relatively simple low-dimensional cases, it does not give any way of finding an appropriate h. Combined with the previous results, however, that do give us good candidate h's near equilibrium points, and provided we can expand these local h's far enough, it may be possible to come up with the global candidate h to satisfy Theorem 12.13. We proceed to give some conditions under which this is possible.

Theorem 12.14 *Let $e_1, \ldots, e_N$ be the points of the Morse specification and suppose we have found corresponding sets $U_1, \ldots, U_N$ and Morse functions h_j, $j = 1, \ldots, N$ such that*

(i) $\cup_{j=1}^N U_j = M^n$,

(ii) the set $\cup_{j=1}^N G_{h_j}^{-1}(S^{n-m-1})$ separates M^n into path-connected components in each of which either $\mathrm{d}h_j(X^*) < 0$ or > 0, where $X^* = \sum_{i=1}^m \alpha_i X_i$ is some combination of the control vector fields with the functions α_i defined in U_i $(1 \le i \le m)$ smooth.

Then the control system (X_0, D) is smoothly stabilizable relative to the Morse function $\sum_{j=1}^N h_j$.

12.5.2 Constant-Rank Control Distribution of Dimension n-1

The smooth controllability problem has been divided into two parts: finding first a subset where control can be used to sweep past the level sets of a member of $\mathcal{F}(\mathcal{M})$ and then using the state dynamics to flow through the remaining set. We have already mentioned (in Sect. 12.4) some genericity results pertaining to this separation. In this final section, we give a related result, this time one that does not arise from genericity considerations, but is a hard constraint imposed by the topology of the situation.

We have seen in Theorem 12.5 that the Gauss map is onto for any compact, connected attractor in $\mathbb{R}^n$. Suppose the control distribution D is of constant rank $n-1$ (we take the largest possible dimension for a nontrivial result, the case of lower dimension being an easy consequence of our theorem). Fix a Morse-Lyapunov function h and fix one of its level sets H_c. Define the set $N_c = \{p \in H_c \,;\, T_pH_c = D(p)\}$.

Theorem 12.15 *Suppose H_c is (strictly) convex. Then the set N_c is not empty for any h and any D, if n is odd.*

Proof. Since the codimension of D is one, we take D to be given by a smooth section α of the cotangent bundle $T^*\mathbb{R}^n$ (identified here with $T\mathbb{R}^n$ using the standard basis). We can then define the Gauss map for D by

$$G_D : \mathbb{R}^n \to S^{n-1} \ , \ p \mapsto \frac{\alpha(p)}{\|\alpha(p)\|} \tag{12.19}$$

Now since H_c is convex, the Gauss map on it, G_{H_c} is a diffeomorphism. Hence $G_D \circ G_{H_c}^{-1}$ is a smooth map from S^{n-1} to itself.

The set N_c will be empty if $G_{H_c}(p) \neq G_D(p)$ for all $p \in H_c$. This is equivalent to saying that a map from the sphere to itself has no fixed point and does not send any point to its antipode. But by a standard result (see Dugundji (1966)) this is not possible if $n-1$ is even. □

12.6 Conclusions

The main objective of this chapter has been to present a totally different approach to the controllability question for nonlinear systems. This approach, by first specifying an equivalence class of desirable control dynamics that we hope to achieve, makes the controllability problem easier to address. For most points in state space it is seen that the controllability problem is trivially verified (relative to some arbitrary Morse function). The remaining points belong to some obstructing set, whose topological features at least are frequently known. The way we solve the problem of bypassing the obstruction is, we believe, a natural generalization of the linear case, interpreted geometrically, and not as a condition involving Lie brackets of vector fields. It is hoped that, by understanding the topology of the problem better, it will be possible to derive existence conditions for Morse-Lyapunov functions that are more general than the ones derived here.

References

Abraham, R., Marsden, J.E., Ratiu, T. 1983, *Manifolds, Tensor Analysis and Applications*, Addison-Wesley, Reading, Massachusetts

Arnol'd, V.I. 1983, *Geometrical methods in the theory of ordinary differential equations,* Springer-Verlag, New York

Bacciotti, A. 1991, *Local stabilizability of nonlinear control systems,* World Scientific Publishers, Singapore

Banks, S. 1988, *Mathematical Theories of Nonlinear Systems,* Prentice-Hall, London

Bott, R. 1988, Morse theory indomitable, Institut des Hautes Etudes Scientifiques Publications Mathématiques, **68**

Byrnes, C.I., Isidori, A. 1991, Asymptotic stabilization of minimum phase systems. *IEEE Transactions on Automatic Control* **36**, 1228–1240

Conley, C. 1978, *Isolated Invariant Sets and the Morse Index*, American Mathematical Society CBMS Series, No.38

Coron, J.-M. 1990, A necessary condition for feedback stabilization. *Systems and Control Letters* **14**, 227–232

Dayawansa, W.P. 1992, Recent advances in the stabilization problem for low dimensional systems. *Proceedings of the IFAC Nonlinear Control Systems Design Symposium, Bordeaux*, 1–8

Dugundji, J. 1966, *Topology,* Allyn and Bacon, Boston

Franzosa, R.D. 1989, The connection matrix theory for Morse decompositions. *Transactions AMS* **311**, 561–592

Guckehneimer, J., Holmes, P. 1983, *Nonlinear oscillations, dynamical systems and bifurcations of vector fields,* Springer Applied Math. Sciences, Vol. 43, Springer, New York

Kappos, E. 1992a, Convex stabilization of nonlinear systems. *Proceedings of the IFAC Nonlinear Control System Design Symposium*, Bordeaux, 424–429

Kappos, E. 1992b, A global, geometrical linearization theory. *IMA Journal of Mathematical Control and Information* **9**, 1–21

Kappos, E. 1986, Large deviation theory for singular diffusions with dissipative drift. UCB/ERL Memo M86/86, University of California, Berkeley

Kawski, M. 1989, Stabilization of nonlinear systems in the plane. *Systems and Control Letters* **12**, 169–175

Krasnosel'skiĭ , M., Zabreiko, P. 1984, *Geometric methods of nonlinear analysis,* Springer-Verlag, Berlin

Salamon, D. 1985, Connected simple systems and the Conley index of isolated invariant sets. *Transactions AMS* **291**, 1–41

Sussmann, H.J. 1973, Orbits of families of vector fields and integrability of distributions. *Transactions AMS* **180**, 171–188

Vidyasagar, M. 1980, *Nonlinear Systems Analysis*, Prentice-Hall, Englewood Cliffs, New Jersey

13. Polytopic Coverings and Robust Stability Analysis via Lyapunov Quadratic Forms

Francesco Amato, Franco Garofalo and Luigi Glielmo

13.1 Introduction

The stability analysis of a linear system subject to uncertain time-varying parameters ranging in a prespecified bounding set, can be performed with the aid of Lyapunov quadratic forms by examining the sign-definiteness of a family of symmetric matrices associated with the so-called Lyapunov derivatives. Robust stability, which means stability ensured independently of the particular realization of the uncertainty, is guaranteed if we can prove the negative definiteness of the whole family.

In the past decade considerable research has been devoted to the problem of determining classes of parameter dependencies for which the stability analysis can be carried out testing a finite number of conditions (see Horisberger and Bélanger (1976), Boyd and Yang (1989), Corless and Da (1988), Yedavalli (1989) and Garofalo et al (1993)).

A general conclusion is that when the convex hull of the image of the dynamical matrix associated with the uncertain system is a *polytope* – this always happens when the matrix depends on the parameters in affine or multiaffine fashion and the parameters range in a hyperrectangle – then the negative definiteness of the set of the Lyapunov derivatives can be checked by performing a *finite number* of tests.

We recall here that a set $\mathcal{A} \subset \mathbb{R}^{n\times n}$ is a polytope if it can be written as

$$\mathcal{A} = \left\{ A \in \mathbb{R}^{n\times n} : A = \sum_{i=1}^{\mu} \lambda_i A_{(i)},\ \sum_{i=1}^{\mu} \lambda_i = 1,\ \lambda_i \geq 0,\ i = 1, \ldots, \mu \right\} \qquad (13.1)$$

We will denote by $\mathcal{A}^v \triangleq \{A_{(i)}, i = 1, \ldots, \mu\}$ the set of vertices of $\mathcal{A}$. The straight line connecting two vertices is said to be an "edge" of $\mathcal{A}$. A hyperrectangle is a particular polytope which generalizes the concept of rectangle to an arbitrary $(n \times n)$-dimensional space. Polytopes in the vector space $\mathbb{R}^n$ are defined in the same way.

When the convex hull of the image is not a polytope — generally this happens when the dependence is nonlinear — we have to resort to brute force gridding. However, this technique, as the following example shows, may fail.

Example 13.1 Consider the family of quadratic forms

$$W(x,p) = -x^T Q(p) x \tag{13.2}$$

where

$$Q(p) = \begin{pmatrix} (p_1-1)^2+1 & 2 \\ 2 & (p_2-1)^2+4 \end{pmatrix} \tag{13.3}$$

and $p = (p_1, p_2)^T \in [0,2]^2$. Since $q_{11}(p) > 0$ for all p, the negative definiteness test reduces to

$$\left((p_1-1)^2+1\right)\left((p_2-1)^2+4\right) - 4 > 0 \quad \forall p \tag{13.4}$$

Clearly the test is verified everywhere in the rectangle except at the point $p_1 = 1$, $p_2 = 1$. Use of brute force gridding will miss this "bad" point with probability one!

To avoid this kind of problem the solution we consider is that of *immersing* the image into a suitable polytope. In other words if $\mathcal{R} \subset \mathbb{R}^{n_p}$ is the parameters bounding set (which is assumed to be a hyperrectangle in the context of this chapter) and $A : \mathcal{R} \to \mathbb{R}^{n \times n}$ is the system matrix, we will show how to construct a polytope $\mathcal{A}$ such that $A(\mathcal{R}) \subseteq \mathcal{A}$. This problem was first considered in De Gaston (1985) and De Gaston and Safonov (1988) for the case of *multiaffine* dependence and then in Sideris and De Gaston (1986) and Sideris and Peña (1988) for the case of *multivariate* dependence. A more general treatment can be found in Garofalo et al (1993b).

More recently in Amato and Garofalo (1993) polytopic coverings techniques have been used to test the robust stability of time-varying systems subject to slowly-varying parameters.

Since immersion introduces conservatism, depending on how the polytope fits the original image, it is necessary to provide techniques which allow one to reduce this conservatism. As we shall see later, the procedures in this chapter solve the problem by introducing fictitious parameters which, on the other hand, increase the number of the vertices of the covering polytope. However, this seems to be a small disadvantage in comparison to the big advantage of testing only a finite number of conditions.

The chapter is organized as follows. In Sect. 13.2 we present some of the applications of polytopic coverings in the field of robust analysis. In Sect. 13.3 the main techniques available to perform these coverings are described. In Sect. 13.4 we consider a more general algorithm, which recovers the others as particular cases. This algorithm is quite general and covers many situations of practical interest. Finally in Section 13.5 some examples are provided.

13.2 Some Applications of Polytopic Coverings to the Robust Stability Problem

13.2.1 Systems Subject to Time-Varying Parameters

Consider a time-varying system of the form

$$\dot{x}(t) = A(p(t))x(t) \tag{13.5}$$

where $p : \mathbb{R}^+ \to \mathcal{R}$ is any Lebesgue measurable function, $A : \mathcal{R} \to \mathbb{R}^{n \times n}$ is continuous and $\mathcal{R} \subset \mathbb{R}^{n_p}$ is a hyperrectangle. Usually it is known that the matrix A is asymptotically stable for a "nominal" value of the parameters, say p_0. Hence, given any symmetric matrix $Q > 0$, there exists a unique symmetric matrix $P > 0$ which is a solution of the algebraic Lyapunov equation

$$A(p_0)^T P + PA(p_0) = -Q \tag{13.6}$$

Now consider the time-invariant Lyapunov quadratic form

$$V(x) = x^T Px \tag{13.7}$$

The derivative of (13.7) along the solutions of system (13.5) is

$$x^T \left[A(p)^T P + PA(p)\right] x$$

Hence a sufficient condition for the exponential stability of (13.5) under any admissible "realization" of the uncertain function p is that

$$L(A(p)) \triangleq -\left[A(p)^T P + PA(p)\right] > 0, \quad \forall p \in \mathcal{R} \tag{13.8}$$

But how does one check the sign-definiteness of a *family* of symmetric matrices? The initial observation is that *a bounded set of symmetric matrices is positive definite if and only if its convex hull is positive definite.* When the convex hull is a polytope, this means that the positive definiteness of the original set is equivalent to the positive definiteness of the vertex matrices of the polytope. As a consequence we can state the following result.

Theorem 13.2 *If the convex hull of $A(\mathcal{R})$ is a polytope, i.e. there exist matrices $A_{(i)}$, $i = 1, \ldots, \mu$, such that*

$$\text{Conv } A(\mathcal{R}) = \text{Conv } \{A_{(i)},\ i = 1, \ldots, \mu\},$$

then the set of matrices $L(A(\mathcal{R}))$ is positive definite iff matrices $L(A_{(i)})$, $i = 1, \ldots, \mu$, are positive definite.

There are a few situations in which Theorem 13.2 turns out to be useful. Obviously it is useful when $A(\cdot)$ is an affine mapping. In this situation the statement of Theorem 13.2 enables us to obtain again the stability robustness conditions found in Kiendl (1985). A more general condition is

$$A(\mathcal{R}) \subseteq \operatorname{Conv} A(\mathcal{R}^v) \tag{13.9}$$

since this obviously implies

$$\operatorname{Conv} A(\mathcal{R}) = \operatorname{Conv} A(\mathcal{R}^v) \tag{13.10}$$

For instance, this happens, as proven in Petersen (1988), when $A(\cdot)$ is *multiaffine*, i.e. affine with respect to each parameter. Hence, application of Theorem 13.2 provides the stability robustness conditions found in Horisberger and Bélanger (1976). The same situation is also found in Theorem 5.2 in Boyd and Yang (1989).

When the convex hull of $A(\mathcal{R})$ is not a polytope, we have to immerse the image into a polytope in order to apply Theorem 13.2.

13.2.2 Systems Subject to Slowly-Varying Parameters

Consider again system (13.5), but suppose we have extra information regarding the rate of variation of parameters, i.e. let $\dot{p}(t) \in \tilde{\mathcal{R}}$, where $\tilde{\mathcal{R}} \subset \mathbb{R}^{n_p}$ is a hyperrectangle centered at the origin. In this way the rate of variation of the i-th parameter is constrained to be bounded, i. e. $|\dot{p}_i(t)| \leq h_i$, $i = 1, \ldots, n_p$. Moreover suppose that $A(p)$ is Hurwitz for all $p \in \mathcal{R}$. In this case the use of a time-invariant Lyapunov function like (13.7) turns out to be conservative, because its derivative does not take into account the information on the rate of variation of parameters.

In Amato and Garofalo (1993) the idea of using *parameter dependent* Lyapunov functions is proposed. Suppose there exists a matrix valued function $P(\cdot) : \mathcal{R} \to \mathbb{R}^{n \times n}$ such that

$$P(p) > 0 \qquad \forall p \in \mathcal{R} \tag{13.11a}$$

$$P(\cdot)\ \text{has continuous first order partial derivatives on}\ \mathcal{R} \tag{13.11b}$$

The derivative along the trajectories of system (13.5) can be written in the form

$$\dot{V}(t,x) = x^T \left(A^T(p(t))P(p(t)) + P(p(t))A(p(t)) + \sum_{i=1}^{n_p} \frac{\partial P(p)}{\partial p_i} \dot{p}_i(t) \right) x \tag{13.12}$$

Since $\mathcal{R} \times \tilde{\mathcal{R}}$ is compact we can conclude that, if for all $(p,q) \in \mathcal{R} \times \tilde{\mathcal{R}}$

$$A^T(p)P(p) + P(p)A(p) + \sum_{i=1}^{n_p} \frac{\partial P(p)}{\partial p_i} q_i < 0 \tag{13.13}$$

the exponential stability of system (13.5) follows.

Since the system we are dealing with is Hurwitz on $\mathcal{R}$, the following choice of $P(\cdot)$ arises quite naturally

$$P(\cdot): p \rightarrow \text{the only positive definite solution of } A^T(p)P + PA(p) = -I. \tag{13.14}$$

In Amato and Garofalo (1993) it is shown that, with this choice of $P(\cdot)$, condition (13.13) leads to the following result.

Theorem 13.3 *System (13.5) is exponentially stable if*

$$\max_{(p,q)\in\mathcal{R}\times\tilde{\mathcal{R}}} \left\| \left[\sum_{i=1}^{n_p} \frac{\partial A}{\partial p_i} q_i\right] \oplus \left[\sum_{i=1}^{n_p} \frac{\partial A}{\partial p_i} q_i\right] \right\| < \frac{1}{\sqrt{n}} \min_{p\in\mathcal{R}} \underline{\sigma}^2(A(p) \oplus A(p)) \tag{13.15}$$

where $\oplus$ denotes the Kronecker sum (see Graham (1981)).

In the same paper is shown that if $\frac{\partial A}{\partial p}(\mathcal{R})$ is a polytope (this is guaranteed if $A(\mathcal{R})$ is a polytope), then the maximum of the left hand side in (13.15) is attained at one of the vertices of $\mathcal{R} \times \tilde{\mathcal{R}}$. In all other circumstances the calculation of the maximum cannot be reduced to a convex problem, hence the alternative to the gridding is the immersion of $\frac{\partial A}{\partial p}(\mathcal{R})$ into a suitable polytope.

Concerning the minimum of the right hand side, it is interesting to note that it is never possible to evaluate its value by checking just a finite number of points, even in the case of affine dependence on parameters. This is due to the fact that $\underline{\sigma}(\cdot)$ is not a convex operator. However, using a polytopic covering it is possible to give an estimate of the minimum. Consider the family of positive definite matrices:

$$\mathcal{V} \triangleq \left\{V \in \mathrm{IR}^{n^2\times n^2} : V = [A(p) \oplus A(p)]^T [A(p) \oplus A(p)], p \in \mathcal{R}\right\} \tag{13.16}$$

Suppose one covers the set $\mathcal{V}$ with a polytope of positive definite matrices

$$\mathcal{H} \triangleq \left\{H \in \mathrm{IR}^{n^2\times n^2} : H = \sum_{i=1}^{n_{\mathcal{H}}} \lambda_i H_{(i)}, \sum_{i=1}^{n_{\mathcal{H}}} \lambda_i = 1, \lambda_i \geq 0, H_{(i)} > 0, i = 1, \ldots, n_{\mathcal{H}}\right\} \tag{13.17}$$

i.e. suppose that $\mathcal{H}$ is such that

$$\mathcal{H} \supseteq \mathcal{V} \tag{13.18}$$

In Amato and Garofalo (1993) it is shown that

$$\min_{p\in\mathcal{R}} \underline{\sigma}^2(A(p) \oplus A(p)) \geq \min\left\{\underline{\sigma}(H_{(i)}), i = 1, \ldots, n_{\mathcal{H}}\right\} \tag{13.19}$$

Inequality (13.19) can be used to establish that a lower bound of $\underline{\sigma}^2(A(p)\oplus A(p))$ can be obtained by examining the minimum singular value of a finite number of matrices.

13.3 Polytopic Coverings: A Survey of the Existing Literature

The problem of covering the image of a function with a polytope was first considered in De Gaston (1985) and De Gaston and Safonov (1988). They dealt with the exact computation of the Multiloop Stability Margin (MSM), which generalizes to multivariable systems the classical concept of gain margin for single input-single output systems.

Consider a stable square transfer matrix $G(s) \in \mathbb{C}^{m\times m}$ and the closed-loop system obtained by interconnecting $G(s)$ with the matrix $k\Delta$, where $k \in [0, +\infty)$ and $\Delta = \text{diag}(\delta_1, \ldots, \delta_m)$.

Suppose $(\delta_1, \ldots, \delta_m)^T \in \mathcal{C} \triangleq [-1, 1]^m$. The closed-loop transfer matrix is $G(s)(I + k\Delta G(s))^{-1}$. For $k = 0$ the closed-loop is stable. As k varies it remains stable until an eigenvalue crosses the imaginary axis for some value of δ. This situation, for a given $\omega \in \mathbb{R}$, is equivalent to the existence of $\tilde{\delta} = (\tilde{\delta}_1, \ldots, \tilde{\delta}_m)^T \in \mathcal{C}$, $\tilde{\Delta} = \text{diag}(\tilde{\delta}_1, \ldots, \tilde{\delta}_m)$, such that

$$f_m(\tilde{\delta}, k) \triangleq \det(I + k\tilde{\Delta}G(j\omega)) = 0. \tag{13.20}$$

The MSM $k_m(\omega)$ is then defined, at each frequency, as the infimum of the values of k for which condition (13.20) holds, i.e.

$$k_m(\omega) = \inf\{k \in [0, +\infty) : \det(I + k\Delta G(j\omega)) = 0 \quad \text{for some } \Delta\} \tag{13.21}$$

The following obvious result holds

Fact 13.4 *For any fixed ω, $k < k_m(\omega)$ if and only if $0 \notin f_m(\mathcal{C}, k)$.*

Now effective algorithms exist to determine if the origin belongs to a given polygon in the complex plane (Anagnost et al (1988)). Thus, even if the image of $f_m(\,\cdot\,, k)$ is not a polygon, since f_m is not affine in the δ_i's, we can determine a polygon which includes it and then apply the above-mentioned algorithms, so as to check condition stated in Fact 13.4 and give an estimate of $k_m(\omega)$.

The key observation is that the function (13.20) is multiaffine on the hypercube $\mathcal{C}$. Under this hypothesis it is possible to apply to $f_m(\,\cdot\,, k)$ the following mapping theorem (Zadeh and Desoer (1963), p. 476).

Theorem 13.5 *Consider the multiaffine function $g : \mathcal{D} \to \mathbb{R}^l$, where $\mathcal{D} \subset \mathbb{R}^m$ is a hyperrectangle. Then $g(\mathcal{D}) \subseteq \text{Conv}\, g(\mathcal{D}^v)$.*

Hence if, for a given k, $0 \notin \text{Conv}\, f_m(\mathcal{C}^v, k)$, then k is a lower bound for k_m. Actually the procedure proposed in De Gaston and Safonov (1988) is based on the following auxiliary result.

Theorem 13.6 *If $k_2 > k_1$, then* $\text{Conv}\, f_m(\mathcal{C}^v, k_2) \supseteq \text{Conv}\, f_m(\mathcal{C}^v, k_1)$.

Therefore starting from $k = 0$ one can increase k until the convex hull of the image of $f_m(\cdot, k)$ includes 0 (as stated above there exist effective procedures to check if the origin belongs to a polygon in the complex plane). However, when this happens, the current value of k still represents a lower bound for k_m, because $0 \in \mathrm{Conv}\, f_m(C^v, k)$ does not imply $0 \in f_m(C, k)$. Hence a procedure to compute k_m exactly is suggested. It is based on an iterative algorithm which subdivides the original C into l hypercubes of smaller dimensions

$$C = \bigcup_{r=1}^{l} C_r \tag{13.22}$$

It is simple to show that for a given k

$$\bigcup_{r=1}^{l} \mathrm{Conv}\, f_m(C_r^v, k) \subseteq \mathrm{Conv}\, f_m(C^v, k) \tag{13.23}$$

and hence the estimate of k_m obtained using $\cup_{r=1}^{l} \mathrm{Conv}\, f_m(C_r^v, k)$, say $\tilde{k}_m^{(l)}$, is less conservative. In De Gaston and Safonov (1988) it is proved that as C is divided ever finer, the union of the convex hulls of the image of the sub-hypercubes converges to the true image of C, and therefore $\tilde{k}_m^{(l)}$ converges to k_m.

A more complicated situation has been analyzed in Sideris and Peña (1988) and Peña and Sideris (1988). They consider the situation in which the function $f_m(\cdot, k)$ depends in a multivariate fashion on δ

$$f_m(\delta, k) = \sum_{\alpha_1, \ldots, \alpha_m} f_{m\alpha_1, \ldots, \alpha_m}(k) \delta_1^{\alpha_1} \cdots \delta_m^{\alpha_m} \tag{13.24}$$

where $f_{m\alpha_1, \ldots, \alpha_m}(k) \in \mathbb{C}$, $\alpha_i \in \mathbb{N}_0$, $i = 1, \ldots, m$ and $\delta \in C$.

It is possible to reduce this problem to the multiaffine case by introducing some fictitious parameters. Let h_i the highest degree of δ_i. Introduce fictitious variables $\delta_{i1}, \ldots, \delta_{ih_i}$ such that

$$\hat{\delta} \triangleq (\delta_{11} \ldots \delta_{1h_1} \ldots \delta_{m1} \ldots \delta_{mh_m}) \in \hat{C} \triangleq [-1, 1]^{\hat{m}} \tag{13.25}$$

where $\hat{m} = \sum_{i=1}^{m} h_i$, and replace in (13.24) each $\delta_i^{\alpha_i}$ with the product $\delta_{i1}\delta_{i2} \cdots \delta_{i\alpha_i}$. In this way we obtain the multiaffine function defined over the hypercube $\hat{C}$

$$\hat{f}_m(\hat{\delta}, k) = \sum_{(i_{11}, \ldots, i_{1h_1}, \ldots, i_{m1}, \ldots, i_{mh_m})^T \in B^{\hat{m}}} \hat{f}_{m i_{11}, \ldots, i_{1h_1}, \ldots, i_{m1}, \ldots, i_{mh_m}}(k) \cdot \delta_{11}^{i_{11}} \cdots \delta_{1h_1}^{i_{1h_1}} \cdots \delta_{m1}^{i_{m1}} \cdots \delta_{mh_m}^{i_{mh_m}} \tag{13.26}$$

where $B \triangleq \{0, 1\}$. Obviously $\hat{f}_m(\hat{C}, k) \supseteq f_m(C, k)$. Again the conservatism related to the immersion can be eliminated by splitting $\hat{C}$ in sub-hypercubes.

The covering procedures so far presented were devoted to solve specific situations. The general problem of how to immerse the image of a function into a polytope has been addressed in Garofalo et al (1993b). The remaining part of this section will be devoted to describe this procedure.

We consider a vector-valued function (obviously, since $\mathbb{R}^{n \times n}$ is isomorphic to $\mathbb{R}^{n^2}$, the theory can be extended in a straightforward way to matrix-valued functions) $a : \mathcal{R} \to \mathbb{R}^n$, where $\mathcal{R} \subset \mathbb{R}^{n_p}$ is a hyperrectangle, under the following assumption.

Assumption 13.7 *There exist known affine functions $\underline{a}, \overline{a}$ such that for all $p \in \mathcal{R}$,*[1]

$$\underline{a}(p) \leq a(p) \leq \overline{a}(p) \tag{13.27}$$

Remember that a polytope in $\mathbb{R}^m$ has exactly 2^m vertices. The following algorithm constructs $2^n 2^{n_p}$ points in $\mathbb{R}^n$ (not necessarily distinct). In Theorem 13.9 it is shown that the convex hull of these points includes $a(\mathcal{R})$.

Algorithm 13.8 *The algorithm is composed of three steps.*

Step 1 Define the hyperrectangle $\mathcal{D} \triangleq \{\delta \in \mathbb{R}^n : \delta_i \in [0,1], i = 1,\ldots,n\}$, and the hyperrectangle $\Omega \triangleq \mathcal{R} \times \mathcal{D}$; compute the vertices $\omega_{(i)}$, $i = 1,2,\ldots,2^n 2^{n_p}$, of Ω;

Step 2 Construct the function

$$a_m(p,\delta) \triangleq (I - \text{diag}(\delta))\underline{a}(p) + \text{diag}(\delta)\overline{a}(p); \tag{13.28}$$

Step 3 Determine the points $a_{m(i)} \triangleq a_m(\omega_{(i)})$, $i = 1,2,\ldots,2^n 2^{n_p}$.

The following holds.

Theorem 13.9

$$a(\mathcal{R}) \subseteq \text{Conv}\,\{a_{m(i)},\ i = 1,2,\ldots,2^n 2^{n_p}\}.$$

Remark. It is interesting to point out that, due to the particular structure of the function (13.28), Algorithm 13.8 works also if we replace the hyperrectangle $\mathcal{R}$ with a polytope $\mathcal{P}$, as proved in Garofalo et al (1993b).

If the function a is continuous, then the affine functions $\underline{a}$ and $\overline{a}$ can be chosen to be constant, e.g. $\underline{a}_i(p) = \min_{p \in \mathcal{R}} a_i(p)$, $\overline{a}_i(p) = \max_{p \in \mathcal{R}} a_i(p)$, $i = 1,2,\ldots,n$. On the other hand it should be clear that the better the functions $\underline{a}, \overline{a}$ fit a, the less conservative the immersion will be.

[1] Inequalities must be intended component-wise.

Generally speaking, the determination of "good" functions $\underline{a}$, $\overline{a}$ is not straightforward and could itself require an optimization algorithm. It can be greatly simplified if *the mapping a is convex and differentiable on $\mathcal{R}$*. Indeed, under this hypothesis, given any point $p_* \in \mathcal{R}$, there exists a (unique) matrix $\Pi(p_*)$, whose rows are the gradients in p_* of the components of a, such that for all $p \in \mathcal{R}$ the following inequality holds (Rockafellar (1970), Theorem 25.1)

$$a(p) \geq a(p_*) + \Pi(p_*)(p - p_*) \tag{13.29}$$

Since $\mathcal{R}$ is a hyperrectangle, each point $p \in \mathcal{R}$ can be expressed as $p = \rho(\lambda)$, where ρ is the linear function $\rho(\lambda) \triangleq \sum_{i=1}^{\mu} \lambda_i p_{(i)}$, and λ belongs to the polytope

$$\mathcal{L} \triangleq \{\lambda \in \mathbb{R}^{2^{n_p}} : \sum_{i=1}^{2^{n_p}} \lambda_i = 1,\, \lambda_i \geq 0,\, i = 1, \ldots, 2^{n_p}\} \tag{13.30}$$

Note that $\mathcal{R} = \rho(\mathcal{L})$. Hence, we can define

$$\hat{a}(\lambda) \triangleq a(\rho(\lambda)), \tag{13.31}$$

$$\underline{a}(\lambda) \triangleq a(p_*) + \Pi(p_*)(\rho(\lambda) - p_*) \tag{13.32}$$

and inequality (13.29) becomes

$$\underline{a}(\lambda) \leq \hat{a}(\lambda) \tag{13.33}$$

On the other hand, from Jensen's inequality (Rockafellar (1970), Theorem 4.3), we have

$$\hat{a}(\lambda) \leq \sum_{i=1}^{2^{n_p}} \lambda_i a(p_{(i)}) \triangleq \overline{a}(\lambda). \tag{13.34}$$

Combining inequalities (13.33) and (13.34) we have

$$\underline{a}(\lambda) \leq \hat{a}(\lambda) \leq \overline{a}(\lambda) \tag{13.35}$$

for each $\lambda \in \mathcal{L}$. Note that both $\underline{a}$ and $\overline{a}$ are affine as required.

As already stated, the goodness of the fitting of the set $a(\mathcal{R})$ by the constructed polytope largely depends on the choice of the functions $\underline{a}$, $\overline{a}$. However this fitting can be improved by immersing the given set in the union of a number of hyperrectangles. In applications this usually results in reduced conservatism.

Let $\mathcal{T}$ be a covering of the hyperrectangle $\mathcal{R}$ into $k(\mathcal{T})$ hyperrectangles, i.e. $\mathcal{T} = \{\mathcal{R}_1, \mathcal{R}_2, \ldots, \mathcal{R}_{k(\mathcal{T})}\}$, $\cup_{r=1}^{k(\mathcal{T})} \mathcal{R}_r = \mathcal{R}$. For the sake of brevity we will call $\mathcal{T}$ a *rectangular* covering, and define $\mathbf{T}$ as the set of all rectangular coverings of $\mathcal{R}$. For *each* hyperrectangle $\mathcal{R}_r \in \mathcal{T}$ one can apply Algorithm 13.8, i.e. first determine affine functions $\underline{a}^{(r)}(\cdot), \overline{a}^{(r)}(\cdot)$ such that

$$\underline{a}^{(r)}(p) \leq a(p) \leq \overline{a}^{(r)}(p), \qquad \forall p \in \mathcal{R}_r \tag{13.36}$$

then construct the multiaffine functions $a_m^{(r)}(\cdot, \cdot)$ according to (13.28). In view of Theorem 13.9, one has

$$
\begin{aligned}
a(\mathcal{R}) = a(\cup_r \mathcal{R}_r) &\subseteq \bigcup_{r=1}^{k(\mathcal{T})} a_m^{(r)}(\mathcal{R}_r \times \mathcal{D}) \\
&\subseteq \bigcup_{r=1}^{k(\mathcal{T})} \text{Conv } a_m^{(r)}(\mathcal{R}_r \times \mathcal{D})
\end{aligned} \tag{13.37}
$$

where the terms in the last union are polytopes computable as illustrated in Algorithm 13.8. In the next theorem it is shown that the RHS of (13.37) approaches its LHS as the covering gets finer and finer, provided the functions $\underline{a}^{(r)}, \overline{a}^{(r)}$ are suitably chosen.

Let

$$
d(\mathcal{T}) \triangleq \max_{r=1,\ldots,k(\mathcal{T})} \text{diam}(\mathcal{R}_r) \tag{13.38}
$$

$$
\sigma(\mathcal{T}) \triangleq \max_{r=1,\ldots,k(\mathcal{T})} \max_{p_1,p_2\in\mathcal{R}_r} \left\| \overline{a}^{(r)}(p_1) - \underline{a}^{(r)}(p_2) \right\| \tag{13.39}
$$

where, for a given set B,

$$
\text{diam}(B) \triangleq \sup_{b_1,b_2\in B} \|b_1 - b_2\| \tag{13.40}
$$

We add the following condition.

Condition 13.10 *The functions $\underline{a}^{(r)}, \overline{a}^{(r)}$ are chosen so that there exists $M_\sigma > 0$ such that, for any rectangular covering $\mathcal{T}$ of $\mathcal{R}$, $\sigma(\mathcal{T}) \leq M_\sigma d(\mathcal{T})$.*

Theorem 13.11 *Let $\{\mathcal{T}_h\}_{h\in\mathbb{N}} \subset \mathbf{T}$ with $\lim_{h\to\infty} d(\mathcal{T}_h) = 0$. If Condition 13.10 is satisfied, then*

$$
a(\mathcal{R}) = \bigcap_{h\in\mathbb{N}} \bigcup_{r=1}^{k(h)} \text{Conv } a_m^{(hr)}(\mathcal{R}_{hr} \times \mathcal{D}). \tag{13.41}
$$

In (13.41) $\mathcal{R}_{hr}$ is the r-th hyperrectangle in $\mathcal{T}_h$; $a_m^{(hr)}$ is the function constructed on $\mathcal{R}_{hr}$ according to Algorithm 13.8.

Condition 13.10 has a non-conventional aspect. The following lemmas give sufficient conditions for its satisfaction.

Lemma 13.12 *Suppose $a(\cdot)$ is Lipschitz on $\mathcal{R}$. Then the (constant) functions $\overline{a}^{(hr)}, \underline{a}^{(hr)}$ whose components are $\overline{a}_i^{(hr)} \triangleq \max_{p\in\mathcal{R}_{hr}} a_i(p)$, $\underline{a}_i^{(hr)} \triangleq \min_{p\in\mathcal{R}_{hr}} a_i(p)$, for $i = 1, 2, \ldots, n$, satisfy Condition 13.10.*

Lemma 13.13 *Suppose $a(\cdot)$ is convex and* continuously *differentiable on $\mathcal{R}$. Then the functions $\overline{a}^{(hr)}, \underline{a}^{(hr)}$ constructed according to (13.32) and (13.34) satisfy Condition 13.10.*

13.4 A More General Algorithm

Suppose that the function $a(p)$ of the previous section can be written as

$$a(p) = \sum_{(i_1,\ldots,i_\nu)^T \in H \subseteq B^\nu} a_{i_1,\ldots,i_\nu} f_1(p)^{i_1} \ldots f_\nu(p)^{i_\nu} \tag{13.42}$$

where $a_{i_1,\ldots,i_\nu} \in \mathbb{R}^n$ and $B \triangleq \{0,1\}$.

In the sequel we will make the following hypothesis.

Assumption 13.14 *For all $j = 1,\ldots,\nu$ if f_j is not multiaffine, there exist multiaffine functions $\underline{f}_j(p)$, $\overline{f}_j(p)$ such that*

$$\underline{f}_j(p) \le f_j(p) \le \overline{f}_j(p), \quad \forall p \in \mathcal{R} \tag{13.43}$$

Let $\mu \le \nu$ the number of non-multiaffine functions in (13.42). The following algorithm constructs $2^{n_p+\mu+n_p(\nu-1)}$ points in $\mathbb{R}^n$. We will show in Theorem 13.16 that the convex hull of these points includes $a(\mathcal{R})$.

Algorithm 13.15 *The algorithm is composed of three steps.*

Step 1 Construct the function $\psi(p,\delta)$ obtained from (13.42) substituting for $f_j(p)$, $j = 1,\ldots,\nu$, the multiaffine function

$$f_j^m(p,\delta_j) \triangleq \begin{cases} (1-\delta_j)\underline{f}_j(p) + \delta_j \overline{f}_j(p) & \text{if } f_j \text{ is not multiaffine} \\ f_j(p) & \text{if } f_j \text{ is multiaffine} \end{cases} \tag{13.44}$$

where

$$\delta_j \in I_j \triangleq \begin{cases} [0,1] & \text{if } f_j \text{ is not multiaffine} \\ \{0\} & \text{if } f_j \text{ is multiaffine} \end{cases} \tag{13.45}$$

Let $\mathcal{D} \triangleq I_1 \times I_2 \times \cdots \times I_\nu$; hence $\psi(p,\delta)$ is defined over $\mathcal{R} \times \mathcal{D}$. Observe that $\mathcal{D}$ is a hyperrectangle with 2^μ vertices.

In Theorem 13.16 it is shown that $\psi(p,\delta)$ can be written

$$\begin{aligned} &\psi(p,\delta) \\ &= \sum_{\substack{(i_1,\ldots,i_{n_p\nu},i_{n_p\nu+1},\ldots,i_{(n_p+1)\nu})^T \in B^{(n_p+1)\nu} \\ i_1+\cdots+i_\nu=1 \\ \vdots \\ i_{n_p\nu+1-\nu}+\cdots+i_{n_p\nu}=1}} \psi_{i_1,\ldots,i_{(n_p+1)\nu}} (p_1)^{i_1} \cdots (p_1^\nu)^{i_\nu} \\ &\quad \cdot \quad \cdots \quad \cdot (p_{n_p})^{i_{n_p\nu+1-\nu}} \cdots (p_{n_p}^\nu)^{i_{n_p\nu}} (\delta_1)^{i_{n_p\nu+1}} \cdots (\delta_\nu)^{i_{(n_p+1)\nu}} \end{aligned} \tag{13.46}$$

where $\psi_{i_1,\ldots,i_{(n_p+1)\nu}} \in \mathbb{R}^n$;

Step 2 Observe that $\psi(p,\delta)$ has the same structure as $a(p)$ in (13.42), considering any term $p_j^{\alpha_j}$ or δ_i as a separate function; *moreover these functions are readily seen to satisfy Assumption 13.14. Observe that in this case the number of non-multiaffine functions is $n_p(\nu-1)$. Hence we can reapply Step 1 to ψ, defining the hyperrectangle $\mathcal{E} \triangleq J_1 \times \cdots \times J_{(n_p+1)\nu}$ where*

$$J_j \triangleq \begin{cases} \{0\} & \text{if } j = 1 + k\nu,\ k = 0,\dots,n_p-1 \\ \{0\} & \text{if } j \geq n_p\nu + 1 \\ [0,1] & \text{otherwise} \end{cases} \tag{13.47}$$

and the function $\phi(p,\delta,\epsilon)$, $\epsilon \in \mathcal{E}$. Observe that the hyperrectangle $\mathcal{E}$ has $2^{n_p(\nu-1)}$ vertices;

Step 3 Define the hyperrectangle $\Omega \triangleq \mathcal{R}\times\mathcal{D}\times\mathcal{E}$ and evaluate the points $\phi_{(i)} \triangleq \phi(\omega_{(i)})$, where $\omega_{(i)}$, $i = 1,\dots,2^{n_p}2^{\mu}2^{n_p(\nu-1)}$, are the vertices of Ω.

Remark. Also if not explicitly stated, it is obvious that, if f_j does not depend on the entire set of parameters $p_1,\dots,p_{n_p}$, the same will be the case for the bounding functions. In other words $\underline{f}_j$, $\overline{f}_j$ depend on the same parameters on which f_j depends.

Theorem 13.16

$$a(\mathcal{R}) \subseteq \operatorname{Conv}\{\phi_{(i)},\ i = 1,\dots,2^{n_p+\mu+n_p(\nu-1)}\}$$

Proof. In (13.42) replace $f_j(p)$ with $f_j^m(p,\delta_j)$ for $j = 1,\dots,\nu$. We obtain

$$\psi(p,\delta) = \sum_{(i_1,\dots,i_\nu)^T \in H \subseteq B^\nu} a_{i_1,\dots,i_\nu} f_1^m(p,\delta_1)^{i_1}\dots f_\nu^m(p,\delta_\nu)^{i_\nu} \tag{13.48}$$

where $(p,\delta) \in \mathcal{R}\times\mathcal{D}$. Now, from the expression of f_j^m we know that for all $p \in \mathcal{R}$ there exists $\delta_j \in [0,1]$ such that $f_j^m(p,\delta_j) = f_j(p)$, and hence for all $p \in \mathcal{R}$ there exists $(\delta_1,\dots,\delta_\nu)^T \in \mathcal{D}$ such that $\psi(p,\delta) = a(p)$. Therefore we can conclude that

$$a(\mathcal{R}) \subseteq \psi(\mathcal{R}\times\mathcal{D}). \tag{13.49}$$

Now in order to express explicitly in (13.48) the dependence on parameters, consider the product of two different factors, f_i^m, f_j^m. We have

$$\begin{aligned} f_i^m(p,\delta_i)f_j^m(p,\delta_j) &= (1-\delta_i)(1-\delta_j)\underline{f}_i(p)\underline{f}_j(p) \\ &+ (1-\delta_i)\delta_j\underline{f}_i(p)\overline{f}_j(p) \\ &+ \delta_i(1-\delta_j)\overline{f}_i(p)\underline{f}_j(p) \\ &+ \delta_i\delta_j\overline{f}_i(p)\overline{f}_j(p) \end{aligned} \tag{13.50}$$

Since $\underline{f}_i$, $\overline{f}_i$, $\underline{f}_j$, $\overline{f}_j$ are multiaffine, (13.50) will contain terms like $p_l^2 p_m^2$, $p_l^2 p_m$ etc., while it is evident that in this expression terms like δ_i^2 cannot appear. When we consider all possible products, we obtain terms like $p_l^{\alpha_l} p_m^{\alpha_m} \cdots p_k^{\alpha_k}$ and, since we have at most ν products, $\alpha_l, \alpha_m, \alpha_k \leq \nu$. By virtue of the previous observation the dependence on the δ_i's will be multiaffine. Therefore we have

$$\psi(p,\delta) = \sum_{\alpha_1=0}^{\nu} \cdots \sum_{\alpha_{n_p}=0}^{\nu} \sum_{(i_{n_p\nu+1},\ldots,i_{(n_p+1)\nu})^T \in B^\nu} \psi_{\alpha_1,\ldots,\alpha_{n_p},i_{n_p\nu+1},\ldots,i_{(n_p+1)\nu}} \cdot p_1^{\alpha_1} \cdots p_{n_p}^{\alpha_{n_p}} \delta_1^{i_{n_p\nu+1}} \cdots \delta_\nu^{i_{(n_p+1)\nu}} \tag{13.51}$$

and hence the structure given in (13.46).

Now it is simple to recognize that when we apply the procedure described in Step 1 to $\psi(p,\delta)$, the resulting function $\phi(p,\delta,\epsilon)$ defined over $\mathcal{R}\times\mathcal{D}\times\mathcal{E}$ (where $\mathcal{E}$ has been defined in Step 2 of Algorithm 13.15), will be multiaffine. Indeed consider two factors in (13.46), say $(p_i^{\alpha_i})^{i_{(i-1)\nu+\alpha_i}}$ and $(p_j^{\alpha_j})^{i_{(j-1)\nu+\alpha_j}}$. The key point is that certainly $i \neq j$ because in any addendum of the summation (13.46) each component of p appears only once. The bounding functions will be affine respectively in p_i and p_j

$$\underline{f}_{(i-1)\nu+\alpha_i} = \underline{a}_i p_i + \underline{b}_i \quad \leq \quad p_i^{\alpha_i} \leq \overline{a}_i p_i + \overline{b}_i = \overline{f}_{(i-1)\nu+\alpha_i} \tag{13.52a}$$

$$\underline{f}_{(j-1)\nu+\alpha_j} = \underline{a}_j p_j + \underline{b}_j \quad \leq \quad p_j^{\alpha_j} \leq \overline{a}_j p_j + \overline{b}_j = \overline{f}_{(j-1)\nu+\alpha_j} \tag{13.52b}$$

When we consider the product between the multiaffine function covering $p_i^{\alpha_i}$,

$$f^m_{(i-1)\nu+\alpha_i} = (1-\epsilon_{(i-1)\nu+\alpha_i})\underline{f}_{(i-1)\nu+\alpha_i} + \epsilon_{(i-1)\nu+\alpha_i}\overline{f}_{(i-1)\nu+\alpha_i} \tag{13.53}$$

and that covering $p_j^{\alpha_j}$,

$$f^m_{(j-1)\nu+\alpha_j} = (1-\epsilon_{(j-1)\nu+\alpha_j})\underline{f}_{(j-1)\nu+\alpha_j} + \epsilon_{(j-1)\nu+\alpha_j}\overline{f}_{(j-1)\nu+\alpha_j} \tag{13.54}$$

the resulting function will present only terms such as $p_i p_j$, $\epsilon_{(i-1)\nu+\alpha_i}\epsilon_{(j-1)\nu+\alpha_j}$, etc.

Obviously the new function $\phi(p,\delta,\epsilon)$ is such that

$$\psi(\mathcal{R}\times\mathcal{D}) \subseteq \phi(\mathcal{R}\times\mathcal{D}\times\mathcal{E}) \tag{13.55}$$

Now, let $\Omega \triangleq \mathcal{R}\times\mathcal{D}\times\mathcal{E}$. Inclusion (13.55) together with (13.49) and the multiaffine nature of ϕ, which enables us to apply Theorem 13.5, yields

$$\begin{aligned} \operatorname{Conv}\phi(\omega_{(i)}) &\supseteq \phi(\Omega) \\ &\supseteq a(\mathcal{R}) \end{aligned} \tag{13.56}$$

Remark. Observe that Algorithm 13.15 yields again Algorithm 13.8 when $a(p)$ is defined over a hyperrectangle and bounding affine functions for the whole vector $a(p)$ exist. Indeed, suppose there exist affine functions $\underline{a}_i(p)$, $\overline{a}_i(p)$, $i = 1, \ldots, n$ such that $\underline{a}_i(p) \leq a_i(p) \leq \overline{a}_i(p)$, $i = 1, \ldots, n$. Let us write $a(p)$ in a form compatible with (13.42)

$$a(p) = \sum_{\substack{(i_1,\ldots,i_n)^T \in B^n \\ i_1+\cdots+i_n=1}} a_{i_1,\ldots,i_n} a_1^{i_1}(p) \cdots a_n^{i_n}(p) \tag{13.57}$$

where $a_{1,0,\ldots,0} = (1, 0, \ldots, 0)^T, \ldots, a_{0,0,\ldots,1} = (0, 0, \ldots, 1)^T$.

Now introducing the vector of fictitious parameters $(\delta_1, \ldots, \delta_n)^T$ and applying Algorithm 13.15, it should be clear that the resulting multiaffine function can be written as

$$\psi(p, \delta) = (I - \mathrm{diag}(\delta))\, \underline{a}(p) + \mathrm{diag}(\delta)\overline{a}(p) \tag{13.58}$$

which is exactly the covering function (13.28).

Obviously the class of functions (13.42) considered in Algorithm 13.15 is considerably larger.

Also in this case the goodness of the fitting of the set $a(\mathcal{R})$ by the constructed polytope can be improved by splitting the set $\mathcal{R}$ into smaller subdomains.

Consider again the rectangular covering $\mathcal{T}$ of the previous section. For each hyperrectangle $\mathcal{R}_r \in \mathcal{T}$ one can apply Algorithm 13.15, i.e. first determine affine functions $\underline{f}_j^{(r)}(\cdot), \overline{f}_j^{(r)}(\cdot)$, $j = 1, \ldots, \nu$ such that

$$\underline{f}_j^{(r)}(p) \leq f_j(p) \leq \overline{f}_j^{(r)}(p), \qquad \forall p \in \mathcal{R}_r \tag{13.59}$$

then construct the multiaffine functions $f_j^{m(r)}(\cdot, \cdot)$ according to (13.44) and then $\phi^{(r)}(\cdot, \cdot, \cdot)$. In view of Theorem 13.16, one has

$$\begin{aligned} a(\mathcal{R}) = a(\cup_r \mathcal{R}_r) \;&\subseteq\; \bigcup_{r=1}^{k(\mathcal{T})} \phi^{(r)}(\mathcal{R}_r \times \mathcal{D} \times \mathcal{E}) \\ &\subseteq\; \bigcup_{r=1}^{k(\mathcal{T})} \mathrm{Conv}\, \phi^{(r)}(\mathcal{R}_r \times \mathcal{D} \times \mathcal{E}) \end{aligned} \tag{13.60}$$

where the terms in the last union are polytopes computable as illustrated in Algorithm 13.15. In the next theorem we state that the RHS of (13.60) approaches its LHS as the covering gets finer and finer, provided the functions $\underline{f}_j^{(r)}, \overline{f}_j^{(r)}$ are suitably chosen.

Let

$$\sigma_j(\mathcal{T}) \triangleq \max_{r=1,\ldots,k(\mathcal{T})} \max_{p_1,p_2 \in \mathcal{R}_r} \left\| \overline{f}_j^{(r)}(p_1) - \underline{f}_j^{(r)}(p_2) \right\|, \; j = 1, \ldots, \nu \tag{13.61}$$

We add the following condition.

Condition 13.17 *The functions $\underline{f}_j^{(r)}, \overline{f}_j^{(r)}$, $j = 1, \ldots, \nu$ are chosen so that there exists $M_{j\sigma} > 0$ such that, for any rectangular covering $\mathcal{T}$ of $\mathcal{R}$, $\sigma_j(\mathcal{T}) \leq M_{j\sigma} d(\mathcal{T})$.*

The next theorem can be proved following the guidelines of Theorem 13.11.

Theorem 13.18 *Let $\{\mathcal{T}_h\}_{h \in \mathbb{N}} \subset \mathbf{T}$ with $\lim_{h\to\infty} d(\mathcal{T}_h) = 0$. If Condition 13.17 is satisfied, then*

$$a(\mathcal{R}) = \bigcap_{h \in \mathbb{N}} \bigcup_{r=1}^{k(h)} \operatorname{Conv} \phi^{(hr)}(\mathcal{R}_{hr} \times \mathcal{D} \times \mathcal{E}). \tag{13.62}$$

In (13.62) $\phi^{(hr)}$ is the function constructed on $\mathcal{R}_{hr}$ according to Algorithm 13.15.

13.5 Examples

13.5.1 Example 1

Consider the time-varying system (13.5) with

$$A(p) = \begin{pmatrix} -6 + p_1 \sin(p_2) + p_1 p_2 & 12 + p_1 \\ -0.5 + p_1 p_2 & -11 + e^{p_1} \cos(p_2) \end{pmatrix} \tag{13.63}$$

where $p(\cdot) = (p_1(\cdot), p_2(\cdot))^T$ is any Lebesgue measurable vector-valued function ranging in $\mathcal{R} \triangleq [-0.5, 0.5]^2$.

Our objective is to check the exponential stability of this system with respect to all admissible "realizations" of the parameters, according to the procedure detailed in section 13.2.1.

Let us choose as Lyapunov matrix the solution of the Lyapunov equation

$$A^T(p_0)P + PA(p_0) = -I \tag{13.64}$$

where $p_0 = (0, 0)^T$. We obtain

$$P = \begin{pmatrix} 0.0787 & 0.0554 \\ 0.0554 & 0.1165 \end{pmatrix} \tag{13.65}$$

Now observe that A is neither multiaffine nor convex. Therefore none of the algorithms discussed in section 13.3 can be applied. However $A(p)$ can be rewritten in the following form

$$\sum_{(i_1,i_2,i_3,i_4,i_5)^T \in B^5} A_{i_1,i_2,i_3,i_4,i_5} (p_1)^{i_1} (p_1 p_2)^{i_2} (\sin p_2)^{i_3} (\cos p_2)^{i_4} (e^{p_1})^{i_5} \tag{13.66}$$

where the non-zero coefficient matrices are

$$A_{0,0,0,0,0} = \begin{pmatrix} -6 & 12 \\ -0.5 & -11 \end{pmatrix} \qquad (13.67a)$$

$$A_{1,0,0,0,0} = \begin{pmatrix} 0 & 1 \\ 0 & 0 \end{pmatrix} \qquad (13.67b)$$

$$A_{0,1,0,0,0} = \begin{pmatrix} 1 & 0 \\ 1 & 0 \end{pmatrix} \qquad (13.67c)$$

$$A_{1,0,1,0,0} = \begin{pmatrix} 1 & 0 \\ 0 & 0 \end{pmatrix} \qquad (13.67d)$$

$$A_{0,0,0,1,1} = \begin{pmatrix} 0 & 0 \\ 0 & 1 \end{pmatrix} \qquad (13.67e)$$

We need to provide bounding functions for $f_3(p) = \sin p_2$, $f_4(p) = \cos p_2$ and $f_5(p) = e^{p_1}$. It is readily verified by graphical inspection that suitable functions are

$$\begin{aligned}
\underline{f}_3(p) &= 0.9689p_2 - 0.0052 && (13.68a)\\
\overline{f}_3(p) &= 0.9689p_2 + 0.0052 && (13.68b)\\
\underline{f}_4(p) &= 0.8776 && (13.68c)\\
\overline{f}_4(p) &= 1 && (13.68d)\\
\underline{f}_5(p) &= p_1 + 1 && (13.68e)\\
\overline{f}_5(p) &= 1.0422p_1 + 1.1276. && (13.68f)
\end{aligned}$$

Now according to Algorithm 13.15 construct the multiaffine functions

$$\begin{aligned}
f_3^m(p) &= (1-\delta_3)\underline{f}_3(p) + \delta_3\overline{f}_3(p) && (13.69a)\\
f_4^m(p) &= (1-\delta_4)\underline{f}_4(p) + \delta_4\overline{f}_4(p) && (13.69b)\\
f_5^m(p) &= (1-\delta_5)\underline{f}_5(p) + \delta_5\overline{f}_5(p) && (13.69c)
\end{aligned}$$

and define $\mathcal{D} \triangleq [0,1]^3$. It is readily seen that the matrix-valued function $\psi(p,\delta)$ obtained replacing $\sin p_2$, $\cos p_2$ and e^{p_1} with the functions $f_3^m(p)$, $f_4^m(p)$ and $f_5^m(p)$ respectively, is multiaffine. Now the multiaffine symmetric matrix-valued function

$$L(\psi(p,\delta)) \triangleq -(\psi(p,\delta)^T P + P\psi(p,\delta)) \qquad (13.70)$$

turns out to be *positive definite* on the $2^2 2^3$ vertices of $\mathcal{R} \times \mathcal{D}$. By virtue of Theorem 13.16 we can conclude that $L(A(p))$ (defined as in (13.8)) is positive definite on $\mathcal{R}$, and hence exponential stability of (13.63) follows.

13.5.2 Example 2

Consider system (13.5) where

$$A(p) = \begin{pmatrix} p_1p_2 - 10 & -p_1^2 \\ p_2^2 & -p_1p_2 - 10 \end{pmatrix} \tag{13.71}$$

and $\mathcal{R} \triangleq [0,2]^2$. Suppose that the parameters have a bounded rate of variation

$$\dot{p} \in \tilde{\mathcal{R}} \triangleq [-0.5, 0.5]^2 \tag{13.72}$$

The characteristic polynomial of $A(p)$ is $s^2 + 20s + 100$, hence $A(p)$ has eigenvalues in -10 for all $p \in \mathcal{R}$. However, exponential stability is not guaranteed since the uncertainties are time-varying.

Now, by application of (13.15), we will try to prove the exponential stability, using the information on the rate of variation of parameters. First observe that

$$\frac{\partial A}{\partial p_1} = \begin{pmatrix} p_2 & -2p_1 \\ 0 & -p_2 \end{pmatrix} \qquad \frac{\partial A}{\partial p_2} = \begin{pmatrix} p_1 & 0 \\ 2p_2 & -p_1 \end{pmatrix} \tag{13.73}$$

In this situation $\frac{\partial A}{\partial p}$ is linear; hence we can compute the maximum of the left hand side of (13.15) by only evaluating the norm at the vertices of $\mathcal{R} \times \tilde{\mathcal{R}}$. We obtain

$$\max_{(p,q)\in\mathcal{R}\times\tilde{\mathcal{R}}} \left\| \left[\sum_{i=1}^{2} \frac{\partial A(p)}{\partial p_i} q_i \right] \oplus \left[\sum_{i=1}^{2} \frac{\partial A(p)}{\partial p_i} q_i \right] \right\| = 6.47 \tag{13.74}$$

Now we have to evaluate the right hand side of (13.15). First note that

$$\begin{aligned}
&(A(p) \oplus A(p))^T (A(p) \oplus A(p)) \\
&= \left(\begin{matrix} 4(p_1p_2 - 10)^2 + 2p_2^4 & -2p_1^3p_2 + 20p_1^2 - 20p_2^2 \\ -2p_1^3p_2 + 20p_1^2 - 20p_2^2 & p_1^4 + p_2^4 + 400 \\ -2p_1^3p_2 + 20p_1^2 - 20p_2^2 & p_1^4 + p_2^4 \\ -2p_1^2p_2^2 & 20p_1^2 - 2p_1p_2^3 - 20p_2^2 \end{matrix} \right. \\
&\qquad \left. \begin{matrix} -2p_1^3p_2 + 20p_1^2 - 20p_2^2 & -2p_1^2p_2^2 \\ p_1^4 + p_2^4 & 20p_1^2 - 2p_1p_2^3 - 20p_2^2 \\ p_1^4 + p_2^4 + 400 & 20p_1^2 - 2p_1p_2^3 - 20p_2^2 \\ 20p_1^2 - 2p_1p_2^3 - 20p_2^2 & 2p_1^4 + 4(p_1p_2 + 10)^2. \end{matrix} \right) .
\end{aligned} \tag{13.75}$$

Equality (13.75) can be rewritten according to (13.42) in the following way

$$\begin{aligned}
&(A(p) \oplus A(p))^T (A(p) \oplus A(p)) \\
&= \sum_{(i_1,\ldots,i_8)\in B^8} A_{i_1,\ldots,i_8} (p_1)^{i_1} (p_1^2)^{i_2} (p_1^3)^{i_3} (p_1^4)^{i_4} (p_2)^{i_5} (p_2^2)^{i_6} (p_2^3)^{i_7} (p_2^4)^{i_8}
\end{aligned} \tag{13.76}$$

where $A_{i_1,\ldots,i_8}$ are suitable matrices.

Bounding functions for p_1^2, p_1^3 and p_1^4 are

$$0 \le p_1^2 \le 2p_1 \tag{13.77a}$$
$$0 \le p_1^3 \le 4p_1 \tag{13.77b}$$
$$0 \le p_1^4 \le 8p_1 \tag{13.77c}$$

The same can be repeated for p_2^2, p_2^3 and p_2^4. Applying Step 1 of Algorithm 13.15 to function (13.75), the resulting function $\psi(p,\delta)$ defined over $\Omega \triangleq \mathcal{R} \times [0,1]^6$ is multiaffine. Hence the polytope $\mathcal{H}$ covering the set $\mathcal{V}$ according to (13.16)-(13.18) coincides with the convex hull of the values of ψ evaluated at the vertices of Ω.
Applying (13.19) we obtain the following estimate

$$\frac{1}{\sqrt{2}}\underline{\sigma}^2(A(p) \oplus A(p)) \geq 6.64 > 6.47, \tag{13.78}$$

and hence the exponential stability of the system is proven.

13.6 Conclusions

In this chapter we have discussed the problem of immersing the image of a given function into a polytope. This has several applications in the field of robust stability analysis of linear systems subject to uncertain time-varying parameters. After a review of the existing literature we have proposed an algorithm which works under quite general assumptions.

Future research will be devoted to extending the class of functions for which the proposed polytopic coverings are applicable.

References

Amato, F., Celentano, G., Garofalo, F. 1992, Stability robustness bounds for linear systems subject to slowly-varying uncertainties. *Proc American Control Conference,* Chicago

Amato, F., Garofalo, F. 1993, A robust stability problem for linear systems subject to time-varying parameters. submitted for publication

Amato, F., Garofalo, F., Glielmo, L., Verde, L. 1993, An algorithm to cover the image of a function with a polytope: applications to robust stability problems. *12th IFAC World Congress*, Sydney, Australia

Anagnost, J. J., Desoer, C. A., Minnichelli, R. J. 1988, Graphical stability robustness tests for linear time-invariant systems: generalizations of Kharitonov's stability theorem. *Proc IEEE Conference on Decision and Control,* Austin, Texas

Barmish, B.R. 1988, New tools for robustness analysis. *Proc IEEE Conference on Decision and Control,* Austin, Texas

Bartlett, A. C., Hollot, C. V., Lin, H. 1988, Root locations of an entire polytope of polynomials: it suffices to check the edges. *Mathematics of Control, Signals, and Systems* **1**, 67–71

Boyd, S., Yang, Q. 1989, Structured and simultaneous Lyapunov functions for system stability problems. *International Journal of Control* **49**, 2215–2240

Corless, M., Da, D. 1988, New criteria for robust stability. *International Workshop on Robustness in Identification and Control*, Turin, Italy

De Gaston, R. R. E. 1985, *Nonconservative calculation of the multiloop stability margin*, Ph.D. Thesis, University of Southern California, California

De Gaston, R. R. E., Safonov, M. G. 1988, Exact calculation of the multiloop stability margin. *IEEE Transactions on Automatic Control* **AC-33**, 156–171

Dorato, P. 1987, *Robust control*, IEEE Press, New York

Dorato, P., Yedavalli, R.K. 1990, *Recent advances in robust control*, IEEE Press, New York

Garofalo, F., Celentano, G., Glielmo, L. 1993a, Stability robustness of interval matrices via Lyapunov quadratic forms. *IEEE Transactions on Automatic Control* **AC-38**, 281–284

Garofalo, F., Glielmo, L., Verde, L. 1993b, Positive definiteness of quadratic forms over polytopes: applications to the robust stability problem. submitted for publication

Graham, A. 1981, *Kronecker product and matrix calculus: with applications*, Ellis Horwood, Chichester

Horisberger, H. P., Bélanger, P. R. 1976, Regulators for linear time-invariant plants with uncertain parameters. *IEEE Transactions on Automatic Control* **AC-21**, 705-708

Kiendl, H. 1985, Totale Stabilitat von linearen regelungssystemen bei ungenau bekannten parametern der regelstrecke. *Automatisierungstechnik* **33**, 379–386

Peña, R. S. S., Sideris, A. 1988, A general program to compute the multivariable stability margin for systems with parametric uncertainty. *Proc American Control Conference*, Atlanta, Georgia

Petersen, I.R. 1988, A collection of results on the stability of families of polynomials with multilinear parameter dependence. *Technical Report EE8801*, University of New South Wales, Australia

Rockafellar, R. T. 1970, *Convex Analysis*, Princeton University Press, Princeton, New Jersey

Safonov, M. G. 1981, Stability margins of diagonally perturbed multivariables feedback systems. *Proc IEEE Conference on Decision and Control*, San Diego, California

Sideris, A. 1989, A polynomial time algorithm for checking the robust stability of a polytope of polynomials. *Proc American Control Conference*, Pittsburgh, Pennsylvania

Sideris, A., De Gaston, R. R. E. 1986 Multivariable stability margin calculation with uncertain correlated parameters. *Proc IEEE Conference on Decision and Control*, Athens, Greece

Sideris, A., Peña, R. S. S. 1988, Fast computation of the multivariable stability margin for real interrelated uncertain parameters. *Proc American Control Conference*, Atlanta, Georgia

Sideris, A., Peña, R. S. S. 1990, Robustness margin calculation with dynamic and real parametric uncertainties. *IEEE Transactions on Automatic Control* **AC-35**, 970–974

Yedavalli, R.K. 1985, Improved measures of stability robustness for linear state space models. *IEEE Transactions on Automatic Control* **AC-30**, 557–559

Yedavalli, R.K. 1989, On Measures of stability robustness for linear state space systems with real parameter perturbations: a perspective. *Proc American Control Conference,* Pittsburgh, Pennsylvania

Zadeh, L. A., Desoer, C. A. 1963, *Linear system theory*, McGraw-Hill, New York

Zhou, K., Khargonekar, P.P. 1987, Stability robustness bounds for linear state-space models with structured uncertainty. *IEEE Transactions on Automatic Control* **AC-32**, 621–623

14. Model-Following VSC Using an Input-Output Approach

Giorgio Bartolini and Antonella Ferrara

14.1 Introduction

Standard VSC techniques have been applied to uncertain systems described in input-output form when the output derivatives of order up to the relative degree of the system can be measured. The stability of the zero dynamics on the sliding manifold (Bartolini and Zolezzi 1988, Sira-Ramírez 1988) is assumed. The reason for this assumption relies on the fact that the *equivalent control* which is the control forcing the state trajectories starting on the sliding manifold to remain on it, depends algebraically on the output derivatives up to order equal to the relative degree. If the order of the available output derivatives is less than the relative degree, the standard procedure fails and more complex structures involving high gain observers should be designed. This topic is actually under investigation, as far as the general nonlinear case is concerned, when there is significant uncertainty in the system dynamics (Walcott and Żak 1987, Utkin 1992).

The linear time-invariant case has been solved, in an adaptive control context, by using dynamic regulators with output, i.e. the plant control signal consisting of a time-varying linear combination of the states of suitable time-invariant linear filters (Monopoli 1974, Landau 1979, Narendra, Lin and Valavani 1980, Narendra and Annaswamy 1989).

The substitution of the continuous adaptation mechanism by discontinuous control laws can be advantageously performed in order to improve robustness and time transient behaviours, as well as to counteract external disturbances (Hsu and Costa 1987, Bartolini and Zolezzi 1988, Hsu 1990, Tao and Ioannou 1991, Bartolini and Ferrara 1992b, Narendra and Boskovic 1992). In particular, when the relative degree of the plant is equal to one, under the assumption that the plant is minimum phase and the dynamic gain is known, even in presence of bounded disturbances, it is possible to use a discontinuous control law of the type

$$u = -\sum_{k=1}^{2h+1} |\theta_{M_k} x_k| \operatorname{sign} e - \Delta \operatorname{sign} e$$

where h and x_k are respectively the order and the states of the linear filters, θ_{M_k} are the components of a vector upperbounding the parameters of the control law in the known plant case (the *ideal control law*), Δ is a number bounding the disturbance, and e is the output error with respect to a reference model. As a result the finite-time convergence to zero of the ouput error can be guaranteed without adaptation of the controller parameters.

When the relative degree is greater than one, it is not possible to reduce to zero the output error, but only to assure the convergence to zero of the error (the so-called *augmented error signal*) obtained with respect to an auxiliary model (Monopoli 1974, Narendra, Lin and Valavani 1980).

In order to synthesize a discontinuous control law asymptotically equivalent to the ideal control Utkin (1978, 1992) avoiding the necessity of performing complex stability analysis, Hsu (1990) proposed some modifications to the control structure. These modifications consist of a cascade of high gain filters with control which is able to steer the augmented error. By means of a sequence of filtering and discontinuous tracking operations, it is possible to obtain a filtered signal which turns out to be asymptotically equivalent to the ideal control law.

In the present chapter the control problem is slightly modified in order to avoid the use of the augmented error signal and to extend the class of uncertain plants controllable by VSC techniques to include some non-minimum phase systems with unknown dynamic gain. More precisely, instead of referring to the classical augmented error signal control scheme, we consider the simplified adaptive controller presented by Bartolini and Ferrara (1991). Such a controller is characterized by the fact that some of the assumptions which are mandatory for the previous scheme, can be relaxed. In particular the complexity of the control scheme is independent of the relative degree of the plant to be controlled, the sign of the high frequency gain can be unknown, and the zeros of the plant, at least in some cases, can be located anywhere in the complex plane. Moreover, recent works have shown the possibility of extending the validity of such a scheme to the unmodelled dynamics case (Bartolini and Ferrara 1992a) and to a wider class of non-minimum phase plants (Bartolini and Ferrara 1992b). These improvements are obtained by building an auxiliary output variable through the introduction of a fixed relative-degree-one compensator in parallel to the plant. Indeed, the use of the parallel compensator allows us to divide the control problem into two separate problems: one aimed at the tracking of a suitable reference model by the auxiliary output variable, the other consisting of plant control via pole assignment. In this chapter the discontinuous control version of the foregoing scheme is considered, showing how adopting a discontinuous parameter adjustment mechanism coupled with a suitable identification algorithm, the control objective can be attained, while maintaining the structural simplicity of the adaptive scheme.

The structure of the chapter will be as follows In the next section some preliminary issues concerning the input/output approach are reported. In Sect. 14.3 the control problem is stated and the underlying linear structure of the proposed controller is introduced. In Sect. 14.4 the discontinuous model-following mechanism is described, while in Sect. 14.5 the solution to the pole assignment problem by means of discontinuous identification is discussed. Finally in Sect. 14.8 a some illustrative examples are provided to complement the theoretical issues.

14.2 Some Preliminary Issues

In the literature the adaptive model reference control of uncertain linear systems has been studied first considering plants with available states, and, in the sequel, extending the results to the more general case of plants in input/output form (Landau 1979, Narendra and Annaswamy 1989). As far as the case of plants with available states is concerned, the adaptive model-following approach can be briefly summarized as follows.

Consider a plant described in state variable form as

$$\dot{x}(t) = Ax(t) + Bu(t) \tag{14.1}$$

where $x(t) \in \mathbb{R}^n$, $u(t) \in \mathbb{R}^1$ for simplicity, and A, B are assumed to be unknown; and a reference model

$$\dot{x}_m(t) = A_m x_m(t) + B_m u_m(t) \tag{14.2}$$

where $x_m(t) \in \mathbb{R}^n$, $u_m(t) \in \mathbb{R}^1$.

The problem is that of choosing the control signal $u(t)$ so that the tracking error $e(t) := x_m(t) - x(t)$ is steered to zero asymptotically in spite of plant uncertainties.

The error equation associated with this objective can be written as

$$\dot{e}(t) = A_m e(t) + (A_m - A)\,x(t) + B_m u_m(t) - Bu(t) \tag{14.3}$$

When the plant parameters are assumed to be known, the control objective is attained, provided that the *matching condition*

$$\operatorname{rank}[B] = \operatorname{rank}[B \qquad A_m - A \qquad B_m] \tag{14.4}$$

is satisfied, and the control signal $u(t)$ is chosen as

$$\begin{aligned} u(t) &= B^{\dagger}\left[(A_m - A)\,x(t) + B_m u_m(t)\right] \\ &= \Theta^{*} X(t) \end{aligned} \tag{14.5}$$

where $B^{\dagger}$ denotes the pseudo-inverse of B, $X^T(t) := [x(t)\ u_m(t)]$ and Θ^{*} is a parameter vector.

When the plant parameters are unknown, the state of the tracking error model converges to zero if the following further conditions are fulfilled:

(i) The control signal $u(t)$ is chosen with the same structure as in the ideal known parameter case, i.e. $u(t) = \Theta^T(t)X(t)$, where $\Theta(t)$ is the time-varying vector of the parameters to be tuned.

(ii) An auxiliary output $e_1(t) = ce(t)$ is generated, so that the overall system satisfies the Kalman-Yacubovich Lemma, i.e.

$$A^T P + PA = -Q \tag{14.6}$$

$$B^T P = c \tag{14.7}$$

(iii) The parameter vector is adapted according to

$$\dot{\Theta}(t) = -\Gamma X(t) e_1(t)$$

where Γ is a suitable gain matrix.

When the plant state is available but the matching condition is violated, as long as it is possible to choose an auxiliary output $y_a(t) = Cx(t)$ such that, in spite of the uncertainties,

$$\det \begin{vmatrix} sI - A & -B \\ C & 0 \end{vmatrix}$$

is Hurwitz, then an input/output representation of the system is obtained which is given by an invertibly stable transfer function with relative degree equal to one. The problem becomes that of steering to zero the error between the auxiliary output signal and the output of a reference model described in terms of the transfer function $W_m(s)$. In practice the availability of the states only allows the generation of the auxiliary output, but it cannot be exploited in setting up the adaptive control mechanism Indeed, the overall system is equivalent to a system with inaccessible states, is minimum phase and has a relative-degree-one transfer function.

For the case of plants with unavailable states consider an input-output descriptionof the form

$$y(t) = \frac{N_p(s)}{D_p(s)} u(t) \tag{14.8}$$

where $s = d/dt$. When the relative degree of the plant is equal to one, the classical approach entails the introduction of two state variable filters placed at the input and at the output of the plant respectively

$$\begin{aligned} \dot{x}_1(t) &= \Lambda_f x_1(t) + b_f u(t) \\ y_1(t) &= h_1(t) x_1(t) \end{aligned} \tag{14.9}$$

$$\begin{aligned} \dot{x}_2(t) &= \Lambda_f x_2(t) + b_f y(t) \\ y_2(t) &= h_2(t) x_2(t) + h(t) y(t) \end{aligned} \tag{14.10}$$

where $x_1(t)$, $x_2(t) \in \mathbb{R}^{n-1}$, $b_f^T = [0 \ldots 1] \in \mathbb{R}^{n-1}$, $u(t)$ and $y(t) \in \mathbb{R}^1$, Λ_f is an $(n-1) \times (n-1)$ matrix in controllable companion form with the elements in the last row equal to the coefficients of the characteristic polynomial of the filters, named $D_f(s)$, which can be arbitrarily chosen. Finally, $h(t)$, $h_1(t)$, $h_2(t)$ are respectively a scalar value and row vectors (time-varying in the unknown parameter case) of the parameters to be adaptively adjusted.

As long as perfect knowledge of plant parameters is assumed, it is possible to determine a constant parameter vector $H^* := [h_1^* \; h_2^* \; h^* \; k^*]$, with h_1^*, h_2^* row vectors, and h^*, k^* scalar values, such that, using a control law of the type

$$u(t) = -H^* \zeta \tag{14.11}$$

where $\zeta^T := [x_1(t)\ x_2(t)\ y(t)\ r(t)]$, and $r(t)$ is a bounded reference input, the controlled plant tracks the output of a suitably chosen reference model

$$y_m(t) = \frac{N_m(s)}{D_m(s)} r(t) = W_m(s) r(t) \tag{14.12}$$

characterized by a strictly positive real (SPR) transfer function. The convergence of the tracking error $y(t) - y_m(t)$ to zero is assured provided that an adaptation mechanism of the type

$$\dot{H}^T(t) = -\Gamma \zeta \left(y(t) - y_m(t) \right) \tag{14.13}$$

where Γ is a suitable gain matrix, and $H(t) := [h_1(t)\ h_2(t)\ h(t)\ k(t)]$, with $k(t)$ a scalar value, is activated.

When the relative degree of the plant is greater than one, it is impossible to choose a reference model with SPR transfer function, but it is possible to assume the existence of an operator $L(s)$ such that $L(s)W_m(s)$ is SPR. According to Monopoli (1974) and Narendra, Lin and Valavani (1980), in this case the controller structure is modified by means of the introduction of an *augmented error signal*

$$e_a(t) = y(t) - y_m(t) + \frac{L(s)N_m(s)}{D_m(s)} \left[L^{-1}(s) H(t) \zeta - H(t) \xi \right] \tag{14.14}$$

where $\xi := L^{-1}(s)\zeta$, and the adaptation mechanism is chosen as

$$\dot{H}^T(t) = -\Gamma \xi e_a(t) \tag{14.15}$$

Recently a link between adaptive control theory and the theory of VSC systems has been established by many authors. The motivation for the use of discontinuous control in an adaptive framework is the possibility of introducing disturbance rejection capabilities and robustness with respect to unmodelled dynamics in the designed controllers.

Among the interesting proposals appeared in the literatures, one of the most significative is constituted by the approach presented by Hsu (1990). The basic idea underlying this approach can be summarized as follows (for the sake of simplicity, we limit ourselves to considering the case of known high frequency gain). The tracking error $y(t) - y_m(t)$ is modified by subtracting the output of an auxiliary model with SPR transfer function, namely $L(s)N_m(s)/D_m(s)$, having as input the difference between the control signal $u_0(t)$ and the plant input filtered by $L^{-1}(s)$. The new error signal, denoted by $e_0(t)$, is described by the following differential equation

$$e_0(t) = \frac{L(s)N_m(s)}{D_m(s)} \left[u_0(t) - H^* \xi \right] \tag{14.16}$$

where the control signal $u_0(t)$ is discontinuous and can be expressed as

$$u_0(t) = -\left| \bar{H} \xi \right| \operatorname{sign} e_0(t) \tag{14.17}$$

with $\bar{H}_i > |H_i^*|$. $e_0(t)$ is exponentially equivalent to $H^*\xi = H^*L^{-1}(s)\zeta$ on the sliding manifold $e_0(t) = 0$.

The signal $H^*\xi$ is assumed to be available at the output of a suitable high-gain filter $H^*\xi \approx u_0'(t) = F(s)u_0(t)$. It remains to evaluate the ideal control signal which is required to act on the plant in order to fulfil the model tracking objective $H^*\zeta$. This can be accomplished by means of the following procedure. Let us assume that $L(s) = \prod_1^N L_i(s)$, with $L_i(s) = (s + a_i)$, and $N = N^* - 1$, N^* being the relative degree. Then

$$\dot{y}_i(t) = -a_i y_i(t) + u_i(t) \; , \qquad e_i = y_i - u_{i-1}'(t) \tag{14.18}$$

$$\xi_i = (s + a_i)\,\xi_{i-1} \; , \qquad \xi_0 = \xi \tag{14.19}$$

$$u_i(t) = -\left|\bar{H}\xi_i\right| \operatorname{sign} e_i(t) \tag{14.20}$$

$$u_i'(t) = F(s)u_i(t) \tag{14.21}$$

and $u_N'(t)$ is equivalent to the ideal control law $H^*\zeta$. Then, the application of $u_N'(t)$ at the input of the plant allows us to solve the control problem in question.

The approach presented in this chapter differs from the above described procedure since it modifies the control problem so that the model tracking is always accomplished by a relative-degree-one system independently of the relative degree of the plant, thereby avoiding the necessity of using the recursive filtering-discontinuous tracking procedure just described. However, to achieve this aim, it is necessary to reduce the control objective from true model tracking to pole assignment, as far as the true plant output is concerned, which also requires the introduction of an indirect adaptation (identification) phase in the averall VSC structure, as will be detailed in the remainder of the chapter.

14.3 The Underlying Linear Structure of the Controller

Let us consider an unknown linear time-invariant single input-single output plant with inaccessible states

$$y_p(t) = \frac{B(s)}{A(s)} u_p(t) = W_p(s)u_p(t) \tag{14.22}$$

where $s = \mathrm{d}/\mathrm{d}t$ in operational notation, $B(s)$ and $A(s)$ are polynomials of degree m and n respectively ($n - m$ is the relative degree), with $A(s)$ monic, and $r(t)$ is a suitable bounded reference input.

The control problem consists of the determination of a control law $u_p(t)$ such that

$$y_p(t) = \frac{B(s)}{A_m(s)} r(t) \tag{14.23}$$

where the roots of polynomial $A_m(s)$ are the poles to be assigned to the plant ($\deg(A_m(s)) = n$, $A_m(s)$ monic), i.e. a pole assignment control problem is to be solved. It should be noted that this control objective cannot be represented in terms of tracking of an arbitrary model since the zeros of the transfer function in (14.24) coincide with those of the unknown plant transfer function $W_p(s)$ and, in general, could be located anywhere in the complex plane. However, we shall prove that the solution to the pole assignment control problem can be reduced to the combined solution to an explicit model tracking problem and to a design problem relevant to the setting up of a suitable feedforward compensator.

Assuming perfect knowledge of the plant parameters, it is possible to design the so-called underlying linear structure of the controller, i.e. the control scheme which would solve the control problem under the assumption that the plant is perfectly known. To this end, we first place in parallel to the plant the fixed first order compensator

$$y_c(t) = \frac{k}{s+a} u_p(t) \tag{14.24}$$

obtaining an overall system (which in the sequel will be called the ***auxiliary plant***) described by

$$y_a(t) = y_c(t) + y_p(t) \tag{14.25}$$

where

$$y_a(t) = \frac{k[A(s) + (1/k)B(s)(s+a)]}{A(s)(s+a)} u_p(t) \tag{14.26}$$

Note that, if $A(s)$ is Hurwitz (determined by root locus evidence), there exists a gain k^* such that, for any $k \in (k^*, \infty)$, the polynomial $[A(s)+(1/k)B(s)(s+a)]$ is Hurwitz. Then, under the assumptions: $(A.1)$ $A(s)$ Hurwitz polynomial, $(A.2)$ k^* known gain, the auxiliary plant described by (14.27) becomes a system with relative degree one, known high frequency gain k, and minimum phase transfer function, even though the original plant $W_p(s)$ had unknown relative degree (greater than one), unknown high frequency gain, and zeros arbitrarily located in the complex plane. When the relative degree of the plant is equal to one, all the above features remain unchanged apart from the knowledge of the high frequency gain. Indeed, in that case, the leading coefficient of the numerator of the auxiliary plant would be $k+b_0$, b_0 being the leading coefficient of $B(s)$. However, if we assume to know a priori some bounds on b_0, then we can choose $|k| > \max|b_0|$, so that the sign of the high frequency gain coincides with the sign of k and again can be arbitrarily fixed. Note that, for the sake of simplicity, in the sequel we assume that the relative degree of the plant is greater than one, since when the relative degree is equal to one, the use of the parallel compensator is only motivated by the possible necessity of making the numerator of the auxiliary plant Hurwitz. However, the case of non-minimum phase relative-degree-one plant can be satisfactorily dealt with by using the approach proposed by Bartolini and Zolezzi (1988).

With reference to the auxiliary plant, a simplified model tracking control scheme can be conceived. The control scheme we set up in order to solve this problem is presented in Fig. 14.1, where the signal $\nu(t)$ is the output error signal representing the difference between the model output, denoted by $y_m(t)$, and the auxiliary plant output $y_a(t)$. As in previous work concerning the adaptive version of this scheme, the controller structure is realized by means of a set of state variable filters, described by the following differential equations

$$\begin{aligned}
\dot{x}_{F_1}(t) &= \Lambda x_{F_1}(t) + b_f u_p(t) && (14.27)\\
y_{F_1}(t) &= F_1 x_{F_1}(t) && (14.28)\\
\dot{x}_{F_2}(t) &= \Lambda x_{F_2}(t) + b_f y_a(t) && (14.29)\\
y_{F_2}(t) &= F_2 x_{F_2}(t) + f_{20} y_a(t) && (14.30)\\
\dot{x}_{F_3}(t) &= \Lambda x_{F_3}(t) + b_f r(t) && (14.31)\\
y_{F_3}(t) &= F_3 x_{F_3}(t) + f_{30} r(t) && (14.32)
\end{aligned}$$

where $x_{F_1}(t)$, $x_{F_2}(t)$, $x_{F_3}(t) \in \mathbb{R}^n$, $b_f^T = [0 \ldots 1] \in \mathbb{R}^n$; Λ is a matrix of dimension $n \times n$ in controllable companion form with the elements in the last row equal to the coefficients of the characteristic polynomial of the filters $D(s)$, which can be arbitrarily chosen; f_{20} and f_{30} are scalar coefficients, so that the transfer functions of the filters can be expressed as

$$\begin{aligned}
&\frac{F_1(s)}{D(s)} && \text{Filter 1}\\
&\frac{F_2(s)}{D(s)} = f_{20} + \frac{F_2'(s)}{D(s)} && \text{Filter 2}\\
&\frac{F_3(s)}{D(s)} = f_{30} + \frac{F_3'(s)}{D(s)} && \text{Filter 3}
\end{aligned}$$

with $\deg(F_1(s)) = n-1$ and $\deg(F_2(s), F_3(s)) = n$ respectively, and $F_3(s)$ monic ($f_{30} = 1$), while $F_1(s)$, $F_2(s)$ is not monic. In (14.29)-(14.33) F_j, $j = 1, \ldots, 3$, are row vectors of dimension n containing the coefficients of the polynomials $F_1(s)$, $F_2'(s)$, $F_3'(s)$. Note that the coefficients of the numerators of the state variable filters are the unknowns of the problem we have to solve. Indeed, they have to be chosen so that the transfer function between $r(t)$ and $y_p(t)$ has poles coinciding with those of the polynomial $A_m(s)$ to be assigned. As anticipated, the scheme in Fig. 14.1 can be viewed as the cascade connection between a pre-filter (namely, $F_3(s)/D(s)$) and a parallel structure aimed at the fulfilment of an explicit model tracking. Then the signal $y_{F_3}(t)$ can be regarded as a *filtered reference input*. With reference to the scheme in Fig. 14.1, the following result can be proved.

Theorem 14.1 *Given the plant $W_p(s)$ in (14.23), and the controller structure (14.29)-(14.33), then there exists a unique control law, expressible as*

$$\begin{aligned}
u_p(t) &= F_3 x_{F_3}(t) + f_{30} r(t) - F_1 x_{F_1}(t) - F_2 x_{F_2}(t) - f_{20} y_a(t)\\
&= \Theta^T X(t) && (14.33)
\end{aligned}$$

where

$$\Theta^T := [F_3 \quad f_{30} \quad -F_1 \quad -F_2 \quad -f_{20}]$$

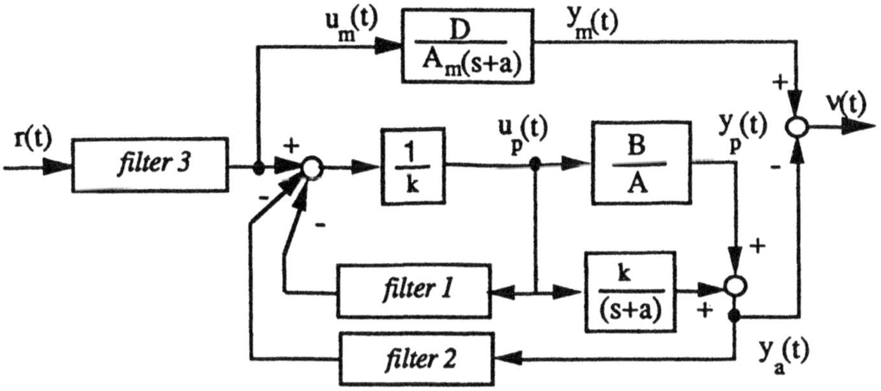

Fig. 14.1. The proposed control scheme

$$X^T(t) := \left[x_{F_3}^T(t) \;\; r(t) \;\; x_{F_1}^T(t) \;\; x_{F_2}^T(t) \;\; y_a(t)\right]$$

such that the control objective (14.24) is attained, while $y_a(t)$ *exactly tracks* $y_m(t)$, *i.e.*

$$y_a(t) = \frac{D(s)}{A_m(s)(s+a)} y_{F_3}(t) \tag{14.34}$$

Proof. Let us calculate the transfer function $T_1(s)$ between the reference input $r(t)$ and the plant output $y_p(t)$

$$T_1(s) = \frac{[B(s)(s+a)D(s)]\, F_3(s)}{P_1(s)D(s)} \tag{14.35}$$

where

$$P_1(s) = (s+a)A(s)\,[F_1(s) + kD(s)] + F_2(s)k\,[A(s) + (1/k)B(s)(s+a)] \tag{14.36}$$

The transfer function $T_2(s)$ between the signals $y_{F_3}(t)$ and $y_a(t)$ is

$$T_2(s) = \frac{\tilde{B}(s)D(s)}{P_1(s)} \tag{14.37}$$

where $\tilde{B}(s) := k\,[A(s) + (1/k)B(s)(s+a)]$ (Hurwitz, by assumption). Thus, the pole assignment control objective can be rewritten as

$$T_1(s) = \frac{B(s)}{A_m(s)} \tag{14.38}$$

while the solution to the tracking problem can be obtained by solving, for any polynomial $[F_1(s) + kD(s)]$, the equation

$$P_1(s) = \tilde{B}(s)A_m(s)(s+a) \tag{14.39}$$

Trivially

$$F_1(s) + kD(s) = \tilde{B}(s) \tag{14.40}$$

$$F_2(s) = (s+a)\,(A_m(s) - A(s)) \tag{14.41}$$

is the unique solution. Moreover, from (14.37), (14.40) and (14.41), it is apparent that the pole assignment requirement is met as long as

$$F_3(s) = \tilde{B}(s) \tag{14.42}$$

This concludes the proof. □

Then, with the choice of polynomials F_j, $j = 1, \ldots, 3$, indicated above, the underlying linear structure is completely determined. Note that if the arbitrary polynomial $D(s)$ (the denominator of the state variable filters) is chosen to be equal to $A_m(s)$ (which is known, since its roots are the poles to be assigned), then the explicit reference model turns out to be a first order strictly proper system, regardless of the plant order and relative degree.

When the plant is affected by parameter uncertainty, in order to fulfil the control objectives indicated in Theorem 14.1, it is necessary to design a control law $u_p(t)$ which solves both the tracking problem (14.36) (by adjusting the parameters of *Filter 1* and *Filter 2*) and the pole assignment problem (14.24) (suitably selecting the parameters of *Filter 3*). Thanks to the particular structure of the proposed scheme, the design of the two parts of the control law (feedback and feedforward) can be easily accomplished in sequence. In the next section we first study the tracking problem by setting up a discontinuous control strategy which assures the convergence to zero of the output error $\nu(t)$. In Sect. 14.5 the design of the feedforward compensator (*Filter 3*) is studied, outlining the necessity of the use of identification in order to determine the parameters which lead to the (asymptotic) solution of the pole assignment problem.

Notes and Comments. From Fig. 14.1 it is easy to see that, when the auxiliary plant exactly tracks its model, the parameters of *Filter 3* do not affect the output error $\nu(t)$, while they have a direct influence on the plant output $y_p(t)$. It is therefore impossible to adopt a tuning mechanism for the components of vector F_3 which is driven, as usual, by $\nu(t)$. The adjustment of F_3 can be performed only if its components are tied to those of one of the parameter vectors of the other filters. Fortunately, in our case there exists a precise relationship between polynomials $F_3(s)$ and $F_1(s)$, i.e. $F_1(s) + kD(s) = F_3(s)$, as indicated in the previous proof.

14.4 Discontinuous Parameter Adjustment Mechanisms

The control scheme presented in the previous section shows interesting features as long as the control law, designed to cope with uncertainties, is discontinuous on the sliding manifold $\nu(t) = 0$, as will be detailed in this section.

Let us keep on considering the plant (14.23), but assume that upper bounds on the modula of the coefficients of the transfer function $W_p(s)$ are known, that is

$$B(s) = b_0 s^m + b_1 s^{m-1} + \ldots + b_m \tag{14.43}$$

$$A(s) = s^n + a_1 s^{n-1} + \ldots + a_n \tag{14.44}$$

with $|b_i| < \beta_i$, for $i = 0, \ldots, m$, $|a_j| < \alpha_j$, for $j = 1, \ldots, n$, where β_i and α_j are known positive numbers. The problem consists of synthesizing a control law $u_p(t)$ such that, in spite of the uncertainties, the output error $\nu(t)$ is steered to zero. In order to solve this problem, the structure of the control law is maintained equal to the one used in the case of perfect knowledge of the plant parameters, i.e.

$$u_p(t) = \Theta^T(t) X(t) \tag{14.45}$$

with the difference that, this time, the parameters contained in $\Theta^T(t) := [\Theta_3(t)\ \theta_{30}(t)\ -\Theta_1(t)\ -\Theta_2(t)\ -\theta_{20}(t)]$, with $\Theta_3(t)$, $\Theta_1(t)$, $\Theta_2(t)$ row vectors of dimension n, and $\theta_{20}(t)$, $\theta_{30}(t)$ scalar values, are time-varying (vector $X(t)$ is the *regressor* (14.35)). The control law (14.47) can be further rewritten as

$$u_p(t) = \Theta^T X(t) + \tilde{\Theta}^T(t) X(t) \tag{14.46}$$

where

$$\tilde{\Theta}^T(t) := [\Theta_3(t) - F_3\ \ \theta_{30}(t) - f_{30}\ \ -\Theta_1(t) + F_1\ \ -\Theta_2(t) + F_2\ \ -\theta_{20}(t) + f_{20}]$$

is the parameter error, and Θ is as in (14.35). Using Theorem 14.1, and applying simple block algebra manipulations, the output error equation can be expressed as

$$\begin{aligned} \nu(t) \quad = \quad & -\frac{D(s)}{A_m(s)(s+a)} \tilde{\Theta}^T(t) X(t) \\ & + \frac{D(s)}{A_m(s)(s+a)} \{[\Theta_3(t) - F_3] x_{F_3}(t) + [\theta_{30}(t) - f_{30}] r(t)\} \end{aligned} \tag{14.47}$$

or analogously as

$$\nu(t) = \frac{D(s)}{A_m(s)(s+a)} \tilde{\Theta}_r^T(t) X_r(t) \tag{14.48}$$

where vectors $\tilde{\Theta}_r^T(t) := [\Theta_1(t) - F_1\ \ \Theta_2(t) - F_2\ \ \theta_{20}(t) - f_{20}]$, and $X_r^T(t) := [x_{F_1}^T(t)\ x_{F_2}^T(t)\ y_a(t)]$ are respectively the *reduced parameter error vector* and the *reduced regressor*. As previously noted, the error equation does not depend on the parameters of the feedforward filter *Filter 3*.

Let us now consider the *filtered output error* $\nu_F(t)$ obtained at the output of a linear filter placed in series to $\nu(t)$

$$\nu_F(t) = \frac{1}{D(s)}\nu(t) = \frac{1}{A_m(s)(s+a)}\tilde{\Theta}_r^T(t)X_r(t) \tag{14.49}$$

Since the states of the filter $1/D(s)$ are accessible, all the derivatives of $\nu_F(t)$ up to the $n+1$-th order turn out to be available. Equation (14.51) can be put in state variable form as

$$\begin{aligned} \nu_F(t) &= \eta_1(t) \\ \dot{\eta}_1(t) &= \eta_2(t) \\ &\vdots \\ \dot{\eta}_{n+1}(t) &= -\sum_{i=0}^{n} a_{i+1}^* \eta_{n+1-i}(t) + \tilde{\Theta}_r^T(t)X_r(t) \end{aligned} \tag{14.50}$$

with

$$\nu(t) = \eta_{n+1} + \sum_{i=1}^{n} d_i \eta_{n+1-i}(t) \tag{14.51}$$

where a_j^*, for $j = 1, \ldots, n+1$, are the coefficients of polynomial $A_m(s)(s+a) = s^{n+1} + a_1^* s^n + \ldots + a_{n+1}^*$, and d_j, for $j = 1, \ldots, n$, are the coefficients of polynomial $D(s) = s^n + d_1 s^{n-1} + \ldots + d_n$. Note that, in (14.52) the signal $\tilde{\Theta}_r^T(t)X_r(t)$, which in the sequel will be denoted by $\tilde{u}_p(t)$, plays the role of control.

The condition $\nu(\eta) = 0$ identifies a sliding manifold for the error model (14.52),(14.53). Then, in order to perform perfect model-following regardless of the uncertainty which affects the plant, standard VSC control theory can be used (Utkin 1978, 1992).

If the *reaching condition*

$$\nu(\eta)\dot{\nu}(\eta) \leq -\gamma^2 |\nu(\eta)| \tag{14.52}$$

where $\gamma \in \mathbb{R}$, is satisfied, then the sliding manifold $\nu(\eta) = 0$ is reached in finite time. So, to fulfil condition (14.54), we have to take into account $\dot{\nu}(\eta)$, which can be derived from (14.52),(14.53) as

$$\begin{aligned} \dot{\nu}(\eta) &= \dot{\eta}_{n+1}(t) + \sum_{i=1}^{n} d_i \dot{\eta}_{n+1-i}(t) \\ &= -\sum_{i=0}^{n} a_{i+1}^* \eta_{n+1-i}(t) + \tilde{u}_p(t) + \sum_{i=1}^{n} d_i \dot{\eta}_{n+1-i}(t) \end{aligned} \tag{14.53}$$

and, for the sake of simplicity, can be rewritten as

$$\dot{\nu}(\eta) = \tilde{u}_p(t) + \Phi(\eta) \tag{14.54}$$

the term $\Phi(\eta)$ being a known linear combination of the states of the error model (14.52),(14.53).

Inequality (14.54) can be rewritten as

$$\begin{aligned} \dot{\nu}(\eta) &\geq \gamma^2 \qquad \text{if} \quad \nu(\eta) < 0 \\ \dot{\nu}(\eta) &\leq -\gamma^2 \qquad \text{if} \quad \nu(\eta) > 0 \end{aligned} \tag{14.55}$$

or, alternatively as

$$\begin{aligned} |\dot{\nu}(\eta)| &\geq \gamma^2 \\ \operatorname{sign} \dot{\nu}(\eta) &= -\operatorname{sign} \nu(\eta) \end{aligned} \tag{14.56}$$

Since $\Phi(\eta)$ is known, it is easy to verify that an additional control signal equal to $\Phi(\eta) + \gamma^2 \operatorname{sign} \nu(\eta)$, fed at the input of the auxiliary plant $\tilde{B}(s)/[A_m(s)(s+a)]$, causes the error equation to become

$$\nu(t) = \frac{D(s)}{A_m(s)(s+a)} \left[\tilde{\Theta}_r^T(t) X_r(t) - (\Phi(\eta) + \gamma^2 \operatorname{sign} \nu(\eta))\right] \tag{14.57}$$

Equation (14.56) is modified to be

$$\dot{\nu}(\eta) = \tilde{u}_p(t) - \gamma^2 \operatorname{sign} \nu(\eta) \tag{14.58}$$

and conditions (14.58) turn out to be reduced to a single relationship

$$\operatorname{sign} \tilde{u}_p(t) = -\operatorname{sign} \nu(\eta) \tag{14.59}$$

It should be noted that, according to these considerations, the control law $u_p(t)$ can be rewritten as

$$\begin{aligned} u_p(t) &= \Theta_r^T(t) X_r(t) + \Theta_3(t) x_{F_3}(t) + \theta_{30}(t) r(t) - (\Phi(\eta) + \gamma^2 \operatorname{sign} \nu(\eta)) \\ &= u_p'(t) + u_p''(t) \end{aligned} \tag{14.60}$$

where $\Theta_r^T(t) := [-\Theta_1(t) \; -\Theta_2(t) \; -\theta_{20}(t)]$, and $u_p'(t) := \Theta_r^T(t) X_r(t)$, while $\tilde{u}_p(t)$ from its definition yields

$$\tilde{u}_p(t) = \Theta_r^T X_r(t) - \Theta_r^T(t) X_r(t) \tag{14.61}$$

with $\Theta_r^T := [-F_1 \; -F_2 \; -f_{20}]$. Then, taking into account the uncertainties of the plant description $W_p(s)$, we have to choose the actual control law $u_p(t)$ so as to make $\tilde{u}_p(t)$ satisfy condition (14.61) (which in turn implies the generation of a sliding motion on the manifold $\nu(\eta) = 0$).

By rewriting (14.63) as

$$\tilde{u}_p = \sum_{i=1}^{2n+1} \theta_{r_i} x_{r_i}(t) - \sum_{i=1}^{2n+1} \theta_{r_i}(t) x_{r_i}(t) \tag{14.62}$$

where θ_{r_i}, $\theta_{r_i}(t)$ are the components of vectors Θ_r and $\Theta_r(t)$ respectively, it is clear that in order to satisfy condition (14.61) it is sufficient that

$$|\theta_{r_i}(t)| > |\theta_{r_i}| \tag{14.63}$$

and

$$\operatorname{sign}\theta_{r_i}(t) = \operatorname{sign} x_{r_i}(t)\ \operatorname{sign}\nu(\eta) \tag{14.64}$$

Conditions (14.65),(14.66) express a rule for the adjustment of the parameter $\theta_{r_i}(t)$, affecting the output error $\nu(t)$, in order to fulfil (14.61). This means in particular that, if an upperbound $\theta_{r_i max}$ of $|\theta_{r_i}|$ is available, the choice $|\theta_{r_i}(t)| = \theta_{r_i max}$ (an adjustment mechanism which only varies the sign of the parameters) is sufficient for our purpose.

In our case, such an upper bound has to be determined on the basis of relationships (14.42),(14.43), which express the correct choice of polynomials $F_1(s)$ and $F_2(s)$ in the known parameter case, and taking into account the uncertainties on the plant parameters. This procedure yields

$$|\theta_{r_i}(t)| = \tilde{b}_{max} + kd_i \quad \text{for } i = 1, \ldots, n \tag{14.65}$$

$$|\theta_{r_i}(t)| = a_{m_{i+1-n}} + a a_{m_{i-n}} + (a+1)\alpha \ \text{ for } i = n+1, \ldots, 2n-1 \tag{14.66}$$

$$|\theta_{r_{2n}}(t)| = a(a_{m_n} + \alpha) \ , \quad |\theta_{r_{2n+1}}(t)| = a_{m_1} + \alpha \tag{14.67}$$

where $\alpha := \max_j a_j, j = 1, \ldots, n$; $\beta := \max_i b_i, i = 0, \ldots, m$; $\tilde{b}_{max} := k\alpha + (a+1)\beta$, with $\tilde{b}_{max}$ being an upper bound on the modula of the coefficients of $\tilde{B}(s)$. Moreover, a is the absolute value of the pole of the fixed compensator placed in parallel to the plant, and a_{m_i}, $i = 1, \ldots, n$, are the coefficients of $A_m(s)$ which are known positive numbers since $A_m(s)$ is Hurwitz by assumption.

Therefore, choosing $|\theta_{r_i}(t)|$ as indicated in (14.67)-(14.69), and switching the signs of $\theta_{r_i}(t)$ according to (14.66), we can build a discontinuous control strategy $u'_p(t)$ which guarantees that $\tilde{u}_p(t)$ satisfies condition (14.61), and consequently the *reaching condition* (14.54). Therefore, the application of such a control results in a sliding motion for the auxiliary plant on the manifold $\nu(\eta) = 0$. In other words the model-following problem which was the subject of this section has been completely solved.

14.5 Pole Assignment via Discontinuous Identification of the Parameters of the Feedforward Filter

From the above discussion it is clear that the parameters of *Filter 3* have not yet been specified, since they are not involved in the error model. The control law $u'_p(t)$ determined in the previous section is aimed at the zeroing of the tracking error $\nu(t)$ which represents the difference between the model output $y_m(t)$ and the auxiliary plant output $y_a(t)$. The pole assignment objective indicated in (14.24) remains to be attained.

The parameters of *Filter 3* play a crucial role. It has been previously outlined how the polynomial $F_3(s)$ is related to the polynomial $F_1(s)$. So it seems natural to conceive a procedure which would allow us to indirectly identify the coefficients of $F_3(s)$ (i.e. the parameters of *Filter 3*) once the *equivalent* parameters of *Filter 1* have been acquired. But the parameters of *Filter 1* are discontinuous functions of the components of the regressor $X_r(t)$ and of the output error $\nu(\eta)$, according to (14.66). We shall use the relevant Filippov solution concept (Filippov 1964), in order to derive, from the available information, a continuous-time parameter vector converging to the ideal parameter vector Θ_r. To this end we prove the following result.

Theorem 14.2 *On the sliding manifold $\nu(\eta) = 0$ the Filippov's equivalent representation of the error model, with a discontinuous control law, consists of a continuous-time system controlled by a continuous-time equivalent control $\tilde{u}_{p_{eq}}(t) = 0$. This in turn implies that $u'_{p_{eq}}(t) = \Theta_r^T X_r(t)$.*

Proof. Rewrite the error model (14.52),(14.53) taking into account (14.67)-(14.69) and the additional input signal $\Phi(\eta) + \gamma^2 \operatorname{sign} \nu(\eta)$, with

$$\dot{\eta}(t) = \Lambda_\eta \eta(t) + b_\eta \left[\tilde{u}_p(t) - \Phi(\eta) - \gamma^2 \operatorname{sign} \nu(\eta)\right] \tag{14.68}$$

where Λ_η is a matrix of dimension $(n+1) \times (n+1)$, in controllable companion form with the elements in the last row equal to the coefficients of the polynomial $A_m(s)(s+a)$, $b_\eta^T = [0 \ \dots \ 1] \in \mathbb{R}^{n+1}$, and

$$\tilde{u}_p(t) = \begin{cases} \tilde{u}_p^+(t) - \gamma^2 & \text{if} \quad \nu(\eta) > 0 \\ \tilde{u}_p^-(t) + \gamma^2 & \text{if} \quad \nu(\eta) < 0 \end{cases} \tag{14.69}$$

By applying Filippov's theory the previous equation turns out to be equivalent to

$$\dot{\eta}(t) = \Lambda_\eta \eta(t) + b_\eta \left\{\lambda \left[\tilde{u}_p^+(t) - \gamma^2\right] + (1-\lambda)\left[\tilde{u}_p^-(t) + \gamma^2\right]\right\} \ , \quad \lambda \leq 1 \tag{14.70}$$

The parameter λ has to be determined so as to make $\dot{\nu}(\eta) = 0$ on the manifold $\nu(\eta) = 0$. This means that

$$\lambda \tilde{u}_p^+(t) + (1-\lambda)\, \tilde{u}_p^-(t) = \tilde{u}_{p_{eq}}(t) = o\left(\gamma^2\right) \tag{14.71}$$

denoting by $o(\cdot)$ the magnitude order. Since γ^2, which determines the velocity of the reaching phase, can be freely chosen, we set $\gamma = \gamma_1$ if $|\nu(\eta)| \geq \varepsilon$, and $\gamma = \gamma_2 = 0$ if $|\nu(\eta)| < \varepsilon$, where ε is an arbitrarily small positive number. Therefore, $\tilde{u}_p(t)$ is equivalent to a null signal and, consequently, $u'_{p_{eq}}(t) = \Theta_r^T X_r(t)$. □

Taking into account Theorem 14.2, we can apply the theory of approximability developed by Utkin (1992) in order to obtain, starting from $u'_p(t)$, the so-called *average control* $u'_{p_{av}}(t)$ as the output of a high (in theory infinite) gain first order filter

$$\tau \dot{u}'_{p_{av}}(t) = -u'_{p_{av}}(t) + u'_p(t) \tag{14.72}$$

The closer $\nu(\eta)$ is to zero the motion of system (14.52), the smaller the time constant τ, and the more negligible the difference between $u'_{p_{av}}(t)$ and $\Theta_r^T X_r(t)$ becomes. Accordingly we can assume that the signal obtained at the output of the high gain filter coincides with the unknown ideal control law $u_{p_{eq}}(t) = \Theta_r^T X_r(t)$. This is not sufficient for our purposes, since we need to extract from the available information (the knowledge of the signal $u'_{p_{av}}(t)$) the ideal parameters of *Filter 1*, the first n components of vector Θ_r, apart from the sign. The problem is then reduced to an identification problem. Note that the necessity of using identification does not affect either the stability or the precision of the feedback part of the controller treated in the previous section.

In order to perform the identification of the unknown parameter vector Θ_r, we consider the scalar product between the regressor $X_r(t)$ and a time-varying vector $\hat{\Theta}_r(t)$ of dimension $2n+1$ as our identification model (this is possible since $X_r(t)$ is accessible), i.e.

$$\hat{y}(t) = \hat{\Theta}_r^T(t) X_r(t) \tag{14.73}$$

while the system to be identified is

$$u'_{p_{av}}(t) = \Theta_r^T X_r(t) \tag{14.74}$$

since $u'_{p_{av}}(t)$, and not $u'_{p_{eq}}(t)$, is the accessible signal. Then, let us define

$$e(t) = \hat{y}(t) - u'_{p_{av}}(t) \tag{14.75}$$

$$\varphi(t) = \hat{\Theta}_r(t) - \Theta_r \tag{14.76}$$

as identification error and parameter error respectively, and choose the adaptation mechanism of integral type

$$\dot{\varphi}(t) = \dot{\hat{\Theta}}_r(t) = -\Gamma X_r(t) e(t) \tag{14.77}$$

where Γ is a diagonal $(2n+1) \times (2n+1)$ gain matrix with positive entries, with the aim of steering the error $e(t)$ to zero (and $\varphi(t) \to 0$).

In order to prove the convergence to zero of the parameter error $\varphi(t)$ the following Lyapunov function needs to be considered

$$V(\varphi(t)) = \frac{1}{2} \|\varphi(t)\|^2 \tag{14.78}$$

Then from (14.79)

$$\dot{V}(\varphi(t)) = -\Gamma \left\|\varphi^T(t) X_r(t)\right\|^2 \tag{14.79}$$

and it follows that the convergence to zero of $\varphi(t)$ is ensured only if the components of vector $X_r(t)$ do not vanish. Since we are interested only in the first n components of vector Θ_r, it is sufficient that only the states of *Filter*

1 do not vanish. More generally, the convergence to zero of $\varphi(t)$ is guaranteed provided that the reference input $r(t)$ is persistently exciting of order greater then or equal to n (instead of $2n+1$ as required if all the parameters contained in Θ_r should be identified). For more details about the problem of persistent excitation, the reader is referred to Narendra and Annaswamy (1989).

As a result, the foregoing procedure through the identification of the parameters of *Filter 1*, indirectly provides (via the relationship $F_1(s) + kD(s) = F_3(s)$) the values of the parameters $[\Theta_3(t)\ \theta_{30}(t)]$. These latter converge asymptotically to the ideal values $[F_3\ f_{30}]$.

The introduction of the estimated parameters in the discontinuous control law (14.62) completes the treatment, since the pole assignment problem turns out to be solved asymptotically.

14.6 Illustrative Examples

In this section some simulation examples are presented in order to demonstrate the role of the identifier in the overall control process. The plant is characterized by the second order transfer function $W_p(s) = 3/(s^2+3)$. The first order compensator $W_c(s) = 1/(s+5)$ is placed in parallel with the plant , while the model to be tracked is $W_m(s) = 5/(s+5)$.

The state variable filters have a common characteristic polynomial $D(s) = s^2+8s+15$. The ideal numerators of the feedback filters (namely, *Filter 1* and *Filter 2*) are given by $F_1(s) = -5s+3$, $f_{20} = 8$, and $F_2'(s) = -12s-60$ respectively.

The first example shows the behaviour of the process of parameter identification applied to the parameters of *Filter 1* when the nominal values of the coefficients of polynomial $F_1(s)$ are assumed to differ from the ideal values -1 and 1, respectively. The persistently exciting reference input is plotted in Fig. 14.2. The behaviour of the tracking error, the auxiliary plant output along with the model output, and the plant output are shown in Figs. 14.3, 14.4 and 14.5 respectively. Finally, in Fig. 14.6 the adjustment of the parameters $\theta_1(t)$ and $\theta_2(t)$, defined here as the difference between the actual and the nominal values, is illustrated. The convergence values are as expected, and they are reached with a good convergence rate.

The second example refers to the same plant, but describes an experiment in which the learning properties of the controlled system with respect to a step reference input are outlined. The nominal values of the coefficients are now assumed to be zero, so that the expected convergence values are -5 and 3, respectively. Figure 14.7 shows a suitably chosen reference input which is given by the sequence of a step signal, a persistently exciting signal, followed by another step. The behaviour of the tracking error, the auxiliary plant output together with the model output and the plant output relevant to this second case, are shown in Figs. 14.8, 14.9 and 14.10 respectively. Finally, the parameters $\theta_1(t)$ and $\theta_2(t)$ are shown in Fig. 14.11.

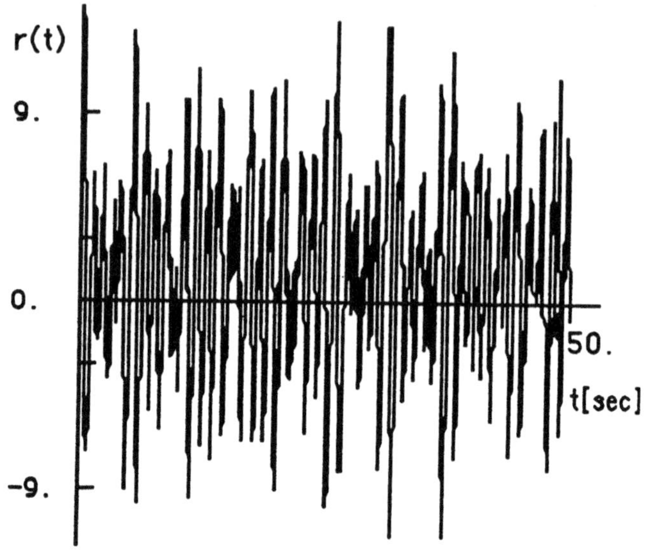

Fig. 14.2. The reference input

The expected values are reached with good precision, which indicates correct design of the feedforward fiter *Filter 3*, and consequent good behaviour of the controlled plant in spite of the uncertainty.

14.7 Conclusions

The model-following problem for uncertain linear SISO systems in input/output form has been discussed. The approach presented is characterized by the use of a discontinuous control law within a control structure inherited from classical adaptive control schemes and simplified, since its complexity turns out to be independent of the relative degree of the plant to be controlled.

The structure of the linear scheme corresponding to the ideal case of perfect knowledge of the plant parameters was first described. The main feature of the design procedure underlying the proposed scheme consists of the fact that a pole assignment objective is pursued by performing explicit model tracking and synthesizing a suitable feedforward compensator.

In the case of an uncertain plant the two stages into which the design process can be divided, were considered in sequence. The tracking problem is easily dealt with by using a discontinuous parameter adjustment mechanism based on the knowledge of the uncertainty bounds of the plant parameters. The pole assignment problem is solved by applying an identification procedure and exploiting the link between the parameters of the feedforward compensator and those of the feedback part of the controller.

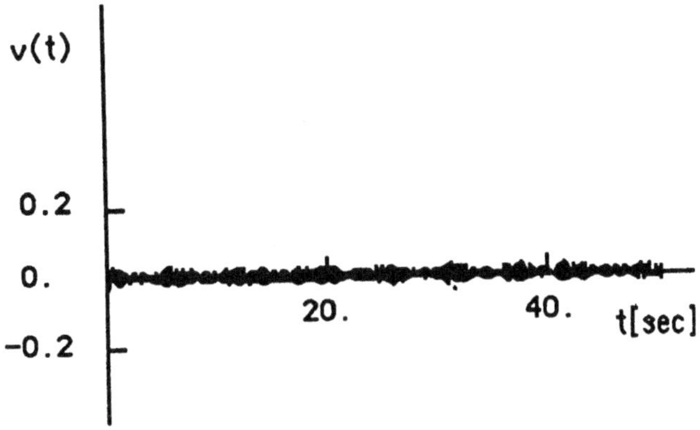

Fig. 14.3. The output error

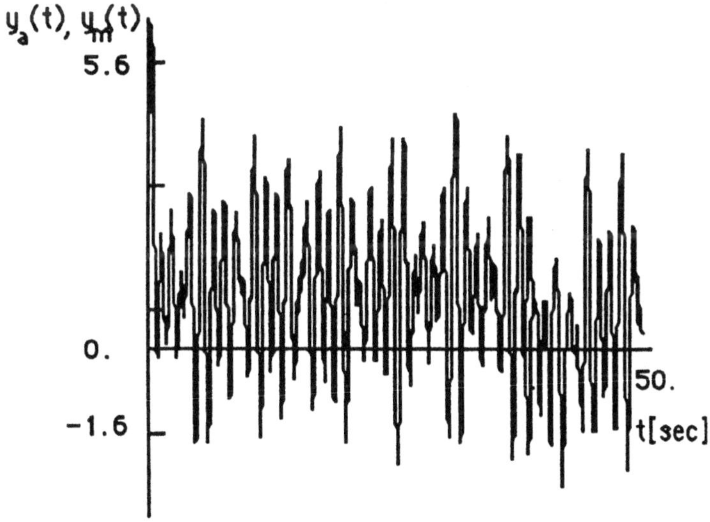

Fig. 14.4. The auxiliary plant output and the model output

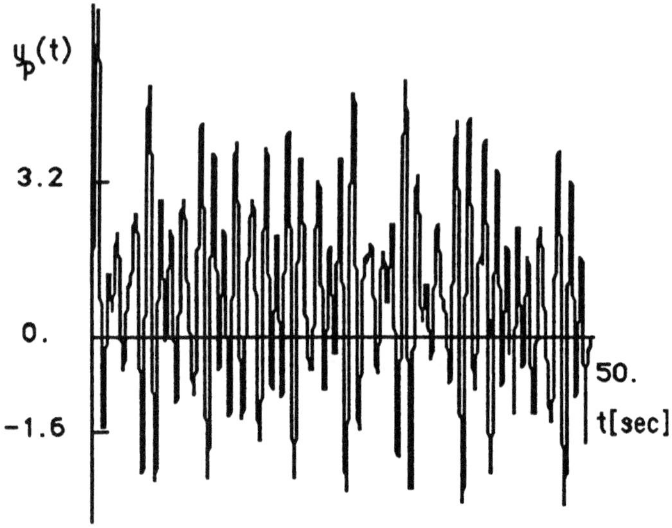

Fig. 14.5. The plant output

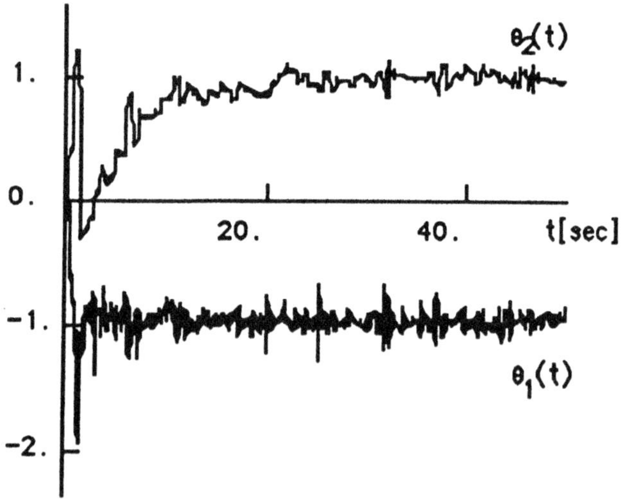

Fig. 14.6. The adjusted parameters

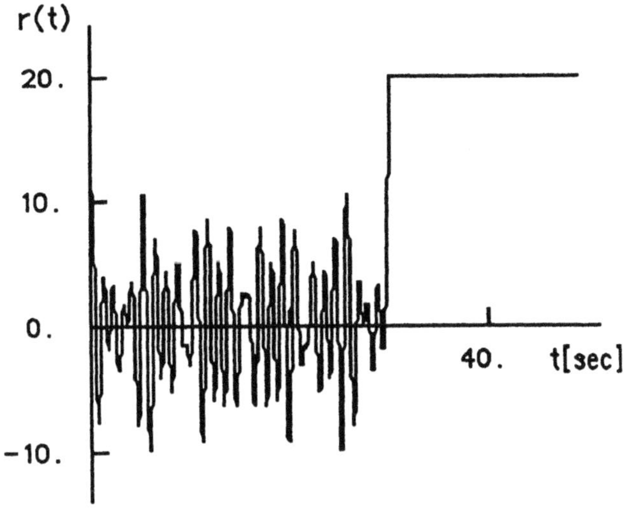

Fig. 14.7. The reference input

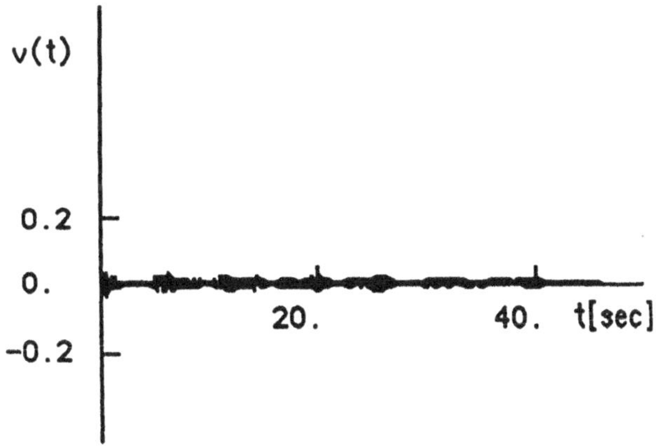

Fig. 14.8. The output error

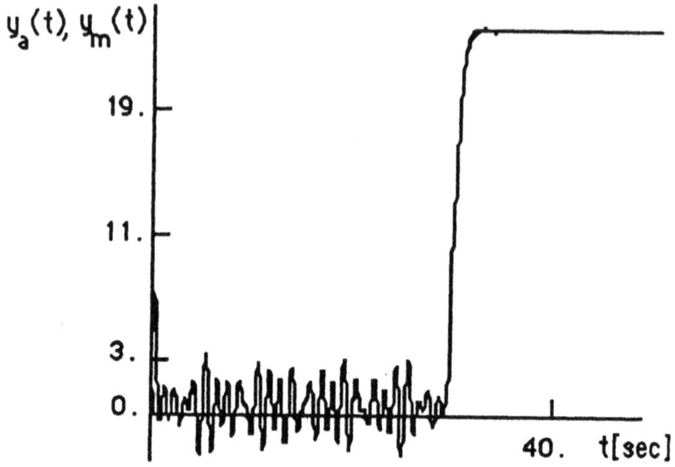

Fig. 14.9. The auxiliary plant output and the model output

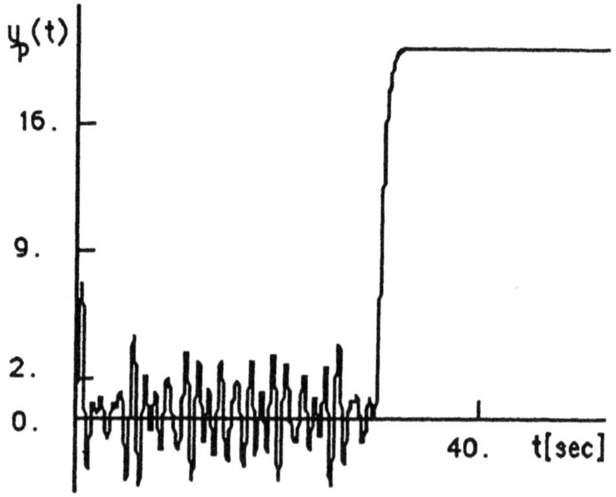

Fig. 14.10. The plant output

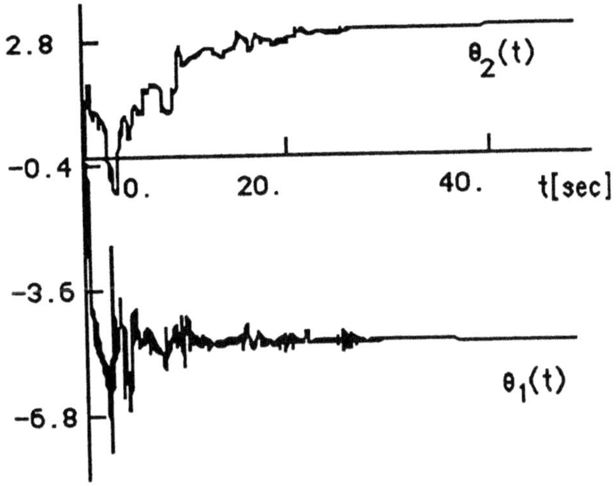

Fig. 14.11. The adjusted parameters

The advantage of the described approach with respect to other proposals in the literature mainly relies on its simplicity and on the fact that it is based on a reduced assumption set about plant uncertainty.

References

Bartolini, G., Ferrara, A. 1991, A simplified direct approach to the problem of adaptive pole assignment. In *New trends in systems theory*, ed. Conte, Perdon, Wyman, Birkhäuser, Berlin, 89–96

Bartolini, G., Ferrara, A. 1992a, Adaptive control of SISO plants with unmodelled dynamics. *International Journal of Adaptive Control and Signal Processing* **6**, 237–246

Bartolini, G., Ferrara, A. 1992b, Pole assignment control of nonminimum phase systems: a combined adaptive control–variable structure approach. *Proc IEEE Conference on Systems Engineering*, Kobe, Japan, 54–59

Bartolini, G., Zolezzi, T. 1988, The VSS approach to the model reference control of nonminimum phase linear plants. *IEEE Transactions on Automatic Control* **33**, 859–863

Bartolini, G., Zolezzi, T. 1989, Asymptotic linearization of unceratin systems by means of approximate sliding modes. *Proc IFAC Symposium on Non Linear System Design*, Capri, Italy

Filippov, A. F. 1964, Differential equations with discontinuous right hand side. *Ann. Math. Soc. Trasl.* **42**, 199–232

Hsu, L. 1990, Variable structure model reference adaptive control using only input and output measurements: the general case. *IEEE Transactions on Automatic Control* **35**, 1238–1243

Hsu, L., Costa, R. R. 1987, Bursting phenomena in continuous systems with s-factor. *IEEE Transactions on Automatic Control* **32**, 84–86

Landau, I. D. 1979, *Adaptive control: the model reference approach*, Dekker, New York

Monopoli, R. V. 1974, Model reference adaptive control with an augmented error signal. *IEEE Transactions on Automatic Control* **19**, 474–484

Narendra, K. S., Annaswamy, A. M. 1989, *Stable adaptive systems*, Prentice-Hall, New Jersey

Narendra, K. S., Boskovic, J. D. 1992, A combined direct, indirect and variable structure method for robust adaptive control. *IEEE Transactions on Automatic Control* **37**, 262–268

Narendra, K. S., Lin, Y. H., Valavani, L. S. 1980, Stable adaptive controller design, part II: proof of stability. *IEEE Transactions on Automatic Control* **25**, 440–448

Sira-Ramírez, H. 1988, Differential geometric methods in variable structure control. *International Journal of Control* **48**, 1359–1391

Tao, G., Ioannou, P. A. 1991, Robust adaptive control of plants with unknown order and high frequency gain. *International Journal of Control* **53**, 559–578

Utkin, V. I. 1978, *Sliding modes and their application in variable structure systems*, MIR Publishers, Moscow

Utkin, V. I. 1992, *Sliding modes in control and optimization*, Springer-Verlag, Berlin

Walcott, B. L., Żak, S. H. 1987, State observation of nonlinear uncertain dynamical systems. *IEEE Transactions on Automatic Control* **32**, 166–170

15. Combined Adaptive and Variable Structure Control

Alexander A. Stotsky

15.1 Introduction

It is well known that Variable Structure Control (VSC) is a powerful tool for solving the problem of the control of uncertain dynamical systems (Utkin 1981). Adaptive algorithms are also widely used. The present chapter is devoted to the study of the design of algorithms including elements and advantages of both techniques; combined VSC and adaptive algorithms.

Combined VSC and direct adaptive algorithms have been intensively studied by Slotine and Coetsee (1986), Andrievsky et al (1988), Hsu (1988) and Stotsky (1991). However, it is known that the main drawback of direct adaptation is slow parameter convergence and, in turn, a slow convergence of the tracking errors to zero (Slotine and Li 1989). Therefore an important problem is the design of combined VSC and indirect adaptive algorithms, since the convergence rate of the estimated parameters, when using indirect algorithms is higher than in direct schemes.

Indirect adaptive algorithms are driven by the prediction error (i.e. the error between the estimated and true parameters), which includes the derivative of the tracking error. To avoid the direct measurement of the derivative, the estimate of the prediction error is obtained via filtering, and indirect algorithms proposed by Middleton and Goodwin (1988), Narendra and Annasvamy (1989) and Miroshnik and Nikiforov (1991), are driven by this estimate. The performance of the system with indirect algorithms depends on how we evaluate the estimate of the prediction error, and hence on the filter parameters.

The indirect algorithms proposed in the above papers, were designed via the Lyapunov function which represents the square norm of the deviation between the adjustable and true parameters. The filter dynamics was not taken into account, and the parameters of the algorithm and the filter were not linked. Also incorrect choice of both algorithm and filter parameters yields a decrease of the performance of the closed system. This necessitates the finding of a Lyapunov function for simultaneous design of the filter and algorithm structures.

It is worth remarking that to use only indirect algorithms for control of uncertain dynamical systems, is not always possible, since in general, the convergence of the prediction error does not imply the convergence of the tracking error.

This chapter is devoted to the design of new combined VSC and adaptive algorithms, based on the Lyapunov design procedure proposed by Stotsky (1993). The advantages of the proposed technique can be summarized as follows:

(i) It allows the design of new combined adaptive direct (indirect) and VSC algorithms.

(ii) A new Lyapunov design procedure allows simultaneous design of filter and adaptive algorithm structures. This, in turn, allows

(a) linking the parameters of the filter with algorithm parameters to improve the performance of the system

(b) deriving both a new type of filtering and an indirect adjustment law, i.e. indirect adaptive algorithms with the relay term driven by the estimate of the prediction error. These algorithms have relevant robust properties when bounded unmeasurable disturbances act on the plant. Note that it is impossible to design the filter and algorithm structures separately in this case.

(iii) The proposed algorithms use various techniques for the improvement of the convergence rate of the tracking errors, including

(a) Proportional-integral and integral-relay adjustment law driven by the tracking error

(b) Proportional and relay terms driven by the estimate of a prediction error

(c) Gain update by the least-squares method.

(iv) The Lyapunov design procedure allows straightforward control performance evaluation with various signals in the closed-loop system.

The proposed Lyapunov design procedure is used for a certain class of linear time-invariant plants where the state vector is available for measurement and for SISO plants with relative degree one using only input-output measurements. A similar design procedure has been used by Panteley and Stotsky (1992) in the trajectory control of robotic manipulators with constraints.

The chapter consists of two parts. The first is devoted to the tracking problem for systems with an implicit reference model. The control goal, the control law and the error model are described in Sect. 15.2. The problem under consideration is to choose a vector of adjustable parameters to ensure the achievement of the control goal.

The adjustment laws can be classified into three groups: direct, indirect and combined algorithms. In Sects. 15.3 and 15.4 direct integral and combined VSC and direct adjustment law are considered. Control performance is evaluated via the boundedness of the tracking error and this motivates the modification of the adjustment law.

Sect. 15.5 is devoted to indirect algorithms. A new Lyapunov design procedure for simultaneous design of the estimate of the prediction error and the adaptive algorithm, is described. Using this procedure a new indirect algorithm with relay term driven by the estimate of the prediction error, is proposed. It is shown that the algorithms have relevant robust properties in the case where bounded unmeasurable disturbances act on the plant. The conditions for the convergence of adjustable parameters to their true values are presented at the end of Sect. 15.5. New combined algorithms are presented in Sect. 15.6 and include the previously described algorithms special cases, yielding an improvement of the convergence rate of the tracking error. Links with the sliding mode are established. Exponential convergence of the Lyapunov function is established under persistent excitation.

The second part of the chapter (Sect. 15.7) is devoted to the control of SISO plants using only input/output measurements. Algorithms similar to those presented in the first part are proposed. The generalized error model is used for algorithm design. It is shown that the new algorithms can be considered to be extensions of the algorithms proposed by Narendra and Annaswamy (1989), Hsu (1990) and Marino (1990). Global exponential stability of the system is ensured with the proposed algorithms under persistent excitation.

15.2 Problem Statement

Consider the linear time-invariant plant in the form

$$\dot{x}(t) = Ax(t) + Bu(t) \tag{15.1}$$

where $x \in R^n$ is the state vector and $u \in R^1$ is the scalar input. Consider the plant in regular *canonical form* (Utkin 1981)

$$\begin{aligned} \dot{x}_1(t) &= A_{11}x_1(t) + A_{12}x_2(t) && (15.2)\\ \dot{x}_2(t) &= A_{21}x_1(t) + A_{22}x_2(t) + bu(t) && (15.3) \end{aligned}$$

where $x_1 \in R^{n-1}$ and $x_2 \in R^1$. Let A_{11} and A_{12} be known constant matrices, b a known constant scalar and A_{21}, A_{22} unknown but constant. The control problem consists of finding a feedback control law which achieves asymptotic stabilization of the tracking error

$$\lim_{t\to\infty} (x(t) - x_d(t)) = 0 \tag{15.4}$$

where $x_d(t)$ is a bounded and continuously differentiable desired trajectory, $x_d(t)^T = (x_{1d}(t), x_{2d}(t))^T$. It should be noted that the control problem in this case requires tracking of n coordinates using a single control action, and hence we should introduce some restrictions on the desired dynamics. Let the desired trajectory x_{1d} satisfy the following differential equation

$$\dot{x}_{1d}(t) = A_{11}x_{1d}(t) + A_{12}x_{2d}(t) \tag{15.5}$$

where A_{11} is a stable matrix. The restriction of the desired dynamics is met in practice, for instance, in the biped locomotion problem and in the problem of aircraft stabilization via aerofoil incidence.

A simple practical example is the problem of aircraft stabilization via the angle of attack. The dynamic equations of an aircraft in the longitudinal plane have the following form (Bukov 1987)

$$\begin{aligned} \dot{\alpha}(t) &= -a_1\alpha(t) + \omega(t) \\ \dot{\omega}(t) &= a_2\omega(t) + a_3\alpha(t) + a_4\delta(t), \end{aligned} \tag{15.6}$$

where $\alpha(t)$ is the angle of attack, $\omega(t)$ is the pitch angle, $\delta(t)$ is the deflection of the elevator, a_1, a_2, a_3 and a_4 are aerodynamical coefficients and a_1 is a positive known constant.

The control problem in this case is to track the desired angle of attack $\alpha_d(t)$ to achieve the control goal (15.4)

$$\begin{aligned} \lim_{t\to\infty} \alpha(t) - \alpha_d(t) &= 0 \\ \lim_{t\to\infty} \omega(t) - \omega_d(t) &= 0 \end{aligned} \tag{15.7}$$

where $\omega_d(t)$ is the desired pitch angle, corresponding to the desired angle of attack, and is defined by

$$\dot{\alpha_d}(t) = \omega_d(t) - a_1\alpha_d(t) \tag{15.8}$$

This example motivates the introduction of the restrictions on the desired dynamics (15.5). It is easy to see that we can solve the tracking problem for $\alpha(t)$ since we have only one control action; therefore the desired pitch angle is uniquely defined by the equation (15.8) and cannot be arbitrarily chosen.

Consider also the auxiliary control goal

$$\lim_{t\to\infty} s(\tilde{x}) = 0 \tag{15.9}$$

where $s = C\tilde{x}(t)$, $C = (C_1, 1)$, C_1 is an $(n-1)$-vector and

$$\tilde{x}(t)^T = (\tilde{x}_1(t), \tilde{x}_2(t))^T, \quad \tilde{x}_i(t) = x_i(t) - x_{id}(t), \quad i = 1, 2 \tag{15.10}$$

It is easy to see that the relationship between the control goals (15.4) and (15.9) is given by the following differential equation

$$\dot{\tilde{x}}_1(t) = (A_{11} - A_{12}C_1)\tilde{x}_1(t) + A_{12}s(t) \tag{15.11}$$

Now the motivation for the introduction of $s(t)$ becomes clear. Indeed when (15.9) is achieved and $s(t)$ is square-integrable, then (15.4) is achieved provided that $(A_{11} - A_{12}C_1)$ is a Hurwitz matrix. A suitable placement of eigenvalues of the matrix $(A_{11} - A_{12}C_1)$ can be made by choosing C_1, if pair (A_{11}, A_{12}) is controllable (Utkin 1981).

It is worth noting that in the present system we have two types of models; implicit and explicit reference models. The implicit reference model is given by the equation (15.11) with $s(t) = 0$ and the explicit reduced order reference model is given by the equations (15.5) where $x_{2d}(t)$ can be seen as the input of the model.

Restrictions on the desired dynamics (15.5) follow from the Erzberger conditions which are well-known in model reference adaptive control (Erzberger 1968). Our first step is to derive the error model with respect to the tracking error $s(t)$. Let the control law have the form

$$u = b^{-1}(\dot{x}_{2d}(t) - \alpha_0 s - C_1(A_{11}\tilde{x}_1 + A_{12}\tilde{x}_2) + f^T\theta), \tag{15.12}$$

where $f^T = (x_1(t), x_2(t))^T$ and $\theta \in R^n$ is a vector of adjustable parameters. Using (15.2) and the control law (15.12) we obtain the error model

$$\dot{s} = -\alpha_0 s + f^T\tilde{\theta} \tag{15.13}$$

where $\tilde{\theta} = \theta - \theta^*$; $\theta^{*T} = (-A_{21}, -A_{22})$, the constant vector of true parameters; and $\alpha_0 > 0$.

Our problem is to find θ to achieve the control objective (15.9). Various techniques are presented for the adjustment of θ. The exposition begins with the direct integral adjustment law presented in Sect. 15.3 where adjustment is achieved by incorporating proportional or relay terms with the aim of performance improvement to be described in Sect. 15.4. As the measure of performance we select the upper bound of the tracking error $s^2(t)$ and accurate bounds of this error are determined via the Lyapunov method. Despite the fact that the procedures for calculating the bounds are different for each algorithm, they are all based on the simple idea of step-by-step improvement of bounds. A comparative analysis of the bounds is the motivation for subsequent modification of the algorithms. In the next two sections we evaluate the control performance and compare the proposed algorithms.

15.3 Direct Integral Adjustment Law

In this section we present the design and performance analysis of the system for a direct integral adjustment law. The design is based upon the Lyapunov function candidate

$$V_1(t) = \frac{1}{2}s^2 + \frac{1}{2\gamma}\tilde{\theta}^T\tilde{\theta} \tag{15.14}$$

where $\gamma > 0$. Taking the derivative of (15.14) along the solutions of (15.13) we get

$$\dot{V}_1(t) = -\alpha_0 s^2 + s f^T\tilde{\theta} + \tilde{\theta}^T\dot{\theta}/\gamma \tag{15.15}$$

In order to cancel the last two terms in $\dot{V}_1(t)$ we choose $\dot{\theta}$ as

$$\dot{\theta} = -\gamma s f \tag{15.16}$$

One obtains $\dot{V}_1 = -\alpha_0 s^2$. Integrating the last identity we have $V_1(t) \leq V_1(0)$ where $V_1(0) = \frac{1}{2}s^2(0) + ||\tilde{\theta}(0)||^2/(2\gamma)$, $\tilde{\theta}(0) = \theta(0) - \theta^*$ and hence

$$\frac{1}{2\gamma}||\tilde{\theta}(t)||^2 \leq V_1(0), \qquad \frac{1}{2}s^2(t) \leq V_1(0) \tag{15.17}$$

Note that the adjustment law $\dot{\theta} = -\gamma s f$ for the error model (15.13) was proposed by Fradkov (1989). The convergence of s(t) to zero can be easily established. Indeed, the boundedness of $s(t)$ implies the boundedness of $\tilde{x}_1(t)$ and in turn, the boundedness of $\tilde{x}_2(t)$. Since the desired trajectories are bounded, we conclude the boundedness of $x_1(t)$ and $x_2(t)$. The boundedness of f and $\tilde{\theta}$ implies the boundedness of $\dot{s}(t)$. Since $s(t)$ is square-integrable we conclude that the control goal (15.9) is achieved. The achievement of the goal (15.4) can be established using (15.11).

However, the analysis presented above does not give any information about transient performance. Our next step is to get an accurate bound for $s^2(t)$ in order to make comparisons. Note that the bound (15.17) is very crude, since it does not include algorithm parameters and it is not suitable for our objective. To get an accurate bound for $s^2(t)$ we need the bound for $||\tilde{x}_1(t)||^2$.

Let us for simplicity choose C_1 such that $(A_{11} - A_{12}C_1) = -\beta I_{n-1}$. To obtain the bound for $||\tilde{x}_1(t)||^2$ consider the positive definite function

$$V_2 = \frac{1}{2}||\tilde{x}_1(t)||^2 \tag{15.18}$$

whose derivative along the solutions of (15.11) is

$$\dot{V_2} \leq -\beta||\tilde{x}_1(t)||^2 + ||A_{12}||||\tilde{x}_1(t)||^2/(2\lambda) + ||A_{12}||\lambda s^2/2, \tag{15.19}$$

where λ is a positive constant. Choosing $\lambda = ||A_{12}||/\beta$ and using the bound (15.17) for $s^2(t)$ we obtain

$$\dot{V_2} \leq -\beta/2||\tilde{x}_1(t)||^2 + ||A_{12}||^2 V_1(0)/\beta \tag{15.20}$$

and

$$\frac{1}{2}||\tilde{x}_1(t)||^2 \leq \frac{1}{2}e^{-\beta t/2}||\tilde{x}_1(0)||^2 + 2||A_{12}||^2 V_1(0)/\beta^2 \tag{15.21}$$

To get an accurate bound for $s^2(t)$ we need the bounds for $||x_1(t)||^2$ and $x_2^2(t)$. The bound for $||x_1(t)||^2$ can be obtained via the bounds of $||\tilde{x}_1(t)||^2$ and $||x_{1d}(t)||^2$.

$$\begin{aligned} ||x_1(t)||^2 &\leq 2||\tilde{x}_1(t)||^2 + 2||x_{1d}(t)||^2 \\ &\leq 2e^{-\beta t/2}||\tilde{x}_1(0)||^2 + 8||A_{12}||^2 V_1(0)/\beta^2 + 2\bar{x}_{1d}^2, \end{aligned} \tag{15.22}$$

where $\bar{x}_{1d}^2$ is an upper bound for $||x_{1d}(t)||^2$. For $\tilde{x}_2(t)$ the following bound

$$\tilde{x}_2(t) \leq 2s^2(t) + 2||C_1||^2||\tilde{x}_1(t)||^2 \tag{15.23}$$

holds. Using (15.23) and (15.21) we obtain the bound for $x_2^2(t)$

$$\begin{aligned} x_2^2(t) \;\leq\; & 8V_1(0) + 4||C_1||^2(e^{-\beta t/2}||\tilde{x}_1(0)||^2 \\ & +4||A_{12}||^2 V_1(0)/\beta^2) + 2\bar{x}_{2d}^2 \end{aligned} \tag{15.24}$$

where $x_{2d}(t)^2 \leq \bar{x}_{2d}^2$. Now we are able to determine an accurate bound for $s^2(t)$. Let us choose the function

$$V_3(t) = \frac{1}{2}s^2(t) \tag{15.25}$$

Taking the derivative along the solutions of (15.13) and using (15.17), (15.22) and (15.24) we obtain

$$\begin{aligned} \dot{V}_3 &\leq -\alpha_0 s^2 + \mid s \mid \|f\| \sqrt{2\gamma V_1(0)} \\ &\leq \left(-\alpha_0 + \sqrt{2\gamma V_1(0)}/(2\lambda)\right) s^2 + \frac{1}{2}\lambda\sqrt{2\gamma V_1(0)}\left(\|x_1\|^2 + x_2^2\right) \\ &\leq -\kappa s^2 + \frac{1}{2}\lambda\ell \end{aligned} \tag{15.26}$$

where

$$\kappa = \alpha_0 - \sqrt{2\gamma V_1(0)}/(2\lambda), \kappa > 0$$

and

$$\begin{aligned} \ell &= \sqrt{2\gamma V_1(0)}(2\|\bar{x}_1(0)\|^2 + 8\|A_{12}\|^2 V_1(0)/\beta^2 + 2\bar{x}_{1d}^2 \\ &\quad +8V_1(0) + 4\|C_1\|^2(\|\bar{x}_1(0)\|^2 + 4\|A_{12}\|^2 V_1(0)/\beta^2) + 2\bar{x}_{2d}^2) \end{aligned}$$

with λ a positive constant. Then the following inequality is valid

$$\frac{1}{2}s^2(t) \leq \frac{1}{2}s^2(0)e^{-\kappa t} + \lambda\ell/(2\kappa) \tag{15.27}$$

It is worth remarking that the bound (15.27) is quite accurate compared to the bound (15.17), since it includes algorithm parameters. This bound can be used for comparative analysis.

As mentioned by Slotine and Coetsee (1986), Landau (1979) and Andrievsky et al (1988), the inclusion of the proportional and/or relay term in the adjustment law improves the convergence rate of the tracking error. The algorithms proposed by Fradkov (1986) are considered in the next section. The section includes control performance evaluation via the boundedness of the tracking error for both proportional-integral and integral-relay adjustment laws.

15.4 Direct Integral and Pseudo-Gradient Adjustment Law

Consider the following Lyapunov function candidate

$$V_4(t) = \frac{1}{2}s^2 + 1/(2\gamma) \parallel \tilde{\theta} + \psi(f,s) \parallel^2 \tag{15.28}$$

where $\psi(f,s) \in R^n$ is the vector function we will describe below. Taking the derivative of $V_4(t)$ with respect to time along (15.13), we obtain

$$\dot{V}_4(t) = -\alpha_0 s^2 - \psi(f,s)^T(fs), \tag{15.29}$$

provided that

$$\frac{\mathrm{d}(\theta + \psi(f,s))}{\mathrm{d}t} = -\gamma(fs) \tag{15.30}$$

Obviously we have to choose the function $\psi(f,s)$ so as to realize the inequality $\psi(f,s)^T(fs) \geq 0$ and hence we can take $\psi(f,s) = \gamma_1 fs$ or $\psi(f,s) = \gamma_2 \mathrm{sign}(fs)$, $\gamma_1, \gamma_2 > 0$. Note that the derivative of $\psi(f,s)$ is not required for algorithm implementation, since we can rewrite (15.30) in the form

$$\theta = -\gamma \int_0^t f(r)s(r)\mathrm{d}r - \psi(f,s) \tag{15.31}$$

The system (15.13) and (15.30) has the following properties: s is bounded and square-integrable, and $(\tilde{\theta} + \psi(f,s))$ is bounded. Algorithms with (15.30) were proposed by Fradkov (1979). One can show that the control goal (15.17) is achieved.

Let us determine an accurate bound for the tracking error $s^2(t)$ in the system (15.17) and (15.30). To this end we need a crude bound for $\|\tilde{\theta}(t) + \psi(f,s)\|^2$. Integrating $\dot{V}_4(t)$ we obtain

$$\frac{1}{2\gamma}\|\tilde{\theta}(t) + \psi(f,s)\|^2 \leq V_4(0) \tag{15.32}$$

where $V_4(0) = \frac{1}{2}s^2(0) + \|\tilde{\theta}(0) + \psi(f(0), s(0))\|^2/(2\gamma)$. Consider the proportional-integral adjustment law $\psi(f,s) = \gamma_1 fs$. Define the function

$$V_5(t) = \frac{1}{2}s^2(t) \tag{15.33}$$

Taking the derivative along the solutions of (15.17) and (15.30) we obtain

$$\begin{aligned} \dot{V}_5 &= -\alpha_0 s^2 + sf^T(\tilde{\theta} + \psi) - \gamma_1 s^2\|f\|^2 \\ &\leq -\alpha_0 s^2 + s^2\|f\|^2/(2\lambda) + \lambda\|\tilde{\theta} + \psi\|^2/2 - \gamma_1 s^2\|f\|^2 \end{aligned} \tag{15.34}$$

Choosing $\lambda = 1/(2\gamma_1)$ and taking (15.32) into account, we get

$$\dot{V}_5 \leq -\alpha_0 s^2 + \gamma V_4(0)/(2\alpha_0\gamma_1) \tag{15.35}$$

and

$$\frac{1}{2}s^2(t) \leq \frac{1}{2}s^2(0)e^{-\alpha_0 t} + \gamma/(2\alpha_0\gamma_1)V_4(0) \tag{15.36}$$

Now we compare the two bounds (15.27) and (15.36). It is easy to see that the convergence rate of the upper bound for algorithm (15.30) is higher than for algorithm (15.16) due to the inclusion of the proportional term. It is rather difficult to compare upper bounds for the region of convergence of these two algorithms, since they depend upon a large number of algorithm parameters; but it is straightforward to see that the upper bound of the region of convergence for algorithm (15.30) does not depend on the upper bound of the desired trajectories, plant parameter A_{12} and algorithm parameter C_1. This allows one to conclude that the inclusion of a proportional term in the algorithm improves the transient performance of the system.

The next step is to show that the inclusion of a relay term improves the transient performance as well. For later convenience let us present (15.30) in the form

$$\theta = \theta_1 - \psi(f, s) \quad (15.37)$$

$$\dot{\theta}_1 = -\gamma f s \quad (15.38)$$

We select $\psi(f, s) = \gamma_2 \operatorname{sign}(fs)$, where γ_2 is a positive design parameter to be chosen as follows. Substituting (15.37) and (15.38) into the error model (15.13) we obtain

$$\dot{s} = -\alpha_0 s + f^T(\theta_1 - \psi(f, s) - \theta^*) = -\alpha_0 s + f^T\tilde{\theta}_1 - f^T\psi(f, s) \quad (15.39)$$

where $\tilde{\theta}_1 = \theta_1 - \theta^*$. To establish an accurate bound for the tracking error $s(t)$ our first step is to obtain the bound for $||\tilde{\theta}_1(t)||$. To this end consider the Lyapunov function candidate

$$V_6(t) = \frac{1}{2}s^2 + \frac{1}{2\gamma}||\tilde{\theta}_1(t)||^2 \quad (15.40)$$

whose derivative along the solutions of (15.39) is

$$\dot{V}_6 = -\alpha_0 s^2 + s f^T\tilde{\theta}_1 - \gamma_2 s f^T \operatorname{sign}(sf) - \tilde{\theta}_1^T f s \leq -\alpha_0 s^2 \quad (15.41)$$

Integrating the last inequality we obtain

$$\begin{aligned} \frac{1}{2\gamma}||\tilde{\theta}_1(t)||^2 &\leq \frac{1}{2}s^2(0) + \frac{1}{2\gamma}||\theta_1(0) - \theta^*||^2 \\ &\leq \frac{1}{2}s^2(0) + \left(||\theta_1(0)||^2 + ||\theta^*||^2\right)/\gamma \\ &\leq \frac{1}{2}s^2(0) + \left(||\theta_1(0)||^2 + \bar{\theta}^2\right)/\gamma \end{aligned} \quad (15.42)$$

where $||\theta^*|| \leq \bar{\theta}$. Finally we have the bound for

$$||\tilde{\theta}_1(t)|| \leq \sqrt{\gamma s^2(0) + 2||\theta(0)||^2 + 2\bar{\theta}^2} \quad (15.43)$$

Note that the last bound contains only known constants.

Now we are able to get an accurate bound for the tracking error $s^2(t)$. Let us once more use the Lyapunov function candidate

$$V_7(t) = \frac{1}{2}s^2 \quad (15.44)$$

Taking the time derivative along (15.39) we obtain:

$$\dot{V}_7(t) \leq -\alpha_0 s^2 + |sf|||\tilde{\theta}_1|| - \gamma_2|sf| \quad (15.45)$$

Substituting the bound (15.43) we obtain

$$\dot{V}_7 \leq -\alpha_0 s^2 + \left(\sqrt{\gamma s^2(0) + 2||\theta(0)||^2 + 2\bar{\theta}^2} - \gamma_2\right)|sf| \tag{15.46}$$

Choosing $\gamma_2 = \left(\gamma s^2(0) + 2||\theta(0)||^2 + 2\bar{\theta}^2\right)^{\frac{1}{2}}$ we obtain the bound

$$\frac{1}{2}s^2(t) \leq \frac{1}{2}s^2(0)e^{-\alpha_0 t} \tag{15.47}$$

Comparing the last bound with the bounds for the tracking error derived for the integral and proportional-integral algorithms (15.27) and (15.36), we conclude that better performance (i.e. the exponential convergence) of the tracking error is achieved for combined VSC and direct adaptive algorithms.

We have not used all the available opportunities for performance improvement, so we can further improve the parameter indentifiability of our adaptive algorithms. Indeed, it is easy to see from (15.45) that the lower the value of $||\tilde{\theta}_1(t)||$, the lower value of the gain γ_2 we have to choose in the algorithm. On the other hand, the lower the gain we use in the adjustment law, the better the performance.

Next we show that better performance can be obtained when the parameters converge to their true values. Suppose at a time instant that there exist positive constant K_1 and K_2 such that the inequality $||\tilde{\theta}_1(t)||^2 \leq K_1 e^{-K_2 t}$ holds instead of (15.43). Then (15.45) has the form

$$\dot{V}_7 \leq -\alpha_0 s^2 + \left(\sqrt{K_1}e^{-K_2/2t} - \gamma_2\right)|sf| \tag{15.48}$$

Selecting $\gamma_2 = \sqrt{K_1}e^{-K_2/2t}$ we have the same bound (15.47), but the gain vanishes with time.

It is worth remarking that in the case where the parameters converge exponentially to their true values, the bound for $s^2(t)$ is better for the combined variable structure and direct adjustment law than for both integral and proportional-integral adaptation.

The next step is the modification of the adaptive algorithms in order to improve their identifiability. The prediction error should be incorporated into the adjustment law. In order to obtain the prediction error, the measurement of $\dot{s}$ is required. In Sect. 15.5 we describe Lyapunov design procedures for obtaining the prediction error estimate and corresponding adjustment laws driven by the prediction error. These results are not related to the main control problem but they will be used in Sect. 15.6 to improve the identifiability of direct adaptive control schemes.

15.5 Prediction Error Estimation and Indirect Algorithms

15.5.1 Lyapunov Design

We derive new Lyapunov design procedures for constructing both the prediction error estimate and an indirect algorithm. This approach allows the development of new robust algorithms which have advantages when bounded unmeasurable disturbances act on the plant.

The first step, using only the tracking error, is to derive the prediction error estimate

$$e_p = \varphi^T \tilde{\theta} \tag{15.49}$$

with $\varphi \in R^n$. Consider the Lyapunov function candidate

$$V(t) = \frac{1}{2}(s - \varepsilon - \varphi^T \tilde{\theta})^2 \tag{15.50}$$

where ε and φ are secondary variables to be chosen below. Taking the derivative of $V(t)$ along the solutions of the error model (15.13), we obtain

$$\dot{V}(t) = (s - \varepsilon - \varphi^T \tilde{\theta})(-\alpha_0 s - \dot{\varepsilon} + f^T \tilde{\theta} - \dot{\varphi}^T \tilde{\theta} - \varphi^T \dot{\theta}) \tag{15.51}$$

The choice

$$\dot{\varepsilon} = -\alpha_0 \varepsilon - \varphi^T \dot{\theta} \tag{15.52}$$

$$\dot{\varphi} = -\alpha_0 \varphi + f \tag{15.53}$$

implies that $\dot{V} = -\alpha_0 V$ and $V(t) = V(0)e^{-\alpha_0 t}$. It is easy to see that $(s - \varepsilon)$ plays the role of a prediction error estimate. Moreover, if $V(0) = 0$, i.e. $s(0) = \varepsilon(0)$ and $\varphi(0) = 0$, then $e_p = s - \varepsilon = \varphi^T \tilde{\theta}$ for all $t \geq 0$.

Our next step is to design the adjustment law using the prediction error estimate (15.49) derived above. Consider the Lyapunov function candidate

$$V(t) = \frac{1}{2}(s - \varepsilon - \varphi^T \tilde{\theta})^2 + \frac{1}{2}\tilde{\theta}^T \tilde{\theta} \tag{15.54}$$

Taking the derivative of $V(t)$ along the solutions of (15.13),(15.52) and (15.53) we have

$$\begin{aligned} \dot{V}(t) &= -\frac{\alpha_0}{2}\left((s - \varepsilon - \varphi^T \tilde{\theta})^2 + (s - \varepsilon)^2 + \tilde{\theta}^T \varphi \varphi^T \tilde{\theta}\right) \\ &\quad + \alpha_0 (s - \varepsilon)\varphi^T \tilde{\theta} + \tilde{\theta}^T \dot{\theta} \end{aligned} \tag{15.55}$$

For

$$\dot{\theta} = -\alpha_0 \varphi (s - \varepsilon) \tag{15.56}$$

we conclude that $(s - \varepsilon - \varphi^T \tilde{\theta})$ is bounded and square-integrable, $\tilde{\theta}$ is bounded, and $(s - \varepsilon)$ and $\varphi^T \tilde{\theta}$ are square-integrable. The use of filtering to derive the prediction error estimate was proposed by Narendra and Annaswamy (1989) and Miroshnik and Nikiforov (1991), and by Middleton and Goodwin (1988) for robot manipulators; but the filter and adaptive algorithm were designed separately. Algorithms (15.52), (15.53) and (15.56) were considered by Narendra and Annaswamy (1989) and Miroshnik and Nikiforov (1991). Miroshnik and Nikiforov (1991) designed these algorithms for a general error model, the filter was designed via a sensitivity approach, and to prove the convergence of the tracking error a restrictive assumption on the boundedness of $\varphi(t)$ was made.

It is worthwhile to note that the function (15.50) was not proposed in the above, and indirect algorithms were designed via the Lyapunov function $V(t) = \frac{1}{2}\|\tilde{\theta}\|^2$. The filter dynamics has not been taken into account although it is easy to see that, if $V(0) \neq 0$, the prediction error estimate decays exponentially. In this case the wrong choice of the algorithm and filter parameters (i.e. α_0 small and the gain in the adjustment law (15.56) large) yields poorer performance. The function (15.54) allows the linking of the filter parameters with the algorithm parameters, and the consideration of the filter dynamics. Also the function (15.54) allows the design of new algorithms with a relay term driven by the prediction error estimate.

Note that indirect adaptive algorithms do not usually use information about the upper bound of the norm of θ^*. However, it is intuitively clear that the more information we use in the adjustment law the better the performance.

15.5.2 Additional Relay Term

Our next step is to design indirect adaptive algorithms which exploit the information of the upper bound of the norm of θ^* and use the relay term driven by the prediction error. The advantages of this algorithm with respect to (15.52), (15.53) and (15.56) will be shown in detail below. The properties of relay algorithms driven by the estimate of the prediction error, are similar to the properties of VSC algorithms; in particular they reduce the influence of bounded unmeasurable disturbances.

Consider the error model (15.13) with the estimator

$$
\begin{aligned}
\dot{\theta} &= -\gamma\alpha_0\varphi\,\mathrm{sign}(s-\varepsilon), && (15.57)\\
\dot{\varepsilon} &= -\alpha_0\varepsilon - \varphi^T\dot{\theta} + \gamma\alpha_0\,\mathrm{sign}(s-\varepsilon) && (15.58)\\
\dot{\varphi} &= -\alpha_0\varphi + f && (15.59)
\end{aligned}
$$

where γ is a design parameter.

Consider the Lyapunov function candidate (15.54) taking the derivative along the solutions of (15.13), (15.57)-(15.59) we obtain

$$
\begin{aligned}
\dot{V}(t) &= -\frac{\alpha_0}{2}\left((s-\varepsilon-\varphi^T\tilde{\theta})^2 + (s-\varepsilon)^2 + \tilde{\theta}^T\varphi\varphi^T\tilde{\theta}\right)\\
&\quad +\alpha_0(s-\varepsilon)\varphi^T\tilde{\theta} + \tilde{\theta}^T(\dot{\theta} + \gamma\alpha_0\varphi\,\mathrm{sign}(s-\varepsilon)) - \gamma\alpha_0|s-\varepsilon|
\end{aligned}
$$

Taking bounds on the right hand side

$$\dot{V}(t) \le -\frac{\alpha_0}{2}\left((s-\varepsilon-\varphi^T\tilde{\theta})^2+(s-\varepsilon)^2+\tilde{\theta}^T\varphi\varphi^T\tilde{\theta}\right)$$
$$+\alpha_0|s-\varepsilon|\left(||\varphi||\left(\bar{\theta}+||\theta||\right)-\gamma\right) \tag{15.60}$$

Choosing $\gamma = ||\varphi||(\bar{\theta}+||\theta||)$, where $||\theta^*|| \le \bar{\theta}$, we conclude that the estimator has properties similar to the one previously derived. However, in the case where the bounded unmeasurable disturbances act on the plant, to reduce the influence of the disturbances, we have to incorporate a relay term in the filter.

15.5.3 Sliding Mode Approach

Next we present another way to obtain the prediction error estimate. This is based on the ***sliding mode***. Consider the following Lyapunov function candidate

$$V_1(t) = (s-\varepsilon_1)^2 \tag{15.61}$$

Taking the time derivative along (15.13) we have

$$\dot{V}_1 = 2(s-\varepsilon_1)(-\alpha_0 s + f^T\tilde{\theta} - \dot{\varepsilon}_1) \tag{15.62}$$

Let us choose $\dot{\varepsilon}_1$ in the form

$$\dot{\varepsilon}_1 = -\alpha_0\varepsilon_1 + \gamma\,\mathrm{sign}(s-\varepsilon_1) \tag{15.63}$$

where $\gamma > 0$ is a design parameter. Then

$$\dot{V}_1 \le 2|s-\varepsilon_1|\left(||f||\left(||\theta||+\bar{\theta}\right)-\gamma\right) \tag{15.64}$$

Choosing $\gamma = ||f||(||\theta||+\bar{\theta})+\beta/2$, $\beta > 0$ we obtain

$$\dot{V}_1 \le -\beta\sqrt{V_1} \tag{15.65}$$

and hence the inequality

$$\sqrt{V_1(t)} \le \sqrt{V_1(0)} - \beta/2t \tag{15.66}$$

is valid. Note, that for $t \ge t^*$, where $t^* \le 2/\beta\sqrt{V_1(0)}$, a sliding mode arises in the system.

Choosing the initial conditions such that $s(0) = \varepsilon_1(0)$, we obtain $V_1(t) = 0$ for all $t \ge 0$. Subtracting (15.63) from (15.13) we have

$$\dot{s} - \dot{\varepsilon}_1 = -\alpha_0(s-\varepsilon_1) + f^T\tilde{\theta} - \gamma\,\mathrm{sign}(s-\varepsilon_1) \tag{15.67}$$

and the ideal sliding equation is

$$f^T\tilde{\theta} = \gamma\,\mathrm{sign}(s-\varepsilon_1) \tag{15.68}$$

which gives the prediction error estimate. Now we can construct the adaptive estimator in the form

$$\dot{\theta} = -\gamma_1 f f^T\tilde{\theta} = -\gamma_1\gamma f\,\mathrm{sign}(s-\varepsilon_1), \tag{15.69}$$

where $\gamma_1 > 0$.

Note that the properties of the estimator (15.69) are similar to the properties of the two previous estimators.

15.5.4 Comparative Analysis of the Two Proposed Indirect Algorithms with Bounded Disturbances

Now let us show the advantages of the algorithm (15.57)-(15.59) with respect to the algorithm (15.52), (15.53) and (15.56).

Suppose that bounded unmeasurable disturbances act on the plant (15.2) additively with the control action. Then using the control law (15.12), one obtains the error model in the form

$$\dot{s} = -\alpha_0 s + f^T \tilde{\theta} + \eta \tag{15.70}$$

where η is the disturbance with $|\eta| \leq \bar{\eta}$ and $\bar{\eta}$ known. Consider the algorithm (15.52), (15.53) and (15.56) with σ-modification which was proposed by Ioannou and Kokotovic (1983). It prevents unbounded growth of adaptation gains when disturbances act on the plant. Let the adjustment law have the form

$$\begin{aligned} \dot{\theta} &= -\alpha_0 \varphi(s-\varepsilon) - \sigma\theta \\ \dot{\varepsilon} &= -\alpha_0 \varepsilon - \varphi^T \dot{\theta} \\ \dot{\varphi} &= -\alpha_0 \varphi + f \end{aligned} \tag{15.71}$$

Let us obtain the bounds for the prediction error estimate and $||\tilde{\theta}||^2$. Consider the Lyapunov function candidate (15.54). Taking the derivative along the solutions of (15.70) and (15.71)

$$\dot{V}(t) = -\alpha_0(s-\varepsilon-\varphi^T\tilde{\theta})^2 + (s-\varepsilon)\eta - \varphi^T\tilde{\theta}\eta - \alpha_0\tilde{\theta}^T\varphi(s-\varepsilon) - \sigma\tilde{\theta}^T\theta \tag{15.72}$$

Applying the inequality

$$\begin{aligned} -\sigma(\theta-\theta^*)^T\theta &\leq \frac{\sigma}{2}\left(-||\theta-\theta_*||^2 + ||\theta_*||^2\right) \\ &\leq \frac{\sigma}{2}\left(-||\theta-\theta_*||^2 + \bar{\theta}^2\right) \end{aligned} \tag{15.73}$$

and taking bounds we obtain

$$\begin{aligned} \dot{V}(t) &\leq -\frac{\alpha_0}{2}\left((s-\varepsilon-\varphi^T\tilde{\theta})^2 + (s-\varepsilon)^2\right) + |s-\varepsilon|\bar{\eta} \\ &\quad -\frac{1}{2}\left(\alpha_0\tilde{\theta}^T\varphi\varphi^T\tilde{\theta} - 2|\varphi^T\tilde{\theta}|\bar{\eta} + \sigma||\tilde{\theta}||^2 - \sigma\bar{\theta}^2\right) \end{aligned} \tag{15.74}$$

Straightforward calculations show that the maximum value of the function $f(z) = (-\alpha_0 z^2/2 + \bar{\eta}z)$ is $f(z_{max}) = \bar{\eta}^2/(2\alpha_0)$, and hence

$$\dot{V}(t) \leq \frac{1}{2}\left(-\alpha_0(s-\varepsilon-\varphi^T\tilde{\theta})^2 - \sigma||\tilde{\theta}||^2 + 2\bar{\eta}^2/\alpha_0 + \sigma\bar{\theta}^2\right) \tag{15.75}$$

Suppose for simplicity that $\alpha_0 = \sigma$. Then the following inequality is valid

$$V(t) \leq V(0)e^{-\alpha_0 t} + \bar{\eta}^2/{\alpha_0}^2 + \bar{\theta}^2/2 \tag{15.76}$$

Now consider the algorithm (15.57)-(15.59) with σ-modification.

$$\begin{aligned}
\dot{\theta} &= -\gamma\alpha_0\varphi\,\mathrm{sign}(s-\varepsilon) - \sigma\theta &(15.77)\\
\dot{\varepsilon} &= -\alpha_0\varepsilon - \varphi^T\dot{\theta} + \gamma\alpha_0\,\mathrm{sign}(s-\varepsilon)\\
\dot{\varphi} &= -\alpha_0\varphi + f
\end{aligned}$$

where $\gamma = ||\varphi||(\bar{\theta} + ||\theta||) + \bar{\eta}/\alpha_0$. Consider the Lyapunov function candidate (15.54), whose derivative along the solutions of (15.70) and (15.77) is

$$\begin{aligned}
\dot{V}(t) &= \frac{1}{2}\left(-\alpha_0(s-\varepsilon-\varphi^T\tilde{\theta})^2 - \alpha_0(s-\varepsilon)^2 + 2(s-\varepsilon)\eta - \alpha_0\tilde{\theta}^T\varphi\varphi^T\tilde{\theta}\right)\\
&\quad -\alpha_0\gamma|s-\varepsilon| - \varphi^T\tilde{\theta}\eta + \alpha_0\tilde{\theta}^T\varphi(s-\varepsilon) - \sigma\tilde{\theta}^T\theta\\
&\leq \frac{1}{2}\left(-\alpha_0(s-\varepsilon-\varphi^T\tilde{\theta})^2 - \alpha_0(s-\varepsilon)^2 - \sigma||\tilde{\theta}||^2 + \bar{\eta}^2/\alpha_0 + \sigma\bar{\theta}^2\right)
\end{aligned}$$

Taking $\alpha_0 = \sigma$ we have $V(t) \leq V(0)e^{-\alpha_0 t} + \bar{\eta}^2/(2\alpha_0^2) + \bar{\theta}^2/2$. Comparing the two bounds it is easy to see that the upper bound of the region of convergence for the algorithm (15.77) is less than for the algorithm (15.71). This is due to the relay term $\gamma\alpha_0\,\mathrm{sign}(s-\varepsilon)$ in the filter. On the other hand this term does not allow the design of a filter for the prediction error separately from the algorithm design.

15.5.5 Convergence of the Parameters

Here we consider the case where φ is persistently exciting, i.e. there exist strictly positive constants δ , β and β_1 such that the following inequality is true for all $t \geq 0$

$$\beta_1 I_n \geq \int_t^{t+\delta} \varphi(r)\varphi(r)^T\,\mathrm{d}r \geq \beta I_n \tag{15.78}$$

Consider the following adjustment law

$$\begin{aligned}
\dot{\theta} &= -\alpha_0\Gamma(t)\varphi(s-\varepsilon) &(15.79)\\
\dot{\Gamma}(t) &= -\alpha_0\Gamma\varphi\varphi^T\Gamma + \alpha_0(\lambda\Gamma - \Gamma^2/k_0),\quad 0 < \Gamma(0) \leq \lambda k_0 I_n &(15.80)
\end{aligned}$$

where λ and k_0 are positive numbers and φ, ε are adjusted as in (15.52) and (15.53). The following properties of the modified least squares estimator (15.80) can be easily proved: (i) $\Gamma(t) \leq \lambda k_0 I_n$ for any $t \geq 0$, (ii) if $\varphi(t)$ is persistently exciting then there exist $\lambda_1 > 0$ and $\lambda_2 > 0$ such that $\lambda\Gamma^{-1}(t) - I_n/k_0 \geq \lambda_1 I_n$ and $\Gamma^{-1}(t) \leq \lambda_2 I_n$ for any $t \geq 0$. The proof is motivated by the work of Lozano et al (1990) and Slotine and Li (1989).

Consider the behaviour of the system (15.13), (15.79) and (15.80) under the condition of persistent excitation (15.78). Let the Lyapunov function candidate be

$$V(t) = \frac{1}{2}(s-\varepsilon-\varphi^T\tilde{\theta})^2 + \frac{1}{2}\tilde{\theta}^T\Gamma^{-1}(t)\tilde{\theta} \tag{15.81}$$

Taking the derivative of $V(t)$ and applying the condition (15.78) we get $\dot{V}(t) \leq -\alpha_1 V(t)$, where $\alpha_1 = \min(\alpha_0, \alpha_0\lambda_1/\lambda_2)$. This implies the convergence of the estimated parameters to their true values.

Note that the main drawback of the above indirect algorithms is that we cannot guarantee that the convergence of the tracking error follows from the convergence of the prediction error. To guarantee the convergence of the tracking and the prediction error simultaneously we need to consider combined direct and indirect algorithms.

15.6 Combined Algorithms

Consider the following adjustment law

$$\begin{aligned}\theta &= -\int_0^t (\Gamma(r)\left[f(r)s(r) + \alpha_0\varphi(r)(s(r)-\varepsilon(r)) + \varphi(r)\psi_1(s(r),\varepsilon(r))\right])\, dr \\ &\quad -\psi(f,s) \qquad (15.82)\end{aligned}$$

where $\psi(f,s) \in R^n$ and $\psi_1(s,\varepsilon) \in R^1$ are functions satisfying the inequalities

$$\psi(f,s)^T(fs) \geq 0 \qquad (15.83)$$

$$\psi_1(s,\varepsilon)(s-\varepsilon) \geq 0 \qquad (15.84)$$

The values $\varphi(t) \in R^n$, $\varepsilon(t) \in R^1$ and the $n \times n$ gain matrix $\Gamma(t)$ are adjusted as follows

$$\begin{aligned}\dot{\varphi}(t) &= -\alpha_0\varphi(t) + f \qquad (15.85)\\ \dot{\varepsilon}(t) &= -\alpha_0\varepsilon(t) + \varphi(t)^T\Gamma(t)\left(fs + \alpha_0\varphi(t)(s(t)-\varepsilon(t)) + \varphi\psi_1(s,\varepsilon)\right) \\ &\quad -f^T\psi(f,s) + \psi_1(s,\varepsilon) \\ \dot{\Gamma}(t) &= -\alpha_0\Gamma\varphi\varphi^T\Gamma + \alpha_0(\lambda\Gamma - \Gamma^2/k_0), \quad 0 < \Gamma(0) \leq \lambda k_0 I_n \qquad (15.86)\end{aligned}$$

where λ and k_0 are positive numbers. The algorithms represented by (15.82) with $\Gamma = \text{const}$, $\alpha_0 = 0$ and $\psi_1(s,\varepsilon) = 0$ were also considered by Andrievsky et al (1988), Fradkov (1979) and Stotsky (1991) in the class of speed gradient algorithms.

To prove the stability of the system (15.13) and (15.82) we use the following Lyapunov function candidate (Stotsky 1993)

$$V(t) = \frac{1}{2}\left(s^2 + (s - \varepsilon - \varphi^T(\tilde{\theta} + \psi(f,s))^2 + ||\tilde{\theta} + \psi(f,s)||^2_{\Gamma^{-1}(t)}\right) \qquad (15.87)$$

whose derivative along the solutions of the system is

$$\begin{aligned}\dot{V}(t) &\leq -\frac{\alpha_0}{2}\left(s^2 + \left(s - \varepsilon - \varphi^T(\tilde{\theta} + \psi(f,s)\right)^2 + (s-\varepsilon)^2\right. \\ &\qquad \left. ||\tilde{\theta} + \psi(f,s)||^2_{(\lambda\Gamma^{-1}(t) - I_n/k_0)}\right) \qquad (15.88)\end{aligned}$$

¿From this we conclude that s is square-integrable, and using the boundedness of the desired trajectories, one can show that $\dot{s}$ is bounded and (15.9) is achieved. The achievement of the control goal (15.4) can be easily established, since the system (15.11) with square-integrable and convergent input, is stable if $\det(pI_{n-1} - (A_{11} - A_{12}C_1))$ is a Hurwitz polynomial. Note that the polynomial $\det(pI_{n-1} - (A_{11} - A_{12}C_1))$ coincides with the numerator of the transfer function $W(p) = C(pI_n - A)^{-1}B$ (Stotsky 1991).

Now consider the case where $\varphi(t)$ is persistently exciting. In this case using the properties of the estimator (15.59) described previously we obtain

$$\begin{aligned}\dot{V}(t) \;\le\; & -\frac{\alpha_0}{2}\Big(s^2 + (s - \varepsilon - \varphi^T(\tilde{\theta} + \psi(f,s))^2 \\ & + \lambda_1/\lambda_2 \|\tilde{\theta} + \psi(f,s)\|^2_{\Gamma^{-1}(t)}\Big)\end{aligned} \tag{15.89}$$

This implies the exponential convergence of the Lyapunov function

$$\dot{V}(t) \le -\alpha_1 V(t) \tag{15.90}$$

where $\alpha_1 = \min(\alpha_0, \alpha_0\lambda_1/\lambda_2)$, and in turn the convergence of the tracking errors and the estimation errors $(\tilde{\theta} + \psi(f,s))$ to zero. Note that, if we choose $\psi(f,s) = \gamma\ \mathrm{sign}(fs)$ and $\psi_1(s,\varepsilon) = \gamma_1\ \mathrm{sign}(s - \varepsilon)$ with $\gamma, \gamma_1 > 0$, we get an interesting identification algorithm with sliding modes. When we choose $\psi(f,s) = \gamma(fs)$ we obtain a proportional plus integral adjustment law, usually designed by the hyperstability approach (Landau 1979).

To conclude the first part of this chapter, we note that the results presented above can be easily extended to the case where the state vector is not completely measured and also to the multi-input case. In the second part we shall apply the ideas described above to the control of SISO plants using input/output measurements only.

15.7 Combined Algorithms for SISO Plants

Consider the tracking error model of SISO plant in the form

$$\dot{e} = Ae + b\omega(x,r)\tilde{\theta}, \qquad e_1 = c^T e \tag{15.91}$$

where $e \in R^n$ is the tracking error; $\omega(x,r)^T \in R^m$ is a known vector function bounded with respect to its arguments; $x \in R^N$ and $r \in R^1$ are the state vector and bounded input respectively; $\tilde{\theta} = \theta - \theta^*$ with $\theta \in R^m$ the vector of adjustable parameters, and θ^* the vector of true parameters; e_1 is the measured scalar output signal; A is an asymptotically stable matrix; and the transfer function $W(p) = c^T(pI_n - A)^{-1}b$ is strictly positive real.

In the design procedure we will use the Yakubovich-Kalman Lemma (Yakubovich 1962).

Lemma 15.1 *If a strictly proper rational function $W(p) = c^T(pI_n - A)^{-1}b$ with Hurwitz matrix A is strictly positive real, then there exists a positive definite matrix P such that*

$$\begin{aligned} A^T P + PA &= -Q \qquad (15.92)\\ Pb &= c \qquad (15.93) \end{aligned}$$

with Q a positive definite symmetric matrix.

The error model (15.91) is used widely in various control problems involving SISO plants (Hsu 1990, Narendra and Valavani 1978, Narendra and Annaswamy 1989) and in the problem of adaptive observer design (Marino 1990).

Our problem is to find the adjustment law to ensure the convergence of the tracking error

$$\lim_{t\to\infty} e(t) = 0 \qquad (15.94)$$

Consider the following adjustment law

$$\frac{\mathrm{d}\,(\theta + \psi(x,r))}{\mathrm{d}t} = -\Gamma(t)\left(\omega(x,r)^T e_1 - \frac{\alpha_0}{2}\varphi^T\varepsilon + \varphi^T c\psi_1(e_1\varepsilon_1)\right) \qquad (15.95)$$

where $\varepsilon \in R^n$ and φ an $n \times m$ matrix satisfies the following set of differential equations

$$\begin{aligned} \dot{\varepsilon} &= A\varepsilon + \varphi\Gamma(t)\left(\omega(x,r)^T e_1 - \frac{\alpha_0}{2}\varphi^T\varepsilon + \varphi^T c\psi_1\right)\\ &\quad + b\omega(x,r)\psi(x,r) + b\psi_1 \qquad (15.96)\\ \dot{\varphi} &= A\varphi + b\omega(x,r) \qquad (15.97) \end{aligned}$$

The gain matrix is adjusted according to

$$\begin{aligned} \dot{\Gamma}(t) &= -\kappa\Gamma\varphi^T\varphi\Gamma + \kappa\lambda(t)\Gamma \qquad (15.98)\\ \Gamma(0) &= \Gamma(0)^T > 0, \qquad \|\Gamma(0)\| \le k_0\\ \lambda(t) &= \lambda_0(1 - \|\Gamma(t)\|/k_0) \end{aligned}$$

where $\lambda_0, k_0 > 0$, $\psi(x,r) \in R^m$ and $\psi_1(e_1,\varepsilon_1) \in R^1$ satisfy the following inequalities

$$\begin{aligned} \psi(x,r)^T(\omega(x,r)^T e) &\ge 0 \qquad (15.99)\\ \psi_1(e_1,\varepsilon_1)(e_1 - \varepsilon_1) &\ge 0 \qquad (15.100) \end{aligned}$$

where $\varepsilon_1 = c^T\varepsilon$, κ is a positive constant we shall choose below; and α_0 is a positive constant such that $0 < \alpha_0 \le \lambda_{\min}(Q)$, where $\lambda_{\min}(Q)$ is the minimal eigenvalue of the matrix Q. The gain of the adjustment law (15.98) was proposed by Slotine and Li (1989) and the adjustment law (15.95) for the error model (15.91) with $\psi = 0$, $\alpha_0 = 0$, $\psi_1 = 0$ and Γ =constant is well known (Narendra and Annaswamy 1989, and Marino 1990). Algorithms similar to (15.95) with $\alpha_0 = 0$, $\psi_1 = 0$, Γ =constant and $\psi(x,r) = \gamma\ \mathrm{sign}(\omega(x,r)^T e_1)$ $(\gamma > 0)$, were also considered by Hsu (1990). We note furthermore that ε plays the role of a prediction error estimate.

The achievement of the control goal (15.94) can be established by mean of the Lyapunov function candidate

$$V(t) = \frac{1}{2}e^T Pe + \frac{1}{2}\left(e - \varepsilon - \varphi(\tilde{\theta} + \psi(x,r)\right)^T P \left(e - \varepsilon - \varphi(\tilde{\theta} + \psi(x,r)\right) + \frac{1}{2}\|\tilde{\theta} + \psi(x,r)\|^2_{\Gamma^{-1}(t)} \tag{15.101}$$

Taking the derivative of $V(t)$ along the solutions (15.91) and (15.95), and using (15.92) and (15.99) one obtains after some simple calculations

$$\dot{V} \leq -\beta e^T Pe/\lambda_{\max}(P) - \alpha_0(e - \varepsilon - \varphi(\tilde{\theta} + \psi(x,r))^T P(e - \varepsilon - \varphi(\tilde{\theta}\psi(x,r))/(4\lambda_{\max}(P)) - \kappa/2\lambda(t)\|\tilde{\theta} + \psi(x,r)\|^2_{\Gamma^{-1}(t)} \tag{15.102}$$

where $\lambda_{\max}(P)$ is the maximum eigenvalue of the matrix P . The positive constants κ and β are chosen to satisfy $\alpha_0/4 - \alpha_0/(4\lambda_1) = \kappa/2$, $\alpha_0/2 - \alpha_0\lambda_1/4 = \beta$, where $\lambda_1 \in (1,2)$. We assume that φ is persistently exciting. In this case there exists $\lambda_2 > 0$ such that $\lambda(t) \geq \lambda_2$ for any $t \geq 0$ (Slotine and Li 1989). Let α_1 be a positive constant such that

$$\alpha_1 = \min\left(\frac{2\beta}{\lambda_{\max}(P)}, \frac{\alpha_0}{2\lambda_{\max}(P)}, \kappa\lambda_2\right)$$

then the inequality $\dot{V} \leq -\alpha_1 V$ is true. This, in turn, implies the exponential convergence of the tracking and estimation errors. It can be easily shown that the control goal (15.94) is achieved also for the case where φ is non-exciting.

The choice of the parameter λ_1 reflects how much attention the adjustment law pays respectively to the tracking error and to the prediction. When choosing the parameter α_0 we should calculate the bounds of the corresponding functions for all frequencies. In practice, however, it is advisable to choose α_0 sufficiently small.

In conclusion we note that the most general form of the algorithms (15.99) is

$$\psi(x,r) = \gamma_1 \operatorname{sign}(\omega(x,r)^T e_1) + \gamma_2\omega(x,r)^T e_1 \tag{15.103}$$

$$\psi_1(e_1,\varepsilon_1) = \gamma_3 \operatorname{sign}(e_1 - \varepsilon_1) + \gamma_4(e_1 - \varepsilon_1) \tag{15.104}$$

with $\gamma_i > 0,\ i = 1,4$.

15.8 Conclusions

A combination of adaptive algorithms and variable structure control for the regulation of a certain class of linear time-invariant plants, has been presented. New algorithms have been developed both for the case where the state vector is available for measurement, and for SISO plants using input/output measurements only. The adaptive algorithms are driven by the tracking and the prediction error. The design is based on a new form of the Lyapunov function and exponential convergence of the tracking and estimation errors in the presence of persistent excitation has been established. The results improve the performance of existing adaptive schemes for this class of system.

References

Andrievsky, B.R., Stotsky, A.A., Fradkov, A.L. 1988, Speed gradient scheme in control and adaptation. *Avtomatika i Telemechanika* **12**, 3–39 (in Russian)

Bukov, V.N. 1987, *Adaptive flight control systems*, Nauka, Moscow, pp 232 (in Russian)

Erzberger, H. 1968, Analysis and design of model reference following systems by state space tecniques.*Proc. JACC*, 572–580

Fradkov, A.L. 1979, Velocity gradient scheme and its application in adaptive control. *Avtomatika i Telemechanika* **9**, 90–101 (in Russian)

Fradkov, A.L. 1986, Integral-derivative speed gradient algorithms. *Doklady Acad. Sci. USSR* **268**, 798–801 (in Russian)

Hsu, L. 1988, Variable structure model reference adaptive control using only input and output measurements. *Proc IEEE Conference on Decision and Control,* , Austin, USA, 2396–2401

Hsu, L. 1990, Variable structure model reference adaptive control (VSS-MRAC) using only input and output measurements: general case. *IEEE Transactions on Automatic Control* **35**, 1238–1243

Ioannou, P., Kokotovic, P. 1983, *Adaptive systems with reduc models*, Lecture Notes Contr. Inform. Sci. **47**, Springer, New York, pp 168

Landau, I.D. 1979, *Adaptive control systems. The model reference approach*, Dekker, New York, pp 406

Lozano, R., Collado, J., Mondie, S. 1990, Model reference robust adaptive control without apriory knowledge of the high frequency gain. *IEEE Transactions on Automatic Control* **35**, 71–78

Marino, R. 1990, Adaptive observers for single output nonlinear systems *IEEE Transactions on Automatic Control* **35**, 1054–1058

Middleton, R.H., Goodwin, G.C. 1988, Adaptive computed torque control for rigid link manipulators. *Systems and Control Letters* **2**, 9–16

Miroshnik, I.V., Nikiforov, V.O. 1991, Convergence rate improvement in gradient procedure. *Izv. Vuzov Priborostroenie* **8**, 76–82 (in Russian)

Narendra, K.S., Valavani, L. 1978, Stable adaptive controller design–direct control. *IEEE Transactions on Automatic Control* **23**, 570-583

Narendra, K.S., Annasvamy, A.M. 1989, *Stable adaptive systems*, Prentice Hall, Englewood Cliffs, NJ, pp 494

Panteley, E.V., Stotsky, A.A. 1992, Trajectory tracking for constrained robot motion: Composite adaptive control design . *Proc. 36th ANIPLA*, Genova,Italy, 563–571

Slotine, J.J.E., Coetsee, J.A. 1986, Adaptive sliding controller synthesis for non-linear systems. *International Journal of Control* **43**, 1631–1651

Slotine, J.J.E., Li, W. 1989, Composite adaptive control of robot manipulators. *Automatica* **25**, 509–519

Stotsky, A.A. 1991, Design of combined variable structure system and reference equation system algorithms. *Proc. First IFAC Symp. on Design Methods of Control Systems* Zurich, Switzerland, 465–469

Stotsky, A.A. 1993, Lyapunov design for convergence rate improvement in adaptive control. *International Journal of Control* **57**, 501–504

Utkin, V.I. 1981, *Sliding modes in optimization and control*, Nauka, Moscow, pp 368 (in Russian)

Yakubovich, V.A. 1962, Solution of some matrix inequalities in systems theory, *Dokl. Akad. Nauk USSR* **143**, 1304–1307 (in Russian)

16. Variable Structure Control of Nonlinear Systems: Experimental Case Studies

D. Dan Cho

16.1 Introduction

The science of control is concerned with modifying the behaviour of dynamical systems so as to achieve desired goals, which may include optimal fuel economy, increased manufacturing productivity, stabilization of magnetic levitation bearing, or landing a lunar module on the moon. It is highly interdisciplinary in nature, drawing from physical principles, mathematical theories, instrumentation and computers, as well as the diverse nature of systems that need to be understood and controlled.

A 1988 white paper from the Society for Industrial and Applied Mathematics entitled, *Future Directions in Control Theory: A Mathematical Perspective*, classifies the process of controlling a dynamical system into the following three fundamental issues:

- Modelling the system based on either physical laws or identification. The goal here is different from other areas of science. The mathematical model is dictated by the control task: it must provide adequate predictive capabilities and yet satisfy the constraints imposed by limitations in mathematical theories and hardware capabilities.
- Synthesizing the control input using mathematical control theories. This must deal successfully with modelling inaccuracies as well as nonlinearities and complexities that are inherent in most physical systems. In addition, this process includes signal processing methodologies such as filtering, prediction and state estimation.
- Experimental research in control. This has the purpose of finding the best control-oriented models as well as testing the validity of new control paradigms. This process also includes developing real-time algorithms and computer codes.

These three processes are not independent; many iterations among the processes are often required to optimize each process.

While many theories and tools are available for synthesizing the control input for linear systems, synthesizing the control input for nonlinear systems is not a trivial task. Variable Structure Control (VSC) theory holds much promise for providing the means to methodically address the performance and robustness trade-off issues of nonlinear systems. Since much of this book is devoted to this subject, the duplication of review and development is avoided in this chapter. It is assumed that the reader is familiar with classical VSC theory and the sliding mode.

Two experimental case studies are presented in this chapter: automotive fuel-injection control and magnetic levitation control. In each of these cases, the nonlinearities in the dynamics of the system are studied and then discussion of the control objectives and the synthesis of the control input is presented. In the first case study, the performance of a digitally-implemented VSC fuel-injection controller is compared to that of a production fuel-injection controller, using a research automobile. In the second case study, the performance of a digitally-implemented VSC levitation controller is compared to that of a linear classical controller, using a laboratory single-axis magnetic levitation system.

16.2 Fuel-Injection Control

16.2.1 Importance of Analytic Control Methodology for Fuel Injection

According to 1988 U.S. Environmental Protection Agency (EPA) data, 80% of carbon monoxide (CO), 33% of hydrocarbons (HC), and 61% of oxidation products of nitrogen (NO_x) pollution in the U.S. result from transportation. One of the most significant causes of this problem is the inability to design good fuel-control algorithms. In the past, the performance of the production Engine Control Module (ECM) has been sufficient to pass all government-mandated emissions standards and to provide adequate drive feel. However, recent U.S. legislation will soon require a much cleaner engine, which in turn will require extensive improvements to the current fuel-injection controllers. In order to fully appreciate that fuel-injection control is a challenging problem, it is first necessary to discuss the highly nonlinear nature engine dynamics and the limitations imposed by sensors. Figure 16.1 depicts a schematic diagram of the engine fuel-injection system. The fuel-injection control software needs to determine how much air is entering the combustion chambers and accordingly command the proper fuel spray rate. Complete combustion of gasoline occurs at the stoichiometric air-to-fuel (AF) ratio. The stoichiometric ratio can vary slightly depending on the gasoline type; here it is taken as 14.64:1. Complete combustion of gasoline results only in the harmless substances carbon dioxide CO_2 and water vapor (H_2O). However, imprecise AF ratio control and other factors result in exhaust gases that contain CO, HC and NO_x.

In order to meet the emission standards, almost all gasoline-engine vehicles in the U.S. utilize three-way catalytic converters. Figure 16.2 depicts the conversion efficiency of a typical three-way catalytic converter. The catalytic converter has a very small band of peak efficiency near the stoichiometric AF ratio. Note that a mere 1% deviation from it results in a 50% degradation in the conversion of one or more pollutants. Thus, the most important objective

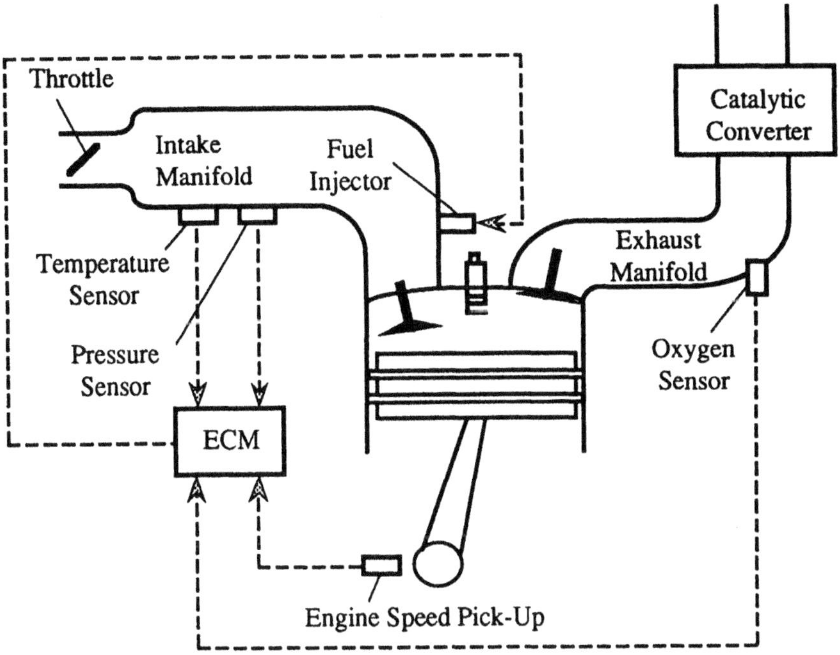

Fig. 16.1. Schematic diagram of fuel-injection systems

of a fuel-injection controller is to maintain the AF ratio at its stoichiometric value through all phases of driving. Note that according to Falk and Mooney's experiments (1980), small oscillations (less than ±0.2 or ±1.4%) of relatively high frequency (at least 1 Hz), do not harm the conversion efficiency because the catalytic converter has a fixed volume, and small and fast oscillations are averaged in the converter.

Referring to Fig. 16.1, in order to maintain a desired AF ratio, a proper amount of fuel rate must be injected in accordance with the mass rate of air entering the combustion chambers, $\dot{m}_{ao}$. Two methods are widely used. The first is the so-called speed-density method:

$$\hat{\dot{m}}_{ao}(t) = \frac{V_e \eta_v m_a(t) \omega_e(t)}{4\pi V_m} \tag{16.1}$$

where $(\hat{\ })$ indicates a model or estimate quantity, V_e is the engine displacement, V_m is the intake manifold volume, η_v is the volumetric efficiency, $m_a(t)$ is the total mass of air in the intake manifold, and $\omega_e(t)$ is the engine speed. Since determining $m_a(t)$ is difficult, using the ideal gas law, (16.1) can be written as

$$\hat{\dot{m}}_{ao}(t) = \frac{V_e \eta_v P_m(t) \omega_e(t)}{4\pi R T_m(t)} \tag{16.2}$$

where $P_m(t)$ is the intake manifold pressure, $T_m(t)$ is the intake manifold temperature, and R is the universal gas constant for air. Then, the desired mass

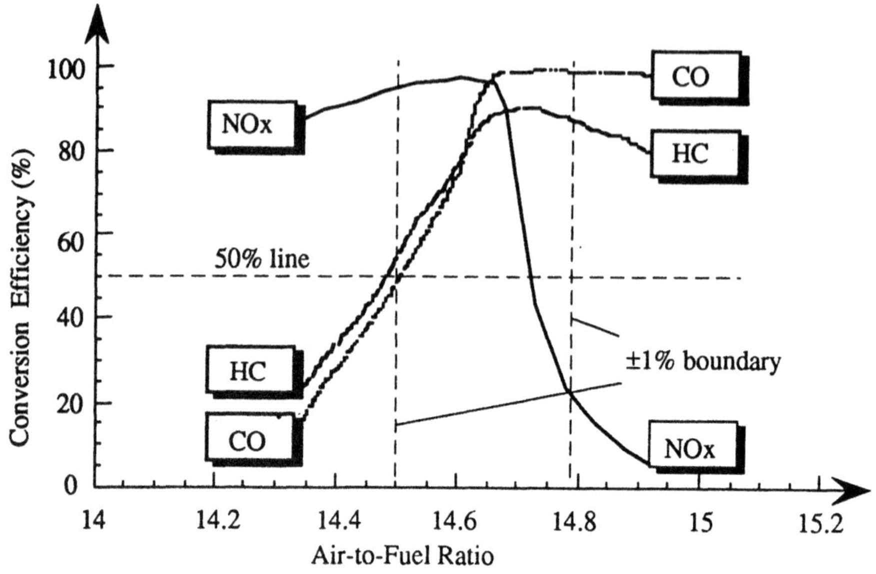

Fig. 16.2. Typical three-way catalytic converter efficiency (Adapted from Falk and Mooney (1980))

rate of fuel entering the combustion chambers can be determined from

$$(\dot{m}_{fo}(t))_{\text{des}} = \frac{\hat{\dot{m}}_{ao}(t)}{\beta} \tag{16.3}$$

where $\dot{m}_{fo}(t)$ is the mass rate of fuel entering the combustion chambers, and β is the desired AF ratio. The second method is called the mass air flow meter method. An air flow rate sensor measures the amount of air entering the intake manifold, $\dot{m}_{ai}(t)$. Then, using the conservation of mass in the intake manifold

$$\dot{m}_{ao}(t) = \dot{m}_{ai}(t) - \dot{m}_a(t) \tag{16.4}$$

Because measuring $\dot{m}_a(t)$ is difficult, a control structure similar to

$$\hat{\dot{m}}_{ao}(t) = \frac{\dot{m}_{ai}(t)}{\tau_m\left(P_m(t), \omega_e(t), \eta_v\right)s + 1} \tag{16.5}$$

is often used, where s is the Laplace operator, and the time constant τ_m varies with manifold pressure, engine speed, and volumetric efficiency. Then, (16.5) can be used in (16.3) to determine the fueling command.

Two major reasons require that a feedback controller be incorporated in both methods. First, the volumetric efficiency η_v is extremely difficult to model accurately. As described in Taylor (1966), Heywood (1988) and Powell (1987), the volumetric efficiency is a complex and nonlinear function of many parameters, including the inlet Reynolds and Mach numbers, manifold and engine geometries, variable valve overlap, intake and exhaust pressures, humidity and temperature. Because of this complexity, the volumetric efficiency cannot be modelled analytically; instead it must be calibrated experimentally. A typical calibration process involves steady-state characterization on a dynamometer for a myriad of load and engine conditions, detailed look-up table generation, and on-vehicle table tuning. Accurately calibrating the volumetric efficiency for all operating conditions to satisfy the stringent emission requirements is not possible. Therefore, feedback control is necessary to compensate for the inaccuracies in modelling and identification.

Second, the typical accuracy of various sensors used in fuel-injection systems are in the 3-5% range (Washino 1989). As shown in Fig. 16.2, 3-5% bias errors completely wipe out the catalyst efficiency. The use of more accurate sensors may improve this situation, but it implies a higher cost, which is highly undesirable.

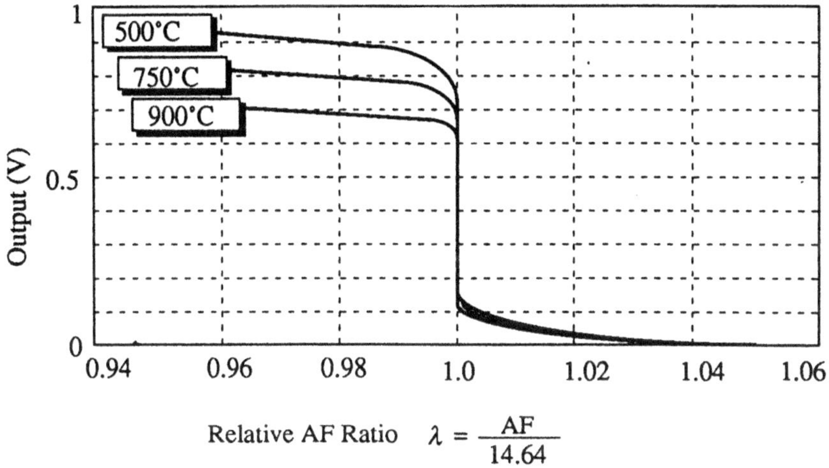

Fig. 16.3. Typical oxygen sensor operating characteristics (adapted from Heywood (1988))

The output signal used for feedback control is generally obtained from an oxygen sensor, which is typically installed just downstream of the exhaust manifold as shown in Fig. 16.1. The oxygen sensor has the binary output characteristics shown in Fig. 16.3. The output voltage abruptly changes at the stoichiometric AF ratio, and consequently, does not provide the desired continuous range output. Most control methods are not readily compatible with such a switching-type feedback sensor. Furthermore, the oxygen sensor location results in the output measurement delay of at least two engine revolutions in addition to the time delay due to the distance between exhaust ports and sensor location. The transport delay varies with engine speed, since two engine revolutions take 150 ms at 800 rpm and 20 ms at 6000 rpm. An ideal sensor for feedback would be a linear AF ratio meter at the inlet, but cost and other constraints make such a sensor impractical.

Popular control methods used in production vehicles include variations of the PI control method. For example, a control structure might be

$$\begin{aligned} \dot{m}_{fc}(t) &= \frac{\hat{\dot{m}}_{ao}(t)}{\beta} - K_p\left(P_m(t),\omega_e(t)\right)\left(V_{O_2}(t)-0.45\right) \\ &\quad - K_i\left(P_m(t),\omega_e(t)\right)\int\left(V_{O_2}(t)-0.45\right)\mathrm{d}\tau \end{aligned} \tag{16.6}$$

where $\dot{m}_{fc}(t)$ is the fuelling command, $\hat{\dot{m}}_{ao}(t)$ can be obtained from a look-up table, $K_p\left(P_m(t),\omega_e(t)\right)$ and $K_i\left(P_m(t),\omega_e(t)\right)$ are manifold pressure and engine speed dependent proportional and integral gains, respectively, $V_{O_2}(t)$ is the oxygen sensor output voltage, and the toggle voltage of 0.45 is selected according to Fig. 16.3. Because engine dynamics cover a wide range, the feedback gains must be determined for many (>100) operating conditions. In addition, the use of the switching feedback sensor make calibrating and tuning these feedback gains vary arduous. Each time a new engine system is developed, the whole process needs to be repeated.

16.2.2 VSC Approach to the Synthesis of Fuel-Injection Control

Using the VSC approach for fuel-injection systems has been proposed by Cho and Hedrick (1988), later extended by Cho (1993a), and demonstrated by Cho and Oh (1993). Below is a recapitulation of the methodology. As formulated in the previous section, the fuel-injection control objective is

$$\dot{m}_{ao}(t) : \dot{m}_{fo}(t) = \beta : 1 \tag{16.7}$$

Equivalently, the objective is to make $S(t) = 0$, where

$$S(t) := \dot{m}_{ao}(t) - \beta\dot{m}_{fo}(t) \tag{16.8}$$

Following the classical VSC approach (Utkin 1978), the $S(t) = 0$ manifold can be made attractive by

$$\dot{S}(t) = -k \operatorname{sgn} S(t) \ , \quad k > 0 \tag{16.9}$$

where sgn $S(t)$ is defined to have a value 1 when $S(t) > 0$ and -1 when $S(t) < 0$.

Substituting (16.8) into the attraction condition (16.9)

$$\dot{S}(t) = \ddot{m}_{ao}(t) - \beta\ddot{m}_{fo}(t) = -k \operatorname{sgn} S(t) \tag{16.10}$$

Various modelling errors and sensor inaccuracies can be combined and expressed as

$$\dot{m}_{ao}(t) = \left(1+\delta_a(t)\right)\hat{\dot{m}}_{ao}(t) \tag{16.11}$$

where $\delta_a(t)$ is an unknown and time-varying modelling error term. Then, differentiating (16.11)

$$\ddot{m}_{ao}(t) = (1+\delta_a(t))\,\ddot{\hat{m}}_{ao}(t) + \dot{\delta}_a(t)\dot{\hat{m}}_{ao}(t) \qquad (16.12)$$

The actual quantity $\dot{m}_{fo}(t)$ cannot be measured or determined accurately. Thus, in order to yield a causal control law, the following simplification of the fuel delivery model is necessary

$$\dot{m}_{fo}(t) = (1+\delta_f(t))\,\dot{m}_{fc}(t) \qquad (16.13)$$

where $\delta_f(t)$ is an unknown time-varying error term that includes the errors in fuel delivery model and injector calibration. The physical justification for this model simplification is that, for the multi-port fuel-injection system under consideration, the time constant between the fuel spray rate at the injector, $\dot{m}_{fc}(t)$, and the fuel rate entering the cylinders, $\dot{m}_{fo}(t)$, is very small. Then, differentiating (16.13)

$$\ddot{m}_{fo}(t) = (1+\delta_f(t))\,\ddot{m}_{fc}(t) + \dot{\delta}_f(t)\dot{m}_{fc}(t) \qquad (16.14)$$

Substituting (16.12) and (16.14) into (16.10)

$$\begin{aligned}(1+\delta_a(t))\,\ddot{\hat{m}}_{ao}(t) \;&+\; \dot{\delta}_a(t)\dot{\hat{m}}_{ao}(t) - \beta(1+\delta_f(t))\,\ddot{m}_{fc}(t)\\ &- \beta\dot{\delta}_f(t)\dot{m}_{fc}(t) = -k\,\mathrm{sgn}\,S(t) \qquad (16.15)\end{aligned}$$

Rearranging (16.15)

$$\begin{aligned}\ddot{m}_{fc}(t) \;&=\; \frac{(1+\delta_a(t))}{\beta}\ddot{\hat{m}}_{ao}(t) + \frac{\dot{\delta}_a(t)}{\beta}\dot{\hat{m}}_{ao}(t) - \delta_f(t)\ddot{m}_{fc}(t)\\ &\quad - \dot{\delta}_f(t)\dot{m}_{fc}(t) + \frac{k}{\beta}\,\mathrm{sgn}\,S(t) \qquad (16.16)\end{aligned}$$

Then the control law

$$\ddot{m}_{fc}(t) = \frac{1}{\beta}\ddot{\hat{m}}_{ao}(t) + k'(t)\,\mathrm{sgn}\,S(t) \qquad (16.17)$$

where

$$\begin{aligned}k'(t) \;&=\; \frac{k}{\beta} + \Big(\frac{\dot{\Delta}_a}{\beta}\dot{\hat{m}}_{ao}(t) + \dot{\Delta}_f\dot{m}_{fc}(t)\Big) + \Big(\frac{\Delta_a}{\beta}|\ddot{\hat{m}}_{ao}(t)| + \Delta_f\,|\ddot{m}_{fc}(t)|\Big)\\ &> \frac{k}{\beta} + \frac{\dot{\delta}_a(t)}{\beta}\dot{\hat{m}}_{ao}(t) - \dot{\delta}_f(t)\dot{m}_{fc}(t)\\ &\quad + \frac{\delta_a(t)}{\beta}\ddot{\hat{m}}_{ao}(t) - \delta_f(t)\ddot{m}_{fc}(t) \qquad (16.18)\end{aligned}$$

can guarantee the attraction condition of the $S(t) = 0$ manifold, even if the four model error terms combine in the worst possible manner. In (16.18) the

Δ_x terms represents the magnitude upper bounds of the $\delta_x(t)$ terms, such that for example,

$$\Delta_a > |\delta_a(t)| \quad \forall t \in [0, \infty] \tag{16.19}$$

The VSC fuel-injection algorithm given in (16.17) and (16.18) is global, nonlinear and completely analytically based. This is in contrast to the conventional methods used to develop production vehicles, which are local, linear and calibration based.

The control gain $k'(t)$ in (16.18) automatically adjusts to operating conditions because the values of $\dot{\hat{m}}_{ao}(t)$, $\ddot{\hat{m}}_{ao}(t)$, $\dot{m}_{fc}(t)$ and $\ddot{m}_{fc}(t)$ vary with operating conditions. The first bracketed term proportionally increases with higher engine speed and load, and the second bracketed term automatically increases (but at an autonomous rate from the first bracketed term) with faster transients. Also, the second bracketed term is close to zero in quasi-steady-state conditions, but during transients this term dominates the aggregate gain. In general, the control gain $k'(t)$ gradually increases with engine speed and load, and exponentially increases during rapid transients. This results in a control law that can provide global performance for all operating conditions without the need for gain scheduling.

Note that the sgn $S(t)$ term in (16.17) is not practical for implementation, since the quantity $S(t)$ defined in (16.8) cannot be easily measured. However, an oxygen sensor is available on production vehicles, which gives a high voltage when the air fuel mixture is rich ($S(t) < 0$) and a low voltage when the mixture is lean ($S(t) > 0$). The switching nature of the oxygen sensor is not well suited for conventional control methods, but it is adequate for a switching-type control method like the VSC approach. Thus, using the oxygen sensor for the feedback signal in (16.17) results in the following control law

$$\ddot{m}_{fc}(t) = \frac{1}{\beta}\ddot{\hat{m}}_{ao}(t) - k'(t)\,\mathrm{sgn}\,(V_{O_2}(t) - 0.45) \tag{16.20}$$

A potential problem here is that the oxygen sensor signal, and hence the feedback signal, is time delayed.

Cho and Hedrick (1987) shows that the use of a time-delayed feedback signal in the VSC method results in increased chatter magnitude and decreased chatter frequency. Consider a generic problem where the control input is formulated based on $\dot{S}(t) = -k\,\mathrm{sgn}\,S(t-\tau)$, where τ is the delay time. Consider initial conditions such that $S(t) > 0$ and $S(t-\tau) > 0$ for $t \leq 0^-$, as shown in Fig. 16.4. Then, for $0 < t < t_r$, $S(t-\tau) > 0$, and $\dot{S}(t) = -k$. Therefore, $S(t)$ approaches the surface defined by $S(t) = 0$ at the rate $-k$. For $t_r < t < (t_r + \tau)$, $S(t) < 0$, but $S(t-\tau) > 0$. Since the feedback signal is obtained from $S(t-\tau)$, both $S(t)$ and hence $S(t-\tau)$ keep moving at the rate $-k$. This results in $S(t)$ moving past the $S(t) = 0$ surface and assuming the value $-k\tau$ at $t = (t_r + \tau)^+$, and the sign of $S(t-\tau)$ changes. For $(t_r + \tau) < t < (t_r + 3\tau)$, $S(t-\tau) < 0$, and $\dot{S}(t) = +k$. Therefore, $S(t)$ reverses direction and moves toward the surface defined by $S(t) = 0$ at the rate $+k$, until $t = (t_r + 3\tau)^+$, when another switch occurs.

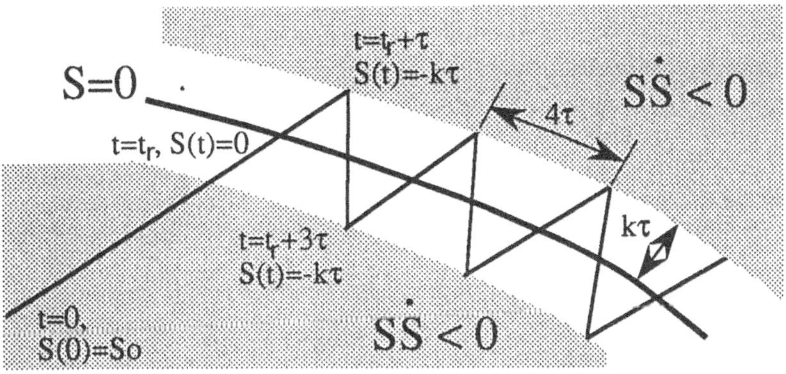

Fig. 16.4. Time-delayed feedback in VSC approach

This process continues with switching occurring at $t = t_r + \tau(2i-1)$ where $i = 1, 2, \cdots, n$. The same results may be obtained for initial conditions such that $S(t) < 0$ and $S(t-\tau) < 0$ for $t \leq 0^-$. Thus, under the time-delayed feedback of $\dot{S}(t) = -k \operatorname{sgn} S(t-\tau)$, the chatter magnitude increases by a factor of 2τ and the frequency decreases by a factor 4τ. The result is still a stable limit cycle about $S(t) = 0$ with a magnitude $2k\tau$. Note that in digital implementation, the finite sample-to-control time has the same effect as the time-delayed feedback.

These effects are undesirable for fuel-injection control, since small fluctuations of high frequencies in AF ratio are preferred. Thus, a small time delay and an accurate model are desirable. These conditions are intuitive and not any different for those methods used in production systems described before. However, in this chapter, test track results will show that these conditions are not overly formidable for the VSC approach. In fact, even with a very rudimentary model (where the only known parameter of the subject engine is its displacement and, hence, relatively large $k'(t)$), the VSC controller can provide performance comparable or superior to that of a production controller.

The actual control command can be determined by integrating (16.17). For example

$$\begin{aligned} \dot{m}_{fc}(t) &= \dot{m}_{fc}(t-t_k) + t_k \ddot{m}_{fc}(t) \\ &= \dot{m}_{fc}(t-t_k) + t_k[\frac{1}{\beta}\hat{\dot{m}}_{ao}(t) - k'(t)\operatorname{sgn}(V_{O_2}(t) - 0.45)] \quad (16.21) \end{aligned}$$

can be employed, where t_k is the time-varying sampling time. The Euler integration method is generally avoided in digital implementation. However, this causes no apparent stability problem here because the integrand contains the feedback term.

Note that the control structure used in (16.21) is a form of dynamic compensation, or dynamic sliding mode controller, adding an integral action. This use of dynamic compensation is very beneficial in reducing the chatter. As discussed in the literature, the VSC approach with the discontinuous sgn function results in output chatter. The problem can be ameliorated by the use of control smoothing approximations, such as the saturation function in Slotine (1984). However, when the feedback signal is obtained from a switching sensor like the oxygen sensor here, such a technique cannot be used. The dynamic sliding mode controller concept and its usefulness in reducing chatter as well as its formalization using the differential algebraic approach are also discussed in Sira-Ramírez (1992).

16.2.3 Implementation and Test Track Results

The designed VSC controller is experimentally evaluated using a research automobile at a test track. The performance is evaluated by comparison testing the VSC controller against a production controller for a number of different drive-by-wire throttle control scenarios.

The research automobile is a 1988 Oldsmobile Cutlass Calais with a Quad-4 (double-overhead-cam, four-cylinder) engine. The production fuel-injection hardware is a multi-port injection type and fires all 4 injectors simultaneously every revolution. The production algorithm is based on the speed-density method. The VSC controller in (16.17), (16.18) and (16.21) is fully compatible with this production hardware configuration. The only difference in the hardware was that the VSC controller was implemented on a 80386 CPU-based portable computer in Turbo Pascal. This implementation takes a fraction of 1 ms of CPU time.

A linear AF ratio sensor was installed in this experiment to accurately assess performance. This sensor provides AF ratio information of 10 to 30. Note that the linear AF ratio sensor was was used for the purpose of evaluating performance only and was not used in the controller as a feedback sensor.

A throttle controller was also developed and implemented for drive-by-wire throttle control. Since the VSC fuel-injection controller uses time-varying sample time based on crank-angle rotation, the throttle controller was implemented on a second portable computer using a fixed sample time. The production cruise control actuator, which is a pneumatic on-off system, was utilized for throttle position control. We have designed several controllers, including a bang-bang controller, a pulse-width-modulated PI controller, and a modified VSC-like controller. Among the options tested, the controller that uses an anticipatory band to switch control action with a switching logic determined in the sliding phase plane provides the best performance. This modified VSC-like controller is described in a paper by Lee, Kim and Cho (1993) and not repeated here.

Only a generic engine model described in Cho and Hedrick (1989) was used to develop the VSC controller. The only model parameter that is pertinent to the research automobile is the engine displacement. As discussed before, an accurate model of a volumetric efficiency is necessary for a fueling system to perform well. However, the volumetric efficiency model used in the VSC controller for the Quad-4 engine here is based on the static data of an overhead-valve, six-cylinder engine in Cho and Hedrick (1989). Thus, modelling errors can be quite significant, and the VSC controller has to be very robust in order to perform well.

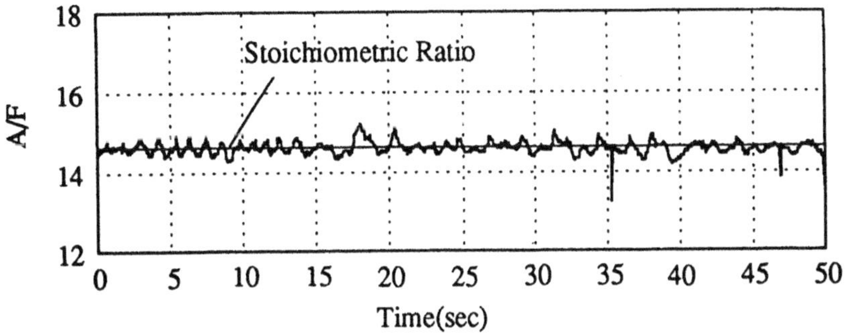

Fig. 16.5. VSC controller performance in rolling test

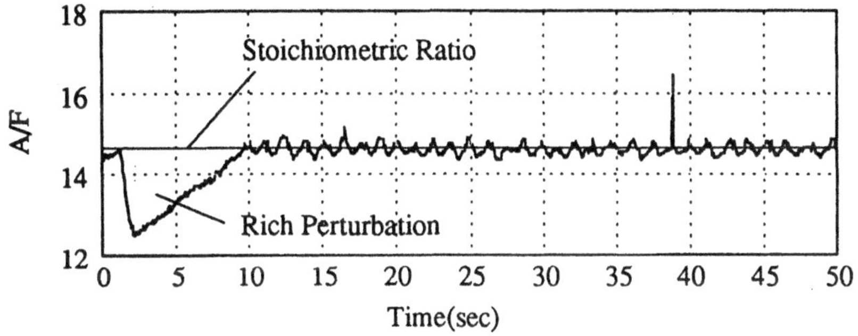

Fig. 16.6. Production controller performance in rolling test

The first test was a rolling test where no throttle or brake was applied. During this test, the engine speed stays near 800 rpm, and the vehicle slowly accelerates to a low speed. Figure 16.5 shows the AF ratio of the VSC controller and Fig. 16.6 shows the AF ratio of the production controller. The sharp spikes in the figures are attributable to sensor noise. In both cases, the performance is good. One exception is the from 1.292 to 9.312 s period (109 engine revolutions) in the production controller case, where the AF ratio goes rich. The peak perturbation in this period reaches 12.50, which represents a 14.62% rich deviation from the desired AF of 14.64. This rich perturbation is difficult to explain. The two controllers were subjected to the identical test procedure under identical conditions on the identical section of the test track, and to the best of our knowledge, there was no arcane engine event to trigger this perturbation. We have repeated the rolling test several times, and the rich perturbation occurred quite frequently and randomly. We do not know the exact reason for this rich perturbation, and thus, it is left unexplained here. However, the rich magnitude is significant, and the duration is long. As a result, the emission properties of the production controller would significantly deteriorate during this period.

In general, the VSC method can result in a high-frequency chatter, which in turn can excite unmodelled dynamics. Furthermore, the time delay in oxygen sensor can further accentuate the chatter problem. Smoothing approximations can be used in place of the discontinuous sgn $S(t)$ to ameliorate the situation, but since the magnitude of the AF ratio is not available from the oxygen sensor, this technique is not applicable here. As discussed earlier, only the dynamic sliding mode controller concept is used to reduce chatter. Comparing Figs. 16.5 and 16.6 (as well as other results presented below), it is evident that the output chatter magnitudes of both VSC and production controllers are comparable. Note that in the VSC case, a smaller feedback gain $k'(t)$ in (16.18) would give a smaller chatter magnitude. Thus, starting out with a more accurate model or incorporating on-line parameter learning capabilities would result in a reduced chatter. In this study, a relatively large $k'(t)$ was necessary, because the only accurate model parameter used was the engine displacement.

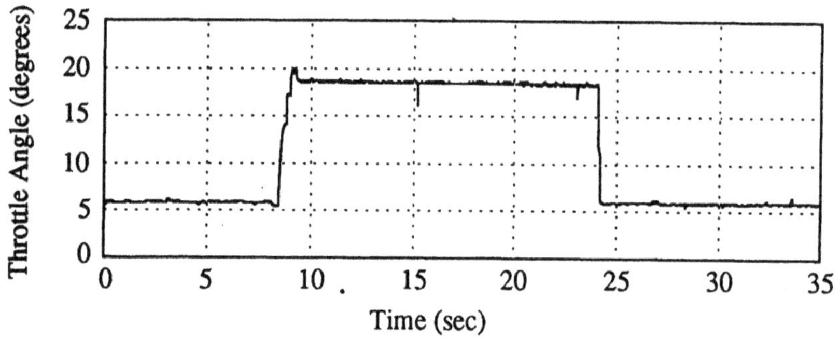

Fig. 16.7. Light throttle driving scenario

Next, a light tip-in/tip-out driving scenario was tested. Throttle time histogram is shown in Fig. 16.7. The throttle command consists of a tip-in command to 20 degrees at $t = 8$ s and a tip-out command at $t = 24$ s.

The corresponding output AF ratios of the VSC and production controllers are shown in Figs. 16.8 and 16.9, respectively. In general, the quasi-steady-state performances of both controllers are good.

In transients, rich tip-in conditions are evident for the VSC case. The throttle tip-in command is given at 8 s, but since the throttle actuator need to build control pressure, the throttle plate does not move until 8.457 s. Following the throttle plate movement, the output AF ratio goes slightly lean, and 4

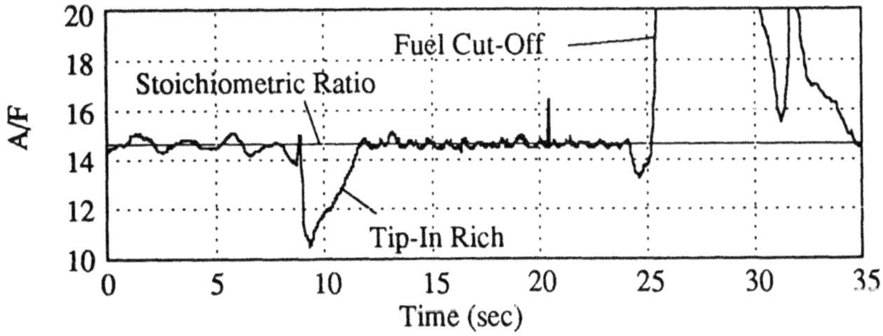

Fig. 16.8. VSC controller performance in light throttle

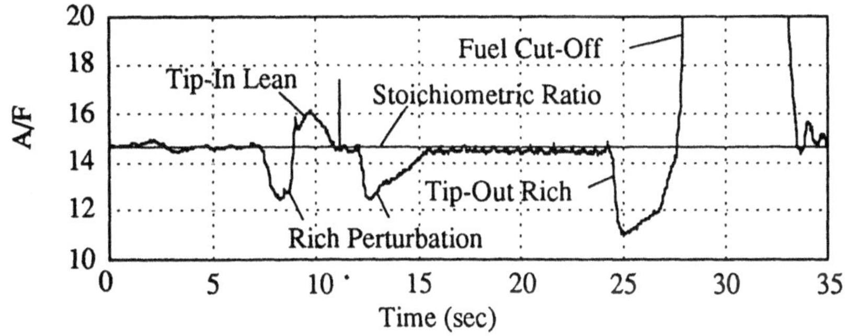

Fig. 16.9. Production controller performance in light throttle

engine revolutions later (8.901 s), reaches a magnitude of 15.01. During these 4 engine revolutions, the feedback gain of the VSC controller rapidly increases, because the $\hat{\dot{m}}_{ao}(t)$ term becomes large. This in combination with the lean signal from the oxygen sensor result in a rapid transient control, and the output goes rich at 8.976 s, and 12 engine revolutions later at 9.934 s, reaches a rich peak of 10.54. By this time, the magnitude of the $\hat{\dot{m}}_{ao}(t)$ term is no longer exceedingly large, and the output remains lean for 89 engine revolutions until 11.640 s. Between the tip-in and tip-out modes, the transmission executes two upshifts (1-to-2 and 2-to-3), but the VSC controller is robust to these transients. In the tip-out mode, commanded at 24 s, the throttle plate starts to move immediately, since no pressure building phase is required in this case. The output goes rich, followed by a fuel cut-off condition. The rich duration lasts for 26 engine revolutions (from 24.183 s to 25.118 s) and reaches a peak magnitude of 13.28. The ensuing fuel cut-off condition is not as harmful for emissions as it might seem. During fuel cut-off, no fuel is being sprayed, and thus, no harmful combustion by-products are generated. Not including the fuel cut-off condition, the total poor transient conditions persist for 127 engine revolutions with a peak rich magnitude of 10.54.

In the production controller case, rich-lean-rich transients are evident in the tip-in mode. The AF ratio goes rich at 7.416 s, reaches a peak of 12.48, and stays rich for 24 engine revolutions until 8.502 s. Note that the actual tip-in command is given at 8.0 s, and the throttle plate does not move until 8.502 s. Thus, this initial rich transient is a rich perturbation, similar to that discussed in the rolling case. Three revolutions after the initial throttle plate movement, at 8.730 s, the output AF ratio reverses its direction, and becomes lean at 8.974 s. This lean transient persists for 67 engine revolutions until 10.865 s and reaches a peak of 16.12. After the lean transient, a rich perturbation is evident again at 12.138 s. This rich perturbation lasts 131 engine revolutions to 15.473 s and reaches a peak of 12.46. Note that the engine speed starts to decrease at 12.788 s, indicating that a 1-2 upshift is in progress, and the shifting transient may have had an effect on the rich AF ratio condition. In the tip-out mode, commanded at 24 s, the output goes initially rich followed by a fuel cut-off condition. The output goes rich at 24.097 s and stays rich for 92 engine revolutions to 27.575 s. The rich peak is 11.02. Not including the two rich perturbations and the fuel cut-off condition, the poor transient conditions due to throttle maneuvers persist for 159 revolutions, and the corresponding lean and rich peaks are 16.12 and 11.02, respectively. In summary, poor transient conditions for the production controller last considerably longer (and hence a larger number of engine revolutions and total bad exhaust gases displaced) than the VSC case.

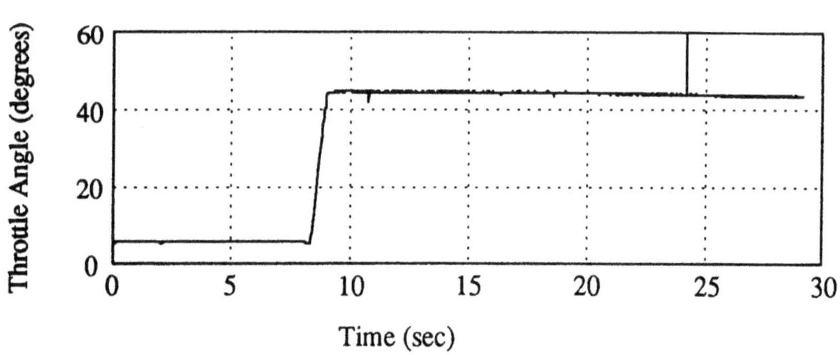

Fig. 16.10. Heavy throttle scenario

The throttle actuator has a maximum actuation limit of 45 degrees. Therefore, the heaviest throttle angle that can be tested using the throttle controller is limited to 45 degrees. Figure 16.10 shows the time histogram of a heavy tip-in/tip-out driving scenario. For this test, the automatic transmission shift lever was left in position 2, so that high engine speeds of up to 5500 rpm can be tested. Figures 16.11 and 16.12 show the AF ratios of the two controllers. In the VSC case, the trend is similar to the light tip-in/tip-out scenario. At the

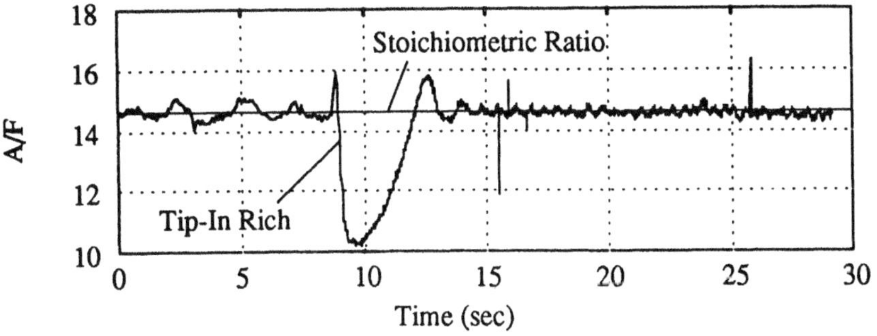

Fig. 16.11. VSC controller performance in heavy throttle

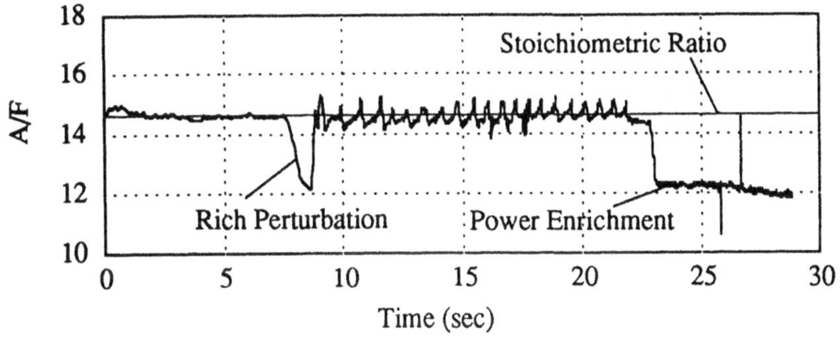

Fig. 16.12. Production controller performance in heavy throttle

tip-in, the AF ratio goes rich, but takes longer to recover to the stoichiometric value. The production controller again shows a rich perturbation before the tip-in. The performance at the actual tip-in is excellent, but soon the production controller goes into the power enrichment mode, where purposely rich AF mixture is supplied to provide better engine torque. During this period, the production of exhaust emissions would significantly increase.

A number of other drive-by-wire scenarios with throttle commands of up to 45 degrees have been tested. The results and conclusions are similar to those presented here and are not included. The results presented in this chapter are time histograms of AF ratio. We have also extensively driven the research automobile and found that the drive feel of the VSC controller is difficult to distinguish from the production controller.

16.2.4 Discussions

The developed VSC fuel-injection control method is fully compatible with production hardware. Based on the demonstrated results, the potential of the developed VSC-based methodology is excellent. This is especially impressive when one considers that these results were achieved within a few days of testing, whereas the production system benefits from years of development and fine tuning. The only model parameter used by the VSC controller that is pertinent to the subject automobile was the engine displacement. Thus, the robust performance demonstrated here is exceptionally good. If more accurate parameters are used in the VSC controller, much better results than those included here would be possible.

16.3 Magnetic Levitation Control

16.3.1 Dynamics of Open-Loop Unstable Magnetic Levitation

Magnetic levitation has been successfully demonstrated for many applications, including frictionless bearings for inertial instruments (Kaplan and Regev 1976, Downer 1980) vibration isolation tables (Dahlen 1985) and high-speed trains (Limbert, Richardson and Wormley 1976, Yamamura and Yamaguchi 1990). Since the magnetic levitation systems are key components in many critical engineering systems, high-performance controllers are very important.

These magnetic levitation systems are open loop unstable, and in general, feedback controllers are used to achieve desired stability and performance. Because of inherent nonlinearities, the governing differential equations are linearized about various operating points, and local feedback controllers are implemented to stabilize perturbations. The most popular methods for designing these controllers are the classical methods such as PID or lead-lag compensators, designed based on the root locus or frequency-response shaping techniques. These classical control methods are well understood for single-input single-output systems and can also consider robustness issues pertaining to modelling uncertainties and disturbances by the use of gain and phase margins, bandwidth limitations, etc. The use of the linear quadratic control method with a Kalman filter for state estimation has also been investigated in Dahlen's thesis (1985); this approach provides similar performance to the classical control methods. In this chapter, the VSC approach is used to synthesize a magnetic levitation controller, and its performance is experimentally compared to that of a classical controller. This chapter presents the results in Cho, Kato and Spilman (1993) and Cho (1993b) as well as some additional results.

Figure 16.13 depicts the schematic diagram of the single-axis magnetic levitation system used in this study. The levitation object was a ping-pong ball with a small permanent magnet bonded to it, to provide an attractive force. A

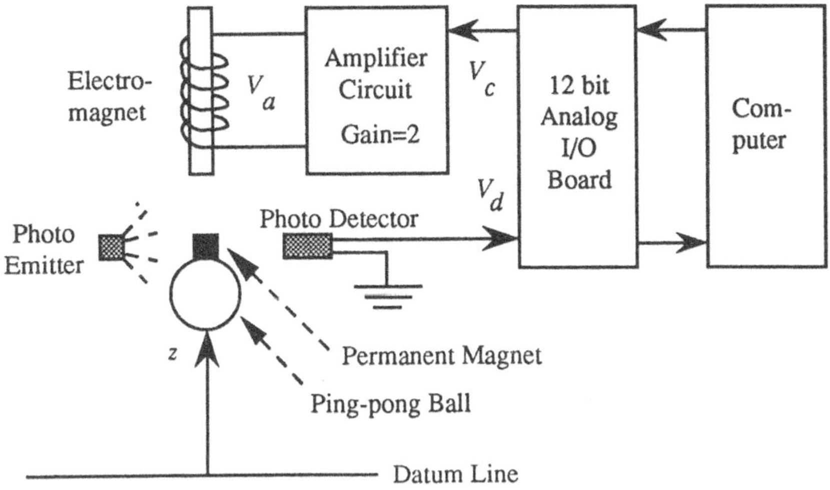

Fig. 16.13. Schematic diagram of magnetic levitation apparatus

photo emitter-detector pair was used to determine the height of the levitation ball. The control computer in this experiment was an IBM PS/2 Model 80, and the sampling rate was 1000 Hz.

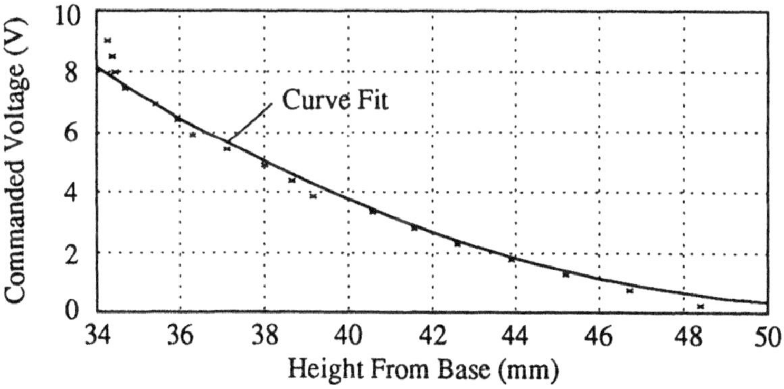

Fig. 16.14. Calibration of electromagnetic force

A force balance analysis in the vertical plane yields the following equation of motion for the levitation ball

$$m\ddot{z}(t) = -mg + F_c(t) \qquad (16.22)$$

where m is the mass of the levitation ball in grams, $z(t)$ is the distance of the levitation ball from a datum point in millimeters, g is gravity, and $F_c(t)$ is the magnetic control force in millinewtons. The magnetic control force between the solenoid and the permanent magnet can be determined by considering the magnetic potential energy. However, analytically determining the magnetic force for the experimental apparatus is extremely difficult, because a combination of an electromagnet (solenoid) and a permanent magnet (levitation ball) is used. Therefore, the magnetic force characteristics as a function of the applied voltage and the height of the levitation ball were calibrated. The calibration experiment consists of resting the levitation ball on an xyz-stage capable of 50×10^{-6}m incremental positioning and determining the minimum voltage required to pick up the levitation ball at various height. The results are depicted in Fig. 16.14. A least squares fit of the data is

$$F_c(t) = \frac{V_c(t)}{a_1 x_1(t)^2 + a_2 x_1(t) + a_3} \tag{16.23}$$

where $V_c(t)$ is the command voltage from the control computer in volts, and

$$a_1 = \frac{0.0231}{mg}, \quad a_2 = \frac{-2.4455}{mg}, \quad a_3 = \frac{-64.58}{mg} \tag{16.24}$$

The results show that the force is inversely proportional to the distance squared. The reason for the nonzero first and zeroth order terms in (16.24) is that the distance $z(t)$ is measured from an arbitrary datum point, rather than from the bottom of the solenoid. The solenoid characteristics change with temperature, and the curve fit in (16.24) can change by up to $\pm 20\%$ when the system has been in operation for a while.

The equation of motion for the magnetic levitation system can be written in controllable canonical form

$$\begin{aligned} \dot{x}_1(t) &= x_2(t) \\ \dot{x}_2(t) &= \hat{b}(x,t)u(t) - g + d(t) \end{aligned} \tag{16.25}$$

where $x_1(t) = z(t)$ and $x_2(t) = \dot{z}(t)$ are the state variables, $u(t) = V_c(t)$ is the control, and $d(t)$ is an unknown disturbance. The model of the force-distance relationship $\hat{b}(x,t)$ is

$$\hat{b}(x,t) = \frac{1}{m(a_1 x_1(t)^2 + a_2 x_1(t) + a_3)} \tag{16.26}$$

The actual force-distance relationship $b(x,t)$ may be expressed as

$$b = \hat{b} + \delta b(t) \tag{16.27}$$

where $\delta b(t)$ is some unknown modelling error. From the calibration data the upper bound of the modelling error, $\delta b(t)_{\max}$, is approximately 20% of the nominal model.

The VSC approach requires full state feedback (i.e. levitation height and its time rate of change), while classical controllers require output feedback

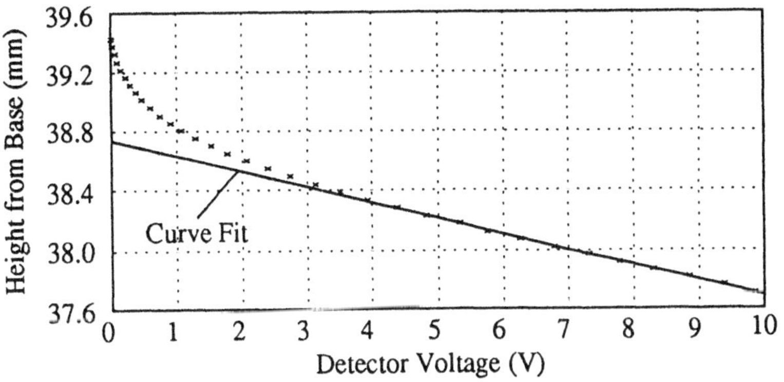

Fig. 16.15. Calibration of position sensor

(i.e. levitation height). A photo emitter-detector pair is used for measuring levitation height, and Fig. 16.15 depicts calibration results. The calibration experiment was performed using the same xyz-stage and from the identical datum point as before. The linear fit of these data is

$$z = -0.0982\, V_d(t) + 38.55 \tag{16.28}$$

where $V_d(t)$ is the detector voltage in volts. Note that the validity of (16.28) is limited to the range of 37.7 mm to 38.5 mm. The sensor accuracy in this range is better than $pm0.02$ mm. The time rate of change measurement required in the VSC controller is obtained by backward differencing the levitation height measurements. This method is very susceptible to sensor noise, which in turn, makes the vertical velocity estimates very noisy. However, since the purpose of this study was to compare various controllers, no addition or modification of hardware was considered.

However, this method is identical to obtaining the derivative action in digitally implementing a classical PD or PID controllers. In our laboratory apparatus, the problems associated with sensor noise were not severe, so no signal conditioning hardware was implemented.

16.3.2 VSC Approach to Levitation Control: Robust and Chatter-Free Tracking

In this section a VSC magnetic levitation controller is synthesized, and its performance in set-point regulation and trajectory following is experimentally evaluated. Again, a detailed review of the VSC theory is not repeated here. The control objective is to maintain the levitation height $x_1(t)$ at some desired height $h_{\mathrm{des}}(t)$, which may be time varying. Define the tracking error as

$$e(t) = x_1(t) - h_{\mathrm{des}}(t) \tag{16.29}$$

A sliding manifold is defined as

$$S(t) := \dot{e}(t) + \lambda e(t) \tag{16.30}$$

where the closed-loop bandwidth λ must be chosen appropriately. The objective is to achieve $S(t) = 0$, and by substituting the equation of motion (16.25) with sliding manifold definition (16.30) into the attraction condition $\dot{S}(t) = -k \operatorname{sgn} S(t)$

$$\frac{u(t)}{m(a_1 x_1^2(t) + a_2 x_1(t) + a_3)} - g - \ddot{h}_{\text{des}}(t) + \lambda(x_2(t) - \dot{h}_{\text{des}}(t)) = -k \operatorname{sgn} S(t) \tag{16.31}$$

and rearranging for the control input $u(t)$

$$u(t) = m(a_1 x_1(t)^2 + a_2 x_1(t) + a_3)(g + \ddot{h}_{\text{des}}(t) - \lambda(x_2(t) - \dot{h}_{\text{des}}(t)) - k \operatorname{sgn} S(t) \tag{16.32}$$

The control law (16.32) results in a chattering solution due to the discontinuous $\operatorname{sgn} S(t)$ function.

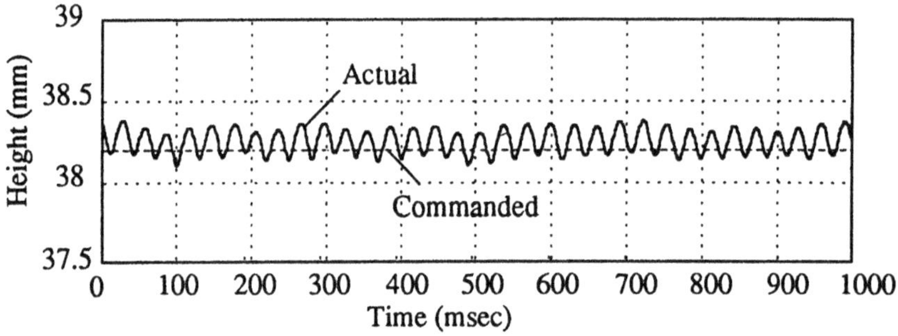

Fig. 16.16. Set-point regulation of classical-VSC controller

Figure 16.16 depicts the experimental results for stabilizing the magnetic levitation system with control law (16.32). The desired levitation height is 38.2 mm. The depicted chatter is highly undesirable, but the control law (16.32) can nevertheless stabilize the system. The chattering levitation height due to $\operatorname{sgn} S(t)$ exhibits a stable limit cycle about the desired equilibrium levitation height. In the experiment the control parameter k is selected to be larger than $\delta b(t)_{\max} u(t)_{\max} + d(t)_{\max}$, where the subscript max denotes the magnitude upper bound of variables. For the 20% upper bound on the modelling error and 10 volt maximum voltage, $\delta b(t)_{\max} u(t)_{\max} = 3750$, and thus, to satisfy the robustness requirements, $k = 5000$ was used in this experiment.

The chattering problem may be improved by the use of control smoothing approximations. One possibility is to replace the infinite gain of $\operatorname{sgn} S(t)$ at $S(t) = 0$ with a finite gain when the magnitude of the $S(t)$ is smaller than some prescribed value ϕ . This can be achieved by replacing the sgn $S(t)$ function with a saturation function described by Slotine (1984)

$$\operatorname{sat}\left(\frac{S(t)}{\phi}\right) := \begin{cases} \operatorname{sgn} S(t) & \text{if} \quad |S(t)| \geq \phi \\ \dfrac{S(t)}{\phi} & \text{if} \quad |S(t)| < \phi \end{cases} \tag{16.33}$$

The boundary layer ϕ must be selected in accordance with the balancing condition give in Slotine (1984)

$$\lambda_A > \frac{k}{\phi} \tag{16.34}$$

where λ_A is the achievable closed-loop bandwidth. As a result of this approximation, the attraction guarantee of the $S(t) = 0$ manifold is possible only when $|S(t)| > \phi$. When $|S(t)| < \phi$, the attraction condition of the $S(t) = 0$ manifold may not be satisfied, due to the presence of modelling errors and disturbances, and the closed-loop dynamics will wander inside the boundaries $S(t) = \pm\phi$. Thus, in general, the control objective $S(t) = 0$ cannot be achieved with the control smoothing approximation (16.33) and the sliding manifold definition (16.30).

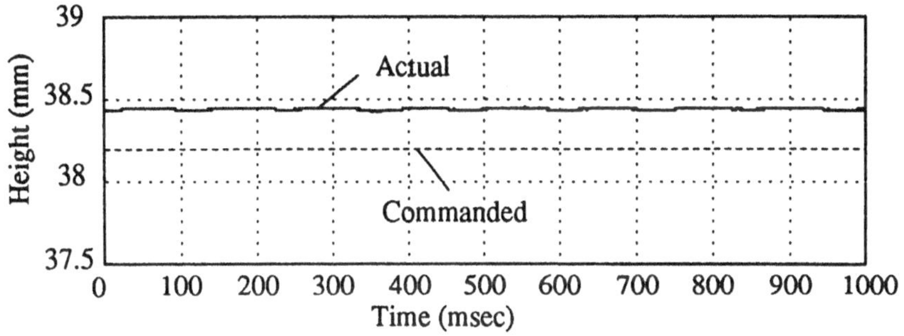

Fig. 16.17. Set-point regulation of smoothing-VSC controller

Experimental results illustrating the tracking error in set-point regulation and 1-Hz trajectory following are depicted in Figs. 16.17 and 16.18. The control parameters were $k = 5000$ and $\phi = 70$. This assumes the bandwidth of soleniod coil to be $\lambda = 100$ rad/s, yielding $\phi > 50 = 5000/100$. The tracking error is due mostly to modelling errors, and the results indicate the presence of parametric errors resulting in a positive bias in the control force, which in turn results in a higher levitation position than desired. For the 1-Hz trajectory, the output error is slightly worse, indicating the presence of additional modelling errors due to unmodelled dynamics.

To understand the origin of these bias tracking errors, it is necessary to consider error dynamics. Inside the boundaries $S(t) = \pm\phi$, the control objective $S(t) = 0$ cannot be achieved if any modelling error is present. This is evident by

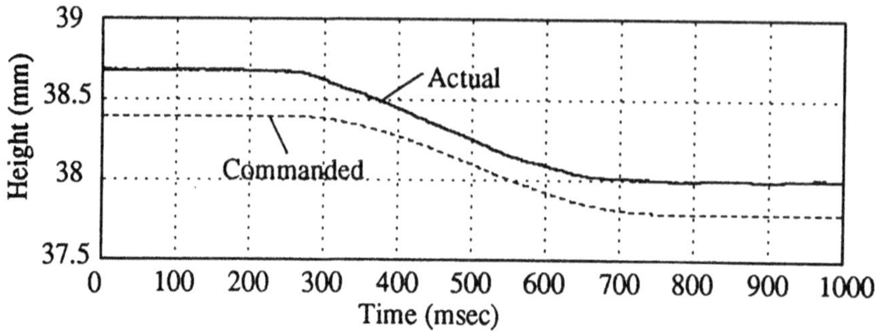

Fig. 16.18. 1-Hz trajectory following of smoothing-VSC controller

substituting the control law (16.32) with sat$(S(t)/\phi)$ to the equation of motion (16.25). Inside the boundaries the closed-loop dynamics are

$$\begin{aligned} \dot{x}_2(t) &= \hat{b}(x,t)u(t) + \delta b(t)u(t) - g + d(t) \\ &= \ddot{h}_{\mathsf{des}}(t) - \lambda(\dot{h}_{\mathsf{des}}(t) - x_2(t)) \\ &\qquad - k\frac{S(t)}{\phi} + \delta b(t)u(t) + d(t) \end{aligned} \tag{16.35}$$

Rearranging (16.35) in terms of closed-loop error dynamics

$$\ddot{e}(t) + \lambda\dot{e}(t) = \delta bu(t) + d(t) - k\frac{S(t)}{\phi} \tag{16.36}$$

Substituting the sliding manifold definition (16.30) into (16.36)

$$\ddot{e}(t) + (\lambda + \frac{\phi}{k})\dot{e}(t) + \frac{\lambda k}{\phi}e(t) = \delta b(t)u(t) + d(t) \tag{16.37}$$

For any nonzero right-hand side no solution of (16.37) gives $e(t) = 0$. Thus, in the presence of any modelling error or disturbance, a nonzero tracking error is unavoidable. As $t \to \infty$, $\ddot{e}(t) = \dot{e}(t) = 0$, and since $k > \delta b(t)_{\mathsf{max}}u(t)_{\mathsf{max}} + d(t)_{\mathsf{max}}$, the minimum tracking guarantee is

$$|e_{ss}| = \left(\frac{\phi}{\lambda}\right)\left(\frac{|\delta b(t)u(t) + d(t)|}{k}\right) < \frac{\phi}{\lambda} \tag{16.38}$$

The minimum tracking guarantee is the worst case scenario. When modelling errors are not severe, the attraction condition may be satisfied even well inside the boundaries $S(t) = \pm\phi$, and a better tracking accuracy can be obtained. Thus, for a given mathematical model of a plant, which contains modelling errors, the trade-off between chattering and tracking accuracy cannot be avoided.

Referring to Figs. 16.17 and 16.18, the errors are approximately 0.24 – 0.28 mm, which is much larger than the sensor accuracy of ±0.02 mm. However, the theoretical minimum tracking guarantee is 0.7 mm ($\phi/\lambda = 70/100$), and the depicted errors are within the theoretical minimum guarantee.

The bias tracking errors can be remedied by modifying the sliding manifold definition (16.30) to include an integral error term

$$S(t) := c_0\dot{e}(t) + c_1 e(t) + c_2 \int e(\tau)\mathrm{d}\tau \tag{16.39}$$

Note that the sliding manifold is of third order, and c_0 cannot be set to zero. If c_0 is set to zero, the sliding manifold definition will not result in a causal input/output relationship; the control $u(t)$ does not appear in the derivative of $S(t)$, and the attraction condition (16.32) cannot be satisfied. With $c_0 = 1$ and the integral sliding manifold definition (16.39), the control law becomes

$$\begin{aligned} u(t) &= \frac{1}{\hat{b}}(g + \ddot{h}_{\mathrm{des}}(t) - c_1(x_2(t) - \dot{h}_{\mathrm{des}}(t)) \\ &\quad - c_2(x_1(t) - h_{\mathrm{des}}(t)) - k \operatorname{sgn} S(t))) \end{aligned} \tag{16.40}$$

Then, inside the boundaries, the closed-loop error dynamics become

$$\ddot{e}(t) + c_1\dot{e}(t) + c_2 e(t) = \delta b(t)u(t) + d(t) - k\frac{S(t)}{\phi} \tag{16.41}$$

Substituting the sliding manifold definition (16.39) into (16.41)

$$\ddot{e}(t) + \left(c_1 + \frac{k}{\phi}\right)\dot{e}(t) + \left(c_2 + c_1\frac{k}{\phi}\right)e(t) + c_2\frac{k}{\phi}\int e(\tau)\mathrm{d}\tau = \delta b(t)u(t) + d(t) \tag{16.42}$$

For stable closed-loop dynamics the condition $(c_1 + k/\phi)(c_2 + c_1 k/\phi) > c_2 k/\phi$ needs to be satisfied. For all positive values of c_1, c_2, k and ϕ, this stability condition is met. Then, for a constant right-hand side, the steady-state solution of (16.42) is $e(t) = 0$ and $\int e(t)\mathrm{d}t \to 0$. Therefore, the control law (16.40) can drive the tracking errors resulting from bias in modelling uncertainties (such as constant and slowly-varying parameter errors) to zero.

Figure 16.19 depicts the experimental results obtained using the control law (16.40) with the control smoothing approximation (16.33) for stabilization at the nominal equilibrium position. The set-point regulation performance is perfect within the sensor accuracy. Figure 16.20 depicts the results for 1-Hz trajectory following. The tracking performance is excellent. The experimental results in Figs. 16.19 and 16.20 were obtained with all control and robustness parameters as before, and $c_1 = 100$ (equivalent to $\lambda = 100$) and $c_2 = 0.2$.

16.3.3 Comparison with Classical Control

The single-axis magnetic levitation system is a single-input single-output system. Using both classical and modern control methodology, a number of linear controllers can be easily synthesized to stabilize the system. Dahlen's thesis

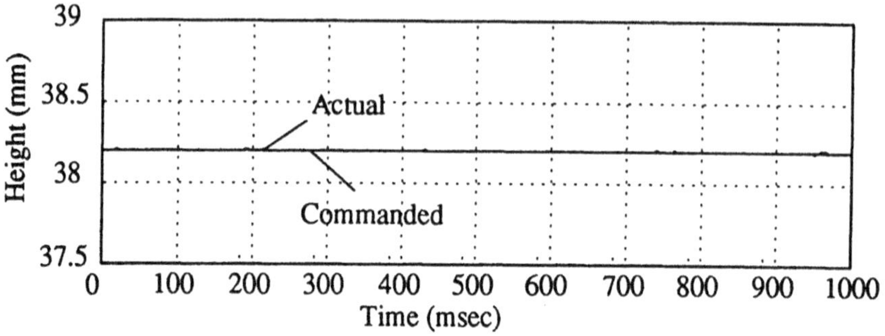

Fig. 16.19. Zero-error regulation with integral-VSC controller

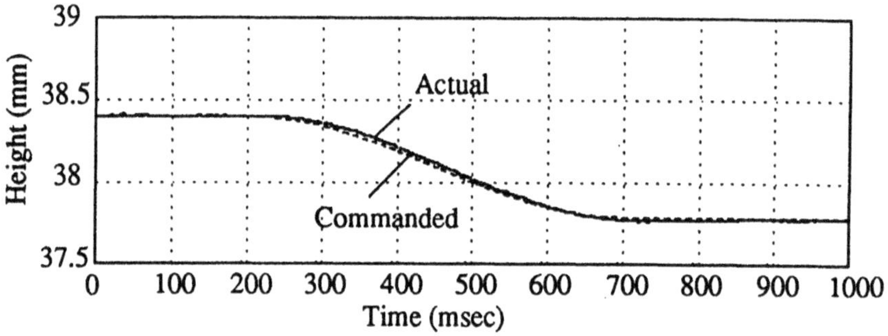

Fig. 16.20. 1-Hz trajectory following with integral-VSC controller

(1985) provides a comparison between a PID controller and a linear quadratic controller with a Kalman filter for state estimation in magnetic vibration isolation tables and shows that both the classical and modern controllers provide comparable performance.

We have also designed and evaluated many different classical controllers. We found that a PI-plus-lead controller provides the best performance. This is a PI controller in cascade with a lead compensator, consisting of two controller zeros and two controller poles. One of the controller poles is placed at the origin to provides zero steady-state error. This PI-plus-lead controller is much more advantageous than a PID controller, consisting of two zeros and one pole, because the extra pole in the PI-plus-lead controller provides the desirable attenuation of high frequency sensor noise. The designed controller is

$$G(s) = \frac{\delta V_c(s)}{\delta z(s)} = K_p\left(1 + \frac{K_i}{s}\right)\frac{(s+z)}{(s+p)} = K_p\frac{(s+K_i)(s+z)}{s(s+p)} \qquad (16.43)$$

To design the controller, the levitation system model described in (16.25) was linearized about a nominal operating point of $z(t) = 38.2$ mm, and the following transfer function was obtained

$$H(s) = \frac{\delta V_c(s)}{\delta z(s)} = \frac{3148}{s^2 - 4551} \tag{16.44}$$

This is an unstable transfer function; the poles are located at ±67.46. The controller parameters were chosen based on both s and z domain root locus analyses. The structure of the plant (one stable pole and one unstable pole) and the controller (two left-half plane zeros and two poles) ensures that, for properly selected controller parameters, the nominal plant is stable. However, this is not true. In reality, a moderately high gain drives the system unstable. This is due to the presence of unmodelled actuator dynamics, which pushes closed-loop poles to the right-half plane. By relating the voltage command to the measured levitation position, we determined that the following transfer function accurately describes the electromagnet RL characteristics

$$\frac{1}{0.0035s + 1} \tag{16.45}$$

The effects of the unmodelled actuator dynamics are quite significant, and the classical controller structure had to be designed with the additional plant pole to achieve good performance. The final parameters of the PI-plus-lead controller were $K_p = 10.5$, $K_i = 38.0$, $p = 500$, and $z = 15.0$. The closed-loop pole and zero locations are $p_i(s) = -22.4, \pm j78.0, -38.8, -27.6$ and -1973, and $z_i(s) = -38.0$ and -15.0. Note that these include the additional pole due to the actuator.

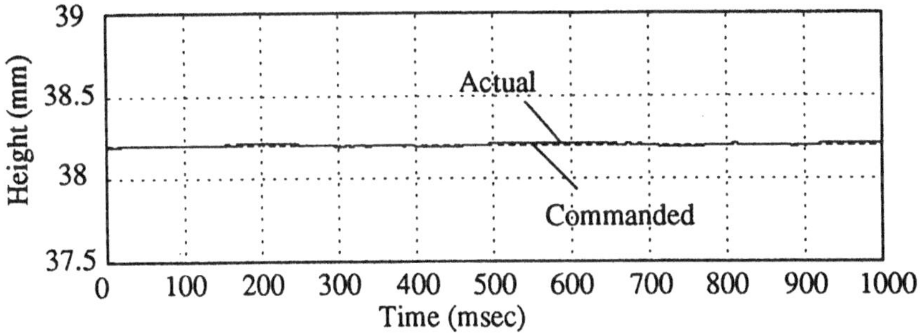

Fig. 16.21. Set-point regulation of classical controller

Figure 16.21 depicts the performance of the PI-plus-lead controller in set-point regulation. The performance is quite good and compares well with the VSC controller case in Fig. 16.19. However, slight oscillations about the desired set point are evident. Figure 16.22 depicts a 1-Hz trajectory following case. The performance is quite good, but the VSC controller case in Fig. 16.20 shows a better performance.

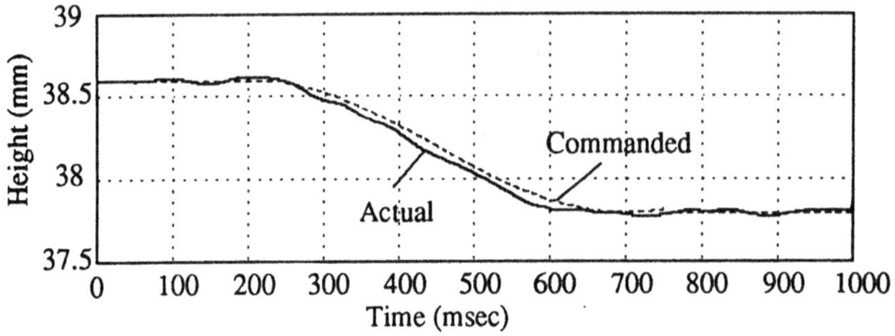

Fig. 16.22. 1-Hz trajectory following of classical controller

It is evident from the large overshoot in Fig. 16.22, as well as the oscillations in Fig. 16.21, that the PI-plus-lead controller does not provide as much damping as the VSC controller. This is mostly attributed to the inherent effects of the controller zeros, which were necessary to stabilize the plant pole in the right-half plane. The effects of linearization at the nominal set-point do not result in large modelling errors that can cause the light damping. In the command range of 38.4 mm and 37.8 mm, the force-distance relationship is quite linear, and the perturbation equation (16.44) does not change appreciably. Throughout the whole range, the plant pole locations change less than 5%.

Performance at various set points of 37.8 mm, 38.0 mm, 38.4 mm and 38.6 mm was also evaluated. As expected, the VSC controller provides perfect regulation in all cases within the sensor accuracy. The classical controller also performs well, even though it was designed based on the linearized plant at 38.2 mm. The performance of the classical controller in all cases are similar to that of the 38.2 mm case shown in Fig. 16.22, which exhibits small-amplitude oscillations about the set point. These results are not included.

Performance for various time-varying trajectories has been investigated. Because of the relatively poor transient characteristics, the classical controller performance degrades rapidly. For a 10-Hz trajectory, the classical controller cannot provide satisfactory performance and results in instability as depicted in Fig. 16.23. For the same trajectory, the VSC case still provides stable performance as depicted in Fig. 16.24. In the VSC case, some excitation of the unmodelled actuator dynamics during the transient is evident. This is due to the fact that the frequency of the desired trajectory (10 Hz) is approaching the bandwidth of the unmodelled actuator (approximately 45 Hz). For a 20-Hz trajectory case the oscillation becomes more pronounced, but the system still reaches a stable equilibrium with a zero steady-state errors shown in Fig. 16.25.

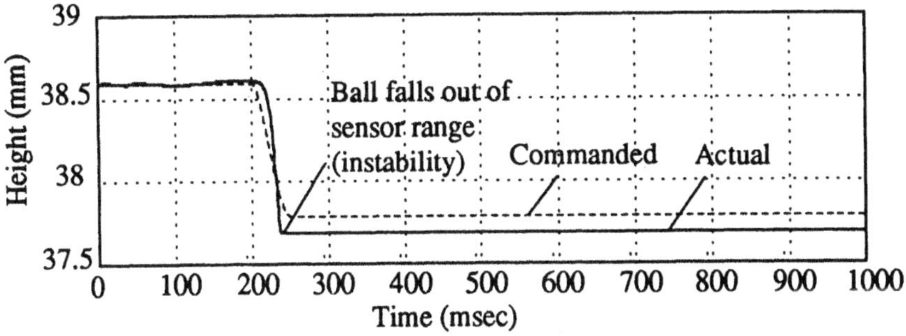

Fig. 16.23. 10-Hz trajectory following of classical controller

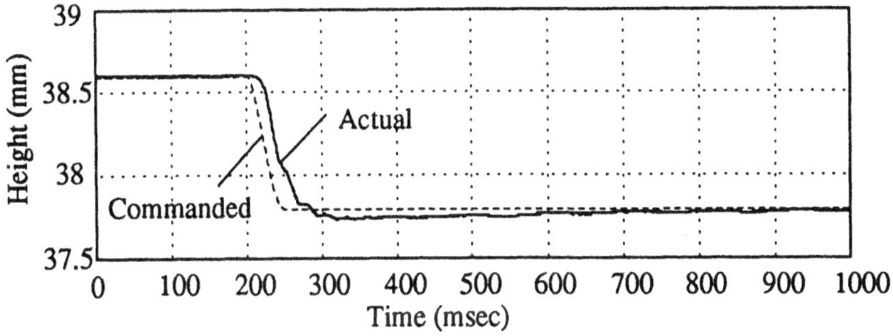

Fig. 16.24. 10-Hz trajectory following of integral-VSC controller

16.3.4 Discussion

Magnetic levitation control is not a trivial problem, because of its nonlinearities and because it is open-loop unstable. This chapter showed that the VSC approach with a control smoothing approximation and integral sliding manifold provides excellent performance and robustness. The classical controller provided performance and robustness levels approaching those of the VSC controller, only when the effects of actuator dynamics were included in the design.

16.4 Conclusions

This chapter started with the premise that the science of control involves the three iterative processes of modelling, control input synthesis, and experimental validation. Control theories provide the means for synthesizing the control input in a methodical and judicious manner, and many such methods exist for linear

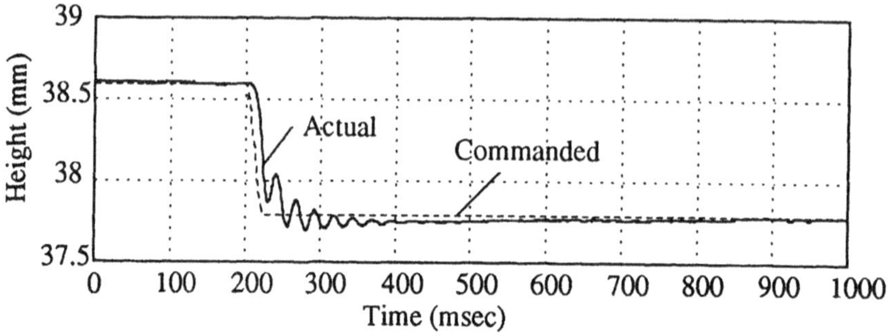

Fig. 16.25. 20-Hz trajectory following of integral-VSC controller

systems. Although many important engineering systems are highly nonlinear, the science of control for nonlinear systems remains much less explored.

This chapter in tandem with many other examples in this book, is an attempt to validate a paradigm for controlling nonlinear systems. The variable structure control concept is well suited for many types of nonlinear engineering systems. The ideal applicability of the VSC approach to engine fuel-injection control and magnetic levitation stabilization problems is experimentally demonstrated. The global performance and robustness properties of these examples are quite good. The problem of chatter and potentially-high control authority does not become a factor in these problems. In the fuel control problem the output chatter of the VSC controller is comparable to or smaller than that of a production controller. In the magnetic levitation problem a control smoothing approximation and an integral sliding manifold are used to provide reduced chatter and improved tracking than a classical controller. It is believed that the VSC approach can be applied to many other important engineering systems to provide better performance levels than those afforded by traditional control methodologies.

16.5 Acknowledgments

The research on fuel-injection control was supported by the Daewoo Motor Company in Korea. The author thanks many collaborators, in particular, Professor J.K. Hedrick and Messrs. H.K. Oh, D. Spilman, Y. Kato, B.J. Lee and Y.W. Kim.

References

Cho, D. 1993a, Automatic control system for IC engine fuel injection, U.S. Patent No. 5,190,020

Cho, D. 1993b, Experimental results on sliding mode control of an electromagnetic suspension. *J. of Mech. Sys. and Sig. Process.* **7**

Cho, D., Hedrick, J.K. 1988, Nonlinear controller design method for fuel-injected automotive engines. *ASME Trans., J. of Eng. for Gas Turbines and Power* **110**, 313–320

Cho, D., Hedrick, J.K. 1989, Automotive powertrain modelling for control. *ASME Trans., J. of Dyn. Sys., Meas. and Control* **111**, 568–576

Cho, D., Oh, H. 1993, Variable structure control of fuel-injection systems. *ASME Trans., J. of Dyn. Sys., Meas. and Control* **115**

Cho, D., Kato, Y., and Spilman, D. 1993, Experimental comparison of classical and sliding mode controllers in magnetic levitation systems. *IEEE Control Sys. Mag.* **13** 42–48

Dahlen, N.J. 1985, *Magnetic active suspension and isolation*, S.M. thesis, Mechanical Engineering, MIT, Cambridge, Mass

Downer, J. R. 1980, *Analysis of single axis magnetic suspension systems*, S. M. thesis, Mechanical Engineering, MIT, Cambridge, Maryland

Falk, C.D., Mooney, J.J. 1980, Three-way conversion catalysts: effects of closed-loop feed-back control and other parameters on catalyst efficiency, *SAE Paper No. 8000462*

Hamann, E., Manger, H., Steinke, L. 1977, Lambda-sensor with Y2O3-Stabilized ZrO2-ceramic for application in automotive emission control systems, *SAE Paper No. 770401*

Heywood, J.B. 1988, *Internal Combustion Engine Fundamentals*, McGraw-Hill, New York

Kaplan, B.Z., Regev, D. 1976, Dynamic stabilization of tuned-circuit levitators. *IEEE Trans. Magnetics* **Mag-12**, 556–559

Lee, B.J., Kim, Y.W., Cho, D. 1993, Anticipatory Control n the Sliding Phase Plane for Throttle Systems, *Proc. of ACC*, 1768-1772 (also to appear in *IEEE Trans. Control Sys. Tech.*)

Limbert, D.A., Richardson, H.H., Wormley, D.N. 1979, Controlled characteristics of ferromagnetic vehicle suspension providing simultaneous lift and guidance. *ASME Trans., J. of Dyn. Sys., Meas. and Control* **101**, 217–222

Powell, J.D. 1987, A review of IC engine models for control system design. Preprint of *Proc IFAC 87*, Munich

Sira-Ramírez, H. 1992, Nonlinear approaches to variable structure control. *Proceedings of Second IEEE Workshop on Variable Structure and Lyapunov Control of Uncertain Dynamical Systems*, Sheffield, UK, 144-155

Society for Industrial and Applied Mathematics 1988, *Future Directions in Control Theory: A Mathematical P erspective, Reports on Issues in the Mathematical Sciences*

Taylor, C.F. 1966, *The Internal Combustion Engine in Theory and Practice* 2nd ed., MIT Press

Utkin, V.I. 1977, Variable structure systems with sliding modes. *IEEE Transactions on Automatic Control* **AC-22**, 212–222

Washino, S. (ed.) 1989, *Japanese Technology Reviews: Automobile Electronics*, Gordon and Breach Science Publishers

Yamamura, S., and Yamaguchi, H. 1990, Electromagnetic levitation system by means of salient-pole type magnets coupled with laminated slotless rails. *IEEE Trans on Vehicular Technology* **39**, 83–87

17. Applications of VSC in Motion Control Systems

Ahmet Denker and Okyay Kaynak

17.1 Introduction

Advances in computer technology and high speed switching circuitry have made the practical implementation of VSC a reality and of increasing interest. The theory of VSC has been well explored over the past two decades and some reports of practical experience have appeared in the literature. To illustrate that VSC theory has reached a sufficiently advanced level to allow its application in various areas, we address here one of its most challenging applications, namely motion control systems. This chapter deals with applications to motor and robot control, the phenomenon of chattering, and the use of different control schemes to alleviate the problem of chatter.

Although VSC is theoretically excellent in terms of robustness and disturbance rejection capabilities, there have been doubts as to its applicability. The theoretical design of VSC which induces the sliding mode does not require accurate modelling; it is sufficient to know only the bounds of the model parameters. When sliding motion occurs, VSC is ideally switched at an infinite frequency. In reality VSC leads to pulse-width amplitude-modulated control signals which contain high and low frequency components. The practical implication of this is that the control is switched at a finite frequency and the corresponding trajectories chatter with respect to the switching plane. Chattering is especially undesirable and can cause excessive wear of mechanical parts and large heat losses in electrical circuits. The high frequency components of the control may also excite unmodelled high frequency plant dynamics which can result in unforeseen instabilities. To eliminate chattering one needs to makethe control input continuous in a region around the sliding surface.

Special emphasis is given here to motion control with examples in motor and robot control. Following a brief introduction on the design techniques, a microprocessor-based sliding mode controller applied to the position control of a dc motor, is described. In order to achieve a parameter and disturbance invariant fast response, the slope of the sliding line is incremented starting from a small initial value. The implemented control law has a switched current feedback term and a switched error velocity term in addition to the normal switched error term. Experimental results are presented showing the invariant nature of the system. Attention is also focussed on reducing chattering and the magnitude of the control effort. With this goal in mind we describe a robot control example which furnishes the VSC with a self-organizing control (SOC) capability. Since in both VSC and SOC the control rule is allowed to change its structure, the idea of combining them is a natural one. The advantage of this combined approach lies in the fact that minimum information of the system is required and modelling becomes much simpler. In the sliding mode self-organizing control (SLIMSOC) scheme, both the control actions and performance evaluation are carried out using the distance from the desired sliding surface and rate of approach to it. An important aspect of this controller is the reduction of the dependency and sensitivity to system uncertainties. It is applicable to systems of any dimension and complexity, even in the presence of random disturbances.

17.2 Design of VSC Controllers

We will not discuss all the intricacies of VSC. Our interest is mainly in its implementation and in the modifications that avoid chattering and non-zero steady state errors. Only a short overview of the design of sliding controllers will be given here.

In all applications the design policy involves the following steps

- selection of the sliding surface such that the sliding system has the desired eigenvalues
- control selection which provides the attractiveness and invariance of the sliding surface.

Control selection can be divided into the following two steps

- selection of a Lyapunov function and its time derivatives such that, if the Lyapunov stability criterion is satisfied, the distances to the sliding surface and their rate of change are of opposite sign
- selection of control to satisfy selected stability criterion.

Consider a system characterized by

$$\dot{x} = f(x,t) + B(x,t)u \tag{17.1}$$

where $x, f \in \mathbb{R}^n$, $u \in \mathbb{R}^m$, $B \in \mathbb{R}^{n \times m}$. The manifold

$$\mathcal{S} = \{x \in \mathbb{R}^n, \quad \sigma_i(x,t) = 0\} \tag{17.2}$$

is the sliding surface.

The selection of a Lyapunov function, on which the control law will be based, is always governed by the requirement that it should be as simple as possible. For the system (17.1) and selected manifold (17.2), consider the Lyapunov function in quadratic form

$$V = \frac{1}{2}\sigma^T \sigma \tag{17.3}$$

$\sigma(x,t) = 0$ will be stable if the first derivative of the Lyapunov function with respect to time can be expressed as

$$\frac{dV}{dt} = -\sigma^T Q \dot{\sigma} \tag{17.4}$$

where Q is a positive definite matrix. The system in the sliding mode satisfies

$$\sigma = 0 \ , \qquad \dot{\sigma} = 0 \tag{17.5}$$

By solving the above equation for the control input, we obtain an expression for $u = \hat{u}$ called the *equivalent control*, (Utkin 1977), which is equivalently the average value of u which maintains the state on the switching surface $\sigma(x) = 0$. However, the existence of the external disturbances and parametric uncertainties in the model make the computation of the exact value of the equivalent control impossible. Instead, only a nominal value can be computed. The application of this nominal value to the system will evidently cause a deviation of the state trajectory from the desired sliding surface. It is due to this reason that the equivalent control is supplemented with a discontinuous term which we will call the *attractive* control, since it ensures the attractiveness of the sliding surface. The attractive control component is determined such that the state is attracted to the sliding surface.

An analogy between feedforward-feedback controller and equivalent-supplementary control is drawn in Meystel (1992) where equivalent control plays a role similar to that of feedforward control in providing the control to track a desired trajectory. The desired trajectory in this case is the user-defined sliding surface itself. The additional term, on the other hand, is similar to the feedback control which tries to eliminate any deviations from the desired trajectory. The actual control u consists of a low frequency (average) component $\hat{u}$ and a high-frequency component u_a

$$u = \hat{u} + u_a \tag{17.6}$$

where u_a is the control which satisfies the following inequality

$$\sigma^T \dot{\sigma} = 0 \tag{17.7}$$

17.3 Application to a Motion Control System

For a set point regulation problem, i.e. for the problem of forcing the system to a desired position p_d with desired velocity $v_d = 0$ from an initial state $p(t_0)$ and $v(t_0)$, (17.1) can be rearranged in error space as given below by defining a new state vector $x_e^T = (e, v)$ where e is the position error

$$\begin{aligned} \dot{e}_i &= v_i \ , \qquad i = 1, ..., n \\ \dot{v}_i &= f_i(e_i + p_{di}, v_i) + b_i(e_i + p_{di})u_i \end{aligned} \tag{17.8}$$

The computation of the equivalent control term (17.6) is done off-line and it requires a priori knowledge about the system. In some cases this is not practical and the control consists only of the term which ensures attractiveness of the sliding surface, and the VSC has to change structure on reaching a set of switching surfaces as in

$$u_i = \begin{cases} u_i^+(p_i, v_i) & \text{if } \sigma_i(e_i, v_i) > 0 \\ u_i^-(p_i, v_i) & \text{if } \sigma_i(e_i, v_i) < 0 \end{cases} \tag{17.9}$$

where u_i is the ith component of u and $\sigma_i(e_i, v_i)$ is the ith component of the m switching hyperplanes $\sigma(ev)$

$$\dot{\sigma}_i = \lambda_i e_i + v_i \ , \qquad \lambda_i > 0 \ , \qquad i = 1, ..., m \tag{17.10}$$

In order to illustrate the disturbance rejection aspect of VSC, the system which is shown in the block diagram in Fig. 17.1 has been considered by Kaynak and Harashima (1985). The state representation of this system is

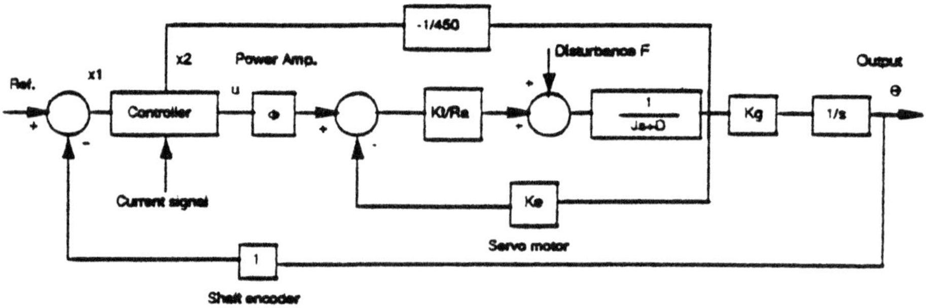

Fig. 17.1. Block diagram of the system

$$\begin{bmatrix} \dot{x}_1 \\ \dot{x}_2 \end{bmatrix} = \begin{bmatrix} 0 & 1 \\ 0 & -a \end{bmatrix} \begin{bmatrix} x_1 \\ x_2 \end{bmatrix} + \begin{bmatrix} 0 \\ -b\phi \end{bmatrix} u(t) + \begin{bmatrix} 0 \\ -1 \end{bmatrix} d(t) \tag{17.11}$$

where

$$\begin{aligned} x_1 &= \text{position error} \\ x_2 &= \dot{x}_1 = -\dot{\theta}_0 = \text{rate of change of error} \\ \phi &= 60^\circ \end{aligned}$$

$K_T = 6.0\text{x}10^{-2}$ V.s/rad
$K_T = 6.0\text{x}10^{-2}$ N.M/A
$R_A = 1.27\ \Omega$
$D = 2.84\text{X}10^{-3}$ n.m.s/rad
$J = 6\text{x}10^{-5}$ kg.m^2
$K_o = 1/450$
$\Phi = 57$
$L = 0.14$ m
$M_o = 9$ kg
$F = K_o L M_o \sin\theta_o = 27.5 \sin\theta_o$ (N.m)

Table 17.1. The parameters of the system

$$
\begin{aligned}
b &= \frac{K_T K_G}{J R_A} = 1.75 \\
a &= \frac{(K_T K_E + D R_A)}{J R_A} = 90 \\
d &= \mp max|\frac{K_G F}{J}|
\end{aligned}
$$

with

$$F = L M_G K_G \sin\theta_0 \tag{17.12}$$

The numerical values of the system parameters are given in Table 17.1. The hardware details of the system used for experimental investigations are shown in Figs. 17.2 and 17.3.

A 24 V 50 W dc servomotor is driven by a PWM power MOSFET chopper operating at 10 kHz. A 10 bit digital shaft encoder is used to sense the output position while a dc tachogenerator coupled directly to the servomotor provides an analog signal for the output speed. Two 10 bit tracking type A/D converters are used to obtain the digital values of the output speed and the motor current. A gear train with a gear ratio of 1/450 is inserted between the motor and the shaft encoder. The mechanical arrangement shown in Fig. 17.3 generates the nonlinearity. The mass on the rod and its distance to the motor shaft can be varied.

For a number of applications the control

$$u = \psi_1 x_1 + k_f \operatorname{sgn} \sigma \tag{17.13}$$

has been proposed (Draženović 1969, Itkis 1976, Utkin 1977, Utkin 1992) where k_f is a constant, and the switching line σ is

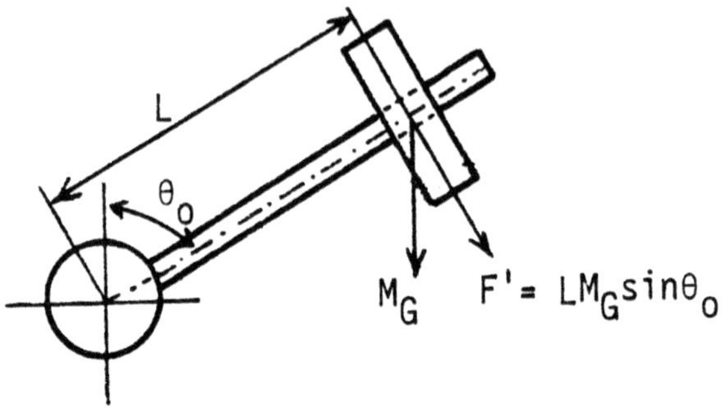

Fig. 17.2. Hardware details of the system

$$\sigma = x_2 + \lambda x_1 \tag{17.14}$$

and

$$\begin{array}{ll} \psi_1 = \alpha_1 & \text{if } \sigma x_1 > 0 \\ \psi_1 = \beta_1 & \text{if } \sigma x_1 < 0 \end{array} \tag{17.15}$$

Alternatively the more general control

$$u = \psi_1 x_1 + \psi_2 x_2 + k_f \operatorname{sgn} \sigma \tag{17.16}$$

can be used. The inclusion of a switched-actuator term allowsthe reduction of the magnitude of the k_f values. In the present system the actuator output can be thought of as the current of the servomotor and the control law

$$u = \psi_1 x_1 + \psi_2 x_2 + \psi_3 i + k_f \operatorname{sgn} \sigma \tag{17.17}$$

$$\begin{array}{ll} \psi_1 = \alpha_1 & \text{if } \sigma x_1 > 0 \\ \psi_1 = \beta_1 & \text{if } \sigma x_1 < 0 \end{array}, \quad \begin{array}{ll} \psi_2 = \alpha_2 & \text{if } \sigma x_2 > 0 \\ \psi_2 = \beta_2 & \text{if } \sigma x_2 < 0 \end{array}, \quad \begin{array}{ll} \psi_3 = \alpha_3 & \text{if } \sigma i > 0 \\ \psi_3 = \beta_3 & \text{if } \sigma i < 0 \end{array} \tag{17.18}$$

is obtained. The first term in (17.17) is the normal proportional term, the second term relaxes the constraints on the slope λ of the sliding line $\sigma = 0$, and the third term ensures that the system does not leave the sliding mode due to the disturbance d. The relay term $k_f \operatorname{sgn} \sigma$ is used to overcome the effects of backlash and coulomb frictional forces. The $\psi_3 i$ term can also be expressed in the form

$$\psi_3 i = (\alpha_3 + \beta_3)\frac{i}{2} + (\alpha_3 - \beta_3)\frac{|i|}{2} \operatorname{sgn} \sigma \tag{17.19}$$

the disturbance d. The relay term $k_f \operatorname{sgn} \sigma$ is used to overcome the effects of backlash and coulomb frictional forces. The $\psi_3 i$ term can also be expressed in the form

$$\psi_3 i = (\alpha_3 + \beta_3)\frac{i}{2} + (\alpha_3 - \beta_3)\frac{|i|}{2} \operatorname{sgn} \sigma \tag{17.19}$$

which reduces to

$$\psi_3 i = a_3 |i| \operatorname{sgn} \sigma \tag{17.20}$$

when $\alpha_3 = -\beta_3$, i.e. it has a similar structure to that of k_f with the difference that when the state is forced to zero, this term also reduces to zero. Equation (17.7) dictates that the slope of the sliding line λ should be chosen so as to satisfy

$$\begin{aligned} \alpha_1 &\leq 0 \\ \beta_1 &\leq 0 \\ b\phi\alpha_2 &\geq \lambda - a \\ b\phi\beta_2 &\leq \lambda - a \end{aligned} \tag{17.21}$$

The slope of the sliding line λ should therefore be chosen accordingly, considering the range of the parameters of the system, a, b and ϕ. In the derivation of the inequalities of (17.21), it is assumed that the control is unrestricted. In the design of a practical system, the fact that the control is limited to a value of $|u| < u_{max}$ should be taken into consideration.

During the experimental investigations, the phase variable state representation (17.11) of the system was expressed in discrete form and a number of digital simulations were carried out (Kaynak and Harashima 1985). The following design values were selected

$$\begin{aligned} &\alpha_1 = 1 \; , \quad \beta_1 = \frac{-1}{2} \; , \quad \alpha_2 = \frac{1}{8} \\ &\alpha_3 = -\beta_3 = \frac{1}{16} \; , \quad k_f = 4 \text{ bits} \end{aligned} \tag{17.22}$$

If the deviations from the sliding line are small, the the solution is approximately

$$x_1(t) = x_1(t_0) \exp\{-\lambda(t - t_0)\} \tag{17.23}$$

In (17.23), t_0 is the time of hitting the sliding line and $1/\lambda$ is the time constant. For a fast response, λ has to be as large as possible, but on the other hand, if λ is large, t_0 will increase. Since the system is invariant only when in a sliding mode, a correspondingly larger part of the trajectory will be sensitive to parameter variations and disturbances. Fig. 17.4 shows the error waveforms for two different conditions. At first, the supply voltage V_s of the chopper is set to 26 V and the mass M is made zero. Afterwards, the supply voltage is increased to 31.2 V, which means that ϕ is increased by 20% and the full disturbance $d = 1$ is applied. The rest is carried out in such a shaft position that the effect of the mass M is additive to the effect of the increase in V_s, i.e. assisting the rotation of the shaft.

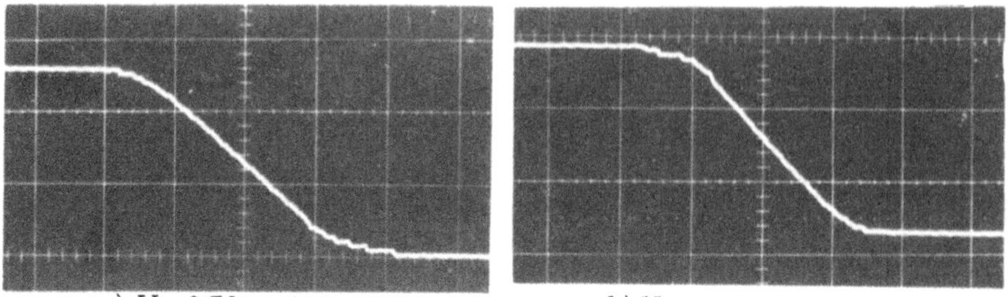

a) V_S=26V and d=0 b) V_S=31.2V and d=1

vert. scale=0.1 rad hor. scale=0.1 s

Fig. 17.4. Error waveforms for two different conditions

A sliding mode throughout the transient response can be ensured if the slope of the sliding line is varied adaptively as suggested in Zinober (1975), starting with a small initial value and increasing it whenever the existence of a sliding mode is verified. The upper trace of Fig. 17.5 shows the error waveform

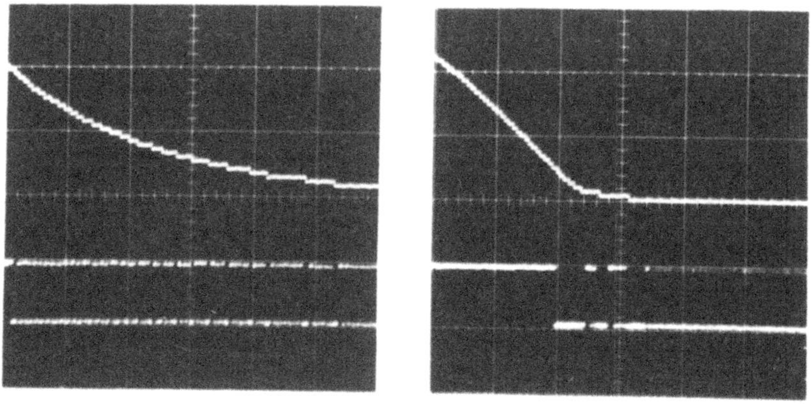

Fig. 17.5. Error waveforms for variable λ

obtained using such an approach for $V_s = 28\,\mathrm{V}$, $d = 1$, $\lambda(n+1) = \lambda(n) + 0.25$, $\lambda(0) = 0.25$ and $\lambda_{\max} = 16$. The lower trace shows the MSB of the digital control voltage, which is the sign bit. It indicates that a sliding mode is ensured throughout the transient response and shows the error waveforms for adaptively varied λ under the same conditions as those for Fig. 17.4 There is very little difference between the oscillograms, i.e. the response is invariant in spite of the large change in ϕ and the disturbance.

In the design and the realization of a sliding mode controller, special attention has to be paid to the switching frequency. The higher it is, the smaller will be the deviations from the ideal sliding line. Furthermore, the devices used to sense the state variables should have high resolution and accuracy. Only then will the motion of the system be almost the same as the one determined by the sliding line.

17.4 Robustness at a Price: Chattering

Ideally the switching of control to eliminate deviations from the sliding surface, occur at infinitely high frequency. But in practice, due to finite switching time, the frequency is not infinitely high. The control is discontinuous across the switching surface and chattering takes place. This effect can be observed in Figs. 17.6 and 17.7. Chattering implies high control activity which can be very harmful to the actuators and may excite the unmodelled dynamics of the system.

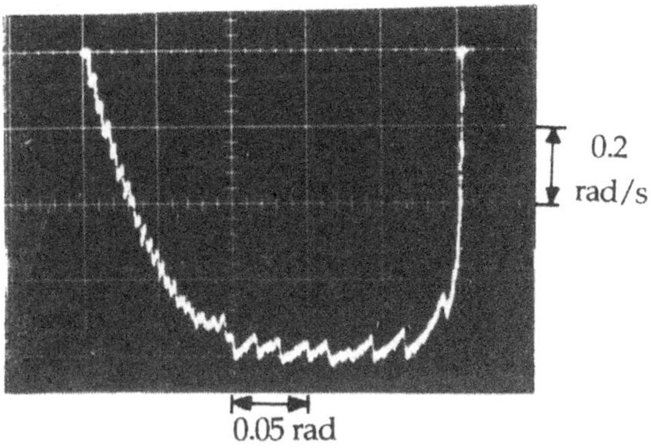

Fig. 17.6. A typical phase-plane trajectory for variable λ

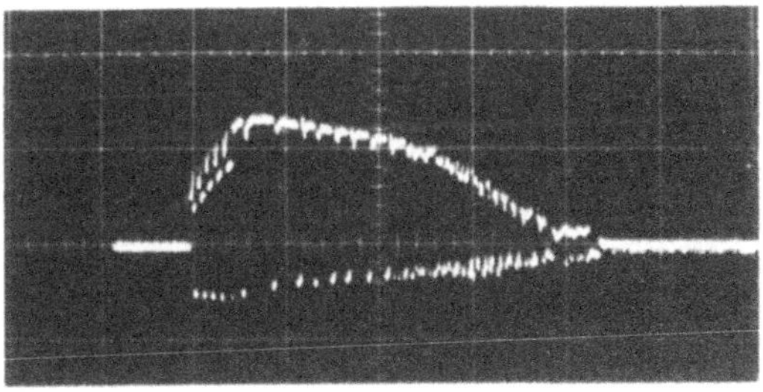

Fig. 17.7. A typical control signal λ

This is a well-known problem and is widely treated in the literature. Different schemes are suggested in order to eliminate chattering. These schemes, in one way or another, try to smooth out the control discontinuity in the vicinity of the sliding surface. One of the first approaches to reduce chattering was to introduce a boundary layer around the sliding surface (Slotine and Sastry

1983, Chang 1991) and use a continuous control within the boundary layer. The relay type function was replaced by a saturation function

$$\operatorname{sgn} \sigma \rightarrow \operatorname{sat} \frac{\sigma}{\Upsilon} \tag{17.24}$$

where $\rightarrow$ means "replace by", and the saturation function sat is

$$\operatorname{sat} \frac{\sigma}{\Upsilon} = \begin{cases} \operatorname{sgn} \sigma & \text{if } \|\sigma\| \leq \Upsilon \\ \sigma/\Upsilon & \text{if } \|\sigma\| \leq \Upsilon \end{cases} \tag{17.25}$$

Fractional interpolation with state dependent offset

$$\operatorname{sgn} \sigma \rightarrow \frac{\sigma}{\|\sigma\| + \delta} \tag{17.26}$$

has been discussed by Burton and Zinober(1986) and Chern and Wu (1991) amongst others. Using an integral transformation with a cone-like boundary layer was proposed by Ning (1989)

$$\operatorname{sgn} \sigma \rightarrow \operatorname{sat} \left(\int_{t_0}^{t} k_i \operatorname{sgn} \sigma \mathrm{d}\tau \right) \tag{17.27}$$

In this approach a cone shaped boundary layer is introduced around the sliding surface. Inside this boundary layer the system has desirable properties and the control law is then selected to guarantee that the state will be attracted to these cones. An approach by Machado and Carvalho (1988) involves periodically redefining the switching surface. From (17.4) and (17.5) the control can be calculated as

$$\begin{aligned} u &= -[GB]^{-1} G f(x,t) + B^{-1} \eta \sigma \\ &= \hat{u} + B^{-1} \eta \sigma \end{aligned} \tag{17.28}$$

which differs from the equivalent control by the term $B^{-1}\eta\sigma$ which is zero if the motion is constrained to the sliding manifold. For the calculation of (17.28) information about the equivalent control is required. Since this representation is not practical, the following form

$$u(t) = u(t^-) + B^{-1}(\eta\sigma + \frac{\mathrm{d}\sigma}{\mathrm{d}t}) \; , \quad t = t^- + \Delta \; , \quad \Delta \rightarrow 0 \tag{17.29}$$

is suggested in Sabanović and Ohnishi (1992). The value of the control at time t is caculated from the value at time $t - \Delta$ of the weighted sum of the control error σ and its rate of change $\dot{\sigma}$. The stability conditions for the selected control can be examined as follows. Using (17.4) and (17.29)

$$\begin{aligned} \frac{\mathrm{d}v}{\mathrm{d}t} &= -\sigma^T Q \dot{\sigma} \\ &= \sigma^T B(\hat{u} - u) \\ &= \sigma^T B(u(t) - u(t - \Delta)) - \sigma^T Q \sigma \end{aligned} \tag{17.30}$$

which for $\Delta \to 0$ becomes $\mathrm{d}v/\mathrm{d}t = \sigma^T \mathrm{d}\sigma/\mathrm{d}t = -\sigma^T Q\sigma$. If Q is a positive definite matrix, the time derivative of the Lyapunov function is negative definite, i.e. the solution $\sigma(x,t) = 0$ is stable and the motion of the system from arbitrary initial conditions will reach the sliding manifold. The resulting control, being continuous, has the potential for eliminating the chatter. The equations of motion with control (17.28) are

$$\frac{\mathrm{d}x_1}{\mathrm{d}t} = f_1(x_1, x_2)$$

$$\frac{\mathrm{d}\sigma}{\mathrm{d}t} + Q\sigma = 0 \tag{17.31}$$

17.5 VSC Design for Robotic Manipulators

It is generally agreed that the dynamics of an n-joint robotic manipulator can be described by n coupled second order nonlinear equations of the form

$$\ddot{x}_i = f_i(x_i, \dot{x}_i) + b_i(x_i)u_i \ , \qquad i = 1,2,3,\ldots,n \tag{17.32}$$

where x_i is the component of the vector of joint angles $x(t) \in \mathbb{R}^n$, and u_i is the the component of the generalized input vector $u(t) \in \mathbb{R}^m$. We need the state to track a reference trajectory in spite of the uncertainty in the system.

In VSC theory this objective corresponds to steering the states of the system in an $(n-m)$-dimensional subspace $\sigma \subset \mathbb{R}^n$ and to maintain the subsequent motion of the state trajectories on this manifold σ which can be defined as

$$\sigma = \bigcap_{i=1}^{m} \sigma_i \tag{17.33}$$

The objective can be reached by making the state variables x_i track the desired state trajectory variables x_{di} . Thus the σ_i can be selected as

$$\sigma_i = \left(\frac{\mathrm{d}}{\mathrm{d}t} + \lambda_i\right) e_i \tag{17.34}$$

where

$$e_i = x_i - x_{di}$$

Having the system track $x_i(t) = x_{di}(t)$, implies making $\sigma_i = 0$ and $\dot{\sigma}_i = 0$. Dropping subscripts for notational clarity, the behaviour of the system with uncertainties is described by the equation

$$\ddot{x} = \hat{f}(x(t)) + \Delta f(x(t)) + \{\hat{b}(x(t)) + \Delta b(x(t)\}u(t) \tag{17.35}$$

where $\hat{f}$ and $\hat{b}$ are the estimated terms of the model and are bounded by some known values

$$|f - \hat{f}| < F$$

and

$$b_{min} < \hat{b} < b_{max} \tag{17.36}$$

The objective is to accomplish tracking in the presence of these uncertainties. Let us choose a Lyapunov function

$$V(\sigma) = \sum_{i=1}^{n} \frac{1}{2}\sigma_i^2 \tag{17.37}$$

which is a measure of the squared distance to the sliding manifold. The controller u_i is chosen such that

$$\frac{1}{2}\frac{\mathrm{d}}{\mathrm{d}t}\sigma^2 < -Q|\sigma| \tag{17.38}$$

where Q is a strictly positive constant. This leads to

$$u = \hat{u} - k \operatorname{sgn} \sigma \tag{17.39}$$

where

$$\hat{u} = \hat{b}^{-1}(-\hat{f} + \ddot{x}_d - \lambda\dot{e})$$

with the values of k and $\hat{b}^{-1}$

$$\begin{aligned} k &\geq \hat{b}^{-1}[\beta(F+Q) + |(\beta-1)|\frac{|\hat{u}|}{\hat{b}^{-1}}] \\ \hat{b}^{-1} &= (b_{min}b_{max})^{-\frac{1}{2}} \end{aligned} \tag{17.40}$$

$\hat{u}$ is equivalently the average value of u which maintains the state on the switching surface $\sigma(x) = 0$ and

$$\beta = (\frac{b_{max}}{b_{min}})^{\frac{1}{2}}$$

From (17.38) this controller satisfies the attractivity condition

$$\sigma\dot{\sigma} < 0 \tag{17.41}$$

When $\sigma = 0$, (17.34) implies that

$$\dot{e} + \lambda e = 0 \tag{17.42}$$

which is asymptotically stable for $\lambda > 0$, and $e \to 0$ as $t \to \infty$. Evidently the $\operatorname{sgn}\sigma(mT)$ term introduces a discontinuity and acts as the main cause of chattering. In addition, the upper bounds lead to an overconservative design thereby yielding unnecessarily high control input values. In order to alleviate these drawbacks, we will try to complement the sliding mode controller with a self-organizing capability.

17.5.1 Merging Sliding Mode and Self-Organizing Controllers

Since both in VSC and SOC the control rule is allowed to change its structure, the idea of combining them is a natural one. A reward for introducing this additional complexity comes from making use of the useful properties of each approach. The advantage of this combined approach lies in reducing the control activity. In (17.39) the equivalent control $\hat{u}$ is the average value of control which maintains the state on the sliding surface.We modify it to be

$$\hat{u} \rightarrow \hat{u} + \Delta\hat{u} \tag{17.43}$$

Substituting (17.43) into $\dot{\sigma}$, we can compute the corrective term as

$$\Delta\hat{u} = \hat{b}^{-1}\dot{\sigma} \tag{17.44}$$

Thus the attractivity control term becomes $\Delta\hat{u} - k\,\mathrm{sgn}\,\sigma$ with the value of k

$$k \geq \hat{b}^{-1}(\beta(F+Q) + |(\beta-1)|\frac{|\hat{u}|}{\hat{b}^{-1}} + |(\beta-1)|.|\dot{\sigma}|)$$

Note that in updating k according to $\dot{\sigma}$, one obtains a less tight switching control than (17.39). Now our supplementary control term u_a becomes

$$u_a = \hat{b}^{-1}(\dot{\sigma} - \beta(F+Q) + |(\beta-1)|\frac{|\hat{u}|}{\hat{b}^{-1}} + |(\beta-1)||\dot{\sigma}|)\,\mathrm{sgn}\,\sigma \tag{17.45}$$

Thus far the control law has been derived using continuous time. However, its application inevitably entails computer implementation. Thus the values corresponding to variations from the desired sliding surface and their rate of change can be more conveniently represented in discrete time by $\sigma(mT)$ and $\dot{\sigma}(mT)$, where T is the sampling period and m is the sample number. $u_a(mT)$ which is to be applied at the instant mT to drive the state trajectories onto the sliding surface, can now be obtained using $\sigma(mT)$ and $\dot{\sigma}(mT)$ as follows

$$u_a(mT) = |(\beta-1)|.|\hat{u}(mT)|\,\mathrm{sgn}\,(\sigma(mT)) + \hat{b}^{-1}K(mT) \tag{17.46}$$

where

$$K(mT) \;=\; \dot{\sigma}(mT) - (\beta(F+Q) + |(\beta-1)|.|\dot{\sigma}(mT)|)\mathrm{sgn}\,\sigma(mT) \tag{17.47}$$

It is obvious that we can calculate the $K(mT)$ values as (17.47).

$$\begin{array}{rl}
K(mT) = -\beta(F+Q) + \beta(\dot{\sigma}(mT)) & \sigma(mT) > 0,\ \dot{\sigma}(mT) < 0 \\
K(mT) = -\beta(F+Q) & \sigma(mT) > 0,\ \dot{\sigma}(mT) = 0 \\
K(mT) = -\beta(F+Q) + |(\beta-2)|(\dot{\sigma}(mT)) & \sigma(mT) > 0,\ \dot{\sigma}(mT) > 0 \\
K(mT) = \dot{\sigma}(mT) & \sigma(mT) = 0,\ \dot{\sigma}(mT) < 0 \\
K(mT) = 0 & \sigma(mT) = 0,\ \dot{\sigma}(mT) = 0 \\
K(mT) = \dot{\sigma}(mT) & \sigma(mT) = 0,\ \dot{\sigma}(mT) > 0 \\
K(mT) = \beta(F+Q) + \beta(\dot{\sigma}(mT)) & \sigma(mT) < 0,\ \dot{\sigma}(mT) < 0 \\
K(mT) = \beta(F+Q) & \sigma(mT) < 0,\ \dot{\sigma}(mT) = 0 \\
K(mT) = \beta(F+Q) + \beta(\dot{\sigma}(mT)) & \sigma(mT) < 0,\ \dot{\sigma}(mT) > 0
\end{array}$$

	$\dot{\sigma}_{min}$ $\dot{\sigma}<0$	$\dot{\sigma}=0$	$\dot{\sigma}>0$ $\dot{\sigma}_{max}$
σ_{max} $\sigma>0$	$K(mT)=-\beta(F+\eta)+\beta\dot{\sigma}(mT)$	$K(mT)=-\beta(F+\eta)$	$K(mT)=-\beta(F+\eta)+\lvert\beta-2\rvert\dot{\sigma}(mT)$
$\sigma=0$	$K(mT)=\dot{\sigma}(mT)$	$K(mT)=0$	$K(mT)=\dot{\sigma}(mT)$
$\sigma<0$ σ_{min}	$K(mT)=\beta(F+\eta)+\beta\dot{\sigma}(mT)$	$K(mT)=\beta(F+\eta)$	$K(mT)=\beta(F+\eta)+\beta\dot{\sigma}(mT)$

Table 17.2. Decision table

It is important to note that (17.47) depends only on $\sigma(mT)$ and $\dot{\sigma}(mT)$. It takes account of the distance from the desired sliding surface and also incorporates the rate of appoach to it.

17.5.2 SLIMSOC

Complementing the sliding mode controller with a self-organizing capability means furnishing the sliding mode controller with a rule-based feature, where the control strategy is improved by the controller itself. We will call the new controller SLIMSOC (SLIding Mode Self-Organizing Controller). SLIMSOC performs two tasks simultaneously, namely

(i) observing $\sigma(mT)$ and $\dot{\sigma}(mT)$ while issuing the appropriate control actions

(ii) using these results to improve the control action further.

These tasks are performed with reference to a rule-based decision table and a performance table as described in Mamdani (1979).

17.5.2.1 Decision Table Three variables $\sigma(mT)$, $\dot{\sigma}(mT)$, and $K(mT)$ form the decision table depicted in Table 17.2,

where the x and y coordinates represent $\dot{\sigma}(mT)$ and $\sigma(mT)$ respectively. $K(mT)$ is shown as an equation in the corresponding entry. The decision table is initially used to observe $\sigma(mT)$ and $\dot{\sigma}(mT)$, and then takes the form of a decision maker, which leads to the control input required from the observed values. In other words, the controller strategy evaluates its own performance at the end of each step and updates itself. The range of values that $\sigma(mT)$ and

$\dot{\sigma}(mT)$ take, determines the boundaries of the decision table. Thus, using the maximum allowable values of the $e(mT)$ and $\dot{e}(mT)$ boundaries of the decision table, the scaling factors are estimated. $\sigma(mT)$ and $\dot{\sigma}(mT)$ are quantized by using the following equations.

$$\begin{aligned} \sigma(mT) &= \mathcal{Q}\left[\sigma(mT) \times \Phi\right] \\ \dot{\sigma}(mT) &= \mathcal{Q}\left[\dot{\sigma}(mT) \times \Psi\right] \end{aligned} \qquad (17.48)$$

where Φ and Ψ are the scaling factors and $\mathcal{Q}[\cdot]$ represents the quantization procedure. Discretization levels which result from the quantization procedure can be chosen according to the desired tracking accuracy. Depending upon the values of $\sigma(mT)$ and $\dot{\sigma}(mT)$, the decision table is partitioned into nine subsections as shown in Table 17.2 where each subsection is associated with a corresponding $K(mT)$ expression.

17.5.2.2 Performance Measure Table In order to improve the control strategy, SLIMSOC evaluates its own performance. The performance criterion should be selected so that the deviation from the desired system behaviour can be directly related to the required improvements in individual control inputs. Recalling that in this approach control inputs depend only upon $\sigma(mT)$ and $\dot{\sigma}(mT)$, it is natural to base the performance on the deviation from the desired sliding surface and its rate of change. In translating this deviation information into a correction of the control action, an important issue to consider is the contribution of the past control inputs to the present situation. In our case control actions at three preceeding time steps are adjusted by a modification term, depending upon the present performance. The number of prior control inputs to be updated is determined by the response time of the system.

In the implementation of SLIMSOC, the performance of the controller is evaluated by using a performance measure table as shown in Table 17.3,

where the rows and columns stand for $\dot{\sigma}(mT)$ and $\sigma(mT)$ respectively, and each entry shows the performance measure value. These values are created by taking into account error sensitivity and control effort limitations which are imposed by the system structure. The zero entries show the cases which require no correction. The entry in the centre of the table is called the set point and corresponds to the ideal case where $\sigma(mT)$ and $\dot{\sigma}(mT)$ are both zero. Performance measure tables are organized in terms of diagonal bands in which the deviations are equally weighted. The zero band stands for tolerable deviations in σ and $\dot{\sigma}$, which will eventually decay to zero, thus ensuring convergence to the set point. The further away we get from the set point, the larger is the performance measure value, thereby requiring a larger modification term. It is important to note that this table is constructed by taking both the distance from the desired sliding surface and the rate of convergence to it. It can be seen from Fig. 17.8 how the decision and performance measure tables are used in SLIMSOC. Let us denote the performance measure value at the mth instant by

$$P(mT) = PM\{\sigma(mT), \dot{\sigma}(mT)\} \qquad (17.49)$$

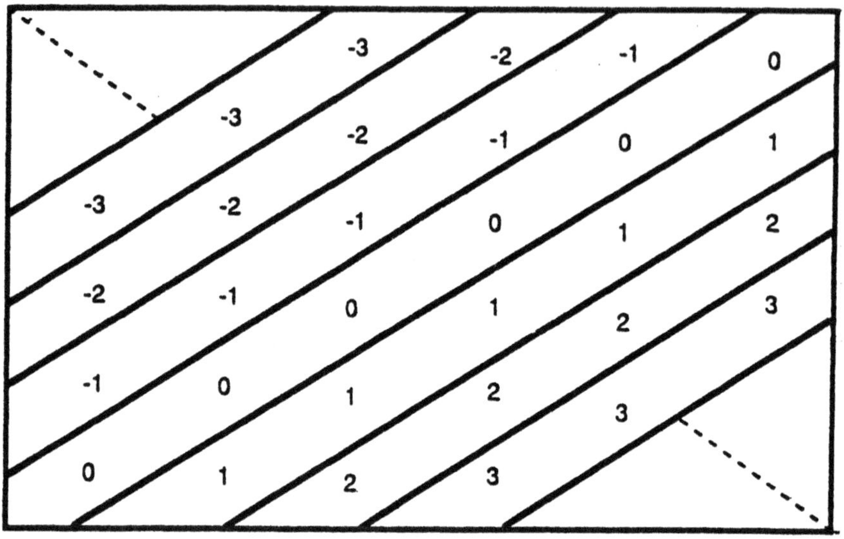

Table 17.3. Performance measure table

where PM represents the corresponding entry of the performance measure table. We will use this to determine the modification terms to the previous control actions

$$K(mT - qT) = K(mT - qT) + \Omega_q P(mT) \ , \qquad q = 1, 2, 3 \tag{17.50}$$

where Ω_q is a weighting factor.

17.5.3 Simulation Results

The performance of SLIMSOC is tested by applying it to the tracking control of the SCARA type robotic manipulator in Fig. 17.9. The angles $\theta_1 = x_{11}$ and $\theta_2 = x_{21}$ are selected according to the D-H conventions (Koivo 1989). The angular velocities are

$$\begin{aligned} \dot{x}_{11} &= \dot{\theta}_1 = x_{12} \\ \dot{x}_{21} &= \dot{\theta}_2 = x_{22} \end{aligned} \tag{17.51}$$

The dynamics of the two-joint manipulator can be expressed as given below:

$$\begin{bmatrix} \dot{x}_{11} \\ \dot{x}_{12} \end{bmatrix} = \begin{bmatrix} 0 & 1 \\ 0 & a_{12} \end{bmatrix} \begin{bmatrix} x_{11} \\ x_{12} \end{bmatrix} + \begin{bmatrix} 0 & 0 \\ 0 & h_1 \end{bmatrix} \begin{bmatrix} x_{21} \\ x_{22} \end{bmatrix} + \begin{bmatrix} 0 \\ b \end{bmatrix} u_1(t)$$

$$\begin{bmatrix} \dot{x}_{21} \\ \dot{x}_{22} \end{bmatrix} = \begin{bmatrix} 0 & 1 \\ 0 & a_{22} \end{bmatrix} \begin{bmatrix} x_2 1 \\ x_2 2 \end{bmatrix} + \begin{bmatrix} 0 & 0 \\ 0 & h_2 \end{bmatrix} \begin{bmatrix} x_{11} \\ x_{12} \end{bmatrix} + \begin{bmatrix} 0 \\ b \end{bmatrix} u_2(t)$$

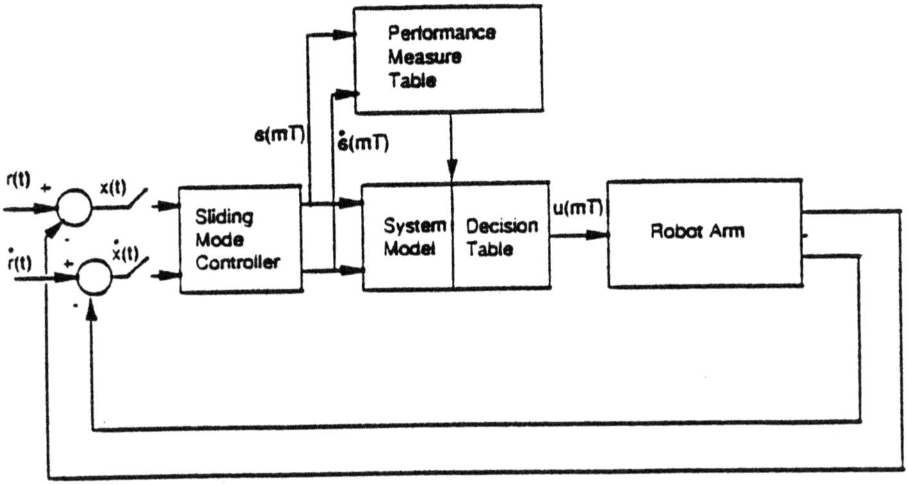

Fig. 17.8. Block diagram of SLIMSOC

where

$$
\begin{aligned}
a_{12} &= C(4\sin(x_{21})x_{22}/3 + 2\sin(x_{21})x_{12} \\
&\qquad + \cos(x_{21})\sin(x_{21})x_{12}) \qquad (17.52) \\
a_{22} &= C(-4\sin(x_{21})x_{12}/3 - 2\sin(x_{21})x_{22}/3 \\
&\qquad - \cos(x_{21})\sin(x_{21})x_{22} - 2\cos(x_{21})\sin(x_{21})x_{12}) \qquad (17.53) \\
h_1 &= 2C\sin(x_{21})x_{22}/2 \\
h_2 &= C(-10\sin(x_{21})x_{12}/3 - 2\cos(x_{21})\sin(x_{21})x_{12}) \\
b &= C
\end{aligned}
$$

with $C = \{16/9 - \cos 2(x_{21})\}^{-1}$. Disturbances were imposed leading to 25% deviation in the parameters. In this configuration motion of the planar and vertical links are decoupled and the control problem in two orthogonal planes can be handled independently. Here we will concentrate on the control of the planar links. Fig. 17.10 displays the outcome of SLIMSOC in tracking a 2D-square in the xy-plane. This involves the simultaneous operation of two SLIMSOC controllers. Each of these controllers treats the link on which it acts, as a single-input single-output system. Each controller improves its own rules in the face of cross-coupling effects experienced by the other system. Thus each controller input is calculated by using $\sigma(mT)$ and $\dot{\sigma}(mT)$ which are dependent on errors experienced by both of the links.

The dual SLIMSOC system is expected to track the outline of a square of side 1 metre which is centred at $(x, y) = (1.4\text{m}, 0.0\text{m})$. It takes 9 seconds to traverse and the sampling time is taken as 10 msec. Fig. 17.10 shows the results for two traverses of the square when using VSC and SLIMSOC controllers. It is important to note that in both cases deviations from the nominal trajectory are bounded by 2%. Thus SLIMSOC results are comparable to VSC with respect to tracking accuracy. However, comparison of Figs. 17.11 and 17.12 exhibits the clear advantage of SLIMSOC over VSC.

It requires much less control effort to obtain similar performance with SLIMSOC. Another advantage of SLIMSOC exhibits itself in Figs. 17.13 and 17.14, which clearly show the reduction of chattering.

The reduction in control effort stems from the fact that with the introduction of the SOC capability, the $K(mT)$ values become much smaller. Fig. 17.15 displays this fact for the $K(mT)$ trajectories. Evidently in the SLIMSOC case these are not only closer to the $\sigma = 0$ zone but also smoother, thus reducing the fundamental cause of chattering.

An overall comparison of the results of SLIMSOC with those of VSC show much better performance both from the control effort and chattering point of view, while tracking accuracy is comparable. It is essential to state that there is a trade-off between control effort and tracking accuracy. Thus, tracking errors can be reduced more but at the cost of higher control inputs.

17.6 Conclusions

In this chapter examples of motor and robotic control have been described which demonstrate that VSC theory can be used effectively in motion control systems. Both experimental and simulation results have shown that VSC is a robust and practical control approach for motion control systems.

References

Burton, J.A., Zinober, A.S.I. 1986, Continuous approximation of variable structure control. *Int. J. Systems Science* **17**, 876–885

Chang, L. 1991, A MIMO sliding control with a first order plus integral sliding condition. *Automatica* **27**, 853–858

Chern, T.L., Wu, Y.C. 1991, Design of integral variable structure controller and application to electrohydraulic velocity servosystems. *Proc IEE* **138D**(5), 439–444.

Draženović, B. 1969, The invariance conditions in variable structure systems. *Automatica* **5**, 287–295

Elmali, H., Olgac, N. 1992, Theory and implementation of sliding mode control with perturbation estimation. *Proc. IEEE Conf. on Robotics and Automation*, Nice, France

Itkis, U. 1976, *Control systems of variable structure*, Wiley, New York

Kaynak, O., Denker, A. 1993, Discrete time sliding mode control in the presence of system uncertainty. *International Journal of Control* **5**, 1177–1189

Kaynak, O., Harashima, F., Hashimoto, H. 1984, Variable structure systems theory applied to sub-time optimal position control with an invariant trajectory. *Trans. IEE of Japan* **104**, 47–52.

Koivo, A. 1989, *Fundamentals for control of robotic manipulators*, Wiley, New York

Mamdani, E., Procyk, T. 1979, A linguistic self organizing process controller. *Automatica* **15**, 15–30

Meystel, A., Nisenzon, Y., Nawathe R. 1993, Merger of rule based and Variable Structure Controller. *Proc. IEEE Regional Conf. on Aerospace Control*, Los Angeles

Ning-Su, L., Chun-bo, F. 1989, A new method for suppressing chattering in variable structure feedback control systems, *Proc. IFAC Symposium on Nonlinear Control Systems Design*, Pergamon Press, Capri, 279–284.

Sabanović, A., Ohnishi, K. 1992, Sliding mode control of robotic manipulators, *Proc. IEEE/RSC Int. Conf. IROS92*, Raleigh

Slotine, J., Sastry, S. 1983, Tracking control of nonlinear systems using sliding surfaces with application to robot manipulators. *International Journal of Control* **38**, 465–492

Slotine, J., Li, W. 1991 *Applied nonlinear control*, Prentice-Hall, New Jersey

Tenreiro Machado, J.A., Martins de Carvalho, J.L. 1988, A smooth variable structure control algorithm for robot manipulators, *Proc. IEE Int. Conf. Control 88*, London, 450–455

Utkin, V.I. 1977, Variable structure systems with sliding mode. *IEEE Transactions on Automatic Control* **22**, 212–222

Utkin, V.I. 1992, *Sliding modes in control optimization*, Springer-Verlag, New York

Zinober, A.S.I. 1975, Adaptive relay control of second-order systems. *International Journal of Control* **21**, 81–98

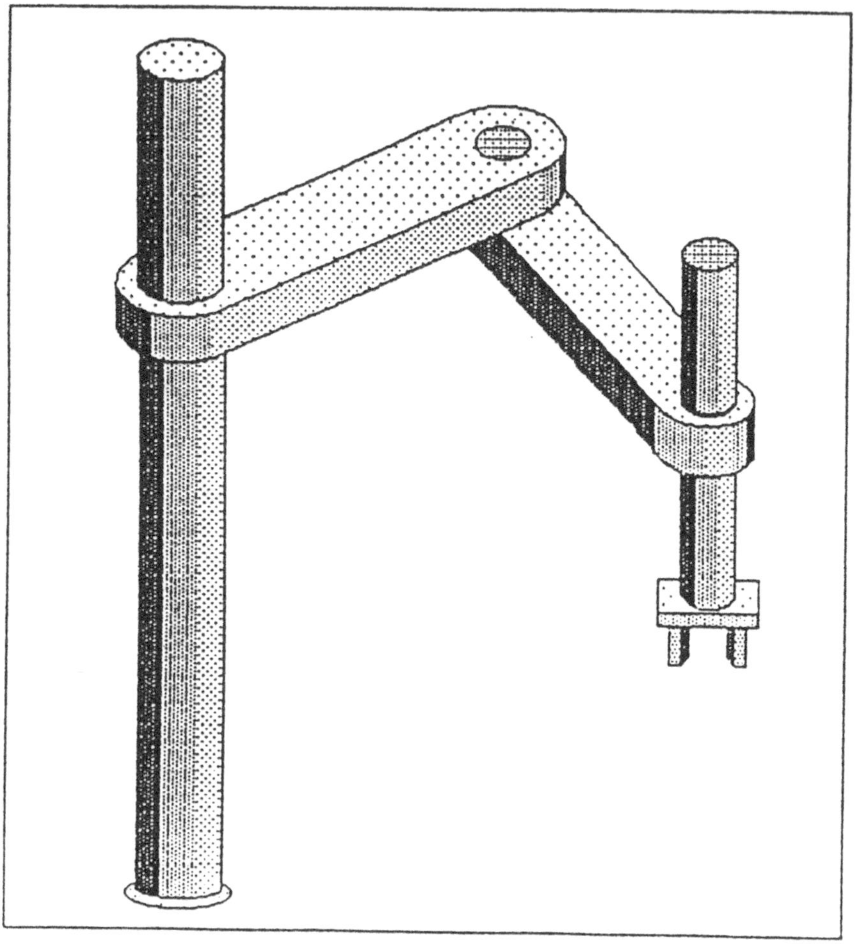

Fig. 17.9. Simulated SCARA configuration

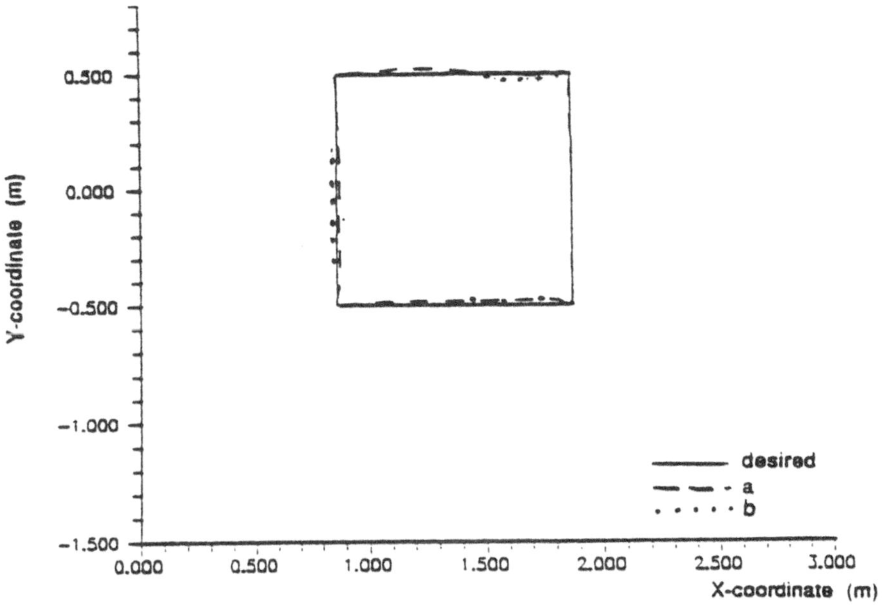

Fig. 17.10. Tracking a square with a) VSC b) SLIMSOC

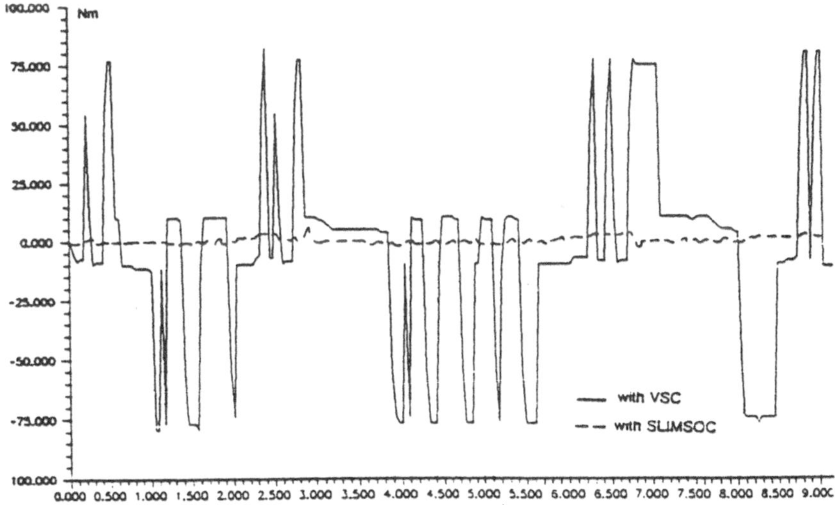

Fig. 17.11. Control input u_1

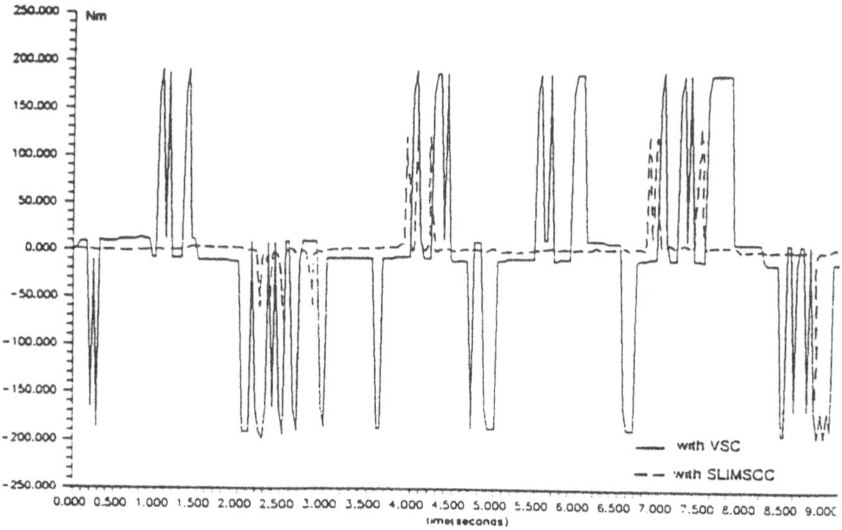

Fig. 17.12. Control input u_2

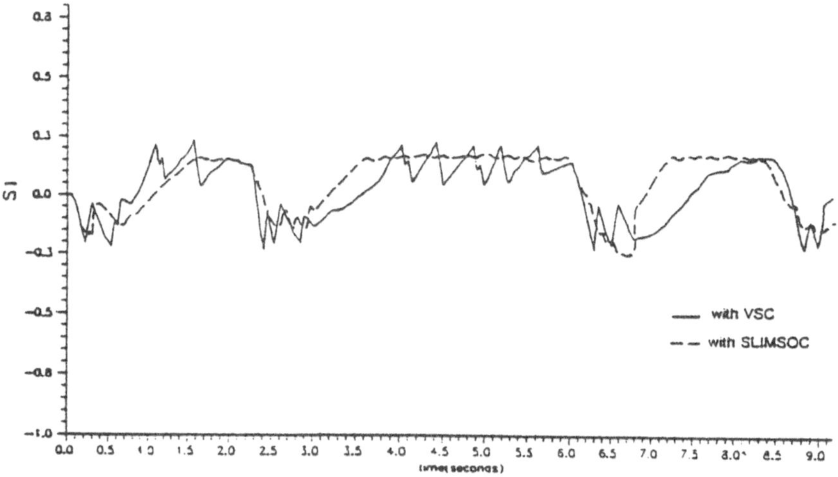

Fig. 17.13. s_1 dynamics

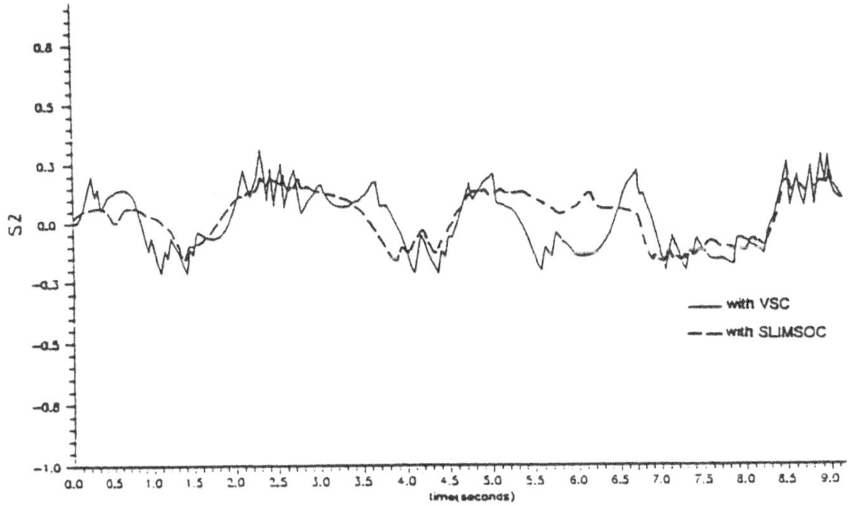

Fig. 17.14. s_2 dynamics

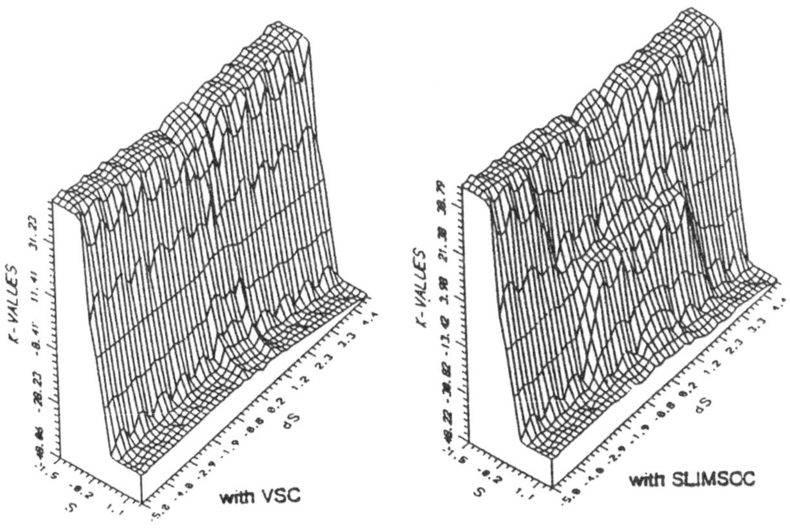

Fig. 17.15. k values

18. VSC Synthesis of Industrial Robots

Karel Jezernik, Boris Curk and Jože Harnik

18.1 Introduction

A Variable Structure Control (VSC) approach is proposed for robust and accurate trajectory tracking of a robotic manipulator with electrical actuators. Decentralized acceleration controllers are used to generate the local switching function. A PI disturbance estimator is proposed to ensure favourable performance. This novel controller gives a zero steady state error and enables each joint to trace the acceleration command. The parameter variation and disturbance insensitive response provided by this control method is demonstrated on a model of a SCARA robot.

The dynamics of an n-link robot mechanism is characterized by a set of highly nonlinear and strongly coupled second-order differential equations

$$D(q)\ddot{q} + C(q,\dot{q}) + G(q) + F(\dot{q}) = \tau \qquad (18.1)$$

where $D(q)$ is the $n \times n$ inertial matrix; $C(q,\dot{q}), G(q)$ and $F(\dot{q})$ are n vectors representing Coriolis and centrifugal forces, the gravity loading, and the friction; $q, \dot{q}$ and $\ddot{q}$ are n vectors of joint angular position, velocity and acceleration; and τ is the n joint torque vector. In general, the matrices D, C, G, F are very complicated functions of q and $\dot{q}$. The fundamental manipulator control problem is to determine the algorithm for generating the joint torque τ, which drives the joint position $q(t)$ to follow closely a desired position trajectory $q^d(t)$. The design of a control algorithm for (18.1) is generally complicated due to the presence of nonlinearity and dynamic coupling (Tarn 1984, Isidori 1989). Even in a well structured industrial setting, the manipulators are subjected to structured and unstructured uncertainties. Structured uncertainty corresponds to the case of a correct dynamic model with parameter uncertainty due to the imprecision on the manipulator link properties, unknown loads, inaccuracies of the torque constants of the actuators, etc. Unstructured uncertainty corresponds to the case of unmodelled dynamics, which results from the presence of the high frequency mode of the manipulator, neglected time-delays, nonlinear friction, etc. The computed torque method is effective for the trajectory control of robotic manipulators (Craig 1988). It has become widely recognized that the tracking performance of the method in high speed operations is often affected by the uncertainties mentioned above. This is especially true for direct drive robots that have no gearing to reduce the dynamic effects.

A severe disadvantage of computed torque control algorithms is that perfect knowledge of the system dynamics is required. The inability to consider the total dynamic model for decoupling and compensation in the control structure, requires robustness of the feedback controller to parameter variations and disturbances. These are the declared features of a variable structure controller in the sliding mode. Many attempts to use VSC in robotics have been reported including Young (1978), Bailey (1987), Wijesoma (1990) and Singh (1990). Exact modelling is not necessary, since it is sufficient that limiting values of model parameters and disturbances, on which basis the control signal is determined, are known. Numerous papers on sliding mode based robot control have selected joint torques as inputs into the system plant as the starting point for the synthesis of the control law. Theoretically, an approach of this kind yields good results.

However, the avoidance of the dynamics of the formation of joint torques may cause a problem. This is particularly true in the case of transistor inverter fed DC or AC motors which use a switching structure with discontinuous torque control. In this case the direct implementation of VSC with discontinuous controllers can result in a hierarchical cascade structure of discontinuous laws (Hashimoto 1988). Since a condition for stable electrical torque tracking is the assurance of a continuous reference torque curve, the use of this cascaded hierarchical structure results in excessive chattering (Jezernik 1990, Jezernik 1991). In order to avoid this problem, some authors have suggested smoothing of the switching law (Xu 1989, Slotine 1983). In this way the discontinuous torque control law is replaced by continuous nonlinear control, which ensures smooth dynamic motion. However, the motion of the system deviates slightly from that achieved by the ideal switching function, and this can result in a steady state error within the boundary layer. The relationship between the boundary layer and the constraints of the the real control input signals is complex, so the control synthesis may be difficult. We have developed an approach which considers the dynamics of torque formation in the synthesis of the control, to be more realistic. In the case of electrical actuators this indicates that armature voltages are *control object inputs* instead of *joint torques.*

One of the underlying assumptions in the design and analysis of VSC systems is that the control can be switched from one value to another infinitely fast. In practical systems, however, it is impossible to achieve the high switching control that is necessary for most VSC designs. There are several reasons for this, including the the presence of finite time delays for control computation and the limitations of physical actuators. Since it is impossible to switch the control at an infinite rate, chattering always occurs in the sliding and steady-state modes of a VSC system. Chattering is almost always objectionable in robotic applications. Here we suggest a new approach to the design of independent VSC joint controllers. Besides the joint acceleration feedback structure and disturbance torque estimation, each controller may possibly comprise elements of computed torque structure. The salient feature of the proposed approach is that the disturbance torque is effectively treated by a computationally straightforward procedure.

18.2 Variable Structure Control Synthesis

For the development of the decentralized control scheme it is convenient to view each joint as a subsystem of the entire manipulator system, with these subsystems interconnected by "coupling torques" representing the inertial coupling terms and the Coriolis, centrifugal, friction and gravity terms in (18.1). The manipulator dynamic model (18.1) is then represented by a collection of n second-order nonlinear scalar differential equations

$$\bar{J}_i(q)\ddot{q}_i \; + \; w_i(q,\dot{q},\ddot{q}) \; = \; \tau_i \tag{18.2}$$

where the subscript i refers to the i-th element. $\bar{J}_i$ is the known varying effective inertia at the i-th joint and is always positive due to the positive-definiteness of D. So $\bar{J}(q)$ can be choosen as a constant diagonal matrix.

Equation (18.2) is the input-output dynamic model of the i-th joint (subsystem) with the joint torque $\tau_i(t)$ as the input and the joint angle $q_i(t)$ as the output. The term w_i, given by (18.3), is treated as a "disturbance torque" by the i-th joint controller ($i = 1, ..., n$) and contains unknown parts of the inertial, gravity, friction, Coriolis and centrifugal torques for the i-th joint, as well as the inertial coupling effects from the other joints

$$w_i = (d_{ii}(q) - \bar{J}_i(q))\ddot{q}_i + \sum_{j=1,\; j\neq i}^{n} d_{ij}(q)\ddot{q}_j + c_i(q,\dot{q}) + g_i(q) + f_i(\dot{q}) \tag{18.3}$$

The dynamics of a DC motor or a DC equivalent of an AC motor with resolved commutation and field generation, can be represented by a first order differential equation

$$L_i\dot{i}_i \; = \; u_i \; - \; e_i \tag{18.4}$$

where L_i is the motor inductance and e_i represents all voltage drops from resistance, back EMF, and for AC motors also equivalent voltages due to inexact commutation, etc. Let us assume a linear connection between the measurable equivalent current i_i and the torque τ_i ($\tau_i = K_{mi}i_i$). The controlled plant will be represent by

$$\begin{bmatrix} \dot{i} \\ \ddot{q} \\ \dot{q} \end{bmatrix} = \begin{bmatrix} 0 & 0 & 0 \\ \bar{J}(q)^{-1}K_m & 0 & 0 \\ 0 & I_{n\times n} & 0 \end{bmatrix} \begin{bmatrix} i \\ \dot{q} \\ q \end{bmatrix} + \begin{bmatrix} 0 & L^{-1} \\ \bar{J}(q)^{-1} & 0 \\ 0 & 0 \end{bmatrix} \begin{bmatrix} w \\ e \end{bmatrix} + \begin{bmatrix} L^{-1} \\ 0 \\ 0 \end{bmatrix} u \tag{18.5}$$

Whenever one seeks to establish a bridge between theory and applications, it no longer suffices to ensure the conditions for the existence of the sliding mode. In real physical systems, such as robots and servodrives, the presence of measuring sensors, idle times due to transistor switchings, idle times due to computer calculation and effects of unmodelled dynamics, cause undesired chattering of the control.

These chatter oscillations are known to result in low control accuracy, high heat losses in electrical power circuits and excessive wear of moving mechanical parts. These phenomena have been considered to be serious obstacles for the application of sliding mode control. In Jezernik (1990, 1991) practical experiments have shown that the chattering caused by unmodelled dynamics may be eliminated by an appropriate choice of the switching function.

In order to obtain smooth mechanical motion of the robot mechanism we prescribe a continuous trajectory with values q_i^d, $\dot{q}_i{}^d$ and $\ddot{q}_i{}^d$. The switching function which determines the mechanical motion is chosen to be second order and a function of angular position, velocity and acceleration errors (for each joint)

$$\sigma_i = (\ddot{q}_i{}^d - \ddot{q}_i) + K_{vi}(\dot{q}_i{}^d - \dot{q}_i) + K_{pi}(q_i^d - q_i) \tag{18.6}$$

where K_{vi} and K_{pi} are constants that determine the damping and the maximum frequency of the decentralized prescribed dynamics of second order. For the practical control implementation the measured quantities are the state variables q_i and $\dot{q}_i$. The acceleration signal $\ddot{q}_i$ is not measurable and can be obtained by double differentiation of the angular position q_i, but is contaminated by the measurement noise to such a degree that it can no longer be used. Consequently, the acceleration signal $\ddot{q}_i$ needs to be replaced by an estimated value which is obtained simply from the differential equation of motion

$$\bar{J}_i(q)\ddot{q}_i = \tau_i - w_i \tag{18.7}$$

where $\bar{J}_i$ is the mean inertia of the robot axes, τ_i is the active measurable drive torque developed by the activator and w_i is the unknown value of the load torque. The expression (18.7) is inserted into the control scheme by replacing the real load torque w_i with an estimated value $\hat{w}_i$. An estimator of reduced order proposed by Jezernik (1990, 1991) is

$$\hat{w}_i = h_i(\dot{q}_i{}^c - \dot{q}_i) \tag{18.8}$$

where h_i is a positive constant linked to the selected dynamics of the asymptotic load observer. The calculated angular acceleration signal $\ddot{q}_i{}^c$ is derived from (18.6), so the condition for the sliding mode operation ($\sigma_i = 0$) of the system is fulfilled

$$\ddot{q}_i{}^c = \ddot{q}_i{}^d + K_{vi}(\dot{q}_i{}^d - \dot{q}_i) + K_{pi}(q_i^d - q_i) \tag{18.9}$$

$$\dot{q}_i{}^c = \int_0^t \ddot{q}_i{}^c \, d\nu \tag{18.10}$$

As a result the control input is based on a modified switching function which contains the estimated acceleration and the estimated disturbance torque as proposed by Jezernik (1991)

$$u_i = \begin{cases} U_i^+ & \sigma_i^e > 0 \\ U_i^- & \sigma_i^e < 0 \end{cases} \tag{18.11}$$

$$\sigma_i^e = i_i^c - i_i = \left[\ddot{q}_i{}^d + K_{vi}(\dot{q}_i{}^d - \dot{q}_i) + K_{pi}(q_i{}^d - q_i) + \frac{\hat{w}_i}{\bar{J}_i}\right]\frac{\bar{J}_i}{K_{mi}} - i_i \quad (18.12)$$

where the desired trajectories of angular position, velocity and acceleration are denoted by the superscript d, and $\hat{w}_i$ is the estimated disturbance torque. The block diagram of the controller with the disturbance torque estimator is shown in Fig. 18.1. The asymptotic observer serves as a bypass for high frequency components, therefore the unmodelled dynamics is not excited.

18.3 Estimation of the Disturbance

From the point of view of controllability, a DC motor with permanent magnet excitation is the most straightforward robotic example. Its motion is governed by a second order equation with respect to angular velocity $\dot{q}_i$ and current i_i with voltage u_i and load torque w_i as a control and a disturbance (18.4) and (18.7).

For the implementation of the control (18.11), angular acceleration $\ddot{q}_i$ is needed. Under the assumptions that the angular velocity $\dot{q}_i$ and current i_i can be measured directly and the load torque varies slowly ($dw_i/dt \sim 0$), a conventional Luenberger reduced order observer may be designed

$$\dot{\hat{z}}_i = \frac{h_i}{\bar{J}_i}(-\hat{z}_i + h_i\dot{q}_i + K_{mi}i_i) \quad (18.13)$$

with h_i constant and $\hat{z}_i = \hat{w}_i + h_i\dot{q}_i$ an estimate of $z_i = w_i + h_i\dot{q}_i$. According to (18.7) and (18.13) the equation for the mismatch $\Delta w_i = w_i - \hat{w}_i$ takes the form

$$\frac{d\Delta w_i}{dt} = -\frac{h_i}{\bar{J}}\Delta w_i \quad (18.14)$$

By a proper choice of gain h_i, the desired convergence rate of Δw_i to zero, or $\hat{w}_i$ to w_i may be obtained provided that the load torque is known, and the acceleration signal $\ddot{q}_i$ may be found from (18.7).

The dynamics of a DC motor is divided into the fast electrical part and the slow mechanical part, which is equivalent to modelled and unmodelled dynamics in robotics. The fast dynamics of the electrical part is defined by the inductance of the motor L_i, which is very much smaller than the moment of inertia J_i. In this case the differential equation (18.4) in the model of DC motor acceleration observer can be ignored and only the differential equation of mechanical motion (18.7) remains to be considered. In connection with the motor current the mechanical variables, angular acceleration and angular velocity, describe the reduced order asymptotic observer of the DC motor (eq. 18.8). If the load torque changes slowly ($dw_i/dt \sim 0$), the observer acceleration signal $\hat{\ddot{q}}$ tracks the real $\ddot{q}$ with the desired dynamics (18.9) in the sliding mode.

We have studied the behavior of two types of reduced order disturbance observers.

a) PI estimator

$$\hat{w}_i = h_i(\dot{q}_i^c - \dot{q}_i) \tag{18.15}$$

$$\dot{q}_i^c = \int_0^t \ddot{q}_i^c \, \mathrm{d}\nu \tag{18.16}$$

$$i_i^c = \left((\ddot{q}_i^c \bar{J}_i + h_i \dot{q}_i^c) - h_i \dot{q}_i\right) / K_{mi} \tag{18.17}$$

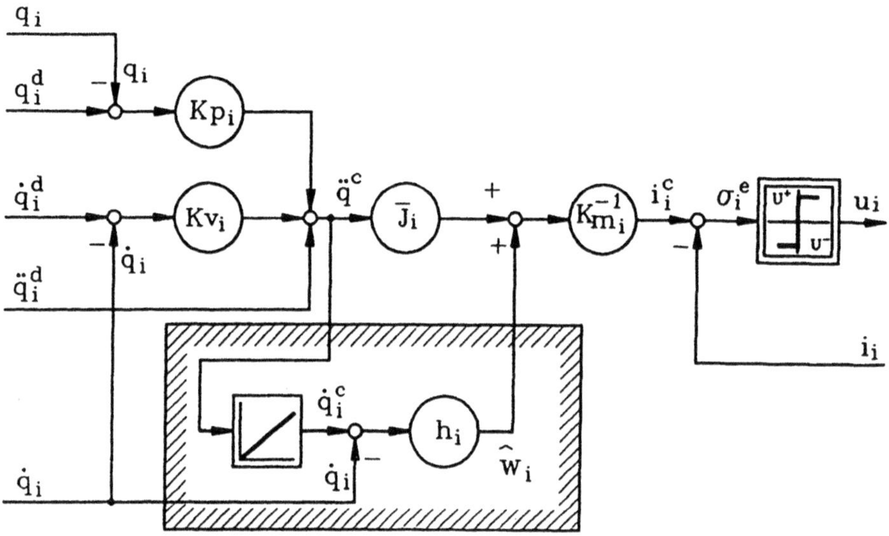

Fig. 18.1. The PI estimator

b) linear disturbance observer (Šabanović 1986)

$$\hat{w}_i = h_i(\dot{\hat{q}}_i - \dot{q}_i) \tag{18.18}$$

$$\dot{\hat{q}}_i = \int_0^t (K_{mi}\, i_i \; - \; \hat{w}_i)/\bar{J}_i \tag{18.19}$$

$$i_i^c = (\ddot{q}_i^c \bar{J}_i + \hat{w}_i)/K_{mi} \tag{18.20}$$

The controlled plant consists of the robot mechanism joint (18.2), the actuator (18.4), the control law (18.11) and the PI disturbance estimator (18.15), which fulfils the sliding mode condition given by (18.21)

$$U_i^+ + \eta_i < U_i^{eq} < U_i^- - \eta_i \ , \qquad \eta_i > 0 \tag{18.21}$$

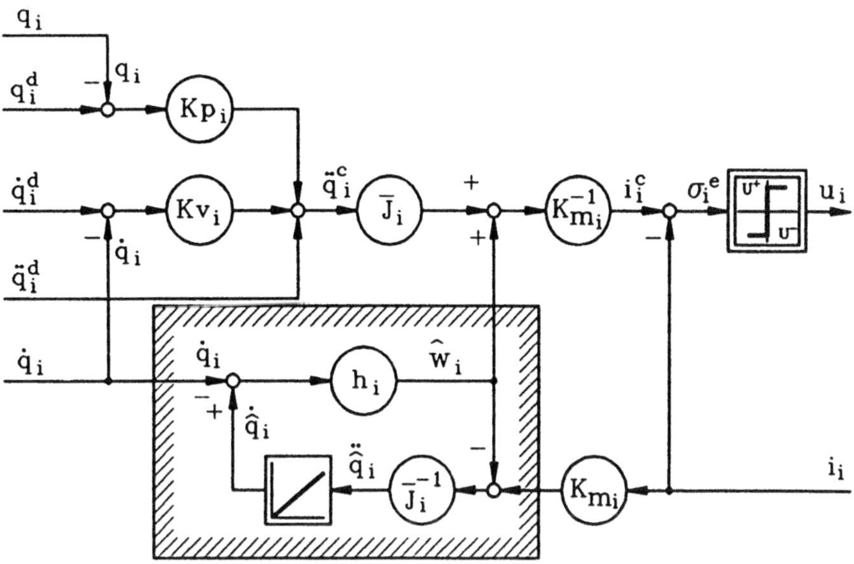

Fig. 18.2. The linear disturbance observer

where the equivalent control U_i^{eq} is defined as the control voltage which assures $\dot{\sigma_i} = 0$ (Utkin 1978)

$$\begin{aligned} U_i^{eq} &= e_i + \frac{L_i}{K_{mi}}[\bar{J}_i(q)\ \dddot{q}_i^{\,d} + \dot{\bar{J}}_i(q,\dot{q})\ddot{q}_i^c + (\bar{J}_i(q)K_{vi} + h_i)(\ddot{q}_i^d - \ddot{q}_i) \\ &\quad + (\bar{J}_i(q)K_{pi} + h_iK_{vi})(\dot{q}_i^d - \dot{q}_i) + h_iK_{pi}(q_i^d - q_i)] \end{aligned} \tag{18.22}$$

The given trajectory is tracked precisely and the initial conditions and disturbances due to indefinitenesses and external influences according to the required dynamics of third order, given by the local tracking error space system representation (18.23), are counteracted.

$$\begin{bmatrix} \dot{x}_i^0 \\ \dot{x}_i^1 \\ \dot{x}_i^2 \end{bmatrix} = \begin{bmatrix} 0 & 1 & 0 \\ 0 & 0 & 1 \\ -h_iK_{pi}/\bar{J}_i & -K_{pi} - h_iK_{vi}/\bar{J}_i & -K_{vi} - h_i/\bar{J}_i \end{bmatrix} \begin{bmatrix} x_i^0 \\ x_i^1 \\ x_i^2 \end{bmatrix} + \begin{bmatrix} 0 \\ 0 \\ -\frac{1}{\bar{J}_i} \end{bmatrix} w_i \tag{18.23}$$

where

$$x_i^0(t) = \int_0^t (q_i^d - q_i)\,\mathrm{d}\nu \tag{18.24}$$

$$x_i^1 = q_i^d - q_i \tag{18.25}$$

$$x_i^2 = \dot{q}_i^d - \dot{q}_i \tag{18.26}$$

The poles of the system are

$$p_{1i} = -\frac{h_i}{\bar{J}_i}\ , \quad p_{2i,3i} = \frac{-K_{vi} \pm \sqrt{K_{vi}^2 - 4K_{pi}}}{2} \tag{18.27}$$

18.4 Simulation Results

Simulations have been done to verify the proposed VSC joint controller to compensate unstructured uncertainties. A two degree of freedom SCARA manipulator was used in the simulation.

The desired trajectory for each joint is

$$q^d(t) = \begin{cases} q(0) + \frac{\Delta q}{2\pi}[\Omega t - \sin(\Omega t)] & 0 \leq \ t \leq t_1 \\ q(t_1) & t > t_1 \end{cases}$$

$$\Delta q = q(t_1) - q(0) \qquad \Omega = \frac{2\pi}{t_1} \tag{18.28}$$

$$\dot{q}^d(t) = \begin{cases} \frac{\Delta q}{t_1}[1 - \cos(\Omega t)] & 0 \leq \ t \leq t_1 \\ 0 & t > t_1 \end{cases} \tag{18.29}$$

$$\ddot{q}^d(t) = \begin{cases} \frac{\Delta q}{t_1}\frac{2\pi}{t_1}\sin(\Omega t) & 0 \leq \ t \leq t_1 \\ 0 & t > t_1 \end{cases} \tag{18.30}$$

The desired trajectory q_1^d is shown in Fig. 18.3. The variation in the moment of inertia ($d_{11}(q)$) from its nominal value to triple the nominal value is presented in Fig. 18.4. The same testing procedure was used for the controller with the PI estimator and the linear observer. The disturbances, i.e. the varied load torque ($w_1(t)$) and its estimated value ($\hat{w}_1(t)$), are presented in Fig. 18.5 for the PI estimator and in Fig. 18.6 for the linear observer. The load torque varies from zero to the nominal value. Figs. 18.7 and 18.8 show the computed current for the PI estimator and linear observer. The tracking errors of the PI estimator and linear observer are compared in Fig. 18.9. The nominal value of joint inertia is $\bar{J}_1 = J_1^{nom}$. Poles in the sliding mode are $p_1 = -500$, $p_2 = p_3 = -25$. The acceleration controller output was calculated every 0.5 ms. The steady state error is compensated and the dynamic error is also asymptotically stable without any restriction for the PI estimator. A major feature of the new controller is its inherent ability to reject payload uncertainty. VSC with the disturbance estimator is also able to solve efficiently the tracking tasks in high speed and direct drive robots.

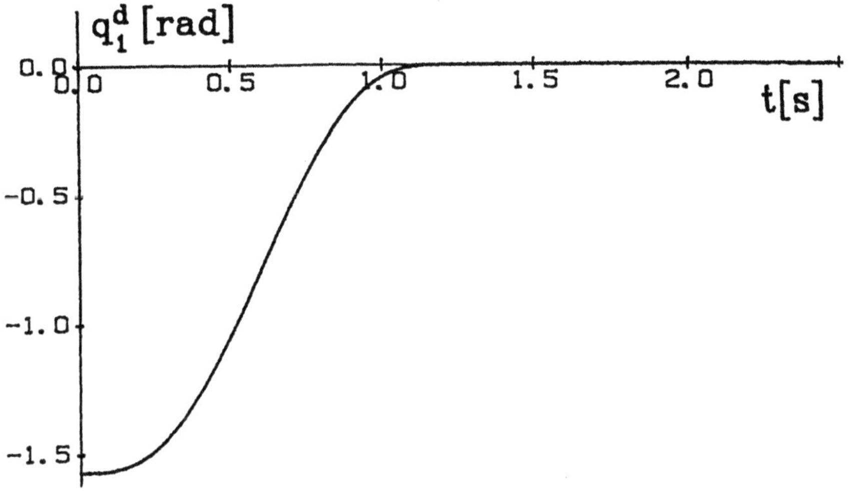

Fig. 18.3. The desired trajectory q_1^d

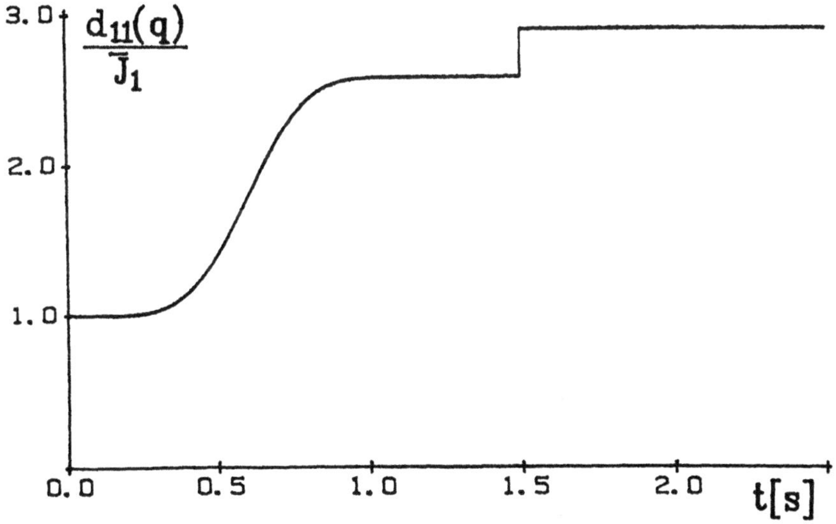

Fig. 18.4. The varied moment of inertia

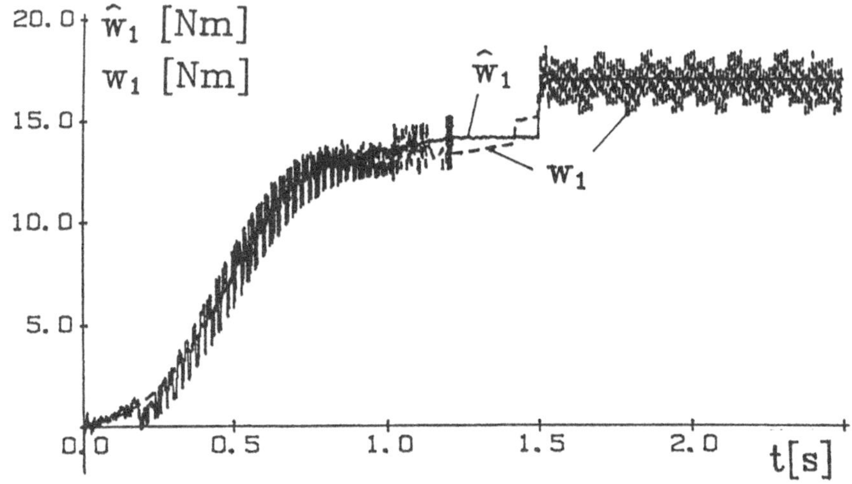

Fig. 18.5. PI estimator: the load torque $w_1(t)$ and its estimated value $\hat{w}_1(t)$

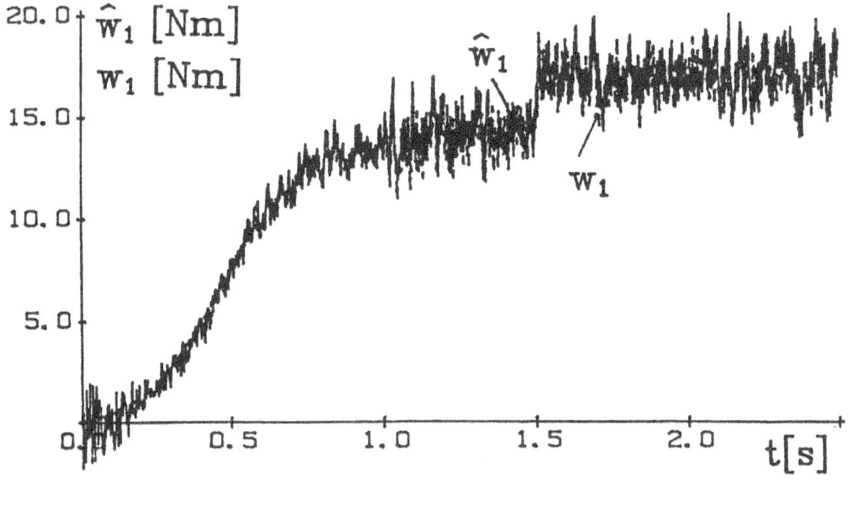

Fig. 18.6. Linear observer: the load torque $w_1(t)$ and its estimated value $\hat{w}_1(t)$

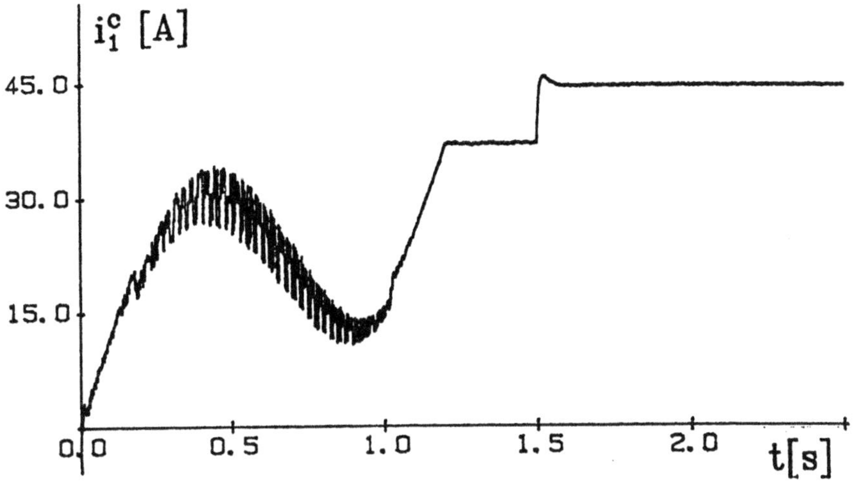

Fig. 18.7. PI estimator: the computed current i_1^c

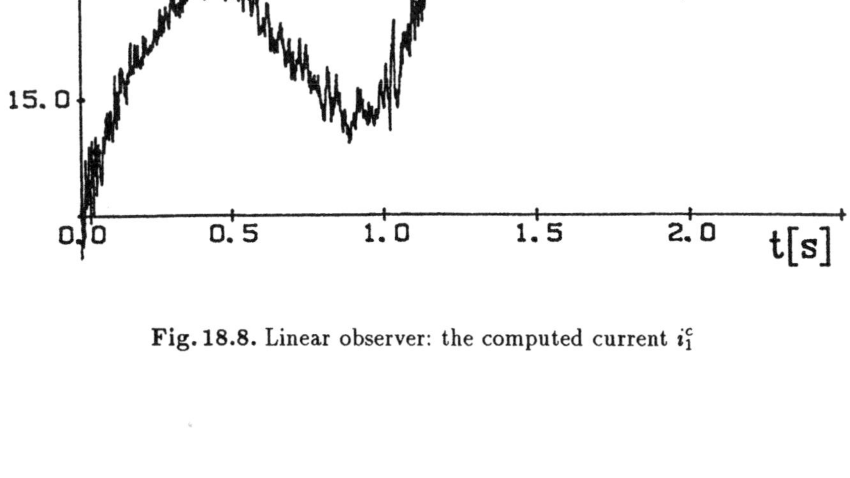

Fig. 18.8. Linear observer: the computed current i_1^c

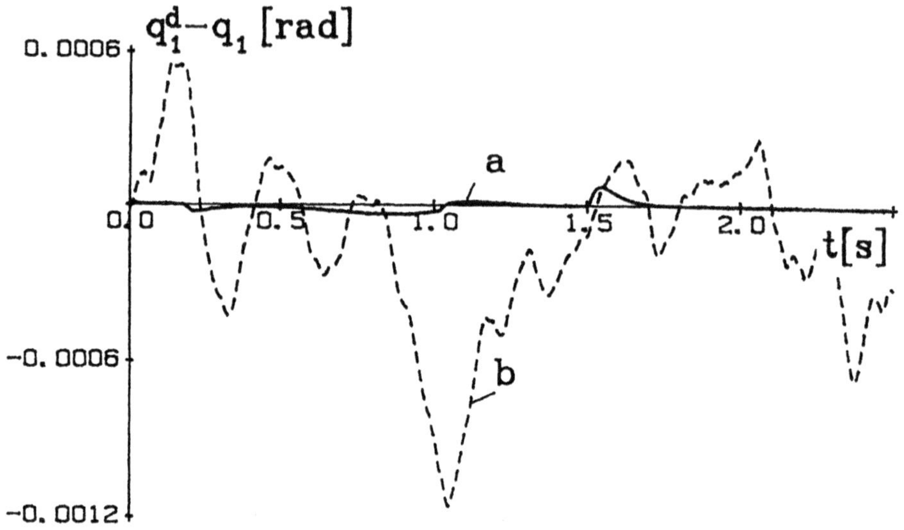

Fig. 18.9. Tracking error: a – PI estimator, b – linear observer.

18.5 Conclusions

The above robot control algorithm consists of acceleration feedback and disturbance torque estimation, and achieves good dynamic performance even in the presence of initial conditions mismatch, parameter perturbations and disturbances. The chattering caused by unmodelled dynamics is eliminated by use of a PI load estimator.

Many authors have used VSC for the control of a robot model with torques functioning as inputs into the system as the starting point for the synthesis. Theoretically such an approach yields good results. However, the dynamics of drive torque generation in real systems results in vibrations because the torque control requires continuous control signals. Due to the structural properties, the direct use of the theory of VSC cannot solve all robotics problems regarding insensitivity to parameter and disturbance variations. It has been found to be necessary to append the on-off controller with an asymptotic observer to estimate the disturbance torque. In this way it is possible to achieve the sliding mode in the vicinity of the desired trajectory by introducing local conditions. However, total insensitivity of the system to disturbances is not possible. Tracking errors are controlled and the dynamic system is asymptotically stable. Better results and a lower tracking error have been achieved with a PI estimator.

References

Bailey, E., Arapostathis, A. 1987, Simple sliding mode control scheme applied to robot manipulator. *International Journal of Control* **45**, 1197–1209

Craig, J.J. 1988, *Adaptive Control of Mechanical Manipulators*, Addison-Wesley, Reading, Maryland

Hashimoto, H., Yamamoto, H., Yanagisawa, S., Harashima F. 1988, Brushless servo motor control using variable structure approach. *IEEE Tr. Ind. App.* **24**, 160–170

Isidori, A. 1989, *Nonlinear Control Systems: An Introduction*, Second Edition, Springer-Verlag

Jezernik, K., Harnik J., Curk B. 1990, Variable structure control of AC servo motors used in industrial robots. *Proc First IEEE International Workshop on variable structure systems and their applications*, Sarajevo, 139–148

Jezernik, K., Curk, B., Harnik, J. 1991, Variable structure field oriented control of an induction motor drive. *4th European conference on power electronics and applications*, Firenze, 2.161–2.166

Singh, S.K. 1990, Decentralized variable structure control for tracking in non-linear systems. *International Journal of Control* **52**, 811–831

Slotine, J.J., Sastry, S.S. 1983, Tracking control of non-linear systems using sliding surfaces, with application to robot manipulators. *International Journal of Control* **38**, 465–492

Tarn, T.J., Begezy, A.K., Isideru, A., Chun, Y.L. 1984, Nonlinear feedback in robot arm control. *Proc IEEE Conference on Decision and Control,* , 736–751

Utkin, V.I. 1978, *Sliding mode and their applications in variable structure systems*, MIR Publishers, Moscow

Wijesoma, S.W. 1990, Robust trajectory following of robots using computed torque structure with VSS. *International Journal of Control* **52**, 935–962

Xu, J.X., Hashimoto, H., Slotine, J.J., Arai, Y., Harashima, F. 1989, Implementation of VSS control to robotic manipulators – smoothing modification. *IEEE Trans. Ind. Electron.* **36**, 321–329

Young, K.D. 1978, Controller design for a manipulator using theory of variable structure systems. *IEEE Trans.Sys., Man. and Cyber.* **SMC-8**, 101–109

Šabanović, A., Bilalović, F. 1986, Sliding mode control of AC drives. *IEEE/AS Ann. Meet.*, Denver

Lecture Notes in Control and Information Sciences

Edited by M. Thoma

1989–1993 Published Titles:

Vol. 135: Nijmeijer, Hendrik; Schumacher, Johannes M. (Eds.)
Three Decades of Mathematical System Theory. A Collection of Surveys at the Occasion of the 50th Birthday of Jan C. Willems.
562 pp. 1989 [3-540-51605-0]

Vol. 136: Zabczyk, Jerzy W. (Ed.)
Stochastic Systems and Optimization. Proceedings of the 6th IFIP WG 7.1 Working Conference, Warsaw, Poland, September 12-16, 1988.
374 pp. 1989 [3-540-51619-0]

Vol. 137: Shah, Sirish L.; Dumont, Guy (Eds.)
Adaptive Control Strategies for Industrial Use. Proceedings of a Workshop held in Kananaskis, Canada, 1988.
360 pp. 1989 [3-540-51869-X]

Vol. 138: McFarlane, Duncan C.; Glover, Keith
Robust Controller Design Using Normalized Coprime Factor Plant Descriptions.
206 pp. 1990 [3-540-51851-7]

Vol. 139: Hayward, Vincent; Khatib, Oussama (Eds.)
Experimental Robotics I. The First International Symposium, Montreal, June 19-21, 1989.
613 pp. 1990 [3-540-52182-8]

Vol. 140: Gajic, Zoran; Petkovski, Djordjija; Shen, Xuemin (Eds.)
Singularly Perturbed and Weakly Coupled Linear Control Systems. A Recursive Approach.
202 pp. 1990 [3-540-52333-2]

Vol. 141: Gutman, Shaul
Root Clustering in Parameter Space.
153 pp. 1990 [3-540-52361-8]

Vol. 142: Gündes, A. Nazli; Desoer, Charles A.
Algebraic Theory of Linear Feedback Systems with Full and Decentralized Compensators.
176 pp. 1990 [3-540-52476-2]

Vol. 143: Sebastian, H.-J.; Tammer, K. (Eds.)
System Modelling and Optimizaton. Proceedings of the 14th IFIP Conference, Leipzig, GDR, July 3-7, 1989.
960 pp. 1990 [3-540-52659-5]

Vol. 144: Bensoussan, A.; Lions, J.L. (Eds.)
Analysis and Optimization of Systems. Proceedings of the 9th International Conference. Antibes, June 12-15, 1990.
992 pp. 1990 [3-540-52630-7]

Vol. 145: Subrahmanyam, M. Bala
Optimal Control with a Worst-Case Performance Criterion and Applications.
133 pp. 1990 [3-540-52822-9]

Vol. 146: Mustafa, Denis; Glover, Keith
Minimum Enthropy H Control.
144 pp. 1990 [3-540-52947-0]

Vol. 147: Zolesio, J.P. (Ed.)
Stabilization of Flexible Structures. Third Working Conference, Montpellier, France, January 1989.
327 pp. 1991 [3-540-53161-0]

Vol. 148: Not published

Vol. 149: Hoffmann, Karl H; Krabs, Werner (Eds.)
Optimal Control of Partial Differential Equations. Proceedings of IFIP WG 7.2 - International Conference. Irsee, April, 9-12, 1990.
245 pp. 1991 [3-540-53591-8]

Vol. 150: Habets, Luc C.
Robust Stabilization in the Gap-topology.
126 pp. 1991 [3-540-53466-0]

Vol. 151: Skowronski, J.M.; Flashner, H.; Guttalu, R.S. (Eds.)
Mechanics and Control. Proceedings of the 3rd Workshop on Control Mechanics, in Honor of the 65th Birthday of George Leitmann, January 22-24, 1990, University of Southern California.
497 pp. 1991 [3-540-53517-9]

Vol. 152: Aplevich, J. Dwight
Implicit Linear Systems.
176 pp. 1991 [3-540-53537-3]

Vol. 153: Hajek, Otomar
Control Theory in the Plane.
269 pp. 1991 [3-540-53553-5]

Vol. 154: Kurzhanski, Alexander; Laseicka, Irena (Eds.)
Modelling and Inverse Problems of Control for Distributed Parameter Systems. Proceedings of IFIP WG 7.2 - IIASA Conference, Laxenburg, Austria, July 1989.
170 pp. 1991 [3-540-53583-7]

Vol. 155: Bouvet, Michel; Bienvenu, Georges (Eds.)
High-Resolution Methods in Underwater Acoustics.
244 pp. 1991 [3-540-53716-3]

Vol. 156: Hämäläinen, Raimo P.; Ehtamo, Harri K. (Eds.)
Differential Games - Developments in Modelling and Computation. Proceedings of the Fourth International Symposium on Differential Games and Applications, August 9-10, 1990, Helsinki University of Technology, Finland.
292 pp. 1991 [3-540-53787-2]

Vol. 157: Hämäläinen, Raimo P.; Ehtamo, Harri K. (Eds.)
Dynamic Games in Economic Analysis. Proceedings of the Fourth International Symposium on Differential Games and Applications. August 9-10, 1990, Helsinki University of Technology, Finland.
311 pp. 1991 [3-540-53785-6]

Vol. 158: Warwick, Kevin; Karny, Miroslav; Halouskova, Alena (Eds.)
Advanced Methods in Adaptive Control for Industrial Applications.
331 pp. 1991 [3-540-53835-6]

Vol. 159: Li, Xunjing; Yong, Jiongmin (Eds.)
Control Theory of Distributed Parameter Systems and Applications. Proceedings of the IFIP WG 7.2 Working Conference, Shanghai, China, May 6-9, 1990.
219 pp. 1991 [3-540-53894-1]

Vol. 160: Kokotovic, Petar V. (Ed.)
Foundations of Adaptive Control.
525 pp. 1991 [3-540-54020-2]

Vol. 161: Gerencser, L.; Caines, P.E. (Eds.)
Topics in Stochastic Systems: Modelling, Estimation and Adaptive Control.
1991 [3-540-54133-0]

Vol. 162: Canudas de Wit, C. (Ed.)
Advanced Robot Control. Proceedings of the International Workshop on Nonlinear and Adaptive Control: Issues in Robotics, Grenoble, France, November 21-23, 1990.
Approx. 330 pp. 1991 [3-540-54169-1]

Vol. 163: Mehrmann, Volker L.
The Autonomous Linear Quadratic Control Problem. Theory and Numerical Solution.
177 pp. 1991 [3-540-54170-5]

Vol. 164: Lasiecka, Irena; Triggiani, Roberto
Differential and Algebraic Riccati Equations with Application to Boundary/Point Control Problems: Continuous Theory and Approximation Theory.
160 pp. 1991 [3-540-54339-2]

Vol. 165: Jacob, Gerard; Lamnabhi-Lagarrigue, F. (Eds.)
Algebraic Computing in Control. Proceedings of the First European Conference, Paris, March 13-15, 1991.
384 pp. 1991 [3-540-54408-9]

Vol. 166: Wegen, Leonardus L. van der
Local Disturbance Decoupling with Stability for Nonlinear Systems.
135 pp. 1991 [3-540-54543-3]

Vol. 167: Rao, Ming
Integrated System for Intelligent Control.
133 pp. 1992 [3-540-54913-7]

Vol. 168: Dorato, Peter; Fortuna, Luigi; Muscato, Giovanni
Robust Control for Unstructured Perturbations: An Introduction.
118 pp. 1992 [3-540-54920-X]

Vol. 169: Kuntzevich, Vsevolod M.; Lychak, Michael
Guaranteed Estimates, Adaptation and Robustness in Control Systems.
209 pp. 1992 [3-540-54925-0]

Vol. 170: Skowronski, Janislaw M.; Flashner, Henryk; Guttalu, Ramesh S. (Eds.)
Mechanics and Control. Proceedings of the 4th Workshop on Control Mechanics, January 21-23, 1991, University of Southern California, USA.
302 pp. 1992 [3-540-54954-4]

Vol. 171: Stefanidis, P.; Paplinski, A.P.; Gibbard, M.J.
Numerical Operations with Polynomial Matrices: Application to Multi-Variable Dynamic Compensator Design.
206 pp. 1992 [3-540-54992-7]

Vol. 172: Tolle, H.; Ersü, E.
Neurocontrol: Learning Control Systems Inspired by Neuronal Architectures and Human Problem Solving Strategies.
220 pp. 1992 [3-540-55057-7]

Vol. 173: Krabs, W.
On Moment Theory and Controllability of Non-Dimensional Vibrating Systems and Heating Processes.
174 pp. 1992 [3-540-55102-6]

Vol. 174: Beulens, A.J. (Ed.)
Optimization-Based Computer-Aided Modelling and Design. Proceedings of the First Working Conference of the New IFIP TC 7.6 Working Group, The Hague, The Netherlands, 1991.
268 pp. 1992 [3-540-55135-2]

Vol. 175: Rogers, E.T.A.; Owens, D.H.
Stability Analysis for Linear Repetitive Processes.
197 pp. 1992 [3-540-55264-2]

Vol. 176: Rozovskii, B.L.; Sowers, R.B. (Eds.)
Stochastic Partial Differential Equations and their Applications. Proceedings of IFIP WG 7.1 International Conference, June 6-8, 1991, University of North Carolina at Charlotte, USA.
251 pp. 1992 [3-540-55292-8]

Vol. 177: Karatzas, I.; Ocone, D. (Eds.)
Applied Stochastic Analysis. Proceedings of a US-French Workshop, Rutgers University, New Brunswick, N.J., April 29-May 2, 1991.
317 pp. 1992 [3-540-55296-0]

Vol. 178: Zolésio, J.P. (Ed.)
Boundary Control and Boundary Variation. Proceedings of IFIP WG 7.2 Conference, Sophia- Antipolis,France, October 15-17, 1990.
392 pp. 1992 [3-540-55351-7]

Vol. 179: Jiang, Z.H.; Schaufelberger, W.
Block Pulse Functions and Their Applications in Control Systems.
237 pp. 1992 [3-540-55369-X]

Vol. 180: Kall, P. (Ed.)
System Modelling and Optimization. Proceedings of the 15th IFIP Conference, Zurich, Switzerland, September 2-6, 1991.
969 pp. 1992 [3-540-55577-3]

Vol. 181: Drane, C.R.
Positioning Systems - A Unified Approach.
168 pp. 1992 [3-540-55850-0]

Vol. 182: Hagenauer, J. (Ed.)
Advanced Methods for Satellite and Deep Space Communications. Proceedings of an International Seminar Organized by Deutsche Forschungsanstalt für Luft-und Raumfahrt (DLR), Bonn, Germany, September 1992.
196 pp. 1992 [3-540-55851-9]

Vol. 183: Hosoe, S. (Ed.)
Robust Control. Proceesings of a Workshop held in Tokyo, Japan, June 23-24, 1991.
225 pp. 1992 [3-540-55961-2]

Vol. 184: Duncan, T.E.; Pasik-Duncan, B. (Eds.)
Stochastic Theory and Adaptive Control. Proceedings of a Workshop held in Lawrence, Kansas, September 26-28, 1991.
500 pages. 1992 [3-540-55962-0]

Vol. 185: Curtain, R.F. (Ed.); Bensoussan, A.; Lions, J.L.(Honorary Eds.)
Analysis and Optimization of Systems: State and Frequency Domain Approaches for Infinite-Dimensional Systems. Proceedings of the 10th International Conference, Sophia-Antipolis, France, June 9-12, 1992.
648 pp. 1993 [3-540-56155-2]

Vol. 186: Sreenath, N.
Systems Representation of Global Climate Change Models. Foundation for a Systems Science Approach.
288 pp. 1993 [3-540-19824-5]

Vol. 187: Morecki, A.; Bianchi, G.; Jaworeck, K. (Eds.)
RoManSy 9: Proceedings of the Ninth CISM-IFToMM Symposium on Theory and Practice of Robots and Manipulators.
476 pp. 1993 [3-540-19834-2]

Vol. 188: Naidu, D. Subbaram
Aeroassisted Orbital Transfer: Guidance and Control Strategies.
192 pp. 1993 [3-540-19819-9]

Vol. 189: Ilchmann, A.
Non-Identifier-Based High-Gain Adaptive Control.
220 pp. 1993 [3-540-19845-8]

Vol. 190: Chatila, R.; Hirzinger, G. (Eds.)
Experimental Robotics II: The 2nd International Symposium, Toulouse, France, June 25-27 1991.
580 pp. 1993 [3-540-19851-2]

Vol. 191: Blondel, V.
Simultaneous Stabilization of Linear Systems.
212 pp. 1993 [3-540-19862-8]

Vol. 192: Smith, R.S.; Dahleh, M. (Eds.)
The Modeling of Uncertainty in Control Systems.
412 pp. 1993 [3-540-19870-9]

GPSR Compliance

The European Union's (EU) General Product Safety Regulation (GPSR) is a set of rules that requires consumer products to be safe and our obligations to ensure this.

If you have any concerns about our products, you can contact us on ProductSafety@springernature.com

In case Publisher is established outside the EU, the EU authorized representative is:

Springer Nature Customer Service Center GmbH
Europaplatz 3
69115 Heidelberg, Germany

Batch number: 10844669

Printed by Printforce, the Netherlands